Novel Metathesis Chemistry: Well-Defined Initiator Systems for Specialty Chemical Synthesis, Tailored Polymers and Advanced Material Applications

NATO Science Series

A Series presenting the results of scientific meetings supported under the NATO Science Programme.

The Series is published by IOS Press, Amsterdam, and Kluwer Academic Publishers in conjunction with the NATO Scientific Affairs Division

Sub-Series

I. **Life and Behavioural Sciences**	IOS Press
II. **Mathematics, Physics and Chemistry**	Kluwer Academic Publishers
III. **Computer and Systems Science**	IOS Press
IV. **Earth and Environmental Sciences**	Kluwer Academic Publishers
V. **Science and Technology Policy**	IOS Press

The NATO Science Series continues the series of books published formerly as the NATO ASI Series.

The NATO Science Programme offers support for collaboration in civil science between scientists of countries of the Euro-Atlantic Partnership Council. The types of scientific meeting generally supported are "Advanced Study Institutes" and "Advanced Research Workshops", although other types of meeting are supported from time to time. The NATO Science Series collects together the results of these meetings. The meetings are co-organized bij scientists from NATO countries and scientists from NATO's Partner countries – countries of the CIS and Central and Eastern Europe.

Advanced Study Institutes are high-level tutorial courses offering in-depth study of latest advances in a field.
Advanced Research Workshops are expert meetings aimed at critical assessment of a field, and identification of directions for future action.

As a consequence of the restructuring of the NATO Science Programme in 1999, the NATO Science Series has been re-organised and there are currently Five Sub-series as noted above. Please consult the following web sites for information on previous volumes published in the Series, as well as details of earlier Sub-series.

http://www.nato.int/science
http://www.wkap.nl
http://www.iospress.nl
http://www.wtv-books.de/nato-pco.htm

Series II: Mathematics, Physics and Chemistry – Vol. 122

Novel Metathesis Chemistry: Well-Defined Initiator Systems for Specialty Chemical Synthesis, Tailored Polymers and Advanced Material Applications

edited by

Y. Imamoglu
Department of Chemistry,
Hacettepe University Beytepe,
Ankara, Turkey

and

L. Bencze
Department of Organic Chemistry,
University of Veszprem,
Veszprem, Hungary

Kluwer Academic Publishers

Dordrecht / Boston / London

Published in cooperation with NATO Scientific Affairs Division

Proceedings of the NATO Study Institute on
Novel Metathesis Chemistry:
Designing Well-Defined Initiator Systems for Specialty Chemical Synthesis, Tailored
Polymers and Advanced Material Applications
Antalya, Turkey
21 September 2002

A C.I.P. Catalogue record for this book is available from the Library of Congress.

ISBN 1-4020-1570-4 (HB)
ISBN 1-4020-1571-2 (PB)

Published by Kluwer Academic Publishers,
P.O. Box 17, 3300 AA Dordrecht, The Netherlands.

Sold and distributed in North, Central and South America
by Kluwer Academic Publishers,
101 Philip Drive, Norwell, MA 02061, U.S.A.

In all other countries, sold and distributed
by Kluwer Academic Publishers,
P.O. Box 322, 3300 AH Dordrecht, The Netherlands.

Printed on acid-free paper

NATO ADVANCED STUDY INSTITUTE

NOVEL METATHESIS CHEMISTRY: WELL-DEFINED INITIATOR SYSTEMS FOR SPECIALTY CHEMICAL SYNTHESIS, TAILORED POLYMERS AND ADVANCED METARIAL APPLICATIONS

8th-21st Sept. 2002 ANTALYA-TURKEY

CO-DIRECTORS:

PROF. IMAMOGLU YAVUZ

Department of Chemistry, Hacettepe University Beytepe, Ankara 06532 TURKEY
imamoglu@hacettepe.edu.tr

PROF. BENCZE LAJOS

Department of Organic Chemistry, University of Veszprem, H-8200, POB 158 Veszprem, HUNGARY
ben016@almos.vein.hu

ORGANISING COMMITTEE:

PROF. Y. IMAMOGLU

Department of Chemistry, Hacettepe University Beytepe, Ankara 06532 TURKEY

PROF. L. BENCZE

Department of Organic Chemistry, University of Veszprem, H-8200, POB 158 Veszprem, HUNGARY

PROF. H. MOL

Istitute of Molecular Chemistry, University of Amsterdam, Nieuwe Achtergracht 166, 1018 WV Amsterdam, NETHERLANDS

PROF. K.B. WAGENER

Departement of Chemistry, University of Florida, POB 117200,Gainsville, Florida 32611-7200, USA.

DR. E. KHOSRAVI

IRC in Polymer Science and technology, University of Durham, South Road Durham DHI 3LE, UK.

TABLE OF CONTENTS

PART 3: INDUSRIAL ASPECTS

PREFACE

The large number and novelty of scientific and technical results reported at the 14th International Symposium on Olefin Metathesis (ISOM 14) meeting has prompted us to initiate the organization of a NATO Advance Science Institue to disseminate the sensational new achievements and experiences.

The main goal of the proposed ASI was to transfer the knowledge accumalated recently in the field of metathesis centred synthesis and engineering of specialty chemicals and new performance materials.

Our lecturers were widely recognized and devoted scientists.

Their plenary lecturers (28) provided the backbone of the scientific program supplemented by special seminars (10) delivered also by distinguished scientists. Posters (22) presented by the young participants students had a great impact on personal relationship and encouraging discussion.

The scientific scope of the ASI features contributions from leading experts on the theory, practice and application of catalytically active comprehensive discussion, the major emphasis was clearly on the production, testing and application of catalytically active transition-metallic species and their application in the production of various speciality polymeric materials. It will serve as a useful reference for graduate students and postdoctoral fellows wishing to get an introduction to the field as well as provide a contemporary overwiev for scientists already working in the field. Accordingly, each tutorial lecturer has been written for such a general audience.

The lecturers were presented in a sequence that is based primatily on the desire to provide a systematic introduction to the material. Fortunately for most readers, the topics were presented in progressively more difficult stages so that those who were not experts in this field could follow the material.

The program was well organized and the coverage was reasonably broad.

Initiating systems for speciality chemical synthesis:

History, mechanism, catalysts and application of alkene metathesis (P. H. Dixneuf); initiation, propagation, chain transfer and termination (L. Bencze), production of Si containig monomers (B. Marciniec) and telechelic macro monomers (L. Bencze); testing the properties of novel and commercialised catalyst in ROMP (E. Khosravi, K. Wagener, H. Mol, B. Merciniec); (B. Çetinkaya, J.C. Mol, J. Vohlidal, F. Verpoort, L. Delaude, P. H. Dixneuf, T. Szymanska-Buzar); development and application of new heterogeneous metathesis catalysts (J. Vohlidal, H. Balcar, K. Weiss, N. Bespalova, E. Sh. Finkelstein); Electrochemically produced tungsten catalsysts (Y. İmamoglu); Late group metals (Pd and Ni) for the polymerisation of norbornene and functionalised norbornenes (A. Bell); A new incite for catalyst studies: X-ray absorption spectroscopy (R. K. Szilagyi).

The range of topics covered in material science:

<u>Alkenamers:</u> Perfect comb graft polymers synthesized via acyclic diene metathesis polymerisation to yield precisely spaced graft sites along an unsaturated polymer backbone; Crystallization Behaviour of branched and functionalised polyethylene (K. Wagener), Well defined crosslinked materials via ROMP (E. Khosravi), Non-crosslinked norbornadiene polymers (L. Bencze); Advanced opto-electronic materials based on addition polymerised norbornene monomers (A. Bell); Synthesis of poly-co-cycloolefins (Y. İmamoglu). Chiral polyolefin possesing amino acids via metathesis

chemistry (K. Wagener); Synthesis and properties highly stereo regular norbornadiene polymers with N and O substituents (L. Delaude and A. F. Noels). Synthesis and properties of poly-co-norbornene-co-norbornadiene (T. Szymanska-Buzar)

Polyacetylenes:

Recent advances in the synthesis of substituted acetylenes, Synthesis and functions of aromatic polyacetylenes; Synthesis and properties of helical substituted polyacetylene, Materials for gas permeation membranes (T. Masuda), Mesomorphous molecular sieves immobilized catalysts for poly-phenylacetylenes (H. Balcar, J. Vohlidal), Synthesis and properties of acyclic diyne metathesis condensation (ADIMET) polymer products, poly(phenylene)ethynylenes (K. Weiss).

Industrial aspects:

Industrial application of olefin metathesis: new industrial routes to important petrochemicals, polymers and specialty chemicals; Metathesis of oleochemical feedstock: toward sustainable chemical industry (J. C. Mol).

Fuels: Metathetic synthesis and application of liquid rocket engine fuels containing strained carbocyclic compounds (E. Sh. Finkelstein).

The topics delivered by scientists are contemporary interest.

The contributing authors generally did an excellent job of pointing out the important issues and exploring the current theoretical and experimental approaches.

Bibliographies referred to were complete and provided an up-to-date review of the progress made in developing the techniques. These should be extremely useful to someone needing references to the primary literature for more theoretical background on the topics discussed.

This ASI is a good starting point for anyone thinking of entering the area of metathesis chemistry related material science research program.

The advanced Study Institute was generously sponsored by the Scientific Affairs Division of NATO and the editor gratefully acknowledges this sponsorship.

We also thank the Members of the Local Organizing Committee for their engagement on a succesful NATO-ASI.

Yavuz İmamoglu   Lajos Bencze

January, 2003

LIST OF PARTICIPANTS

Philip ALMOND

179 Chemistry Building
Department of Chemistry
Auburn University
Auburn, Alabama 36849, USA

Cemil AYDOĞDU

Hacettepe University
Faculty of Education,
Department of Primary Education
06532 Beytepe, Ankara, TURKEY

Hynek BALCAR

Department of Physical Chemistry
and Macromolecular Chemistry
Faculty of Science,
Charles University, Albertov 6,
128 43 Prague 2, CZECH REPUBLIC

Andrew BELL

APT, LLC
9921 Brecksville Road
Brecksville OH 44141, USA

Lajos BENCZE

Department of Organic Chemistry
University of Veszprém
P.O. Box. 158,
H-8210 Veszprém, HUNGARY

Natalia BESPALOVA

Topchiev Institute of Petrochemical Synthesis
Russian Academy of Sciences
Leninsky Prospect 29,
117912 Moscow B-71, RUSSIA

Emine BOZ

Hacettepe University
Department of Chemistry
06532 Beytepe, Ankara, TURKEY

Carsten BRANDT

Institut für Technische
und Makromolekulare Chemie
Universität Hamburg, Bundesstrasse 45,
D-20146 Hamburg, GERMANY

Victor BYKOV

Topchiev Institute of Petrochemical Synthesis
Russian Academy of Sciences
Leninsky Prospect 29,
117912 Moscow B-71, RUSSIA

Tom CASTLE

IRC In Polymer Science and Technology
Chemistry Department,
University of Durham
Durham DH1 3LE, UK

Dariusz CHADYNIAK

Faculty of Chemistry
Adam Mickiewicz University
Grunwaldzka 6,
60-780 Poznan, POLAND

Olga V. CHOUVALOVA

Topchiev Institute of Petrochemical Synthesis
Russian Academy of Sciences
Leninsky Prospect 29,
117912 Moscow B-71, RUSSIA

Bob de CLERCQ

Department of Inorganic
and Physical Chemistry
Ghent University, Krijgslaan 281 (S3),
9000 Ghent, BELGIUM

Robert CSONKA

Department of Organic Chemistry
University of Veszprém
P.O. Box. 158,
H-8210 Veszprém, HUNGARY

Izabela CZELUSNIAK

Faculty of Chemistry
University of Wroclaw
14 F. Joliot-Curie Street,
50-383 Wroclaw, POLAND

Tuba ÇAKIR

Hacettepe University
Department of Chemistry
06532 Beytepe, Ankara, TURKEY

Bekir ÇETİNKAYA

Ege University, Faculty of Science
Chemistry Department, Bornova
35100 İzmir, TURKEY

Engin ÇETİNKAYA

Ege University, Faculty of Science
Chemistry Department, Bornova
35100 İzmir, TURKEY

Sevil ÇETİNKAYA

Hacettepe University
Department of Chemistry
06532 Beytepe, Ankara, TURKEY

Lionel DELAUDE

Center For Education and
Research On Macromolecules
Institut De Chimie (B6a),
University of Liege, Sart-Tilman,
B-4000 Liege, BELGIUM

Okan DERELİ

Hacettepe University
Department of Chemistry
06532 Beytepe, Ankara, TURKEY

Pierre H. DIXNEUF

Institut de Chimie de Rennes
Bat 10 C, Campus de Beaulieu
35042 Rennes Cedex, FRANCE

Bülent DÜZ

Hacettepe University
Department of Chemistry
06532 Beytepe, Ankara, TURKEY

Eugene FINKELSHTEIN

Topchiev Institute of Petrochemical Synthesis
Russian Academy of Sciences
Leninsky Prospect 29,
117912 Moscow B-71, RUSSIA

Steffen FISHER

Institut für Technische
und Makromolekulare Chemie
Universität Hamburg, Bundesstrasse 45,
D-20146 Hamburg, GERMANY

Ildiko GANSZKY

Department of Organic Chemistry
University of Veszprém
P.O. Box. 158,
H-8210 Veszprém, HUNGARY

Stephania GARBACIA

Institut de Chimie de Rennes
Bat 10 C, Campus de Beaulieu
35042 Rennes Cedex, FRANCE

Georgy GRANCHAROV

IRC In Polymer Science and Technology
Chemistry Department,
University of Durham
Durham DH1 3LE, UK

Emin GÜNAY

Ege University, Faculty of Science
Chemistry Department, Bornova
35100 İzmir, TURKEY

David HAIGH

IRC In Polymer Science and Technology
Chemistry Department,
University of Durham
Durham DH1 3LE, UK

Olga ILIEVA

Bulgarian Academy of Sciences
Institute of Polymers, ACAD, G.
Bonchev Str. Bl. 103 A,
1113 Sofia, BULGARIA

Fırat İLKER

University of Massachusetts
Department of Polymer Science
& Engineering, 120 Governors Drive,
Amherst MA, 01003-4530, USA

Yavuz İMAMOĞLU

Hacettepe University
Department of Chemistry
06532 Beytepe, Ankara, TURKEY

Magdalena JANKOWSKA

Faculty of Chemistry
Adam Mickiewicz University
Grunwaldzka 6,
60-780 Poznan, POLAND

Christo JOSSIFOV

Bulgarian Academy of Sciences
Institute of Polymers, ACAD, G.
Bonchev Str. Bl. 103 A,
1113 Sofia, BULGARIA

Radostina KALINOVA

Bulgarian Academy of Sciences
Institute of Polymers, ACAD, G.
Bonchev Str. Bl. 103 A,
1113 Sofia, BULGARIA

Solmaz KARABULUT

Hacettepe University
Department of Chemistry
06532 Beytepe, Ankara, TURKEY

Ezat KHOSRAVI

IRC In Polymer Science and Technology
Chemistry Department, University of Durham
Durham DH1 3LE, UK

Rafet KILIÇASLAN

Ege University, Faculty of Science
Chemistry Department, Bornova
35100 İzmir, TURKEY

Malgorzata KUJAWA-WELTEN

IRC In Polymer Science and Technology
Chemistry Department, University of Durham
Durham DH1 3LE, UK

Robert KURDI

Department of Organic Chemistry
University of Veszprém
P.O. Box. 158,
H-8210 Veszprém, HUNGARY

Jerome LE NOTRE

Institut de Chimie de Rennes
Bat 10 C, Campus de Beaulieu
35042 Rennes Cedex, FRANCE

Grzegorz HRECZYCHO

Faculty of Chemistry
Adam Mickiewicz University
Grunwaldzka 6,
60-780 Poznan, POLAND

Elnur MAMEDOV

Azerbaijan National Academy of Sciences
Department of Chemical Sciences
370001 Baku,
Istiglaliyyat Street 10, AZERBAIJAN

Bogdan MARCINIEC

Faculty of Chemistry
Adam Mickiewicz University
Grunwaldzka 6,
60-780 Poznan, POLAND

Toshio MASUDA

Department of Polymer Chemistry
Graduate School of Engineering,
Kyoto University
Kyoto 606-8501, JAPAN

Hans MOL

Institute of Molecular Chemistry
University of Amsterdam
Nieuwe Achtergracht 166,
1018 WV Amsterdam, NETHERLANDS

Abel MUHARREMOV

Baku State University
Faculty of Chemistry
Organic Chemistry Division,
Baku, AZERBAIJAN

Ani NEDELCHEVA

Bulgarian Academy of Sciences
Institute of Polymers, ACAD, G.
Bonchev Str. Bl. 103 A,
1113 Sofia, BULGARIA

Tom OPSTAL — Department of Inorganic and Physical Chemistry, Ghent University, Krijgslaan 281 (S3), 9000 Ghent, BELGIUM

Piotr PAWLUC — Faculty of Chemistry, Adam Mickiewicz University, Grunwaldzka 6, 60-780 Poznan, POLAND

Dmitry REDKIN — Topchiev Institute of Petrochemical Synthesis, Russian Academy of Sciences, Leninsky Prospect 29, 117912 Moscow B-71, RUSSIA

Zakir RIZAYEV — Azerbaijan National Academy of Sciences, Department of Chemical Sciences, 370001 Baku, Istiglaliyyat Street 10, AZERBAIJAN

Ju. ROGAN — Topchiev Institute of Petrochemical Synthesis, Russian Academy of Sciences, Leninsky Prospect 29, 117912 Moscow B-71, RUSSIA

Katrin SATTLER — Laboratorium Fur Anorganishe Chemie, University of Bayreuth, D-95440 Bayreuth, GERMANY

Felix SCHELIGA — Institut für Technische und Makromolekulare Chemie, Universität Hamburg, Bundesstrasse 45, D-20146 Hamburg, GERMANY

Fatma SEVİN — Hacettepe University, Department of Chemistry, 06532 Beytepe, Ankara, TURKEY

J. SVOBODA — Department of Physical Chemistry and Macromolecular Chemistry, Faculty of Science, Charles University, Albertov 6, 128 43 Prague 2, CZECH REPUBLIC xix

Robert SZILAGYI — Chemistry Department
Stanford University
Stanford CA, USA

Teresa SZYMANSKA-BUZAR — Faculty of Chemistry
University of Wroclaw
14 F. Joliot-Curie Street,
50-383 Wroclaw, POLAND

Cyril THURIER — Institut de Chimie de Rennes
Bat 10 C, Campus de Beaulieu
35042 Rennes Cedex, FRANCE

Gergely TOTH — Department of Organic Chemistry
University of Veszprém
P.O. Box. 158,
H-8210 Veszprém, HUNGARY

Hayati TÜRKMEN — Ege University, Faculty of Science
Chemistry Department, Bornova
35100 İzmir, TURKEY

Canan ÜNALEROĞLU — Hacettepe University
Department of Chemistry
06532 Beytepe, Ankara, TURKEY

Francis VERPOORT — Department of Inorganic
and Physical Chemistry
Ghent University, Krijgslaan 281 (S3),
9000 Ghent, BELGIUM

Jiry VOHLIDAL — Department of Physical Chemistry
and Macromolecular Chemistry
Faculty of Science,
Charles University, Albertov 6,
128 43 Prague 2, CZECH REPUBLIC

Kenneth B. WAGENER — Department of Chemistry,
University of Florida
P.O. Box 117200, Gainsville
Florida 32611-7200, USA

Hayley WAN — Organic Chemistry
Department of Chemistry,
University of Durham
Durham DH1 3LE, UK

Karin WEISS

Laborartorium fur Anorganische Chemie
University of Bayreuth
D-95440 Bayreuth, GERMANY

Christine WIRT-PFEIFER

Laboratorium Fur Anorganishe Chemie
University of Bayreuth
D-95440 Bayreuth, GERMANY

Birgül ZÜMREOĞLU-KARAN

Hacettepe University
Department of Chemistry
06532 Beytepe, Ankara, TURKEY

THE ALKENE METATHESIS RUTHENIUM CATALYST SAGA

David Sémeril, Pierre H. Dixneuf
Institut de Chimie de Rennes, UMR 6509 Université de Rennes - CNRS, Organométalliques et Catalyse, Campus de Beaulieu, 35042 Rennes, France

Abstract

The history of the alkene metathesis ruthenium catalyst discovery is presented and illustrated with catalyst precursor preparation methods and with references to applications in fine chemistry and polymerisation. These precursors involve well-defined and in situ prepared ruthenium-alkylidenes, -vinylidenes and -allenylidenes containing bulky electron-rich phosphine or imidazolylidene and imidazolinylidene ligands.

Key words: Ruthenium catalysts, alkylidenes, vinylidenes, allenylidenes, alkene metathesis, enyne metathesis.

INTRODUCTION

The catalytic olefin metathesis reaction has known in the last years a considerable development and it has become an important tool in organic[1] macromolecules[1] or polymer chemistry[2]. The "olefin metathesis" expression was introduced for the first time by Calderon[3] in 1967, but was actually discovered as soon as 1955 by Anderson and Mercking[4]. It consists in a simultaneous cleavage and formation of two double carbon-carbon bonds (Scheme 1).

(Scheme 1)

Alkene metathesis reactions actually gather several subtypes such as in organic reactions : Ring Closing Metathesis (RCM), Ring Opening Metathesis (ROM), Cross Metathesis (CM)[1] or in polymerization reactions : Ring Opening Metathesis Polymerization (ROMP)[2] and Acyclic Diene Metathesis (ADMET)[5] (Scheme 2).

1

Y. Imamoglu and L. Bencze (eds.), Novel Metathesis Chemistry: Well-Defined Initiator Systems for Specialty Chemical Synthesis, Tailored Polymers and Advanced Material Applications, 1–21.

(Scheme 2)

The first example of norbornene polymerization using $TiCl_4/EtMgBr$ was reported in 1955 by Anderson and Merckling[4] and as the first "olefin disproportion", the old name for cross metathesis, was published in 1964 by Banks and Bailey[6]. These first catalytic systems were *in situ* generated starting from a transition metal derivative for example tungsten chloride[3b, 7], rhenium oxide[8] and an alkylating agent ($SnMe_4$, $MgMe_2$, Cp_2TiMe_2, $PbEt_4$...) to generate a metal-carbene. This active species was observed by Muetterties[9] starting from WCl_4O and $MgMe_2$ (Scheme 3).

(Scheme 3)

The low cost and the easy preparation of these catalytic systems rapidly led to their use in important industrial applications. Following the SHOP process[10] (Shell Higher Olefin Process) the alkene metathesis was used for the preparation of middle olefins (C_{11}-C_{14}) starting from shorter and longer chains. Meanwhile, the harsh reaction conditions, a strong Lewis acid is often required, and the toxicity of alkylating agent, for example tin derivatives, limit the utility of such systems and motivated the search of more efficient catalysts.

The first applications of RCM in organic chemistry, using $WCl_6/SnMe_4$, was reported by Villemin[11] in 1980 for the synthesis of exaltolide.

In the 1990's, the *in situ* generated catalytic systems were replaced by well-defined molybdenum catalyst precursor (**1**) developed by Schrock[12] and tungsten complex (**2**) developed by Basset[13]. Asymmetric versions[14] of molybdenum complexes were synthesized by Grubbs and Schrock–Hoveyda groups.

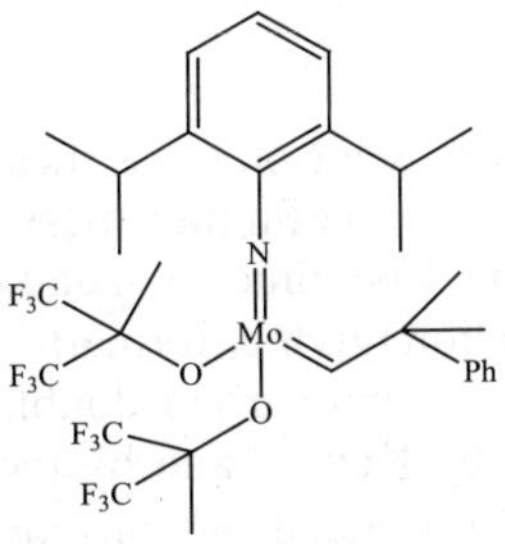

1 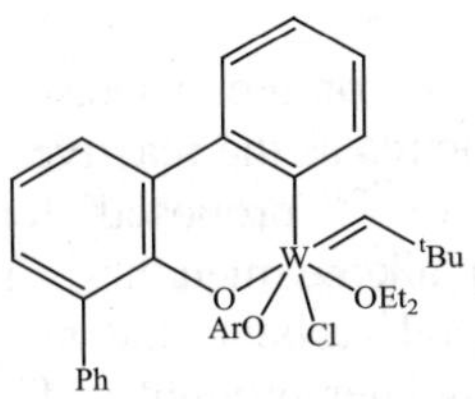   2

Although molybdenum complexes are very active, they are also sensitive to water and solvent impurities. This is a reason why many groups attempted to develop catalytic systems based on ruthenium, less sensitive and well tolerant toward organic functions : alcohol, acetal, ether, silyl ether, ketone, aldehyde, amide, ester[1e]…

After a description of alkene metathesis mechanism, this review will describe the development of ruthenium alkene metathesis catalysts, starting from $RuCl_3$(hydrate) to well defined ruthenium alkylidene complexes containing phosphine or aminocarbene ligand to recent *in situ* prepared multicomponent alkene metathesis catalyst. The key-date of the alkene metathesis catalyst saga show an important acceleration in catalyst precursors discoveries within the last decade and the increasing use of ruthenium catalysts (Scheme 4).

1955	- discovery of olefin metathesis[4]
1955-1990	- use of *in situ* catalytic system[3, 6, 7, 8, 11]
1965	- $RuCl_3$(hydrate)[15]
1970	- proposition of mechanism by Chauvin[16]
1990	- synthesis of Mo-alkylidene[12]
1992	- synthesis of W-alkylidene[13]
1992	- synthesis of Ru-carbene $RuCl_2(=CH\text{-}CH=CR_2)(PPh_3)_2$[34]
1995	- use of diazoalkane to generate *in situ* Ru-carbene[18]
	- $(PCy_3)_2RuCl_2=CHPh$[19]
1996	- asymmetric Mo complex[14a]
1998	- synthesis of Ru-vinylidene[20]
	- synthesis of Ru-allenylidene[21]
	- use of aminocarbene ligand[22]
2000	- Ru-aminocarbene catalyst *in situ* generated[23]
2001	- asymmetric Ru complex[24]

(Scheme 4)

ALKENE METATHESIS MECHANISM

In 1968, Calderon identified, from his work on cross metathesis of deuterated butene, the double bonds as the reactive center in olefin metathesis[25]. Two years later, Chauvin and Herisson[16a] proposed, for the first time, a metal-carbene as the key catalytic species in alkene metathesis reaction giving formally a reversible 2 + 2 cycloaddition of metal carbene and alkene carbon-carbon double bond (Scheme 5). This proposition was later brought to force by Katz[26] and by the characterization of metallacyclobutane intermediates[27] and olefin π-metal carbene complexes[28].

$$R^1HC=CHR^1 \;+\; [M]=CHR^2 \;\rightleftharpoons\; \begin{array}{c} R^1HC-CHR^1 \\ | \qquad\quad | \\ [M]-CHR^2 \end{array} \;\rightleftharpoons\; \begin{array}{c} CHR^1 \\ \| \\ [M] \end{array} \;+\; \begin{array}{c} CHR^1 \\ \| \\ CHR^2 \end{array}$$

(Scheme 5)

$$[M]=CH_2$$

(Scheme 6)

The catalytic cycle for diene RCM is represented in Scheme 6. during the catalytic cycle, a loss of an olefin, generally ethylene, is observed.

ALKENE METATHESIS RUTHENIUM CATALYST PRECURSORS

A.   RuCl₃(HYDRATE)

RuCl₃(hydrate) has been used since 1965 as metathesis initiator for ROMP of norbornene,[15] but induction periods were often long and only a small amount of ruthenium became active[29]. The induction period was reduced when $Ru(tos)_2(H_2O)_6$ complex (tos= p-toluenesulfonate) was used[30]. In this case, the supposed active species was a ruthenium-alkylidene obtained from an alkene[31]. The use of ethyl diazoacetate increased the activity with this new catalytic system, the polymerization of less strained cycloolefins as cyclooctene became possible[32]. Benzylic alcohol was used by Basset[33] to generate from $RuCl_3$(hydrate) a ruthenium-hydride which gave with an olefin the ruthenium-alkylidene, active species for ROMP of norbornene.

These observations and hypothesis of the key role of ruthenium-alkylidene intermediates in alkene metathesis motivated the search of well-defined stable ruthenium-alkylidene catalysts.

B.   RUTHENIUM-PHOSPHINE COMPLEXES

In the 1990's, well-defined ruthenium complexes containing a carbene, vinylidene or allenylidene ligand, were synthesized and used with success in olefin metathesis reactions. The Grubbs type ruthenium-carbene catalysts and their various modifications will be first presented following by relevant ruthenium-phosphines complexes.

1.   Grubbs ruthenium-carbene catalysts

1.1.   Ruthenium-vinylcarbene complexes

In 1992, Grubbs synthesized for the first time a well-defined ruthenium-carbene complex[34] **3a** bearing two triphenylphosphine ligands and a carbene ligand, starting from $RuCl_2(PPh_3)_3$ or $RuCl_2(PPh_3)_4$ and 3,3-

(Scheme 7)

diphenylcyclopropene (Scheme 7)[17, 34].

The complex **3a** was first used in polymerization *via* ROMP process of norbornene[34] and after substitution of triphenylphosphines by tricyclohexylphosphines (PCy_3), the resulting complex **3b** became very active either for ROMP or for fine chemistry *via* RCM reaction[17]. The reactivity of catalyst **3** increased with the increasing of both phosphine basicity and bulkiness: $PPh_3 << P^iPr_3 < PCy_3$.

As 3,3-diphenylcyclopropene is unstable and difficult to synthesize on large scale, different carbene precursor as propargylic chloride[35] or propargylic alcohol[36] were used.

1.2.   Ruthenium-carbene complexes

In 1995, Grubbs[19] synthesized a new type of ruthenium-carbene **4** (Y= H, NMe_2, OMe, Me, F, Cl and NO_2) using phenyldiazomethane as carbene precursor (Scheme 8).

(Scheme 8)

Kinetic studies[37a] showed that with **4**, the activation step was faster compared to catalysts **3**. The activation rate depends also on the phenyl substituent: the more efficient catalyst found was **4a** (Y= H). The halogen influence was studied and, contrary to phosphines, the catalyst reactivity increased when the halogen size decreased and electron withdrawing character increased: $Cl > Br \gg I$[37b].

The use of diazoalkane as carbene source is dangerous, especially on large-scale reaction. For this reason, several groups developed new synthetic route for the synthesis of catalyst **4a**. For example, one way to obtain ruthenium-carbene involves the use of 1,1-dichloroalkane and a ruthenium-zero precursor (scheme 9). The dichloroalkane reacts with the Ru(0) by oxidative addition and after chloride α-elimination give the Ru(II)-carbene[38].

R= H, Ph

(Scheme 9)

Various one step reactions using an acetylenic derivative were developed by Werner[39] starting from $RuCl_3.3\,H_2O$ or by Schaaf et Hafner[40] using $[RuCl_2(COD)]_n$ as ruthenium source. Vinyl chloride[35a] and benzyl ylide[41] were used as carbene precursor to generate a ruthenium-carbene.

1.3. Modification of Grubbs complex

With the aim to obtain a better reactivity or to adapt the ruthenium-carbene **4a** to specific catalytic conditions, several modifications on complex **4a** were reported. For example, in 1998 an organometallic moiety was used as ligand in **4a** to generate bimetallic carbene complexes **5**[42]. These complexes were efficient for ROMP or RCM. Their activity depends on the second metal nature and increases in the order $Rh > Os > Ru$.

5a   5b   5c

The introduction of a Schiff base ligand on complex **4a** was performed by Grubbs[43] and Verpoort[44] to give complexes **6** (Scheme 10). These complexes are soluble in polar solvents and could be used for RCM in ionic liquid[45] or in methanol[43]

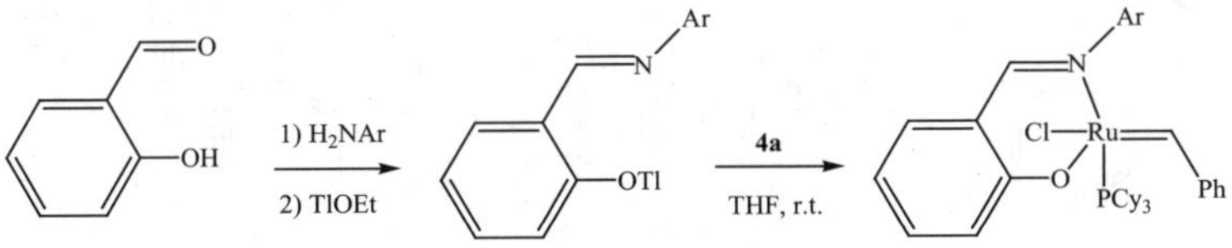

(Scheme 10)

The substitution of tricyclohexylphosphines by water-soluble phosphines gave complexes 7[46] which catalyzed ROMP of functional norbornene[46a, b] or RCM[46d] in protic solvent.

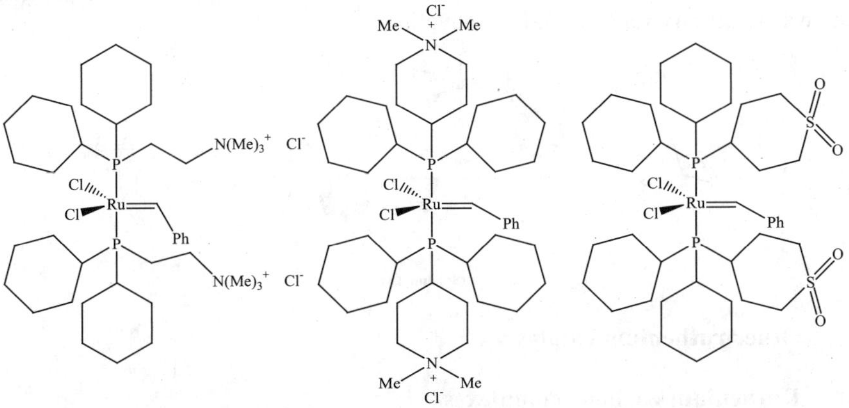

7a   7b   7c

Grubbs showed that during the catalytic cycle, complex **4** lost a phosphine ligand[37b] with the aim to obtain more efficient catalysts, he synthesized[47] 14-electron complexes containing only one PCy_3 (**8**) (Scheme 11). These new catalysts became more reactive than catalysts **4** only when Brønsted acid (HCl) was added to exchange alcoxide ligand by chloride.

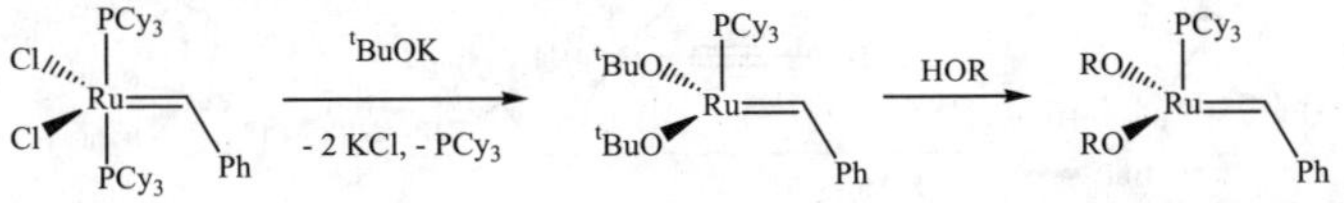

(Scheme 11)

Another type of 14-electron precursor was developed by Hoveyda[48] in 1999, this complex **9** possessed a hemilabile ligand instead of the second phosphine. The complex was obtained directly by cross metathesis between the carbene and 1-isopropoxy-2-vinyl-benzene (Scheme 12). This ruthenium-carbene complex was very robust and could be reused after reaction by filtration on silica and was later supported on a dendrimer[49] on polyethylene glycol polymer[50] and on polystyrene[51].

(Scheme 12)

Barrett immobilized Grubbs complex on polystyrene and showed that it was possible to use the complex **10** in ROMP of norbornene[52a] (Scheme 13). The complex **10** could be reused only twice in RCM of ethyl diallylmalonate but the presence of hexene was necessary for the catalyst recycling[52b].

(Scheme 13)

2.   Other ruthenium catalysts

2.1.   Ruthenium-carbene complexes

Hofmann[53] synthesized a trans diphosphine bisruthenium complex (**11**) starting from propargylic chloride and trimethysilyltriflate as chloride abstractor (Scheme 14). This catalyst was an efficient catalyst precursor in ROMP of cyclooctene.

(Scheme 14)

A way to generate a ruthenium-carbene is to use *in situ* a diazoalkane and a ruthenium-arene as ruthenium source. For example, Noels[18, 54] used as soon as 1995 the $RuCl_2(PCy_3)(p\text{-cymene})$ complex and trimethylsilyldiazomethane for ROMP of cyclic olefins. The *in situ* generated active species was close to Grubbs active species (scheme 15).

(Scheme 15)

On a similar way, Herrmann[55] *in situ* generated ruthenium-carbene using bisallylruthenium(IV) derivatives (Scheme 16).

(Scheme 16)

Rieger[56] used $RuCl_2(\eta^1:\eta^6-R_2PR')$ complexes (**12**) as ruthenium source in addition to a chloride abstractor, a silver salt, and diazoalkane. These catalytic systems were efficient in ROMP of norbornene and the polymerization kinetic depends on the nature of the silver salt employed.

Leitner[57] used a diphosphine bis-methallyl ruthenium complex (**13**) for ROMP of norbornene, it this case, the *in situ* formation of a ruthenium-carbene starting from the bis-methallyl ligands was supposed, based on liberation of 2,4-dimethylpenta-1,4-diene during the polymerization and inhibition of the reaction in compressed CO_2.

2.2. Ruthenium-vinylidene complexes

Instead of using a carbene ligand, Ozawa[20, 58] synthesized vinylidene complexes (**14**), starting from $[RuCl_2(p\text{-cymene})]_2$ and an acetylenic derivative (Scheme 17). These complexes were used either in RCM or in ROMP reactions.

(Scheme 17)

Van Koten[59] used the 2,6-bis[(dimethylamino)methyl]pyridine-ruthenium-vinylidene (**15**) in ROMP of norbornene. In this polymerization, the molecular weight and the polydispersity showed the presence of different actives species.

Ozawa observed that the vinylidene complex $Cp^*RuCl(PPh_3)(=C=CHPh)$ (**16**) was able to polymerize norbornene however in lower yield than with other catalysts[20a].

Kurosawa[60] used the cationic arene-ruthenium precursor **17** to generate *in situ* either a ruthenium-vinylidene, *via* addition of phenylacetylene for RCM, or a supposed ruthenium-hydride *via* reaction with triethylamine for cycloisomerization reaction.

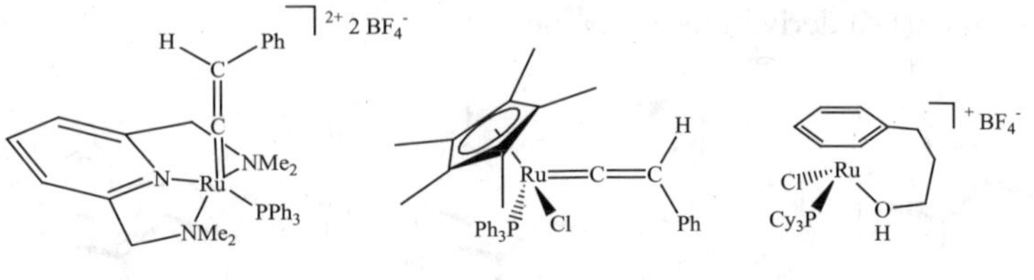

15   16   17

2.3.   Ruthenium- allenylidene complexes

In 1998, Dixneuf and Fürstner reported the synthesis and the use of a ionic ruthenium-allenylidene complex **18** for RCM of dienes[21, 61,62] leading to a variety of macrocycles in the same range of activity as the Grubbs catalyst. Nevertheless, the reactivity and the selectivity of the reaction drastically depend on the nature of escorting anion[63]. In all cases, the reactivity of these complexes could be increased with a UV irradiation at room temperature just before heating[64a]. These ruthenium-allenylidene catalysts were applied to enyne metathesis[64] and ROMP[65].

With complexes **18**, metathesis reactions could be performed in ionic liquids[66, 65b] of the type 1-butyl-3-methylimidazolium due to the cationic nature of these complexes which allowed a perfect solubility.

The synthesis of catalyst precursors **18** was carry out in two or three steps starting from the commercially available $[RuCl_2(p\text{-cymene})]_2$. In all cases, the first step was the formation of the neutral derivative $RuCl_2(PCy_3)(p\text{-cymene})$ and for the complex with PF_6^- as counter anion, the second step was the reaction with $HC{\equiv}CCPh_2OH$ in the presence of $NaPF_6$ to give the ruthenium-allenylidene **18a** [21]. With other counter anions (BF_4^- $CF_3SO_3^-$), the complexes were obtained using a silver salt to abstract chloride ligand. The cationic 16-electron intermediate reacted then with propargylic alcohol to give the allenylidene complexes **18**[63] (Scheme 18).

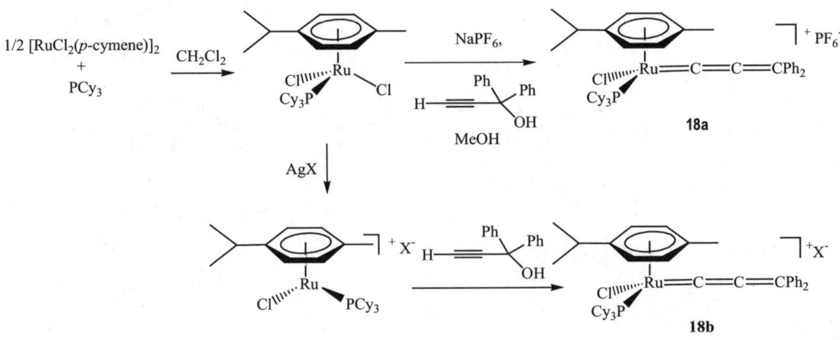

(Scheme 18)

An immobilised version of the ruthenium-allenylidene complex **18a** on polystyrene was recently synthesized by Kobayashi[67]. This precatalyst **19** could be reused several times in RCM but only after a reactivation process.

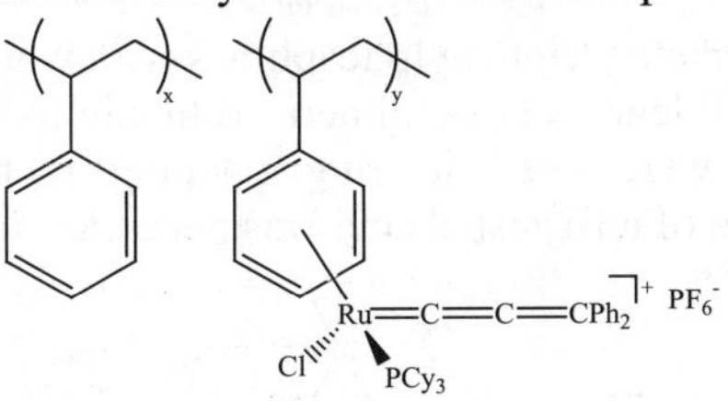

19

The decoordination of the *p*-cymene ligand was expected in the case of ruthenium-allenylidene **18** to give a highly coordinatively unsaturated ruthenium species[64a]. The decoordination of the arene ligand requires a thermal or photochemical activation. With the aim to facilitate the activation step, arene free ruthenium-allenylidenes were synthesized starting from $RuCl_2(dmso)_4$[68]. Cationic $[RuCl(=C=C=CPh_2)(PCy_3)_x(dmso)_2]OTf$ (x= 1 or 2) and neutral $RuCl_2(=C=C=CPh_2)(PCy_3)(dmso)_2$, or $RuCl_2(=C=C=CAr_2)(PCy_3)_2(dmso)$ complexes, were tested in ROMP of norbornene and cyclooctene. Moweover the reactivity of these complexes was lower than with precatalyst **18b** (X⁻= TfO⁻)[65].

Hill's group[69] synthesized, starting from $RuCl_2(PPh_3)_3$ and $HC{\equiv}CCPh_2OH$, a ruthenium-indenylidene (**20**) (Scheme 19) which was first supposed to be a ruthenium-allenylidene. Nolan[70] showed that during the reaction, the allenylidene ligand rearranged to give an indenylidene, by arene electrophilic substitution with the Ru=C carbon atom. The complex **20** was further used in RCM[71].

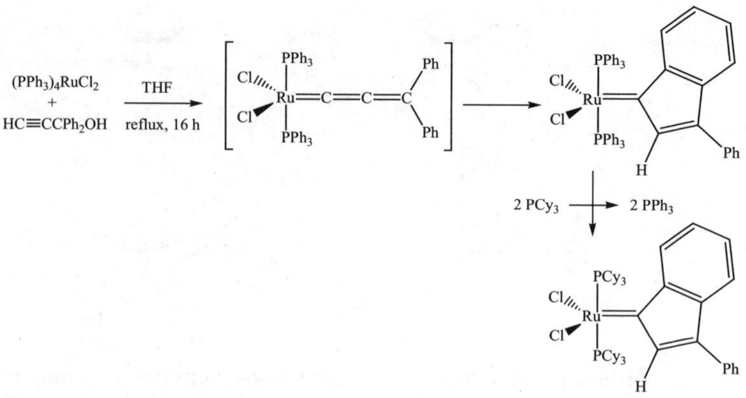

(Scheme 19)

Peruzzini[72] exchange triphenylphosphines with water-soluble phosphines and obtained ruthenium-vinylidene (**21**) or dimeric ruthenium-allenylidene (**22**) (Scheme 20). These complexes were used in ring opening metathesis of cyclopentene, nevertheless, the presence of a Brønsted acid was necessary to activate the catalyst.

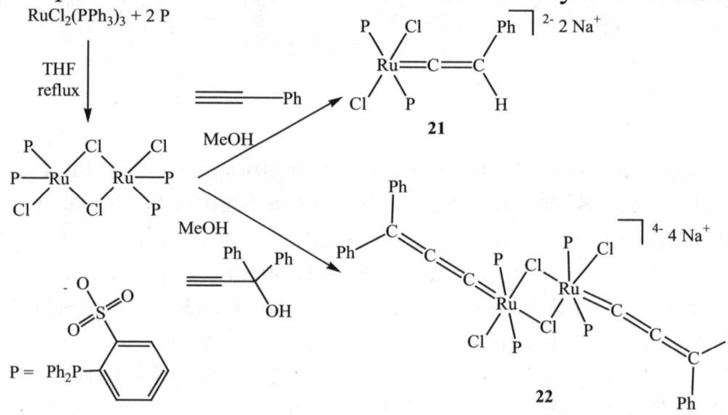

(Scheme 20)

Werner[73] synthesized a cationic ruthenium-allenylidene complex with two phosphines containing a hemilabile methoxy **23**, which was less efficient than precatalysts **18** both in ROMP of cyclooctene and RCM of *N,N*-diallyltosylamide.

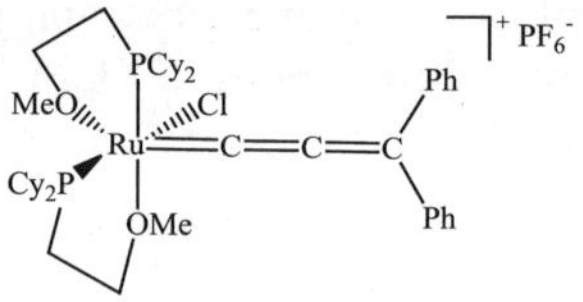

2.4.   Other ruthenium complexes

Several arene-ruthenium precatalysts were used in metathesis reactions as ruthenium source. Lindner[74] reported a series of arene-ruthenium hydride bearing a hemilabile ether **24** for ROMP of norbornene.

$$R = CH_3,$$

24

An *in situ* generated catalytic system, composed of the dimeric [$RuCl_2(p$-cymene)]$_2$ and tricyclohexylphosphine, was used by Fürstner[75] in RCM. The catalyst was generated by neon light and was sensitive to the catalytic conditions.

A catalytic system, *in situ* generated starting from $RuCl_3$ or $RuBr_3$ as ruthenium source, 2-butyne diacetate and PCy_3, was reported by Nubel[76] and used in cross metathesis of α-olefins. The reaction was carried at 80-90°C under an atmosphere of hydrogen. In these conditions, the olefins were not hydrogenated or isomerized. Meanwhile the nature of the active species was unknown.

A bisruthenium complex was synthesized by Mol[77] using silver trifluoroacetate and Grubbs complex **4a** to obtain the $Ru_2(=CHPh)_2(CF_3CO_2)_2(\mu$-$CF_3CO_2)_2(PCy_3)_2(\mu$-$H_2O)$ where the two ruthenium atoms were bridged by two trifluoroacetate and one water molecule. This catalyst was used in cross metathesis of internal alkenes.

C   RUTHENIUM-IMIDAZOLYLIDENE AND IMIDAZOLINYLI-DENE COMPLEXES

The complexation of Wanzlick type carbenes[78], possessing a 1,3-disubstituted imidazol-2-ylidene (**25a**) or 1,3-disubstituted-4,5-dihydroimidazol-2-ylidene (**25b**) moiety, to the ruthenium give a generation of ruthenium metathesis catalysts. These aminocarbene ligands were shown especially by Lappert[79] to afford a large variety of electron-rich carbene-metal complexes.

25a   **25b**

1.   Grubbs aminocarbene-ruthenium catalysts and derivatives

In 1998, Herrmann[22] substituted on Grubbs catalyst **4a** one or two tricyclohexylphosphines by imidazol-2-ylidene carbene. These new mono **26**[22, 80] or bimetallic **27**[80] catalysts showed a higher reactivity, compared to **4a**, both in ROMP or RCM[81] reactions.

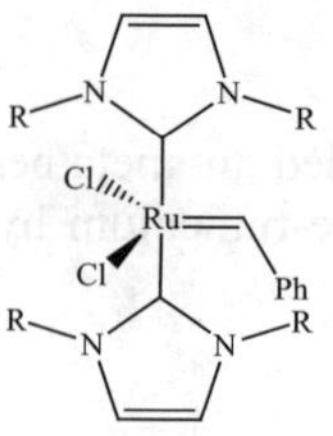

14

26a

R = Cy, ^{i}Pr

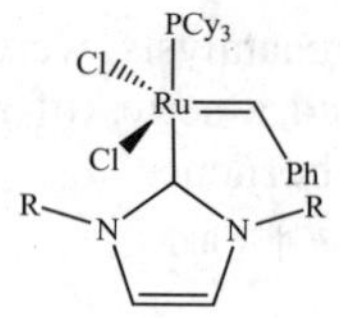

26b

R = Cy, (R,R)-CH(CH$_3$)Ph,
(R,R)-CH(CH$_3$)naphthyl

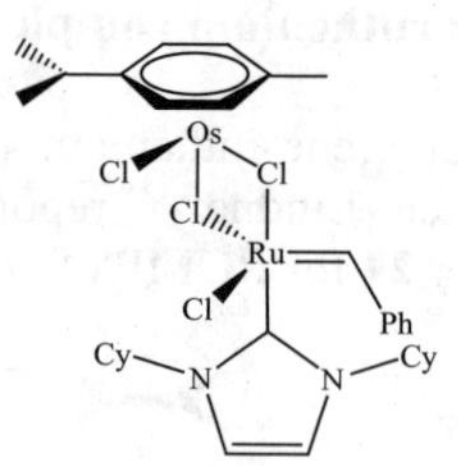

27

These aminocarbene-ruthenium complexes were obtained by

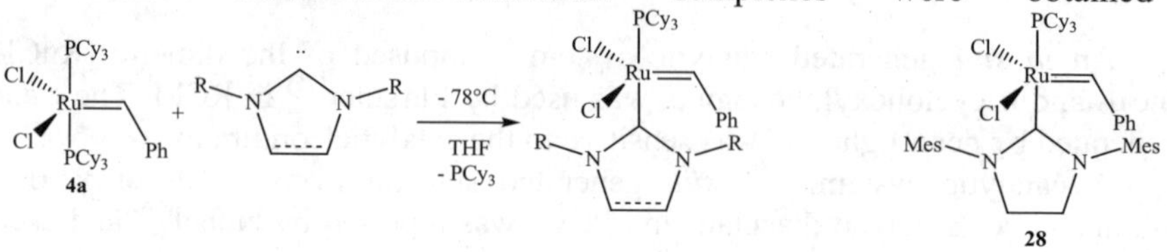

4a

28

exchange of the phosphine by a carbene on **4a** in THF at -78°C (Scheme 21)[80b].

Nolan[82] and Fürstner[83] studied the influence of the substituent on the aminocarbene and an increasing of the reactivity was found when bulky and aromatic substituents (mesityl or 2,6-diisopropylphenyl) were used with respect to alkyl or aromatic but non-bulky substituents.

Non-aromatic carbene, more electron-rich, were used by Grubbs[84] to generate a more reactive catalyst, in this case, the mesityl substituted aminocarbene (**28**) was even more efficient that molybdenum complex **1** in RCM[84a] and ROMP[84b]. The new precatalyst **28** was especially active in cross metathesis[84,85] and its activity could be improved using microwave in liquid ionic[86].

By changing the nature of the phosphine ligand on complex **28**, Grubbs[87] synthesized a series of complexes containing a triarylphosphine ligand. These catalysts were almost two orders of magnitude more active for olefin metathesis reactions than the previous complex **28**.

Enantioselective RCM catalyzed by ruthenium catalyst was realized by Grubbs[24] using a chiral carbene (**29**). In this case, the stereoselectivity transfer of the ligand to the metal center was increased by ortho-substituents of the aryl groups and by an *in situ* substitution of chloride by iodide. Another chiral complex (**30**) was reported by Hoveyda[88] and used in asymmetric ring opening metathesis in air.

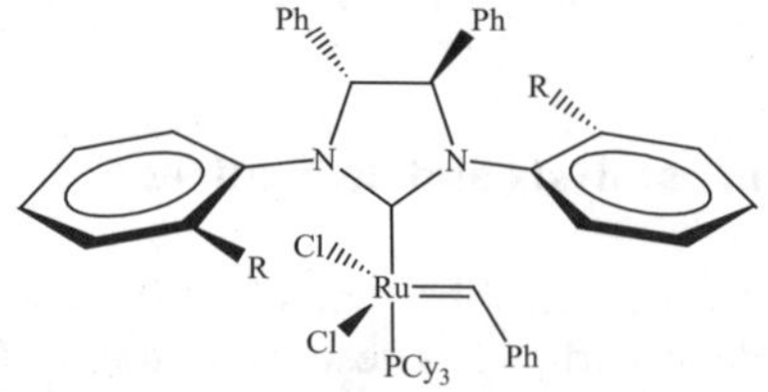

R= Me, ^{i}Pr

29

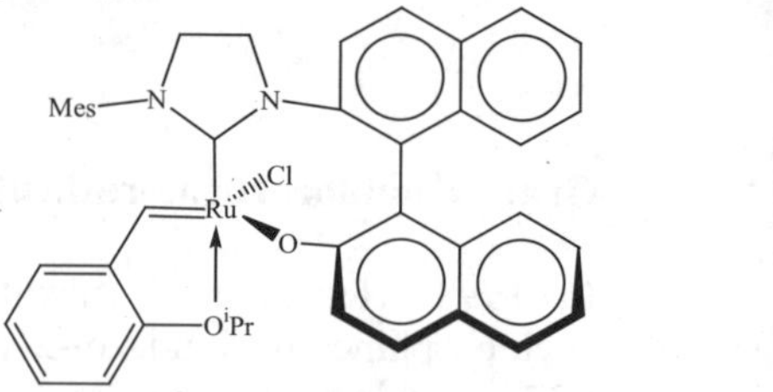

30

Herrmann[89] synthesized complexes bearing a pyridinylalcoholate ligand (**31**) which acted as a "dangling" ligand during catalysis (Scheme 22). These catalysts showed rather low activity at room temperature due to the pyridinyl stabilization but at elevated temperatures their performance could be compared to that of the catalyst **28** in ROMP of norbornene and cyclooctene.

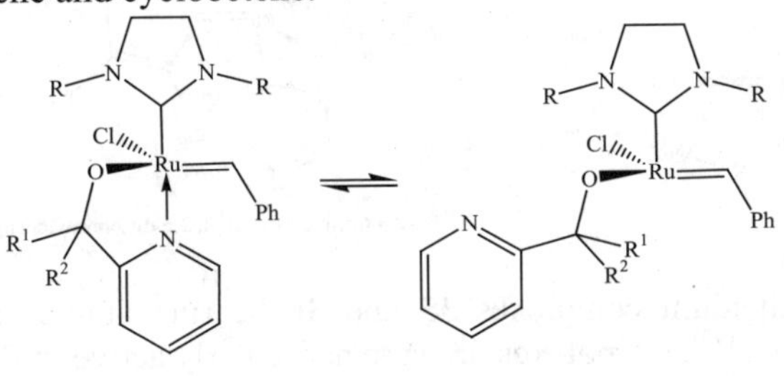

31

(Scheme 22)

Blechert[90] modified complex **28** to obtain precatalyst **32a** (Scheme 23). Recently, the reactivity of catalyst **32a** was

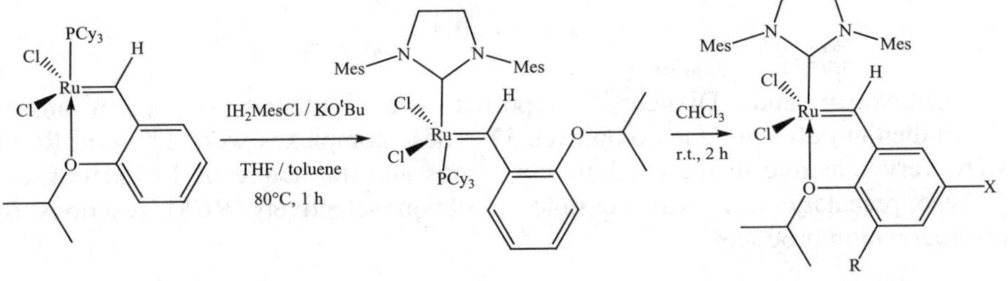

32a R= H, X= H
32b R= Ph, X= H
32c R= H, X= NO$_2$

(Scheme 23)

increased by the introduction of a phenyl group near the *iso*-propyl ether (**32b**)[90b] or by introduction of a nitro group[91] (**32c**). In the presence of a strong electron-withdrawing group at the 2-isopropoxystyrene, the high reactivity could be explained by a decrease of the electron density of the oxygen atom which reduces its chelating ability, thus facilitating formation of the 14-electron active species. The complex **32a** could be attached to dendrimers and they could be easily reused[49].

Aminocarbene complexes were also immobilized. For example, the precatalyst **28** was directly microencapsulated in polystyrene by Gibson[92] or by Barrett[93] and attached by Blechert[94] on a polystyrene-support.

2. Other aminocarbene-ruthenium catalysts

Aminocarbene ligands **25** were introduced on ruthenium complexes already synthesized with tricyclohexylphosphines. Nolan[70] showed that starting from [RuCl$_2$(*p*-cymene)]$_2$ and depending on the synthetic way, it was possible to obtain complexes **33** or **34** with an allenylidene or an indenylidene ligand. Complex **33** was not active in RCM contrary to complexes **34** which were particularly efficient in the formation of trisubstituted double bonds.

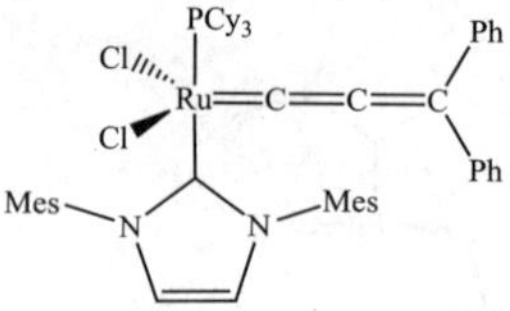

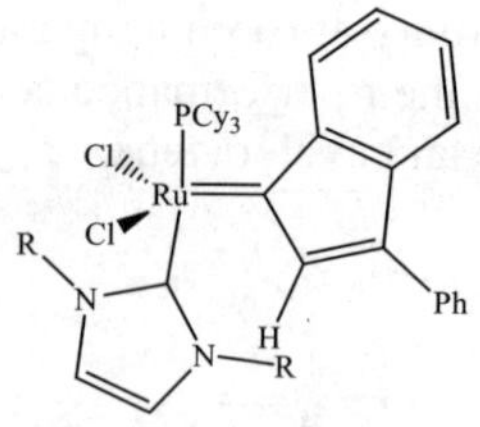

R= 2,4,6-trimethylphenyl, 2,6-diisopropylphenyl

33   **34**

Arene-ruthenium complexes **35** and **36** bearing aminocarbene ligand were synthesized by Nolan[95]. Complexes **35** were particularly active in ROMP[96].

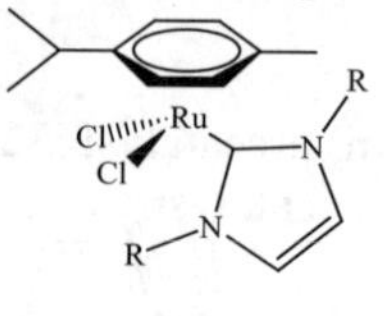

35a R= Mesityl
35b R= 2,6-diisopropylphenyl   **36**

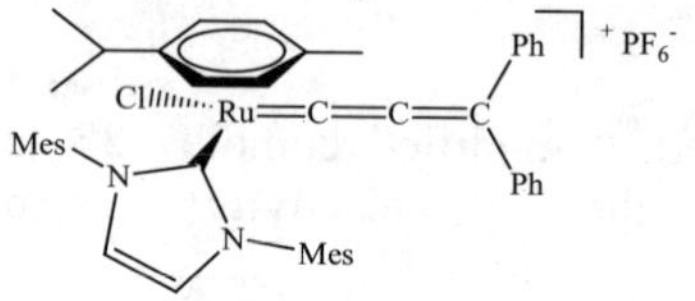

Çetinkaya and Dixneuf[97] reported the synthesis of (η^1:η^6-arene-carbene)ruthenium-allenylidene complexes **37**. These complexes were active in RCM but were very sensitive to the catalytic conditions and the nature of 1,6-diene used. With these precatalysts, it was possible to obtain selectively RCM reactions or cycloisomerization products.

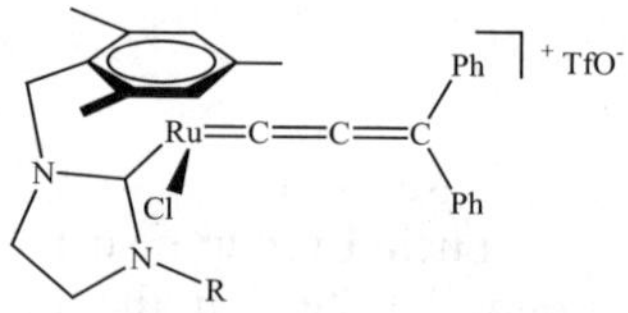

R= CH$_2$CH$_2$OCH$_3$, CH$_2$Mes

37

3.   *In situ* generated catalytic systems

The synthesis of the aminocarbene complexes require several organometallic steps and purification under inert atmosphere with anhydrous solvents. To remedy these drawbacks, *in situ* syntheses of catalytic systems were developed. Grubbs[23] generated *in situ* complex **28** starting from complex **4a**, imidazolium salt and ^{t}BuOK. The free tricyclohexylphosphine was then protonated to avoid its recoordination to the ruthenium center. No difference of reactivity was observed between this *in situ* generated catalyst and isolated catalyst **28**.

Grubbs[98] *in situ* generated a ruthenium-vinylidene complex starting from the [RuCl$_2$(p-cymene)]$_2$, bis(1,3-mesitylimidazol-2-ylidene and *tert*-butylacetylene (Scheme 24), with non aromatic carbene, no conversion was observed in RCM.

(Scheme 24)

In 2001, the Rennes group[66b, 99] generated a three-component catalytic system using the [RuCl$_2$(*p*-cymene)]$_2$, 1,3-dimesitylimidazolium or 1,3-dimesityl-4,5-dihydroimidazolium chloride as carbene precursor and cesium carbonate in the molar ration 1 / 2 / 4. This catalytic system was efficient in enyne metathesis even with substituted double[99a] or triple bond[99].

This *in situ* generated catalytic system, using non-aromatic carbene precursor, gave with 1,6-dienes a dichotomic behavior[100]. When the reaction was carried out under an atmosphere of argon or nitrogen, only the cycloisomerization product was observed. In the presence of acetylene the resulting catalytic system selectively catalyzed the metathesis reaction, probably by formation of a ruthenium-vinylidene derivative.

This three component catalytic system was used by Noels[96], using imidazolium or 4,5-dihydroimidazolium salts, and by Dixneuf[101], using 4,5-dihydroimidazolium salt, in ROMP of norbornene and cyclooctene.

CONCLUSION

The alkene metathesis reaction has become, within the last few years, an unavoidable tool for the formation of carbon-carbon double bonds and for the design of useful molecules, macrocycles, polymers, and immobilized catalysts. Although Molybdenum-alkylidene catalysts are also efficient, the increasing importance of alkene metathesis is certainly due to the synthesis of very active ruthenium catalysts, more tolerant towards organic functions and less sensitive to water or impurities.

This review shows the evolution of ruthenium catalysts for metathesis reactions starting from RuCl$_3$(hydrate) to well-defined ruthenium-carbene complexes. The use of various electron-rich and sterically hindered ligands, such as phosphines or aminocarbenes coordinated to a ruthenium alkylidene, constitutes a key for these well-defined ruthenium catalyst precursors to reach catalytic efficiency. Moreover, alkylidene analogues such as vinylidene and especially allenylidene are now commonly used. A new trend using *in situ* prepared catalysts is appearing. It is based upon the attempt to *in situ* form the catalytic species assumed to arise from well-defined catalysts.

REFERENCES

(1)   a- Grubbs R. H., Miller S. J., Acc. Chem. Res., 1995; 28: 446-452; b- Schmalz H. G., Angew. Chem., Int. Ed. Engl., 1995; 34: 1833-1836; c- Schuster M., Blechert S., Angew. Chem., Int. Ed. Engl., 1997; 36: 2036-2056; d- Ed.: A. Fürstner, *Alkene Metathesis in Organic Synthesis*. Berlin: Springer, 1998; e- Armstrong S. K., J. Chem. Soc., Perkin Trans. 1, 1998; 371-388; f- Grubbs R. H., Chang S., Tetrahedron, 1998; 54: 4413-4450; g- Pariya C., Jayaprakash K. N., Sarkar A., C. Chem. Rev., 1998; 168: 1-48; h- Ivin K. J., J. Mol. Catal A: Chem., 1998; 133: 1-16; i- Fürstner A., Angew. Chem., Int. Ed. Engl., 2000; 39: 3012-3043; j- Trnka T. M., Grubbs R. H., Acc. Chem. Res., 2001; 34: 18-29; k- Jafarpour L., Nolan S. P., J. Organomet. Chem., 2001; 617: 17-27

(2)   a- Ivin K. J., Mol J. C., *Olefin Metathesis and Metathesis Polymerization*. San Diego: Academic Press, 1997; b- Frenzel U., Nuyken O., J. Polym. Sci. Part A: Polym. Chem., 2002; 40: 2895-2916; c- Buchmeiser M. R., Chem. Rev., 2000; 100: 1565-1604

(3)   a- Calderon N., Chem. Eng. News, 1967; 45: 51-53; b- Calderon N., Chen H. Y., Scott K. W., Tetrahedron Lett., 1967; 34: 3327-3329

(4)   Andreson A. W., Merckling M. G., US-A 2721189, 1955

(5)   Schwendeman J. E., Church A., Wagener K. B., Adv. Synth. Catal., 2002; 344: 597-613

(6)   Banks R. L., Bailey G. C., Ind. Eng. Chem. Process Res. Dev., 1964; 3: 170-173

(7)   a- Tsuji J., Hashiguchi S., Tetrahedron Lett., 1980; 21: 2955-2958; b- Tsuji J., Hashiguchi S., J. Organomet. Chem., 1981; 218: 69-80; c- Bosma R. H. A., van den Aardweg F., Mol J. C., J. Chem. Soc., Chem. Commun., 1981; 1132-1133

(8)   Junga H., Blechert S., Tetrahedron Lett., 1993; 34: 3731-3732

(9)   Muetterties E. L., Band E., J. Am. Chem. Soc., 1980; 102: 6572-6574

(10)   Freitas E. R., Gum C. R., Chem. Eng. Prog., 1979; 75: 73

(11)   Villemin D., Tetrahedron Lett., 1980; 21: 1715-1718

(12)   Schrock R. R., Murdzek J. S., Bazan G. C., Robbins J., DiMare M., O'Regan M., J. Am. Chem. Soc., 1990; 112: 3875-3886

(13)   a- Couturier J. L., Paillet C., Leconte M., Basset J. M., Weiss K., Angew. Chem., Int. Ed. Engl., 1992; 31: 628-631; b- Couturier J. L., Tanaka K., Leconte M., Basset J. M., Ollivier J., Angew. Chem., Int. Ed. Engl., 1993; 32: 112-115

(14)   a- Fujimura O., de la Mata F. J., Grubbs R. H., Organometallics, 1996; 15: 1865-1871; b- Alexander J. B., La D. S., Cefalo D. R., Hoveyda A. H., Schrock R. R., J. Am. Chem. Soc., 1998; 120: 4041-4042; c- La D. S., Alexander J. B., Cefalo D. R., Graf D. D., Hoveyda A. H., Schrock R. R., J. Am. Chem. Soc., 1998; 120: 9720-9721; d- Hoveyda A. H., Schrock R. R., Chem. Eur. J., 2001; 7: 945-950

(15)   a- Michelotti F. W., Keaveney W. P., J. Polym. Sci., 1965; A3: 895-905; b- Rinehart R. E., Smith H. P., Polym. Lett., 1965; 3: 1049-1052

(16)   a- Herisson J. L., Chauvin Y., Makromol. Chem., 1970; 141: 161-176; b- Chauvin Y., Commereuc D., Revue de L'Institut Français du Pétrole, 1974; 29: 73-85

(17)   a- Nguyen S. T., Grubbs R. H., J. Am. Chem. Soc., 1993; 115: 9858-9859; b- Fu G. C., Nguyen S. T., Grubbs R. H., J. Am. Chem. Soc., 1993; 115: 9856-9857

(18)   Stumpf A. W., Saive E., Demonceau A., Noels A. F., J. Chem. Soc., Chem. Commun., 1995; 1127-1128

(19)   Schwab P., France M. B., Ziller J. W., Grubbs R. H., Angew. Chem., Int. Ed. Engl., 1995; 34: 2039-2041

(20)   a- Katayama H., Yoshida T., Ozawa F., J. Organomet. Chem., 1998; 562: 203-206; b- Katayama H., Ozawa F., Organometallics, 1998; 17: 5190-5196

(21)   Fürstner A., Picquet M., Bruneau C., Dixneuf P. H., Chem. Commun., 1998; 1315-1316

(22)   Weskamp T., Schattenmann W. C., Spiegler M., Herrmann W. A., Angew. Chem., Int. Ed. Engl., 1998; 37: 2490-2493

(23)   Morgan J. P., Grubbs R. H., Org. Lett., 2000; 2: 3153-3155

(24)   Seiders T. J., Ward D. W., Grubbs R. H., Org. Lett., 2001; 3: 3225-3228

(25)   Calderon N., Ofstead E. A., Ward J. P., Judy W. A., Scott K. W., J. Am. Chem. Soc., 1968; 90: 4133-4140

(26)   a- Katz T. J., McGinnis J., J. Am. Chem. Soc., 1975; 97: 1592-1594; b- Katz T. J., McGinnis J., Altus C., J. Am. Chem. Soc., 1976; 98: 606-608; c- Katz T. J., Rothchild R., J. Am. Chem. Soc., 1976; 98: 2519-2526

(27)   a- Howard T. R., Lee J. B., Grubbs R. H., J. Am. Chem. Soc., 1980; 102: 6876-6878; b- Kress J., Osborn J. A., Greene R. M. E., Ivin K. J., Rooney J. J., J. Am. Chem. Soc., 1987; 109: 899-901; c-

Bazan G. C., Oskam J. H., Cho H. N., Park L. Y., Schrock R. R., J. Am. Chem. Soc., 1991; 113: 6899-6907

(28) a- Kress J., Osborn J. A., Angew. Chem., Int. Ed. Engl., 1992; 31: 1585-1587; b- Tallarico J. A., Bonitatebus Jr. P. J., Snapper M. L., J. Am. Chem. Soc., 1997; 119: 7157-7158

(29) Novak B. M., Grubbs R. H., J. Am. Chem. Soc., 1988; 110: 960-961

(30) Novak B. M., Grubbs R. H., J. Am. Chem. Soc., 1988; 110: 7542-7543

(31) France M. B., Grubbs R. H., McGrath D. V., Paciello R. A., Macromolecules, 1993; 26: 4742-4747

(32) France M. B., Paciello R. A., R. H. Grubbs, Macromolecules, 1993; 26: 4739-4741

(33) Mutch A., Leconte M., Lefebvre F., Basset J. M., J. Mol. Catal. A : Chem., 1998; 133: 191-199

(34) Nguyen S. T., Johnson L. K., Grubbs R. H., J. Am. Chem. Soc., 1992; 114: 3974-3975

(35) a- Wilhelm T. E., Belderrain T. R., Brown S. N., Grubbs R. H., Organometallics, 1997; 16: 3867-3869; b- Amoroso D., Snelgrove J. L., Conrad J. C., Drouin S. D., Yap G. P. A., Fogg D. E., Adv. Synth. Catal., 2002; 344: 757-763

(36) Harlow K. J., Hill A. F., Welton T., J. Chem. Soc., Dalton Trans., 1999; 1911-1912

(37) a- Grubbs R. H., Ziller J. W., J. Am. Chem. Soc., 1996; 118: 100-110; b- Dias E. L., Nguyen S., Grubbs R. H., Ziller J. W., J. Am. Chem. Soc., 1997; 119: 3887-3897

(38) a- Belderrain T. R., Grubbs R. H., Organometallics, 1997; 16: 4001-4003; b- Olivan M., Caulton K. G., J. Chem. Soc., Chem. Commun., 1997; 1733-1734

(39) Wolf J., Stüer W., Grünwald C., Werner H., Schwab P., Schulz M., Angew. Chem., Int. Ed. Engl., 1998; 37: 1124-1126

(40) van der Schaaf P. A., Kolly R., Hafner A., Chem. Commun., 2000; 1045-1046

(41) Gandelman M., Rybtchinski B., Ashkenazi N., Gauvin R. M., Milstein D., J. Am. Chem. Soc., 2001; 123: 5372-5373

(42) Dias E. L., Grubbs R. H., Organometallics, 1998; 17: 2758-2767

(43) Chang S., LeRoy J., Wang C., Henling L. M., Grubbs R. H., Organometallics, 1998; 17: 3460-3465

(44) De Clercq B., Verpoort F., Adv. Synth. Catal., 2002; 344: 639-648

(45) Bayer AG, EP 1035093A2, 2000

(46) a- Mohr B., Lynn D. M., Grubbs R. H., Organometallics, 1996; 15: 4317-4325; b- Lynn D. M., Mohr B., Grubbs R. H., J. Am. Chem. Soc., 1998; 120: 1627-1628; c- Rölle T., Grubbs R. H., Chem. Commun., 2002; 1070-1071; d- Kirkland T. A., Lynn D. M., Grubbs R. H., J. Org. Chem., 1998; 63: 9904-9909

(47) Sanford M. S., Henling L. M., Day M. W., Grubbs R. H., Angew. Chem., Int. Ed. Engl., 2000; 39: 3451-3453

(48) Kingsbury J. S., Harrity J. P. A., Bonitatebus P. J., Hoveyda A. H., J. Am. Chem. Soc., 1999; 121: 791-799

(49) Garber S. B., Kingsbury J. S., Gray B. L., Hoveyda A. H., J. Am. Chem. Soc., 2000; 122: 8168-8179

(50) Yao Q., Angew. Chem., Int. Ed. Engl., 2000; 39: 3896-3898

(51) Dowden J., Savovié J., Chem. Commun., 2001; 37-38

(52) a- Barrett A. G. M., Cramp S. M., Roberts R. S., Org. Lett., 1999; 1: 1083-1086; b- Ahmood M., Barrett A. G. M., Braddock D. C., Cramp S. M., Procopiou A., Tetrahedron Lett., 1999; 40: 8657-8662

(53) a- Hansen S. M., Rominger F., Metz M., Hofmann P., Chem. Eur. J., 1999; 5: 557-566; b- Hansen S. M., Volland M. A. O., Rominger F., Eisenträger F., Hofmann P., Angew. Chem., Int. Ed. Engl., 1999; 38: 1273-1276; c- Adlhart C., Volland M. A. O., Hofmann P., Chen P., Helv. Chim. Acta, 2000; 83: 3306-3311

(54) Demonceau A., Stumpf A. W., Saive E., Noels A. F., Macromolecules, 1997; 30: 3127-3136

(55) a- Wache S., Herrmann W. A., Artus G., Nuyken O., Wolf D., J. Organomet. Chem., 1995; 491: 181-188; b- Herrmann W. A., Schattenmann W. C., Nuyken O., Glander S. C., Angew. Chem., Int. Ed. Engl., 1996; 35: 1087-1088

(56) Abele A., Wursche R., Klinga M., Rieger B., J. Mol. Catal. A : Chem., 2000; 160: 23-33

(57) Six C., Beck K., Wegner A., Leitner W., Organometallics, 2000; 19: 4639-4642

(58) a- Katayama H., Urushima H., Ozawa F., J. Organomet. Chem., 2000; 606: 16-25; b- Katayama H., Yonezawa F., Nagao M., Ozawa F., Macromolecules, 2002; 35: 1133-1136

(59) del Rio I., van Koten G., Tetrahedron Lett., 1999; 40: 1401-1404

(60) Miyqki Y., Onishi T., Ogoshi S., Kurosawa H., J. Organomet. Chem., 2000; 616: 135-139

(61) Fürstner A., Liebl M., Lehmann C. W., Picquet M., Kunz R., Bruneau C., Touchard D., Dixneuf P. H., Chem. Eur. J., 2000; 6: 1847-1857

(62) a- Osipov S. N., Artyushin O. I., Kolomiets A. F., Bruneau C., Dixneuf P. H., Synlett, 2000; 7: 1031-1033; b- Osipov S. N., Artyushin O. I., Kolomiets A. F., Bruneau C., Picquet M., Dixneuf P. H., Eur. J. Org. Chem., 2001; 3891-3897

20

(63) Picquet M., Touchard D., Bruneau C., Dixneuf P. H., New J. Chem., 1999; 23: 141-143

(64) a- Picquet M., Bruneau C., Dixneuf P. H., Chem. Commun., 1998; 2249-2250; b- Sémeril D., Le Nôtre J., Bruneau C., Dixneuf P. H., Kolomiets A. F., Osipov S. N., New J. Chem., 2001; 25: 16-18

(65) a- Castarlenas R., Sémeril D., Noels A. F., Demonceau A., Dixneuf P. H., J. Organomet. Chem., 2002; 663: 235-238; b- Csihony S., Fischmeister C., Bruneau C., Horváth I. T., Dixneuf P. H., New J. Chem., 2002; 25: 1667-1670

(66) a- Sémeril D., Olivier-Bourbigou H., Bruneau C., Dixneuf P. H., Chem. Commun., 2002; 146-147; b- Sémeril D., Bruneau C., Dixneuf P. H., Adv. Synth. Catal., 2002; 344: 585-595

(67) Akiyama R., Kobayashi S., Angew. Chem., Int. Ed. Engl., 2002; 41: 2602-2604

(68) Alaoui Abdallaoui I., Sémeril D., Dixneuf P. H., J. Mol. Catal A : Chem., 2002; 182-183: 577-583

(69) Harlow K. J., Hill A. F., Wilton-Ely J. D. E. T., J. Chem. Soc., Dalton Trans., 1999; 285-291

(70) a- Schanz-H. J., Jafarpour L., Stevens E. D., Nolan S. P., Organometallics, 1999; 18: 5187-5190; b- Jafarpour L., Schanz H. J., Stevens E. D., Nolan S. P., Organometallics, 1999; 18: 5416-5419

(71) a- Fürstner A., Hill A. F., Liebl M., Wilton-Ely J. D. E. T., Chem. Commun., 1999; 601-602; b- Fürstner A., Thiel O. R., J. Org. Chem., 2000; 65: 1738-1742; c- Fürstner A., Guth O., Düffels A., Seider G., Liebl M., Gabor B., Mynott R., Chem. Eur. J., 2001; 7: 4811-4820

(72) Saoud M., Romerosa A., Peruzzini M., Organometallics, 2000; 19: 4005-4007

(73) Jung S., Brandt C. D., Werner H., New J. Chem., 2001; 25: 1101-1103

(74) Lindner E., Pautz S., Fawzi R., Steimann M., Organometallics, 1998; 17: 3006-3014

(75) Fürstner A., Ackermann L., Chem. Commun., 1999; 95-96

(76) Nubel P. O., Hunt C. L., J. Mol. Catal. A : Chem., 1999; 145: 323-327

(77) Buchowicz W., Mol J. C., Lutz M., Spek A. L., J. Organomet. Chem., 1999; 588: 205-210

(78) a- Wanzlick H. W., Schönherr H. J., Angew. Chem., Int. Ed. Engl., 1968; 7: 141-142; b- Herrmann W. A., Köcher C., Angew. Chem., Int. Ed. Engl., 1997; 36: 2162-2187; c- Bourissou D., Guerret O., Gabbaï F. P., Bertrand G., Chem. Rev., 2000; 100: 39-92; d- Herrmann W. A., Angew. Chem., Int. Ed. Engl., 2002; 41: 1291-1309

(79) a- Cardin D. J., Çetinkaya B., Lappert M. F., Chem. Rev., 1972; 72: 545-574; b- Çetinkaya B., Dixneuf P. H., Lappert M. F., J. Chem. Soc., Dalton Trans., 1974; 1827-1833

(80) a- Weskamp T., Kohl F. J., Hieringer W., Gleich D., Herrmann W. A., Angew. Chem., Int. Ed. Engl., 1999; 38: 2416-2419; b- Weskamp T., Kohl F. J., Herrmann W. A., J. Organomet. Chem., 1999; 582: 362-365

(81) a- Scholl M., Trnka T. M., Morgan J. P., Grubbs R. H., Tetrahedron Lett., 1999; 40: 2247-2250; b- Ackermann L., Fürstner A., Weskamp T., Kohl F. J., Herrmann W. A., Tetrahedron Lett., 1999; 40: 4787-4790

(82) a-Huang J., Schanz H. J., Stevens E. D., Nolan S. P., Organometallics, 1999; 18: 5375-5380 ; b- Huang J., Stevens E.D., Nolan S.P., Petersen J.L., J. Am. Chem. Soc., 1999; 121 :2674-2678

(83) a- Fürstner A., Krause H., Ackermann L., Lehmann C. W., Chem. Commun., 2001; 2240-2241; b- Fürstner A., Ackermann L., Gabor B., Goddard R., Lehmann C. W., Mynott R., Stelzer F., Thiel O. R., Chem. Eur. J., 2001; 7: 3236-3253

(84) a- Scholl M., Ding S., Lee C. W., Grubbs R. H., Org. Lett., 1999; 1: 953-956; b- Bielawski C. W., Grubbs R. H., Angew. Chem., Int. Ed. Engl., 2000; 39: 2903-2906; c- Chatterjee A. K., Grubbs R. H., Org. Lett., 1999; 1: 1751-1753

(85) Chatterjee A. K., Grubbs R. H., Angew. Chem., Int. Ed. Engl., 2002; 41: 3172-3174

(86) Mayo K. G., Nearhoof E. H., Kiddle J. J., Org. Lett., 2002; 4: 1567-1570

(87) Sanford M. S., Love J. A., Grubbs R. H., Organometallics, 2001; 20: 5314-5318

(88) van Veldhuizen J. J., Garber S. B., Kingsbury J. S., Hoveyda A. H., J. Am. Chem. Soc., 2002; 124: 4954-4955

(89) Denk K., Fridgen J., Herrmann W. A., Adv. Synth. Catal., 2002; 344: 666-670

(90) a- Gessler S., Randl S., Blechert S., Tetrahedron Lett., 2000; 41: 9973-9976; b- Wakamatsu H., Blechert S., Angew. Chem., Int. Ed. Engl., 2002; 41: 2403-2405

(91) Grela K., Harutyunyan S., Michrowska A., Angew. Chem., Int. Ed. Engl., 2002; 41: 4038-4040

(92) Gibson S. E., Swamy V. M., Adv. Synth. Catal., 2002; 344: 619-621

(93) Ahmed M., Arnauld T., Barrett A. G. M., Braddock D. C., Procopiou P. A., Synlett, 2000; 1007-1009

(94) Randl S., Buschmann N., Connon S. J., Blechert S., Synlett, 2001; 10:1547-1550

(95) Jafarpour L., Huang J., Stevens E. D., Nolan S. P., Organometallics, 1999; 18: 3760-3763

(96) a- Delaude L., Demonceau A., Noels A. F., Chem. Commun., 2001; 986-987; b- Delaude L., Szypa M., Demonceau A., Noels A. F., Adv. Synth. Catal., 2002; 344: 749-756

(97) Çetinkaya B., Demir S., Özdemir I., Toupet L., Sémeril D., Bruneau C., Dixneuf P. H., New J. Chem., 2001; 25: 519-521

(98)   Louie J., Grubbs R. H., Angew. Chem., Int. Ed. Engl., 2001; 40: 247-249
(99)   a- Ackermann L., Bruneau C., Dixneuf P. H., Synlett, 2001; 397-399; b- Sémeril D., Cléran M., Bruneau C., Dixneuf P. H., Adv. Synth. Catal., 2001; 343: 184-187; c- Sémeril D., Cléran M., Jimerez Perez A., Bruneau C., Dixneuf P. H., J. Mol. Catal A : Chem., 2002; 190: 9-25, d- Le Nôtre J., Bruneau C., Dixneuf P. H, Eur. J. Org. Chem., 2002; 3816-3820
(100)   Sémeril D., Bruneau C., Dixneuf P. H., Helv. Chim. Acta, 2001; 84: 3335-3341
(101)   a- Castarlenas R., Alaoui Abdallaoui I., Mernari B., Guesmi S., Sémeril D., Dixneuf P. H., New J. Chem., 2003; 27: 6-8; b- Castarlenas R., Fischmeister C., Dixneuf P.H., New J. Chem, 2003; 27:215-217

RUTHENIUM CARBENES AS CATALYSTS FOR ALKENE METATHESIS

Cédric Fischmeister, Ricardo Castarlenas, Christian Bruneau, Pierre H. Dixneuf
Institut de chimie de Rennes, UMR 6509 CNRS-Université de Rennes, Campus de Beaulieu

Abstract

Two families of ruthenium-based olefin metathesis catalysts and their use in organic and polymer synthesis are presented. Ruthenium-allenylidene complexes $[Ru=C=C=CR_2(Cl)(PR_3)(arene)]^+ X^-$ are efficient pre-catalysts for the Ring Closure of dienes and enynes but they are also able to polymerise strained and unstrained cyclic olefins under mild conditions. These complexes have also shown interesting recycling capacity in ionic liquid solvents. The *in-situ* generated catalysts readily prepared from easily available reagents: a ruthenium complex, an imidazolium salt and a base promote alkene and enyne metathesis. A two-component system for the polymerization of norbornene and cyclooctene is also presented.

Key words: Olefin metathesis, ruthenium catalysts, enyne, ROMP, ionic liquids.

Introduction

Although alkene metathesis is a catalytic reaction discovered as early as 1955[1], it has found only during the last decade a leading role in synthetic methodology due to its power to break and make carbon-carbon double bonds according to the general equation;

$$R^1HC=CHR^1 + R^2HC=CHR^2 \rightleftharpoons 2\ R^1HC=CHR^2$$

Alkene metathesis has acquired a key role in the formation of functional cyclic and macrocyclic molecules *via* ring closing metathesis reactions (RCM)[2] (Figure 1). This catalytic reaction is now selectively promoting the combination of simple olefins with functional alkenes *via* the cross metathesis reaction [2], [3] Alkene metathesis has found useful applications in material science *via* the polymerisation of cyclic olefins (ROMP) or acyclic dienes (ADMET) reactions [4] (Figure 1). The intramolecular, atom economy cyclisation rearrangement of enynes into 1-alkenyl cycloalkenes constitutes an important contribution to fine chemistry as the formed 1,3-diene moieties allows functionalisations such as Diels-Alder reaction. [5] It is related to alkene metathesis as it is promoted by alkene metathesis catalysts, [6] and thus called enyne metathesis reaction (Figure 1), however other catalysts promote the same transformation *via* a non related metathesis mechanism. [7]

Y. Imamoglu and L. Bencze (eds.), Novel Metathesis Chemistry: Well-Defined Initiator Systems for Specialty Chemical Synthesis, Tailored Polymers and Advanced Material Applications, 23–42.
© 2003 *Kluwer Academic Publishers. Printed in the Netherlands.*

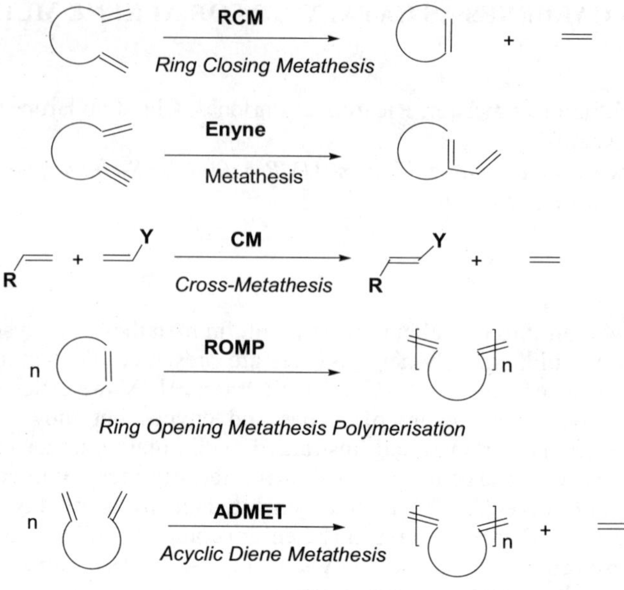

Figure-1. *Olefin metathesis and related processes*

This tremendous impact of alkene metathesis during the 1990 decade is certainly due to the discovery of efficient, well-defined coordinatively unsaturated carbene metal catalysts, such as alkylidene-molybdenum [8] and ruthenium. [9] These discoveries directly result from the elucidation of the alkene metathesis mechanism by Chauvin [10] as early as 1970, soon after the first preparation of carbene metal complex by E.O. Fischer. [11] Chauvin was the first to propose an alkene metathesis mechanism involving a carbene-metal catalyst (Figure 2).

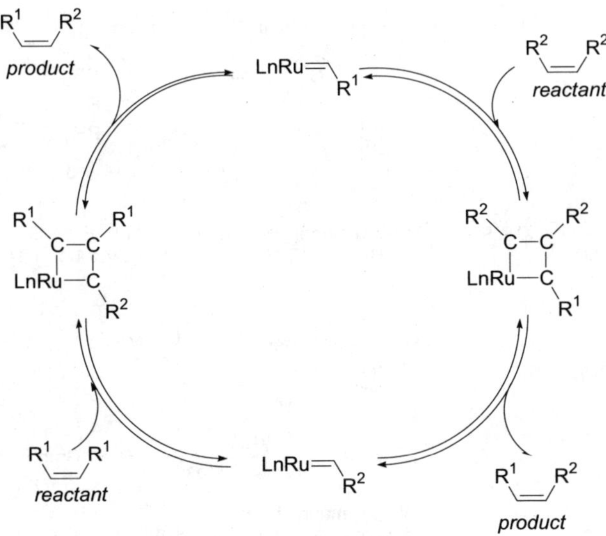

Figure 2: *The Chauvin mechanism*

The Chauvin mechanism involves a 2+2 addition of an alkylidene metal moiety $M=CHR^1$ with the alkene double bond in equilibrium with an equivalent $M=CHR^2$ moiety and the new alkene $R^1CH=CHR^2$.

Twenty years after the elucidation of the mechanism, during which time the validity of the Chauvin's proposal was recognised, several groups attempted to produce well-defined coordinatively unsaturated alkylidene transition metal complexes (Figure 3). The first one, an alkylidene molybdenum derivative **1**, was reported by Schrock in 1990[8] followed by Basset for a related tungsten system **2**. [12] Another series involving ruthenium alkylidene complexes $Ru=CH-CH=CR_2$ **3** [9] and then $Ru=CHPh$ **4**, **5** and **6** [13], [14] was discovered by Grubbs and shown to be at the same time, active and highly tolerant toward functional groups.

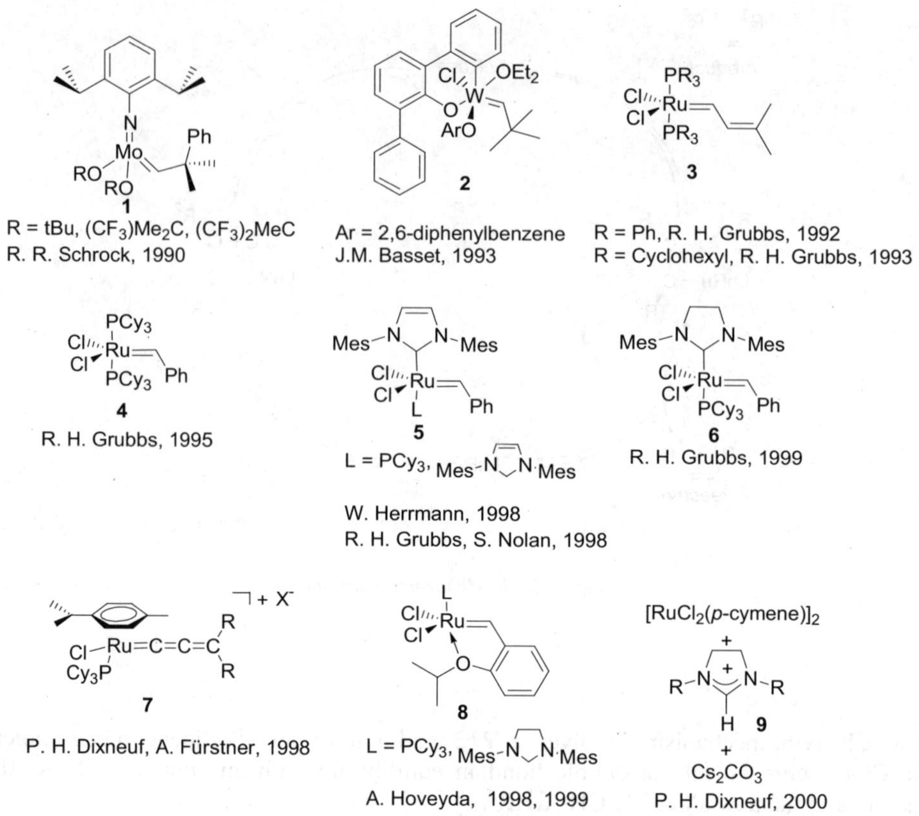

Figure-3. *Olefin metathesis catalysts*

Dixneuf and Fürstner groups showed that ionic, 18 electron ruthenium-allenylidenes **7** were also good alkene metathesis catalyst precursors. [15], [16] Recently the modification of the Grubbs catalysts by introduction of heterocyclic electron-rich carbene ancillary ligands by Herrmann [17], Nolan [18] and Grubbs [19] **5** led to efficient ruthenium catalysts, whereas chelating carbene ruthenium catalysts **8** were introduced by Hoveyda [20], Blechert [21] and Fürstner [22] with a chelation *via* a carbonyl group. The later mixed imidazol(in)ylidene alkylidene ruthenium catalysts led the Rennes group to develop easily *in-situ* prepared catalyst from a ruthenium source and an imidazol(in)ylidene precursors **9** which are promoting alkene metathesis reactions for both fine chemistry and polymer science (Figure 3).

The objective of this review is to present the main lines of the Rennes contribution in the field of alkene and enyne metathesis for fine chemistry and ROMP polymerisation, involving both allenylidene-ruthenium precursors and *in-situ* prepared ruthenium catalysts.

Alkene metathesis catalysts for fine chemistry

Ruthenium-allenylidene in catalysis

Cationic arene ruthenium-allenylidene complexes are the result of the activation of propargylic alcohol by ruthenium(II) complexes [23-25] (Scheme 1). These complexes display many advantages as they can be easily prepared in high scale from commercially available and non toxic reagents. The synthesis pathways (Scheme 1) make possible the synthesis of a large number of complexes by modifying the arene, the phosphine and the allenylidene ligands as well as the escorting anions (X). [15], [16], [26]

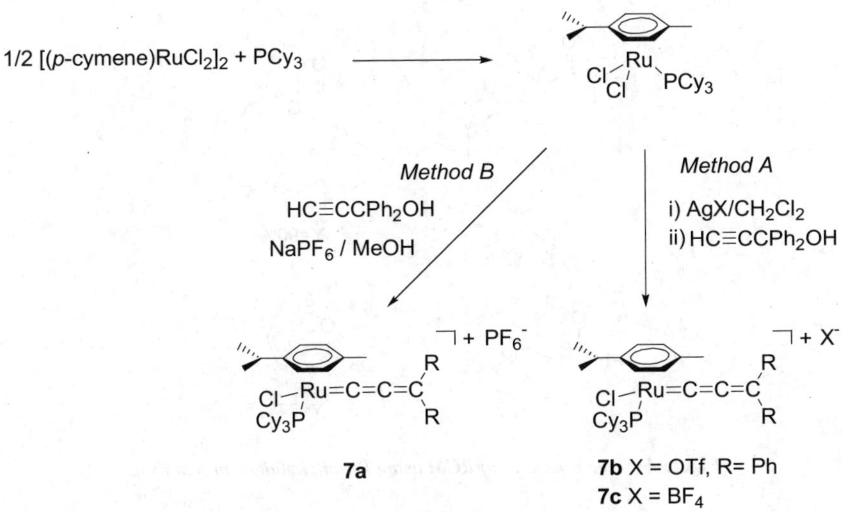

Scheme-1. *Ruthenium allenylidene complex synthesis*

Among these parameters the anion (X$^-$) appeared to be of major importance both for the catalytic activity and selectivity (Scheme 2 and Table 1).

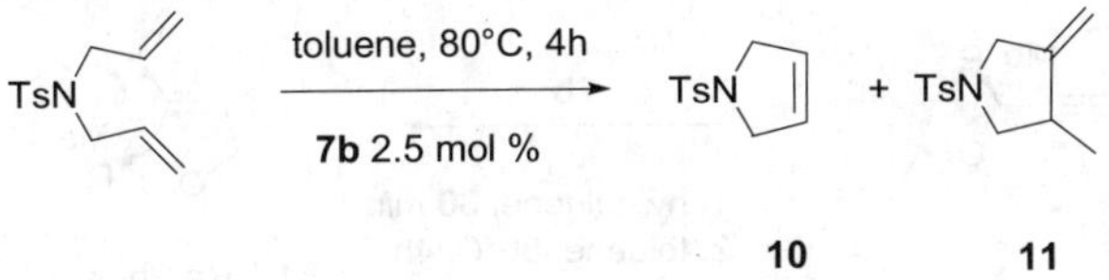

Scheme-2. *RCM and cycloisomerisation of diallytosylamine*

As illustrated in Scheme 2 a cycloisomerisation side-product **11** can be generated along with the expected ring closed compound **10**. When BF$_4$ is the counter-anion this side product is the major whereas only traces are obtained when a triflate anion (OTf) or PF$_6$ are used (table 1).

Table-1. *Counter-anion influence*

Anion	Yield of **10**(%)	Yield of **11**(%)
PF_6	91	-
BPh_4	91	-
BF_4	31	43
OTf	90	3

Ring Closing Metathesis of dienes

The ruthenium allenylidene precatalysts have been extensively used for the RCM of various substrates showing an activity in the range of that of $RuCl_2(=CHPh)(PCy_3)_2$ **4** and a very good tolerance for polar functionalities and heteroatoms (Figure 4).[16]

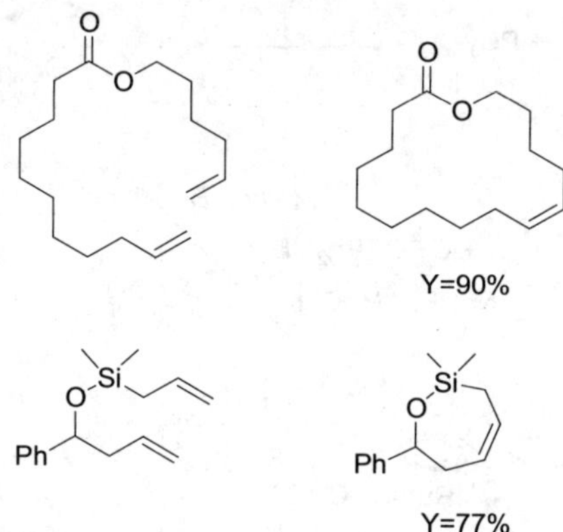

Figure-4. *Some examples of RCM using Ru-allenylidene precatalysts*

Enyne Metathesis

Ruthenium allenylidene catalysts have also shown good activities for enyne metathesis reactions. An illustration is the straightforward synthesis of 3-vinyl-2,5-dihydrofurans in the presence of precatalyst **7b** either by classical heating or with a preliminary UV activation of the precatalyst (Scheme 3).[15]

Scheme-3. *Synthesis of furane derivatives by enyne metathesis*

An interesting feature of the enyne metathesis products is the conjugated diene that can be used in subsequent transformation in particular Diels-Alder reaction type. This possibility has been used to synthesize new fluorinated dehydroproline derivatives of major importance in some biological processes (Scheme 4). [27]

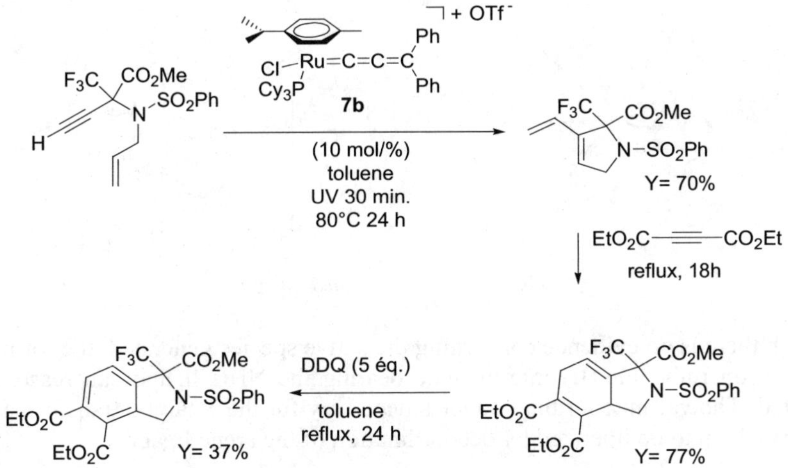

Scheme-4. *Fluorinated dehydropoline analogs by enyne metathesis*

In-situ generated ruthenium catalysts

Historically the first catalysts for olefin metathesis were *in-situ* generated from a metal source and a ligand source. Later on, well-defined complexes enabling a better understanding of the reaction mechanism and the ligand influence have attracted much attention. As a result, organometallic chemists have designed very sophisticated catalysts or precatalysts that are very active but sometimes requiring costly multistep synthesis. For these reasons it was necessary to move to catalyst offering a good compromise between activity, accessibility and cost. With this in mind, new catalytic systems that can be *in-situ* generated from easily available precursors have been investigated. Actually the approach is to *in-situ* produce the catalytic species arising from an isolated, stable precursor. Thus a commercially available ruthenium complex $[RuCl_2(p\text{-cymene})_2]_2$ has been used in combination of a *N*-heterocyclic carbene (NHC) precursor and a base (Figure 5).

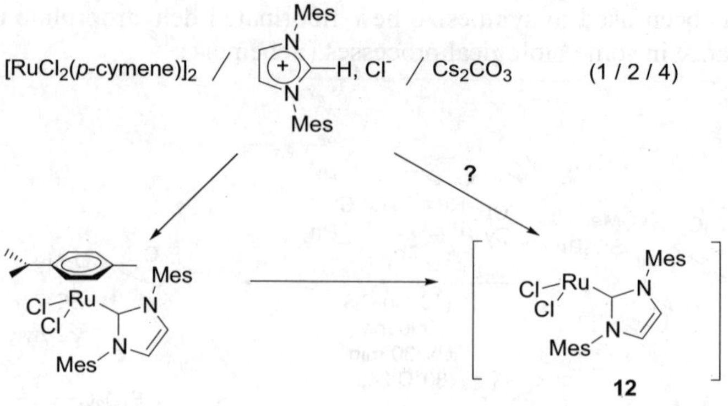
[RuCl2(p-cymene)]2
Mes
N
H, Cl-
Cs2CO3
N
Mes
(1 / 2 / 4)
?
Mes
Cl
Ru
Cl
N
N
Mes
12

Figure-5. *In-situ generated catalyst*

Although there is no evidence concerning the active species generated, the formation of a 18 electron ruthenium(II) intermediate bearing the NHC ligand can reasonably be postulated. The vacant coordination sites necessary for the catalytic transformation are then very likely to be liberated by decoordination of the arene ligand.

Enyne metathesis

This system shows a very good activity for enyne metathesis as illustrated by the synthesis of furan derivatives. [(28), (29)] This first example (Scheme 5) illustrates the important reactivity enhancement with regard to the ruthenium-allenylidene precatalyst.

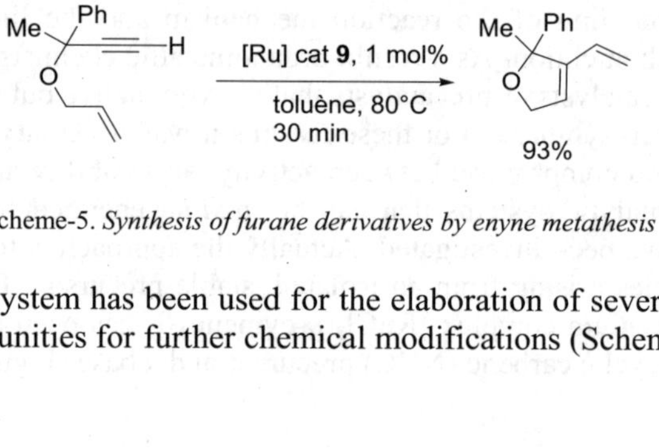
Me
Ph
H
O
[Ru] cat 9, 1 mol%
toluène, 80°C
30 min
Me
Ph
O
93%

Scheme-5. *Synthesis of furane derivatives by enyne metathesis*

This new catalytic system has been used for the elaboration of several cyclic siloxanes offering new opportunities for further chemical modifications (Scheme 6). [(29)]

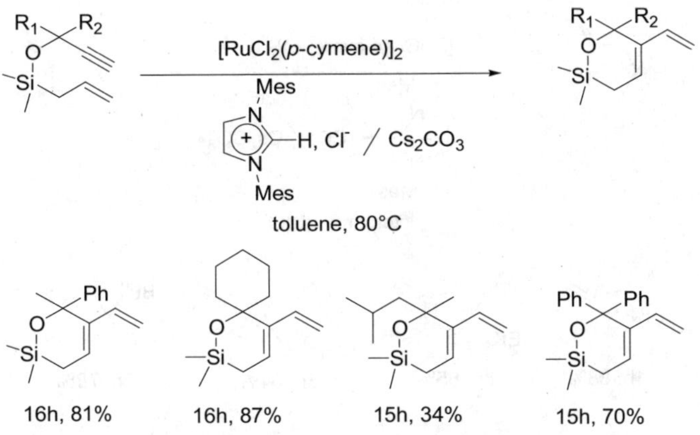

Scheme-6. *Cyclic siloxanes by enyne metathesis*

Indeed, the conjugated diene can be used in a Diels-Alder reaction (Scheme 7) and the siloxane bond reactivity also allows the synthesis of a large variety of products such as diols and internal alkenes.

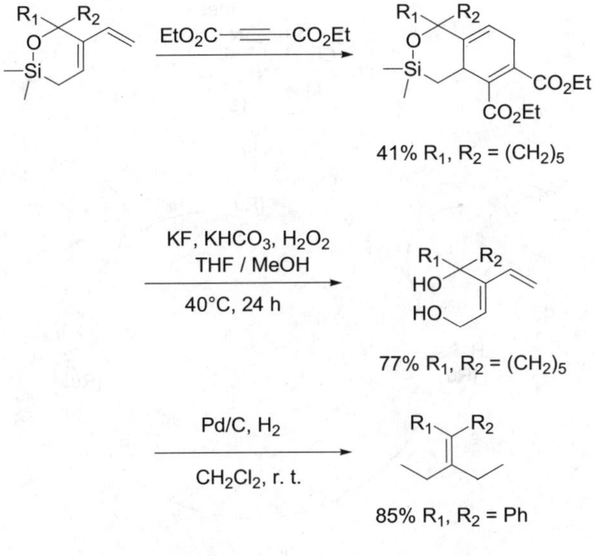

Scheme-7. *Chemical modifications of cyclic siloxanes*

Diene Cycloisomerisation *vs* Ring closing metathesis

Following the discovery of high activity of the *in-situ* generated system for enyne metathesis a similar system using an imidazolinium salt was tested for the RCM of dienes. Instead of the RCM products cycloisomerisation compounds were obtained in very good yields (Scheme 8).[30]

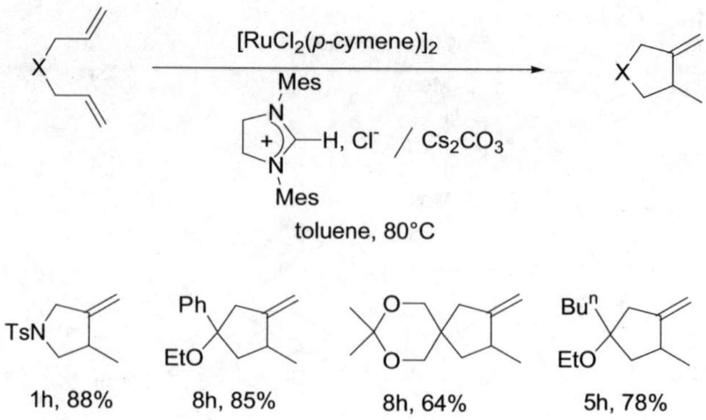

Scheme-8. *Cycloisomerisation of dienes*

The proposed mechanism is expected to involve the ruthenium species **13** likely to undergo an oxidative coupling with the diene. β-H elimination followed by reductive elimination release the cycloisomerisation product (Figure 6).

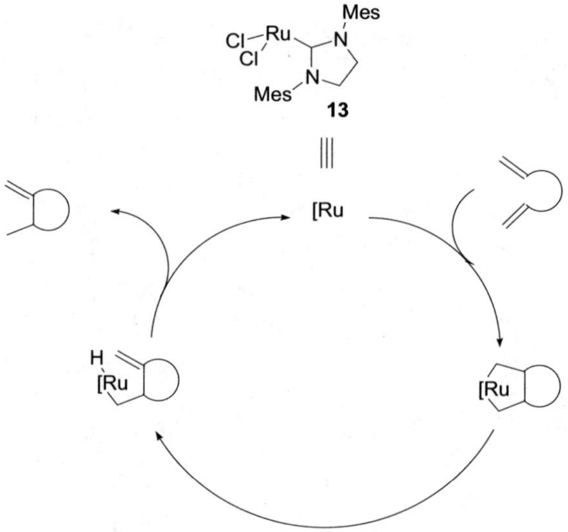

Figure-6. *Proposed mechanism for the cycloisomerisation of dienes*

It is believed that alkylidene ligands are necessary to get olefin metathesis. Thus the outcome of the reaction can be attributed to the absence of such a ligand. Therefore the same reactions have been performed under an acetylene atmosphere since it is well known that ruthenium(II) complexes are able to activate terminal acetylene leading to vinylidene species [31] $Ru=C=CH_2$ which can promote olefin metathesis reactions [32]. The reactions have then been carried out under acetylene atmosphere and as expected the reaction product turned out to be the Ring Closing Metathesis one (Scheme 9). [30]

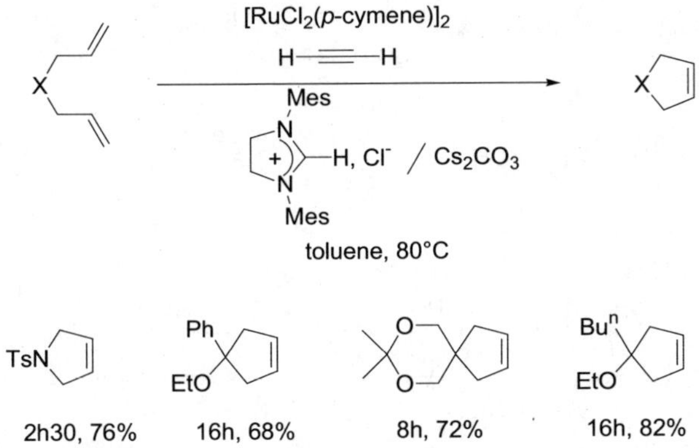

Scheme-9. *RCM of dienes under acetylene atmosphere*

As mentioned the crucial step would be the formation of the ruthenium-vinylidene complex **14** by reaction of the highly unsaturated ruthenium species **13** with acetylene (Figure 7). Reaction of this intermediate with one alkene of the substrate would then generate an alkylidene complex able to perform the RCM reaction.

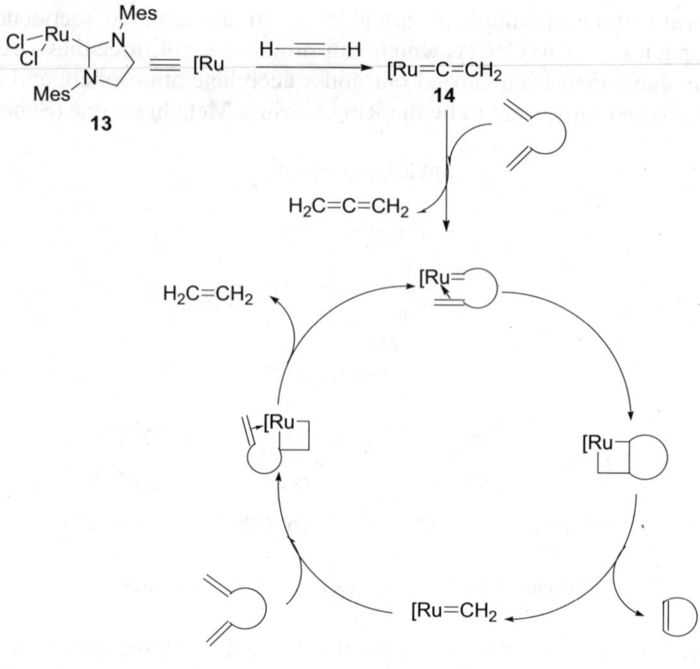

Figure-7. *RCM of dienes under acetylene atmosphere*

As a consequence a single catalytic system is able to perform two different transformations simply by switching the reaction atmosphere!

Alkene metathesis polymerisation

Olefin polymerisation by homogeneous transition metal catalyst has attracted much attention in the last fifty years since Ziegler and Natta discovery in the early 50's. Alkene metathesis polymerisation processes (ROMP and ADMET) constitute the best way to produce linear polymers with C=C bonds regularly displayed along the polymer backbone (Figure 1). These double bonds can further be functionalised or hydrogenated [33] to yield polyolefins. Ruthenium based olefin metathesis catalysts tolerate a large variety of polar functionalities and do not require drastic conditions . In particular the Grubbs catalysts have been efficiently used to polymerise cyclic olefins (ROMP) [4] or dienes (ADMET). [4] In the next part are presented recent developments in the field of polymer synthesis by the ROMP process achieved by the Rennes team using two catalyst families: Ruthenium-allenylidenes and *in-situ* generated systems.

Ruthenium allenylidene catalysts

The search for more active ruthenium-allenylidene precatalysts has led to the design of a new class of arene-free ruthenium allenylidene complexes. A large number of complexes bearing DMSO ligands have been synthesized for this reason (Scheme 10). [34]

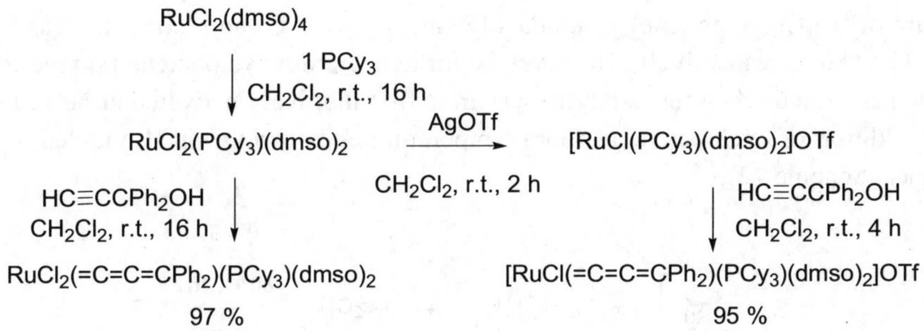

$$RuCl_2(dmso)_4$$

(1 PCy_3, CH_2Cl_2, r.t., 16 h)

$$RuCl_2(PCy_3)(dmso)_2 \xrightarrow[CH_2Cl_2, \text{r.t., 2 h}]{AgOTf} [RuCl(PCy_3)(dmso)_2]OTf$$

$HC{\equiv}CCPh_2OH$, CH_2Cl_2, r.t., 16 h

$HC{\equiv}CCPh_2OH$, CH_2Cl_2, r.t., 4 h

$$RuCl_2(=C=C=CPh_2)(PCy_3)(dmso)_2 \qquad [RuCl(=C=C=CPh_2)(PCy_3)(dmso)_2]OTf$$

97 %　　　　　　　　　　　　　　　　95 %

Scheme-10. *Arene-free ruthenium-allenylidene complexes*

These complexes polymerise norbornene and cyclooctene but their activity did not reach that of the arene ruthenium complex **7b** (Scheme 11). With this catalyst norbornene is readily polymerised at room temperature. [35] Under the same conditions the less strained cyclooctene monomer requires longer reaction time.

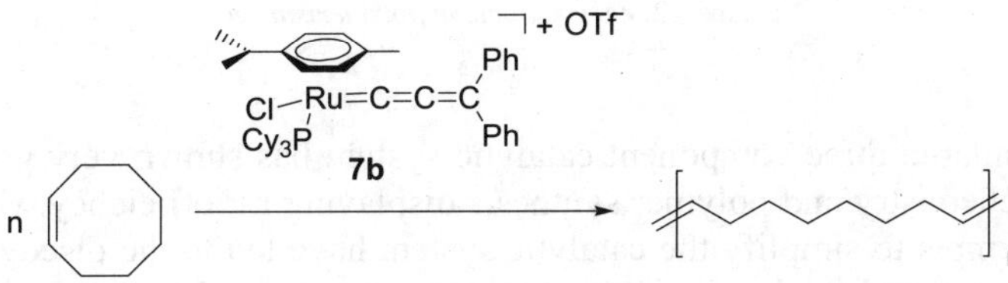

Scheme-11. *ROMP of cyclooctene*

Alternatively a preliminary thermal or photochemical activation of the catalyst has allowed the fast room temperature polymerisation of cyclooctene within 10 min (Table 2).

Table-2. *Cyclooctene polymerisation*

Catalyst	Conditions	Polym. Yield (%)	Mn x 10^{-3}	PD*i*
7b	15h r.t	95	143	1.9
7b[a]	UV[a]	99	143	1.8
7b[b]	10 min r.t[b]	97	151	1.7

[a] **7b** and cyclooctene were stirred at r. t. for 2h and 2h at r.t. under UV irradiation. [b] **7b** was heated for 25 min. at 60°C and allowed to cool down to r. t. before addition of cyclooctene.

In-situ generated ruthenium catalysts

The use of the three components catalytic system earlier found to be very efficient for fine chemistry transformations has been extended to polymer synthesis. Both norbornene and cyclooctene can be polymerised in very good yields when a mixture of $[RuCl_2(p\text{-cymene})_2]_2$, imidazolinium salt, Cs_2CO_3 and monomer was heated at 60°C or 80°C respectively. However, room temperature cyclooctene polymerisation can be performed when the catalytic system is first thermally activated at 80°C for 1h. Then addition of cyclooctene at room temperature and stirring for 1h yielded 85% of polymer (Scheme 12).[36]

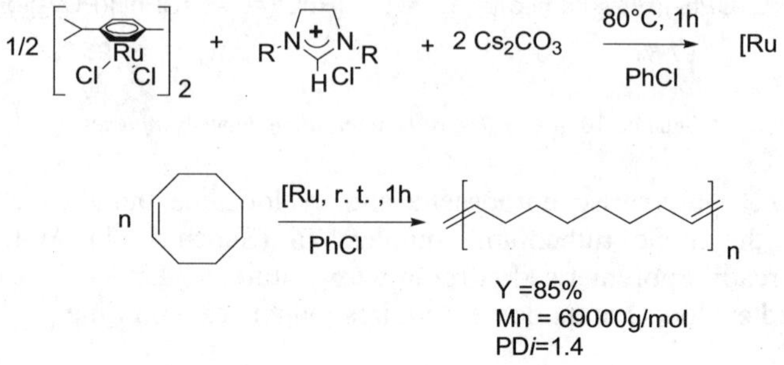

Scheme-12. *r.t. cyclooctene polymerisation*

The easily available three component catalytic system has shown very good activities for both fine chemistry and polymer synthesis displaying r.t. efficiency when correctly activated. Attempts to simplify the catalytic system have led to the discovery of a new catalytic system now based only on two components. Indeed it was found that when using (bis-methallyl)ruthenium diphosphines **15** with imidazolinium salts, the use of a base is no longer required (Scheme 13).[37]

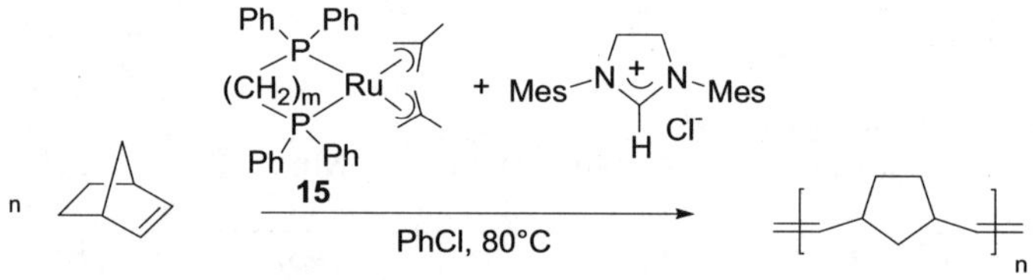

Scheme-13. *Two component catalytic system*

Another interesting feature was found when varying the length of the alkyl chain (m) of the diphosphine. This appeared to be a way to get a reasonable control of the molecular weight of the polymer without loosing the catalyst activity (table 3). Finally this system also performs the room temperature of cyclooctene provided a preliminary thermal activation step is performed.

Table-3. *Norbornene polymerisation*[a]. *Influence of the diphosphine*

m	Polym. Yield (%)	Mn x 10^{-3} [b]	PD*i*[c]
1	99	610	1.3
2	99	636	1.3
4	98	308	1.7
6	99	374	1.5

[a] General conditions : the catalyst was in-situ prepared from 1.5×10^{-5} mol of **15**, $1.5 \; 10^{-5}$ mol of imidazolinium salt in 2.5 ml of chlorobenzene ([monomer]/[Ru] = 300). [b] Determined by GPC in THF calibrated with polystyrene standards. [c] Polydispersity index Mw/Mn

Alkene metathesis in ionic liquids

One major problem for both fine chemistry, ROMP polymerisation and generally for any reaction catalysed by organometallic complex is the recovery or recycling of the catalysts for both cost and environmental issues. The most common way involves the heterogenising [38] of a homogeneous catalyst by anchoring it onto a solid substrate. More recently, unconventional solvents have attracted a lot of attention in catalysis due to their ability to dissolve organometallic species. Among these solvents are supercritical fluids, [39] perfluoroalkanes [40] and ionic liquids. [41], [42] The later seemed to be of high interest since good to excellent solubilities of the ionic allenylidene precatalysts could be foreseen in an ionic solvent. Ionic liquids are defined as salts that melt at or near room temperature. They are composed of a cation such as, ammonium, phospnonium, imidazolium or pyridinium that can be associated to coordinating or non-coordinating anions Their interest is due to their low vapor pressure, high thermal stability and capacity to dissolve either organic or organometallic species. In addition, their non-miscibility with some organic solvents facilitates the extraction of the reaction products. [41], [42]

Ring Closing Metathesis

Apart from a patent by Bayer[43] there was no report on olefin metathesis in ionic liquids when the ruthenium-allenylidene complex was first tested for the RCM of diallytosylamine in imidazolium based ionic liquids. [44] This first attempt has revealed the activity of this precatalyst for the RCM reaction (Scheme 14).

Scheme-14. *RCM in an ionic liquid*

While the best conversions were obtained with the complex **7b** the influence of Y (ionic liquid anion) appeared to be very important in terms of conversion as well as selectivity (table 4).

Table-4. *Influence of the ionic liquid anion*

Solvent	Time (h)	Conversion (%)	10	11	16
[bmim][BF$_4$]	5	60	23	31	6
[bmim][PF$_6$]	5	85	83	2	-
[bmim][OTf]	2.5	100	97	3	-
[bmim][OTf]	18	100	100	-	-

General conditions: (0.5 mmol), (2.5mol%), [bmim][X] 2ml, 80°C. Determined by GC/MS.

A series of products has been synthesized in very good yield with the precatalyst **7b** (Figure 8). In all cases the extraction of the organic product was simplified by using a biphasic media toluene-ionic liquid.

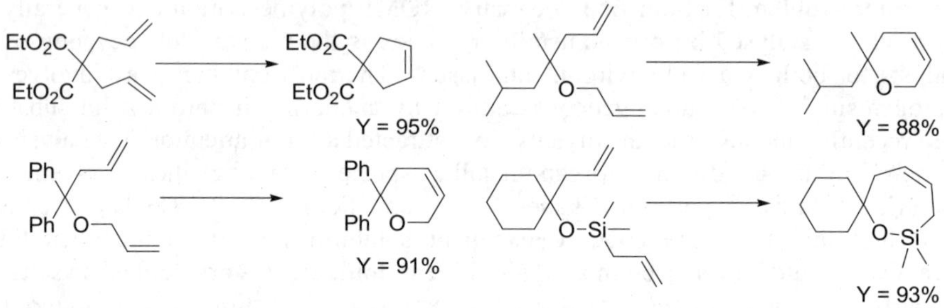

Figure-8. *5, 6 and 7 membered (hetero)cycles by RCM in ionic liquids*

The possibility of recycling the catalyst arising from **7b** in [bmim][OTf] was attempted at room temperature for the RCM of diallyltosylamine. The first experiment gave 100% conversion after 18h and extraction of the organic product with toluene. A second loading of substrate was stirred for an extra 18h to give lower 86% conversion and a third experiment showed a high decrease of the activity since only 33% conversion were reached. In the meantime, Buijsman reported on the same reaction in an ionic liquid catalyzed by the second generation Grubbs catalyst **6**. [45] As observed with the allenylidene precatalyst the conversion was found to dramatically decrease on the third catalyst use. The above modest initial results nevertheless show the feasibility of alkene metathesis in ionic liquids, but give evidence that developments are linked to the parallel discovery of stable metathesis catalysts in a given ionic liquid.

Ring Opening Metathesis Polymerisation

The ROMP polymerisation of norbornene has been carried out with the precatalyst **7b** in [bdmim][PF$_6$]. [46] This ionic liquid has a methyl substituent at the 2-position in order to prevent any possibility of metal protonation [47] potentially responsible of catalyst decomposition (Scheme 15).

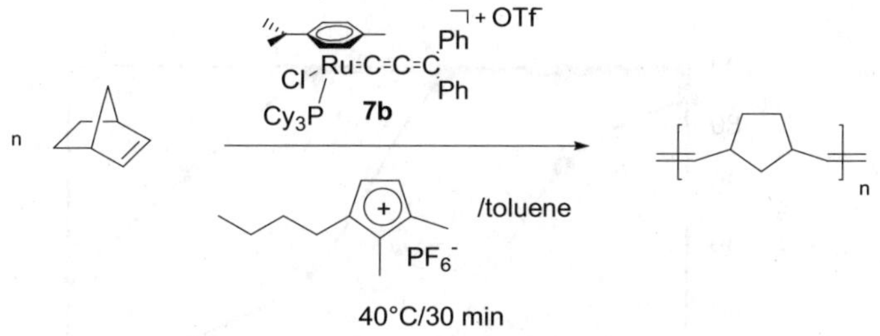

Scheme-15. *ROMP of norbornene in an ionic liquid*

In this case it has been shown that a co-solvent is necessary during the reaction due to the poor solubility of the growing polymer chains. For this reason the experiments have been carried out in a biphasic mixture of ionic liquid and toluene and the optimal experimental conditions have been found to be 30 minutes at 40°C for the polymerisation of norbornene. [46] It has been possible to use four consecutive times the catalysts without loss of activity and even the seventh use gave good polymer yield provided a longer reaction time was used. (Table 5).

Table-5. *Catalyst recycling*

Run	T (h)	Polym. Yield (%)	$Mn \times 10^{-3a}$	PDi^{b}
1	0.5	96	114	1.9
2	0.5	99	168	1.9
3	0.5	98	207	1.8
4	0.5	97	-	-
5	0.5	64	-	-
6	0.5	27	303	1.5
7	2	82	260	1.6

[a] Determined by GPC in THF calibrated with polystyrene standards. [b] Polydispersity index Mw/Mn

These recycling data were compared to that obtained with the 1st and 2nd generation Grubbs catalysts **4** and **6** under the same reaction conditions. As depicted in Figure 9 the ruthenium-allenylidene precatalyst **7b** offers better recycling properties. This has been attributed to its better solubility into the ionic liquid whereas the neutral Grubbs' catalysts were found to be weakly soluble into this media and easily extracted.

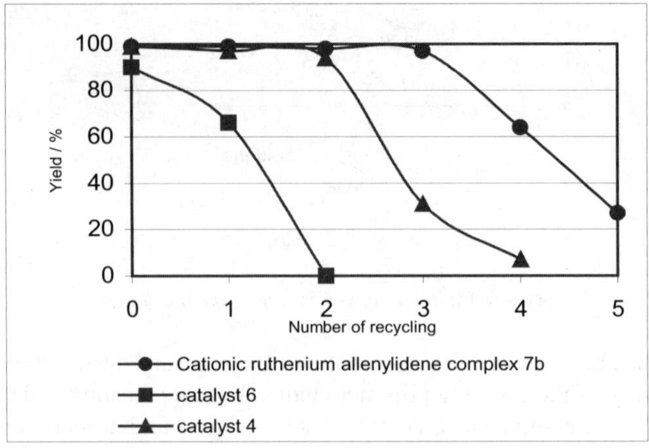

Figure-9. *Comparison of complexes **7b**, **4** and **6** recycling possibilities*

Finally it has also been shown that the ionic liquid used in the first set of experiment could be reused simply by loading a new batch of ruthenium-allenylidene complex allowing to recover the very good recycling characteristic initially observed. The design of more active catalysts able to operate in ionic liquid media remains a challenge.

Conclusion

This review shows various applications of two families of alkene metathesis catalysts. The ionic ruthenium-allenylidene precatalysts have been used in organic transformations and polymer synthesis. They have also been used in ionic liquid solvent with very good recycling properties in particular for the ROMP of norbornene. These results are of particular interest for cost decreasing and environment tolerant processes and open the way for the elaboration of new recyclable catalysts. *In-situ* generated catalytic systems involving a ruthenium source and a *N*-heterocyclic carbene very easy to prepare display very good activities for both organic or polymer synthesis under mild conditions.

References

(1)   Anderson A. W., Merckling M. G., US-A 2721189,1955
(2)   For a recent review on olefin metathesis, see: Fürstner A., Angew. Chem., Int. Ed., 2000; 39: 3012-3043
(3)   Crowe W. E., Zhang Z. J., J. Am. Chem. Soc., 1993; 115: 10998-10999
(4)   Buchmeiser M. R., Chem. Rev., 2000; 100: 1565-1604
(5)   a- Saito N., Sato Y., Mori M., Org. Lett., 2002; 4: 803-805. b- Manas M. M., Pleixats R., Santamaria R., Synlett; 2001:1784-1786.
(6)   a- Poulsen C. S., Madsen R., Synthesis, 2003; 1-18. b- Mori M., Kitamura T., Sato Y., Synthesis, 2001; 654-664
(7)   Fürstner A., Szillat H., Stelzer F., J. Am. Chem. Soc., 2000; 122: 6785-6786
(8)   Schrock R. R., Murdzek J. S., Bazan G. C., Robbins J., DiMare M., O'Regan M., J. Am. Chem. Soc., 1990; 112: 3875-3886

(9)	a- Nguyen S. T., Johnson, L. K., Ziller J. W., J. Am. Chem. Soc., 1992; 114: 3974-3975. b- Nguyen S. T., Grubbs R. H., J. Am. Chem. Soc., 1993; 115: 9858-9859

(10)	Hérisson J.-L., Chauvin Y., Makromol. Chem. 1970; 141: 161-176

(11)	Fisher E. O., Maasböl A., Angew. Chem. Int. Ed. Engl., 1964; 3: 580-582

(12)	Couturier J. L., Paillet C., Leconte M., Basset J-M., Ollivier J., Angew. Chem., Int. Ed. Engl., 1993; 32: 112-115

(13)	Schwab P., France M. B., Ziller J. W., Grubbs R. H., Angew. Chem., Int. Ed. Engl., 1995; 34: 2039-2041

(14)	Trnka T. M., Grubbs R. H., Acc. Chem. Res., 2001; 34: 18-29

(15)	Fürstner A, Picquet M., Bruneau C., Dixneuf P. H., Chem. Commun., 1998; 1315-1316

(16)	Fürstner A., Liebl M., Lehmann C. W., Picquet M., Kunz R., Bruneau C., Touchard D., Dixneuf P. H., Chem. Eur. J., 2000; 6: 1847-1857.

(17)	Weskamp T., Schattenmann W. C., Spiegler M., Herrmann W. A., Angew. Chem., Int. Ed. Engl., 1998; 37: 2490-2493

(18)	Huang J., Stevens E. D., Nolan S. P., Petersen J. L., J. Am. Chem. Soc., 1999; 121: 2674-2678

(19)	a- Scholl M., Trnka T. M., Morgan J. P., Grubbs R. H., Tetrahedron Lett., 1999; 40: 2247-2250. b- Scholl M., Ding S., Lee C. W., Grubbs R. H., Org. Lett., 1999; 1: 953-956

(20)	a- Harrity J. P. A., La P. A., Cefalo D. R., Visser M. S., Hoveyda A. H., J. Am. Chem. Soc., 1998; 120: 2343-2351. b- Kingsburry J. S., Harrity J. P. A., Bonitatebus P. J., Hoveyda A. H., J. Am. Chem. Soc., 1999; 121: 791-799. c- Garber S. B., Kingsburry J. S., Gray B. L., Hoveyda A. H., J. Am. Chem. Soc., 2000; 122: 8168-8179

(21)	a-Gessler S., Randl S., Blechert S., Tetrahedron Lett., 2000; 41: 9973-9976. b- Wakamatsu H., Blechert S., Angew. Chem. Int. Ed., 2002; 41: 2403-2405

(22)	Fürstner A., Thiel O. R., Lehmann C. W., Organometallics, 2002; 21: 331-335

(23)	Selegue J. P., Organometallics, 1982; 1: 217-218

(24)	a Touchard D., Haquette P., Daridor A., Romero A., Dixneuf P. H., Organometallics, 1998; 17: 3844-3852 and references therein. b Touchard D., Pirio N., Toupet L., Fettouhi M., Ouahab L., Dixneuf P. H., Organometallics, 1995; 14: 5263-5272

(25)	Esteruelas M. A., Gomez A. V., Lopez A. M., Modrego J., Oñate E., Organometallics, 1997; 16: 5826-5835

(26)	Picquet M., Touchard D., Bruneau C., Dixneuf P. H., New J. Chem., 1999; 141-143

(27)	Sémeril D., Le Nôtre J., Bruneau C., Dixneuf P. H., Kolomiets A. F., Osipov S. N., New J. Chem., 2001; 25: 16-18

(28)	Ackermann L., Bruneau C., Dixneuf. P. H., Synlett; 2001; 3: 397-399

(29)	Sémeril D., Bruneau C., Dixneuf P. H., Adv. Synth. Catal., 2002; 344: 585-595

(30)	Sémeril D., Bruneau C., Dixneuf P. H., Helv. Chim. Acta, 2001; 84: 3335-3341

(31)	a- Bruneau C., Dixneuf P. H., Acc. Chem. Res., 1999; 32: 311-323. b- Puerta M. C., Valerga P., Coord. Chem. Rev., 1999; 193-195: 977-1025

(32)	a- Katayama H., Yoshida T., Ozawa F., J. Organomet. Chem., 1998; 562: 203-206. b- Katayama H., Ozawa F., Organometallics, 1998; 17: 5190-5196

(33)	a- Bielawski C. W., Louie J., Grubbs R. H., J. Am. Chem. Soc., 2000; 122: 12872-12873. b- Drouin S. D., Zamanian F., Fogg D. E., Organometallics, 2001; 20: 5495-5497

(34)	Alaoui-Abdallaoui I., Sémeril D., Dixneuf P. H., J. Mol. Catal A : Chem., 2002; 182-183: 577-583

(35)	Castarlenas R., Sémeril D., Noels A. F., Demonceau A., Dixneuf P. H., J. Organomet. Chem., 2002; 663: 235-238

(36)	Castarlenas R., Alaoui-Abdalaoui I., Sémeril D., Mernari B., Guesmi S., Dixneuf P. H., New J. Chem, 2003; 27:6-8

(37)	Castarlenas R., Fischmeister C., Dixneuf P. H., New J. Chem, 2003; 27: 215-217

(38)	Choplin A., Quignard F., Coord. Chem. Rev., 1998; 178-180: 1679-1702

(39)	For a review on homogeneous catalysis in supercritical fluids, see : Jessop P. G., Ikariya T., Noyori R., Chem. Rev., 1999; 99: 475-493

(40)	For recent reviews on perfluoroalkanes, see: a- Sandford G., Tetrahedron, 2003; 59: 437-454. b- Horváth I. T., Acc. Chem. Res., 1998; 31: 641-650

(41)	For recent reviews on Ionic Liquids, see: a- Wasserscheid P., Keim W., Angew. Chem. Int. Ed., 2000; 39: 3772-3789. b- Welton T., Chem. Rev., 1999; 99: 2071-2083

(42)	For recent reviews on catalysis in Ionic Liquids, see: a- Olivier-Bourbigou H., Magna L., J. Mol. Catal. A: Chem., 2002; 182-183, 419-437. b- Zhao D., Wu M., Kou Y., Min E., Catal Today, 2002; 74: 157-189. c- Dupont J., De Souza R. F., Suarez P. A., Chem. Rev., 2002; 102: 3667-3692d- Sheldon R., Chem. Commun., 2001; 2399-2407

(43) Gürtler C., Jautelat M., Bayer AG., EP 1035093A2, 2000
(44) Sémeril D., Olivier-Bourbigou H., Bruneau C., Dixneuf P. H., Chem. Commun., 2002; 146-147
(45) Buijsman R. C., Van Vuuren E., Sterrenburg J. G., Org. Lett., 2001; 3: 3785-3787
(46) Csihony S., Fischmeister C., Bruneau C., Horváth I. T., Dixneuf P. H., New J. Chem., 2002; 25: 1667-1670
(47) Examples of imidazolium oxidative addition to Ni, Pt and Pd have been reported. McGuiness D. S., Cavell K. J., Yates B. F., Skelton B. W., White A. H., J. Am. Chem. Soc., 2001; 123: 8317-8328

METATHETICAL CONVERSION AND SILYLATIVE COUPLING OF ALKADIENES AND CYCLOALKENES WITH MONO- AND DI-VINYLSILICON COMPOUNDS

BOGDAN MARCINIEC

Department of Organometallic Chemistry, Faculty of Chemistry,
Adam Mickiewicz University, Grunwaldzka 6, Poznań, 60-780 Poland

Abstract

New catalytic routes for synthesis of vinylene-siloxylene-alkenylene copolymers via ADMET copolymerization of divinyltetraethoxydisiloxane with dienes (e.g., 1,9-decadiene, dihexenyl- ether and 1,4-divinylbenzene) and via tandem ROM/CM polymerization of cyclooctene with divinyltetraethoxydisiloxane as well as via silylative coupling (SC) polycondensation of divinyldisiloxanes with dienes (e.g. divinylbenzene) has been overviewed . Monovinyl(trialkoxy- and trisiloxy)silanes undergo (in the presence of Grubbs catalyst) cross-metathesis with 1,9-decadiene and/or tandem ROM/CM with cyclooctene to give, in excess of olefin, bis(silyl)dienes.

Keywords: ADMET copolymerization, SC polycondensation, cross-metathesis, Grubbs catalyst, divinyldisiloxanes, vinylsilane, ruthenium complexes

Although until the late 1980s very little information on effective metathesis conversion of organosilicon compounds had been reported, the use of molybdenum (Schrock) catalyst and ruthenium (Grubbs) carbene complexes as catalysts tolerating functional groups in substrates, have opened new synthetic opportunities in organosilicon chemistry. Silicon-containing dienes undergo two types of metathetical transformation [1].
- intramolecular Ring Closing Metathesis (RCM) leading to cyclic compounds;
- intermolecular Acyclic Diene Metathesis (ADMET) polymerization to yield linear polyenes see Scheme 1

Scheme 1

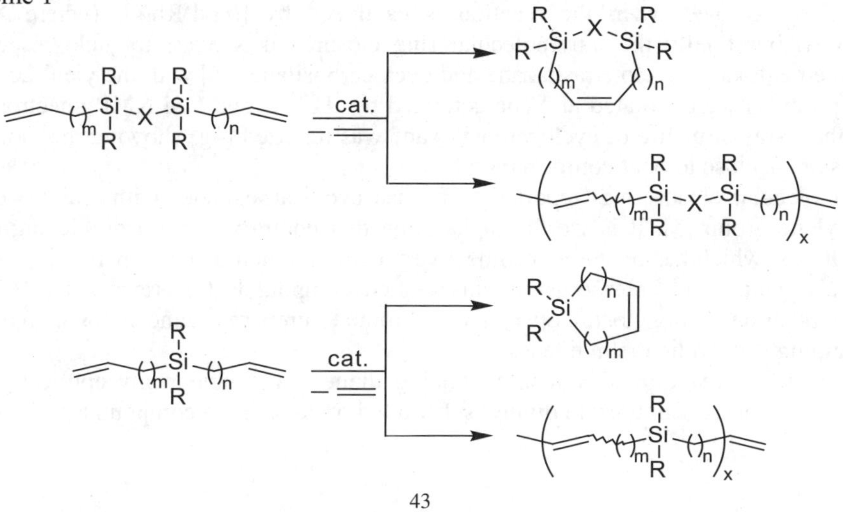

Y. Imamoglu and L. Bencze (eds.), Novel Metathesis Chemistry: Well-Defined Initiator Systems for Specialty Chemical Synthesis, Tailored Polymers and Advanced Material Applications, 43–49.

According to the hitherto knowledge, divinylderivatives of organosilicon compounds which are of fundamental industrial importance, are inert to RCM and ADMET polymerization [2]. These compounds undergo efficient cross-metathesis condensation (to yield a mixture of linear oligomers (mostly if ruthenium complexes are used as catalysts) and cyclic dimers and trimers (particularly if rhodium complexes are used), according to Scheme 2 [3].

Scheme 2

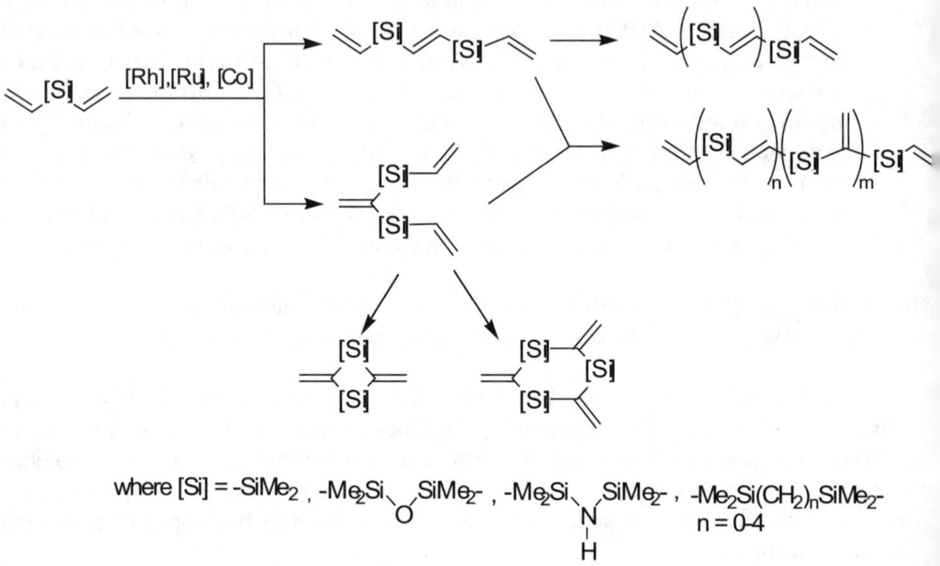

where [Si] = -SiMe$_2$, -Me$_2$Si$\diagdown$O$\diagup$SiMe$_2$- , -Me$_2$Si$\diagdown$N(H)$\diagup$SiMe$_2$- , -Me$_2$Si(CH$_2$)$_n$SiMe$_2$- n = 0-4

The mechanism of catalysis involves the insertion of vinyl-silicon dienes into M-H and M-Si bonds, followed by β-Si and β-H elimination, to yield ethene and two isomeric *trans* and *gem-* bis(vinylsilyl)ethenes, respectively [3,4]. *Gem*-dimeric products can be effectively formed when the reaction is catalyzed by [(cod)RhX]$_2$ (where X=Cl, OSiMe$_3$) but finally the intramolecular ring closure takes place to yield respective cyclotetrasiloxane, cyclotetrasilazane and cyclocarbosilanes [4] with the yield 20-25%. The products were isolated and characterized by ^{1}H, ^{13}C and ^{29}Si NMR spectroscopy and the X-ray structure of cyclocarbosiloxane was reported [4g] showing the molecule to assume a close to boat conformation.

Intramolecular closure was reported to furnish cyclocarbosilanes with one *exo*-cyclic methylene group [5]. It is worth emphasizing that contrary to the ring closing diene metathesis, which has recently become a very common method of preparation of *endo*-cyclic organic (and heteroorganic) unsaturated compounds (for review see [6]) the above mentioned ring closure opens a novel route to prepare organosilicon compounds containing *exo*-cyclic methylidenes.

RCM reaction of silicon-containing dienes has found many applications in organic synthesis. Contrary to divinyl substituted organosilicon compounds, vinyl silyl alkenyl derivatives, e.g. acyclic vinyl silyl alkenyl ether, particularly in the presence of molybdenum and ruthenium alkylidene catalyst, undergo a cyclization to yield 6-8 and 10-membered rings, which followed by oxidative cleavage of cyclic siloxy alkenes [7] (Scheme 3) or via the palladium catalyzed cross-coupling with aryl iodides [8] (Scheme 3b) give hydroxy alkenes

Scheme 3

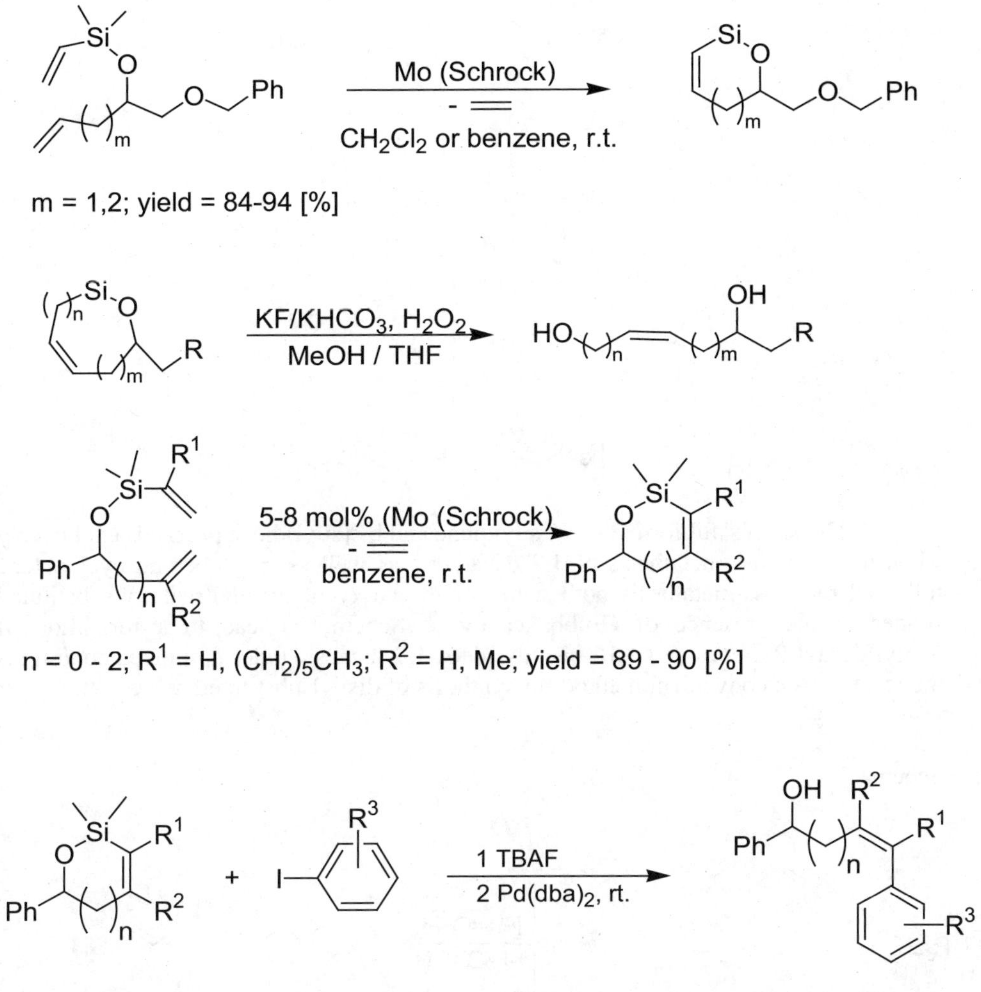

Mo (Schrock)
- =
CH_2Cl_2 or benzene, r.t.

m = 1,2; yield = 84-94 [%]

$KF/KHCO_3$, H_2O_2
MeOH / THF

5-8 mol% (Mo (Schrock))
- =
benzene, r.t.

n = 0 - 2; R^1 = H, $(CH_2)_5CH_3$; R^2 = H, Me; yield = 89 - 90 [%]

1 TBAF
2 $Pd(dba)_2$, rt.

Results of the efficient cross-metathesis of monovinylsubstituted silicon compounds with alkenes reported earlier (see paper in this book) permitted initiation of experiments on metathetical transformation involving reactions of alkadienes and cycloalkenes with vinylsilanes. These two reactions compete with ROMP and ADMET polymerisation of organic substrates. Scheme 4 illustrates all the competitive-consecutive reactions, which can occur in this system.

Scheme 4

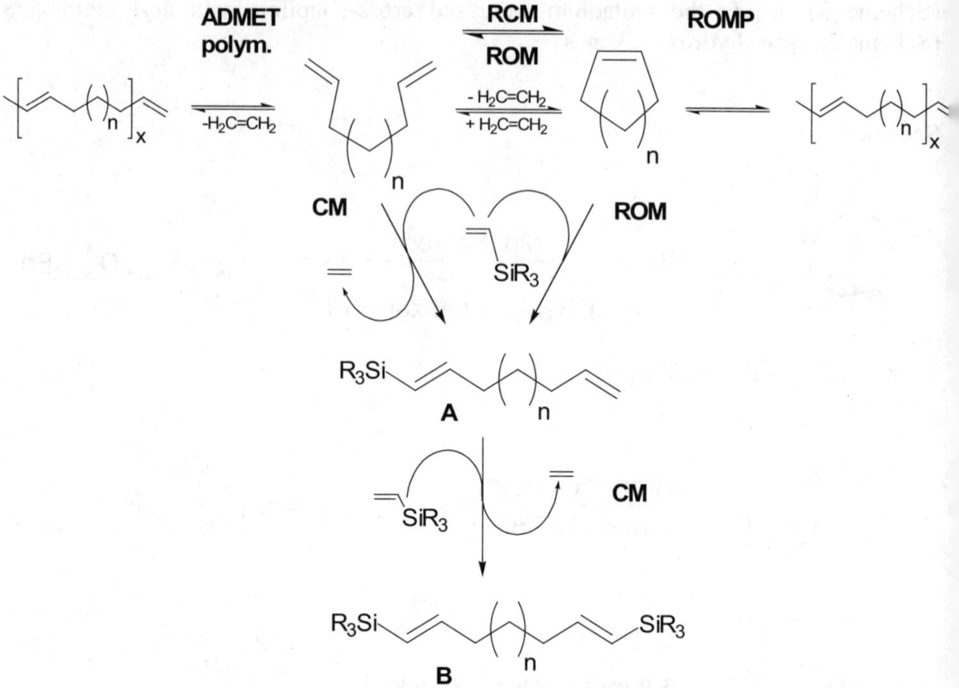

The successful formation of silyldiene and bis(silyl)diene proceeds exclusively when n = 4. Cross-metathesis of 1,9 decadiene as well as ring opening cyclooctene followed by cross-metathesis both with vinyltrialkoxy- and vinyltrisiloxy-substituted silanes in the presence of Grubbs catalysts (Scheme 5) lead to a formation of 1(silyl)deca-1,9-dienes and 1,10-bis(silyl)deca-1,9-dienes. Under optimized conditions the reaction is a convenient method for synthesis of disilylsubstituted dienes [9].

Scheme 5

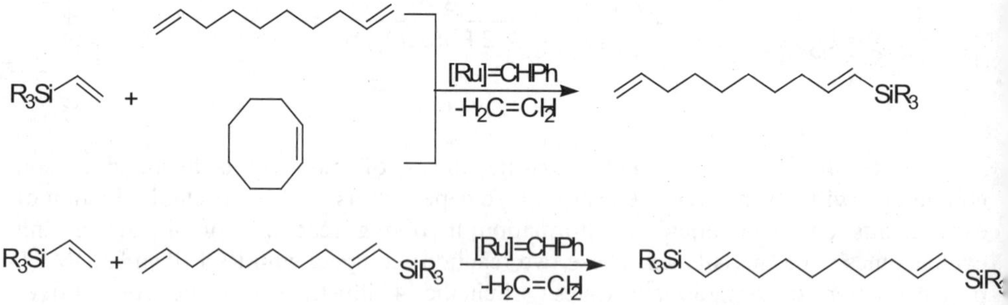

No products were observed when cyclopentene, cyclohexene, cyclododecene or indene were used as reagents. For vinylsilanes containing at least one methyl substituent no metathesis was observed either with 1,9-decadiene or with cyclooctene which is due to decomposition of Grubbs catalyst in the presence of vinylmethylsilanes as reported earlier [10].

High efficiency of the cross-metathesis of 1,9-decadiene with vinylsilanes indicates the possibility of an effective run of the ADMET copolymerisation of divinylsilanes with diene. This supposition was fully confirmed by the effective ADMET copolymerization of divinyltetraethoxydisiloxane with 1,9-decadiene leading, in the presence of Grubbs initiator, to the formation of siloxylene-vinylene-alkenylene polymer [11].

This copolymer was isolated and characterized by 1H , 13C NMR spectroscopy and GPC analysis (Mn = 18000, PDI = 2.9). However, quantitative analysis of olefinic region of the spectrum and signals from the ethoxy groups at silicon enabled a calculation of the molar ratio of both building blocks in the copolymer (siloxane : diene = 1 : 1.7). so the following scheme can be proposed to illustrate the formation of the copolymer

Scheme 6

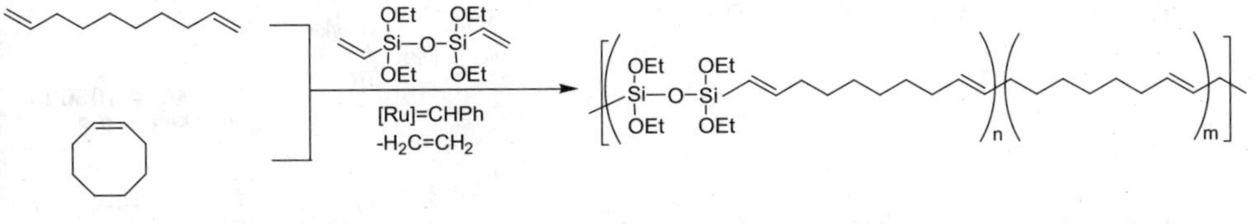

As analogous polymer was obtained when cyclooctene was used instead of diene [9] $M_w = 4500$, PDI = 3.20

Also an earlier attempt of ADMET copolymerization of divinyldimethylsilane with 1,9-decadiene in the presence of Schrock tungsten catalyst [12] (M_n = 7800, M_w/M_n = 2.1, 75% *trans*) felt to obtain much less (only 6 mol %) of vinylene-silylene unit in the polymer.

Moreover, divinylsiloxanes undergo copolymerization with dihexenyl ether (catalyzed by Grubbs catalyst) but also no evidence for the consecutive vinylsilane linkage has been found [13].

Scheme 7

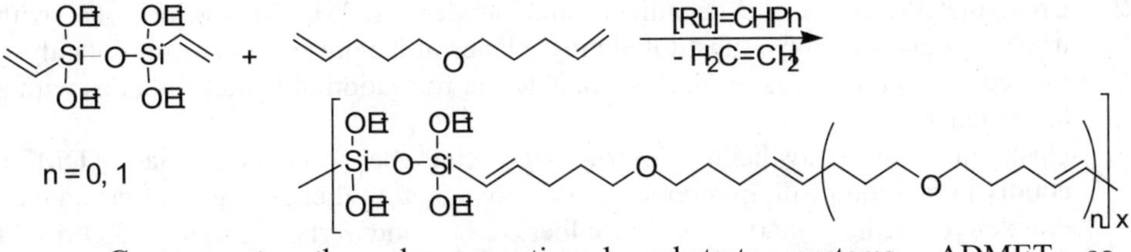

Contrary to the above-mentioned substrate systems, ADMET co-polymerization of divinyltetraethoxysilane with 1,4-divinylbenzene performed in the presence of Grubbs catalyst produces siloxylene-vinylene-p-phenylene copolymers (M_w = 9100, PDI = 4.0) with perfect consecutive vinylsiloxane linkage according to the following equation (Scheme 8) [14]

48

Scheme 8

On the other hand, the corresponding reaction of the silylative co-polycondensation of

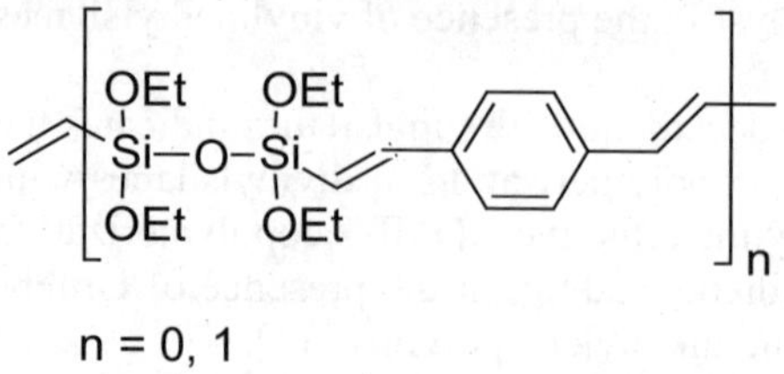

n = 0, 1

divinylsiloxanes ethoxy or methyl groups at silicon with 1,4-divinylbenzene catalyzed by Ru-H complex furnishes similar copolymers but with no consecutive vinylene-siloxylene ligands.

Scheme 9

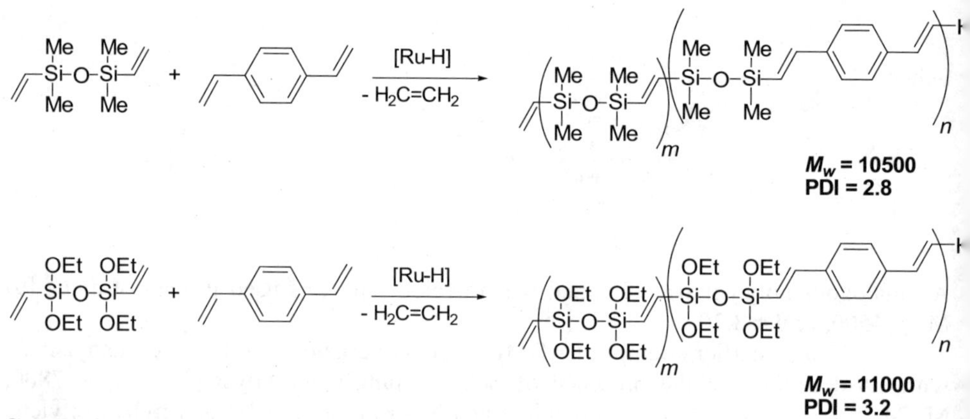

$M_w = 10500$
PDI = 2.8

$M_w = 11000$
PDI = 3.2

Conclusions

1. In view of the inactivity of metallacarbenes in ADMET polymerization of divinylsilicon compounds, the latter can undergo effective silylation coupling (poly)condensation to yield, in the presence of appropriate catalysts, either, e.g. linear polymers - silylene-(siloxylene,silazalene)- vinylenes, silylene-alkylene-vinylenes, silylene-arylene-vinylenes or the respective cyclic compounds containing exo-cyclic methylenes, e.g. cyclocarbosiloxane.

2. Cross-metathesis of 1,9-decadiene and tandem ROM/CM cyclooctene with trialkoxy- and trisiloxy -substituted vinylsilanes in the presence of Grubbs catalyst, carried out in an excess of olefins, leads to the formation of bis(silyl)dienes with a high yield.

3. Unsaturated organosilicon co-polymers can be produced via ADMET copolymerization of divinyltetraethoxydisiloxane with dienes (e.g., 1,9-decadiene, dihexenyl- ether and 1,4-divinylbenzene) and via tandem ROM/CM polymerization of cyclooctene with divinyltetraethoxydisiloxane as well as via silylative coupling (SC) polycondensation of divinyldisiloxanes with dienes (e.g. divinylbenzene). All the new transformations open new convenient routes to synthesis of vinylene-siloxylene-alkenylene polymers.

Acknowledgement

My warmest thanks are due to the co-workers whose names appear in the references. Our recent research was partly supported by State Committee for Scientific research in Poland, projects No. K012/T09/2000 and K026/T09/2001.

References

1. Marciniec, B. and Pietraszuk, C. *Curr. Org. Chem.* (2003) (in press).
2. Schrock, R.R., DePue, R.T., Feldman, J., Scheverin, C.J., Dewan, S.C. and Liu, A.H. (1988) *J. Am. Chem. Soc.* **110**, 1423.
3. Marciniec, B. In *Ring Opening Metathesis and Related Chemistry.* Khosravi, E. and Szymanska-Buzar, T. Eds.; Kluwer Acad. Publ.: Dordrecht, (2002) p.331-340
4. (a) Wakatsuki, Y., Yamazaki, H., Nakano, M. and Yamamoto, Y. (1991) *J. Chem. Soc. Chem. Commun.* 703; (b) Marciniec, B. and Pietraszuk, C. (1997) *Organometallics.* **16**, 4320; (c) Marciniec, B. (2000) *Molecular Crystal & Liquid Crystals.* **353-354**, 173; (d) Marciniec, B. and Lewandowski, M. (1996) *J. Polym. Sci. Part A, Polym. Chem.* **34**, 1443; (e) Marciniec, B. and Lewandowski, M. (1995) *J. Inorg. & Organometal. Polym.* **15**, 115; (f) Marciniec, B. and Malecka, E. (1999) *Macromolecular Rapid Commun.* **20**, 475; (g) Marciniec, B., Lewandowski, M., Bijpost, E., Malecka, E., Kubicki, M. and Walczuk-Gusciora, E. (1999) *Organometallics.* **18**, 3968; (h) Marciniec, B. and Malecka, E. *Macromolecules.* (submitted for publication); (i) Marciniec, B., Malecka, E., Majchrzak, M. and Itami, Y. (2001) *Macromol. Symp.* **174**, 137.
5. Mise, T., Takaguchi, Y., Umemiya, T., Shimizu, S. and Wakatsuki, Y. (1998) *Chem. Commun.* 699.
6. (a) *Alkene Metathesis in Organic Synthesis*, Fürstner, A. Ed., Springer: Berlin, **1998**; (b) Ivin, K.J.; Mol. J.C. *Olefin Metathesis and Metathesis Polymerization*, Academic Press: San Diego, **1997**; (c) Fürstner, A. (2000) *Angew. Chem., 112*, 3140; *Angew. Chem., Int. Ed.* **2000**, *39*, 3012.
7. Chang, S. and Grubbs, R.H. (1997) *Tetrahedron Lett.* **38**, 4757.
8. Denmark, S.E. and Yang, S.M. (2001) *Org. Lett.* **3**, 1749.
9. Pietraszuk, C., Marciniec, B. and Jankowska, M. (2002) *Adv. Synth. Catal.* **344**, 789.
10. (a) Pietraszuk, C., Marciniec, B. and Fischer, H. (2000) *Organometallics.* **19**, 913; (b) Majchrzak, M., Itami, Y., Marciniec, B. and Pawluc, P. (2000) *Tetrahedron Lett.* **41**, 10303.
11. Malecka, E., Marciniec, B., Pietraszuk, C., Church, A.C. and Wagener, K. (2002) J. Mol. Catal. **190**, 27.
12. Wagener, K.B. and Smith, Jr. D.W. (1991) *Macromolecules.* **24**, 6073.
13. Malecka, E., Pawluc, P. and Marciniec, B. (unpublished results).
14. Marciniec, B. and Majchrzak, M. *J. Organometal. Chem.* (submitted for publication).

SYNTHESIS OF ORGANOSILICON REAGENTS OF VINYLSILICON FUNCTIONALITY VIA CROSS-METATHESIS AND SILYLATION OF ALKENES

BOGDAN MARCINIEC

Department of Organometallic Chemistry, Faculty of Chemistry,
Adam Mickiewicz University, Grunwaldzka 6, Poznań, 60-780 Poland

Abstract

The catalytic reactions of two types occurring between the same parent substances, i.e. silylative coupling (also called trans-silylation, silyl group transfer) and cross-metathesis of functionalized (such as p-substituted styrenes, vinyl- and allyl-substituted hetero(O,N,S,B)alkenes with vinylsubstituted silicon compounds (vinyltrisubstituted silanes, vinylsilsesquioxanes, vinylcyclosiloxanes, vinylcyclosilazanes, tris(dimethylvinylsilyl)benzene) have been tested as possible universal route toward synthesis of well-defined molecular compounds with vinylsilicon functionality. Both types of mostly stereo- and/or regio-selective reactions occur in the presence of ruthenium, rhodium and iridium complexes containing or generating the M-H and/or M-Si bonds (the silylative coupling (**I**)) and, in the presence of Grubbs complexes, in the case of trialkoxy- and trisiloxysubstituents at silicon (the cross-metathesis (**II**)).

Keywords: cross-metathesis, silylative coupling, vinylsilane, vinylsiloxane, Grubbs catalyst, silicometallics

Organometallic compounds (organometallics) are defined as materials with one or more bonds (ionic or covalent, localized or delocalized) between carbon atoms of an organic group or molecules and the atoms from the main group of the elements, transition, lanthanide or actinide metals e.g. Fig. 1a.

Fig. 1 a

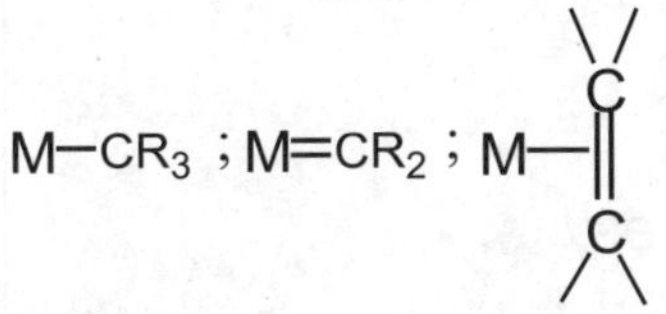

$$M-CR_3 \; ; \; M=CR_2 \; ; \; M-\overset{\overset{\diagup}{C}}{\underset{\underset{\diagdown}{C}}{\|}}$$

Fig. 1b

$$M-SiR_3 \; ; \; M=SiR_2 \; ; \; M-\overset{\overset{\diagup}{C}}{\underset{\underset{\diagdown}{Si}}{\|}} \; ; \; M-\overset{\overset{\diagup}{Si}}{\underset{\underset{\diagdown}{Si}}{\|}}$$

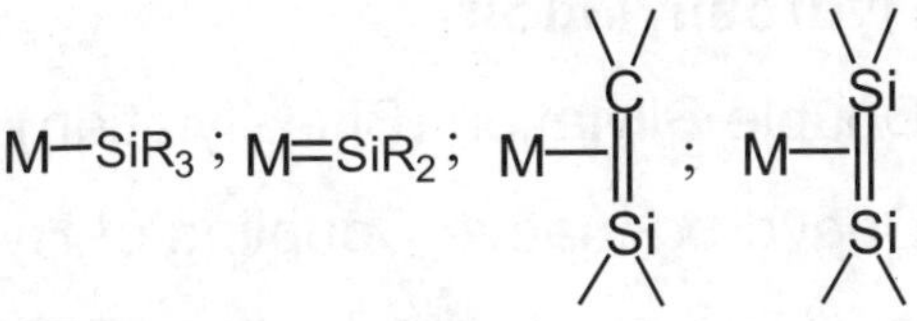

The making and breaking of the metal-carbon bonds play a fundamental role in the homogeneous and heterogeneous catalyses.

Organic derivatives of silicon besides other metalloids (boron, germanium, arsenic), are also included in the class of organometallic compounds, and the Si-C bond is the basis of organosilicon monomers and polymers of great industrial application.

On the other hand, if such metalloids (p-block) replace the carbon atom in the metal-carbon bonding then they form the real metal-nonmetal bond. Such compounds are the subjects of a new field of study called "inorganometallic chemistry" [1].

In particular, compounds containing d-block-p-block element bonds, which although closely related to organometallics yet distinctly differ from them, can play a

Y. Imamoglu and L. Bencze (eds.), Novel Metathesis Chemistry: Well-Defined Initiator Systems for Specialty Chemical Synthesis, Tailored Polymers and Advanced Material Applications, 51–64.
© 2003 *Kluwer Academic Publishers. Printed in the Netherlands.*

significant role as intermediates in transformations of *p*-block compounds. Silicon is the most common and useful element among *p*-block elements, so silicometallic (or silaorganometallic) compounds include the derivatives with metal-silicon bonds, e.g. Fig. 1b.

The triangle (see Fig.2) explains the difference between the organometallic, organosilicon and silicometallic compounds. The reactivity of the latter, and in particular the transition metal silicon bond, is a key point in most conversions of silicon compounds catalyzed by transition metal (TM) complexes.

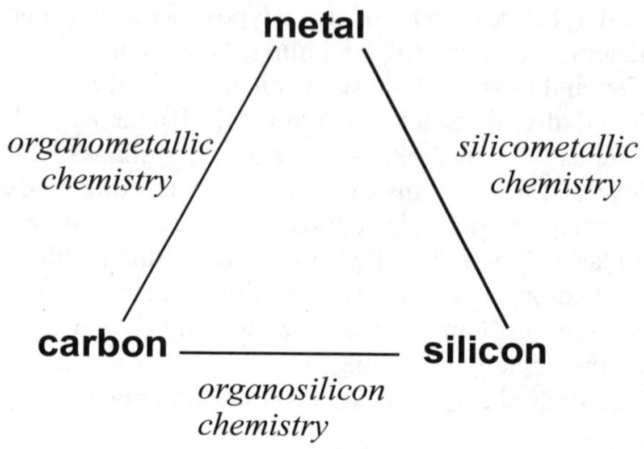

Fig. 2

Therefore, although in catalytic conversions of (organo)silicon compounds only hydrosilylation reactions are well-know as processes of industrial importance [2], in the last - two decades other reactions of silicon compounds catalyzed by TM complexes have been revealed and spectacularly developed [2]. Most of them occur via a mechanism involving metal-silicon and metal-hydrogen bonds, only occasionally accompanied (or sided) by metal-carbon bonds (see Scheme 1).

Hydrosilylation

Double Silylation (Bis-Silylation)

Dehydrogenative Coupling of Hydrosilanes

Dehydrogenative Coupling of Olefins with Hydrosilanes

Silylcarbonylation, Silyl(hydro)formylation

Metathesis of Silicon-containing Olefins

Silylative Coupling of Olefins with Vinylsilanes

Scheme 1. Silicometallics *vs.* Organometallics catalyzed transformation of silicon compounds

The aim of the lecture is to present recent achievements in the synthesis of organosilicon reagents involving alkenylsilanes of vinyl-silicon functionality i.e. of the general formula shown below (Fig.3), on the basis of two catalytic reactions occurring between the same initial substances (i.e. silylative coupling called also silyl group transfer or *trans*-silylation) (**I**) and cross-metathesis (**II**) of olefins with monovinyl-substituted silicon compounds.

Fig.3.

I proceeds via intermediates with TM-silicon bonds (i.e. silicometallics) whereas **II**- via TM carbene species (i.e. organometallics). Both reactions give the same or similar isomeric (Scheme 2) vinylsilanes, which have become one of the most common class of silyl reagents used in organic synthesis.

Silylative coupling (*trans*-silylation) (**I**)

Silylative coupling (*trans*-silylation) (**I**)

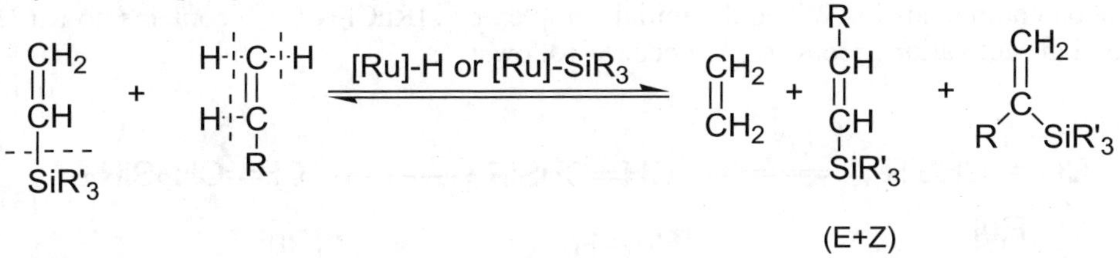

$$CH_2\!=\!CH\!-\!SiR'_3 \; + \; H_2C\!=\!CH\!-\!R \xrightarrow{\;[Ru]-H \text{ or } [Ru]-SiR_3\;} CH_2\!=\!CH_2 \; + \; CH(R)\!=\!CH\!-\!SiR'_3 \; + \; CH_2\!=\!C(R)\!-\!SiR'_3$$

(E+Z)

Cross-metathesis (**II**)

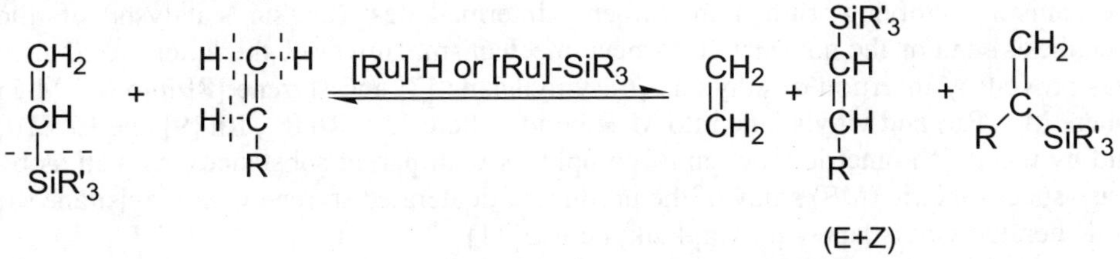

$$CH_2\!=\!CH\!-\!SiR'_3 \; + \; H_2C\!=\!CH\!-\!R \xrightarrow{\;[Ru]-H \text{ or } [Ru]-SiR_3\;} CH_2\!=\!CH_2 \; + \; CH(SiR'_3)\!=\!CH\!-\!R \; + \; CH_2\!=\!C(R)\!-\!SiR'_3$$

(E+Z)

Scheme 2

The silylative coupling of vinyltrisubstituted silanes with a variety of olefins (**I**) catalysed by ruthenium and rhodium complexes was efficiently developed to lead to the novel and well-known silicon-containing olefins.

In 1984 we reported the first very effective example of metathesis (disproportionation) of vinylsubstituted silicon compounds which could be catalyzed by Ru and Rh complexes. It has allowed the synthesis of a series of unsaturated silicon compounds according to the following equations, with the yield predominantly higher than 70% [3,4].

Disproportionation [3]

$$2\ R_3SiCH=CH_2 \xrightarrow[-\ CH_2=CH_2]{} R_3SiCH=CHSiR_3\ +\ (R_3Si)_2C=CH_2 \tag{1}$$

where: R = Me, Ph, OR'

Co-disproportionation [4]

$$R_3SiCH=CH_2\ +\ CH_2=CH(CH_2)_nR' \xrightarrow[-\ CH_2=CH_2]{} R_3SiCH=CH(CH_2)_nR' \tag{2}$$

where: R = Me, Ph, OR''; R' = Me, m= 3-15; R = Ph, m=0

$RuCl_2(PPh_3)_3$, $RuHCl(PPh_3)_3$, $[RuCl_2(CO)_3]$, $RuH_2(PPh_3)_4$, $RhCl_3$ x n H_2O, $RuCl_3$ x n H_2O, $RhCl(PPh_3)_3$, $HSiEt_3$, $HSi(OEt)_3$, $HSiPh_3$ as well as $LiAlH_4$ were used as co-catalysts.

All experimental results as well as the report of Seki et al. on catalysis of metathesis (disproportionation) of vinylsilanes by $Ru_3(CO)_{12}/HSiPh_3$ and $RuHCl(CO)(PPh_3)_3$ [5] have provided convincing evidence for the thesis that formation of Ru-H bond is a crucial stage in the initiation of catalytically active species in all the ruthenium complex-catalyzed metathesis (and co-metathesis) of vinyl substituted silicon compounds [3]. When the initial complex e.g. $[RuCl_2(CO)_3]_2$ contains no Ru-H bond the activation process could occur as follows:

$$\underset{[Ru]}{CH_2=CHSiR_3} \rightleftharpoons \underset{[Ru]-H}{CH=CHSiR_3} \longrightarrow \underset{[Ru]}{CH=CH_2SiR_3} \tag{3}$$

However, the obtained results did not allow us to distinguish between the reaction mechanism involving ruthenium-carbene intermediates (classic catalysts of the metathesis) and/or the non-metallacarbene mechanism. However, the latter mechanism was proved by insertion of ethylene [6], vinylsilane [7] and styrene [8] into the M-Si where M = Ru) and vinylsilane into M-H bond (where M = Ru [8], Rh [9] and Co [10] and by the stoichiometric reaction of complexes with parent substances, as well as by mass-spectrometric (MS) study of the product of deuterated styrene with vinylsilane [8] or deuterated vinylsilane with vinyl alkyl ether [11].
A mechanistic scheme of this new type of silyl olefin conversion, studied in detail for the reaction with styrene, involves the migratory insertion of the olefin into M-Si bond and vinylsilane into M-H bond followed by β-H and β-Si elimination to give E-alkylsilylethene (and ethylene), see Scheme 3 [12].

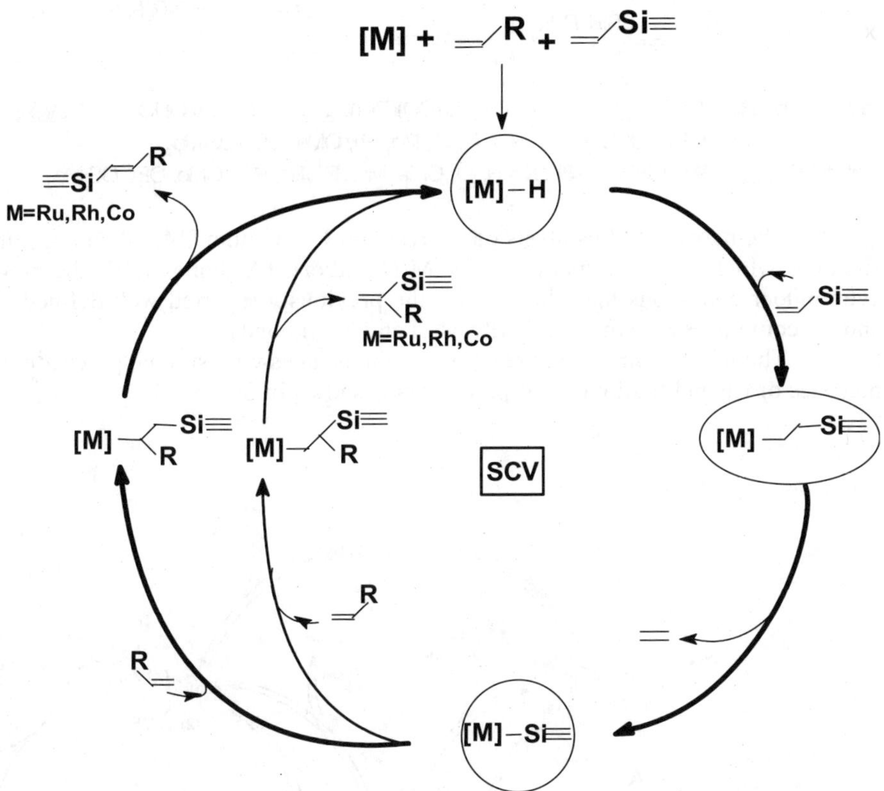

Scheme 3. General mechanism of the silylative coupling of olefins with vinyltrisubstituted silanes catalyzed by TM complexes containing M-H and/or M-Si bonds

On the other hand, although well-defined or in situ initiated metallacarbenes are inactive in self-metathesis of vinylsubstituted silanes and siloxanes due to steric hindrance stimulating non-productive metathesis of decomposition to bis(silyl)substituted metallacyclobutane, we revealed recently a high catalytic activity of Grubbs catalyst (in cross-metathesis of vinyltrialkoxysilanes and vinyltrisiloxysilanes with 1-styrene [13].

In our recent study of the stereo- and regioselective coupling reaction (**I**) [14-16] and cross-metathesis (**II**) [18] of alkenes with vinylsilanes, we used *p*-substituted styrenes and new catalytic systems - rhodium [14,15] and iridium [16] siloxides, RuHCl(CO)PR$_3$ + CuCl co-catalyst as well as Grubbs catalyst [18]. Both reactions **I** and **II** give stereoselectively the same products - *trans*-silylstyrenes.

An efficient, stereo- and regioselective synthesis of silylstyrenes was observed when styrene or *p*-substituted styrenes reacted with vinylsilanes in the presence of both ruthenium (or rhodium) complexes containing or generating the M-H bond and Grubbs type carbene complex. Both reactions often proceed quantitatively according to (4). In all cases *E*-styrylsilanes and siloxanes were exclusively formed. The two reactions yielding the same products often proceed quantitatively according to the equation:

$$X\text{-}C_6H_4\text{-}CH=CH_2 + CH_2=CH\text{-}SiR^1R^2R^3 \xrightleftharpoons[- CH_2=CH_2]{cat.} X\text{-}C_6H_4\text{-}CH=CH\text{-}SiR^1R^2R^3 \qquad (4)$$

cat. = [{(cod)Rh(μ-OSiMe$_3$)}$_2$] [14,15]; [RuHCl(CO)(PPh$_3$)$_3$]+CuCl [17], [RuHCl(CO)(PCy$_3$)$_2$]+CuCl [17]
X = H, Cl, Br, Me, OMe; R^1, R^2,R^3 = H, Cl, Me, Ph, OMe, OEt, OSiMe$_3$
cat. = [(PCy$_3$)$_2$Cl$_2$Ru(=CHPh)] [18] ; X = H, Cl, OMe, Me; R^1, R^2, R^3 = OMe, OEt, OSiMe$_3$

The new type of catalysts of the reaction **I** - rhodium [14,15] and iridium [16] siloxides which do not contain initially M-H and/or M-Si bonds leads the process at even milder conditions than those with the previously reported well-defined Ru, Rh and Co complexes containing initially M-H and M-Si bonds.

The mechanisms of the reaction of vinylsilanes with styrene proceeding in the presence of Rh and Ir siloxides as precursors is shown in Scheme 4.

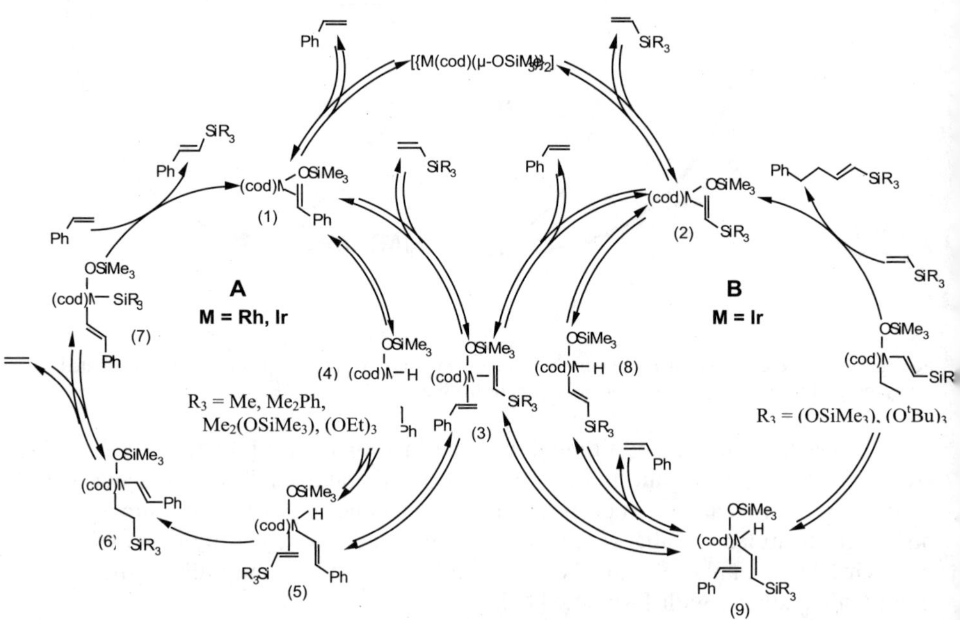

Scheme 4. Catalytic activattion of =C-H of styrene and vinylsilane by Rh and Ir siloxides

The reaction occurs according to the non-metallacarbene mechanism involving the M-H (M = Rh, Ir). 16e intermediate (**1**) generated *in situ* is followed by oxidative addition of the =C-H bond of initiating the well-known process (see Scheme 3). But when metal-siloxide complex is used (Scheme 4, cycle **A**) , the catalytic cycle does not involve a migratory insertion of alkene into the M-Si bond since the final step of the product formation occurs via reductive elimination of the product [19]. The reaction seems to be a convenient tool for synthesis of substituted vinylsilanes, particularly bis(silyl)ethenes [14] and silylstyrenes [15].

However, contrary to rhodium-siloxide, iridium-siloxide complex catalyses also hydrovinylation (co-dimerization) of styrene with $CH_2=CHSi(OR')_3$ (where R = $SiMe_3$, *t*-Bu) (see Scheme 4, cycle **B**) i.e. to get E-1-silyl-4-phenyl-1-butene as a main product [16].

$$Ph{-}{=} \;+\; {=}{-}SiR_3 \quad \xrightarrow{[Ir]} \quad Ph\diagdown\diagup\diagdown SiR_3$$

$$(5)$$

Apparently, steric hindrance of three siloxy substituents at silicon stops the silylative coupling and promotes the hydrovinylation process.

The comprehensive catalytic study of the reaction of vinylsubstituted silanes and siloxanes with heteroatom-functionalized alkenes, catalyzed by ruthenium complexes, has allowed us to propose two general synthetic routes.

The reaction of vinylsubstituted silanes with vinylsubstituted hetero (O, N, S, B) organic compounds proceeds effectively according to the non-metallacarbene mechanism as a silylative coupling (**I**) and yields, under optimum conditions (usually 5-fold excess of alkene; 80-110°C and in the presence of ruthenium complexes containing or generating Ru-H and/or Ru-Si bonds), 1-silyl-2-heteroatom substituted ethenes with high preference for *E*-isomer:

$$R_3Si{-}{=} \;+\; {=}{-}X \quad \xrightarrow[{-\,=}]{[Ru]} \quad R_3Si\diagup{=}{-}X \qquad (6)$$

[Ru] = [RuHCl(CO)(PPh$_3$)$_3$], [RuCl(SiMe$_3$)(CO)(PPh$_3$)$_2$], [RuCl$_2$(PPh$_3$)$_3$], [RuHCl(CO)(PCy$_3$)$_2$]
SiR$_3$ = SiMe$_3$, SiMe$_2$Ph, Si(OEt)$_3$; X = OR'; R' = Et, Pr, Bu, *t*-Bu, C$_6$H$_{13}$, SiMe$_3$; Yield = 18-73% [10-20];
SiR$_3$ = SiMe$_3$, SiMe$_2$(OEt); X = COOMe; Yield 59 and 70% resp. [21]

X = (2-oxopyrrolidin-1-yl), (phthalimido-N-yl), (formamido) Yield = 12-97% [22]

X = (1,3,2-dioxaborolan-2-yl) Yield = 72-78% [23]

On the other hand, the reactions of vinylsilanes with allyl-substituted heteroorganic compounds in the presence of Grubbs catalyst, are examples of successful cross-metathesis and yield, under optimum conditions (10-15 excess of vinylsilane, 40°C), 1-silyl-2-O,N or S, B-substituted propenes) according to the following equation

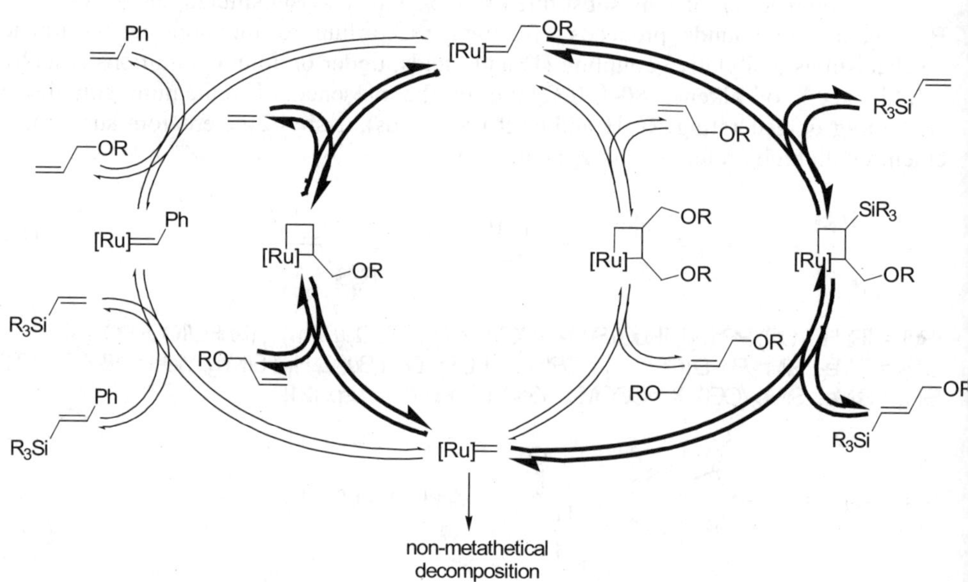

(7

Y = SiR'$_3$; R' = Me, OEt [18]
    OR' ; R' = Et, Bu, Cy, Ph, CH$_2$Ph, glycidyl, SiMe$_3$; R = Me, Et, SiMe$_3$, Yield 50-95% [24
    OCOR; R = Me, Et, Pr, Yield = 87-89% [21];
    NMePh, N(CO)MePh; R = Et; Yield = 57-63% [22]
    SCMe$_3$; R = Et, Yield = 57% [25]

Vinylboranes undergo also cross-metathesis with vinylsilanes in the presence of Grubbs catalyst [23].

A general mechanisms for the cross-metathesis of vinyltri(alkoxy, siloxy)silanes with heteroatom containing olefins in the presence of ruthenium carbene is shown in Scheme 5:

Scheme 5

Mechanistic implication of a general cross-metathesis of vinyl-silicon with allyl-substituted heteroorganic compounds were studied in detail for the reaction with allyl alkyl ether [24].

The detailed NMR study of the stoichiometric reaction of Grubbs catalyst with allyl-n-butyl ether has brought the information on the individual steps of the catalytic cycle. The preliminary reaction resulted in formation of styrene and 3-(alkoxy)propylidene complex 2. This process is accompanied by a series of subsequent reactions as illustrated in Scheme 6.

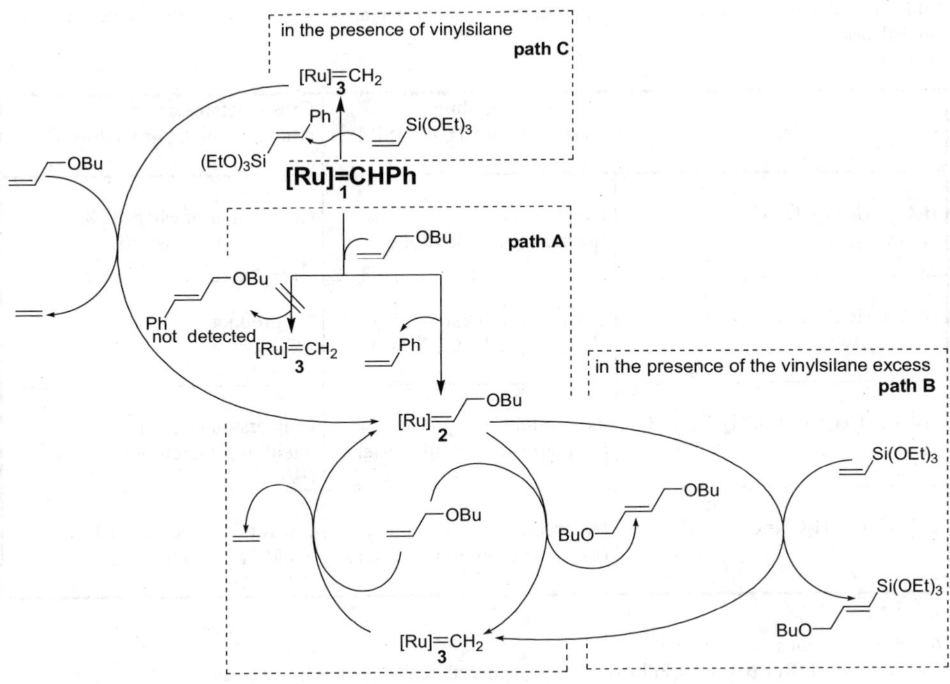

Scheme 6

The catalytic process must be carried out using an excess of vinylsilane in order to avoid the formation of allyl ether homo-metathesis product. Therefore, according to the scheme this process leads to effective formation of the $[Ru]=CH_2$ (3) complex, which unfortunately readily undergoes non-metathetical decomposition. So an addition of some amount of the allyl ether during the reaction would enhance the final yield of the cross-metathesis product, which has been confirmed experimentally [24].

In order to illustrate the different role of the two reactions, i.e. silylative coupling (**I**) catalyzed by Ru-H initiated complex and cross-metathesis (**II**) catalyzed by Grubbs catalyst the compilation of the results for unsaturated esters with vinylsubstituted silanes is shown in Table 1. While effective cross-metathesis is observed for the reaction of trialkoxy- and trisiloxy-substituted vinylsilanes with allyl esters, and methyl butenoate [21] the silylative coupling is observed for methyl acrylate [21].

Table 1. Silylative coupling vs. cross-metathesis of unsaturated esters and methyl butenoate with vinylsilanes

Olefin	Silylative coupling[1] reaction with $CH_2=CHSiMe_3$	Cross-metathesis[2] reaction with $CH_2=CHSi(OEt)_3$
$H_2C=CHCH_2COOCH_3$ methyl-3-butenoate	No product (isomerization of allyl ester)	Conversion of ester=82% Yield of products=70% (E/Z=4/1)
$H_2C=CHCOOCH_3$	Conversion of ester=73% Yield of products=72% (E)	No product
$H_2C=C(CH_3)-COOCH_2CH=CH_2$	No product (isomerization of allyl ester)	Conversion of ester =91% Yield of product=90% (E/Z=3/1)
$H_3C-COOCH_2CH=CH_2$	No product (isomerization of allyl ester)	Conversion of ester =100% Yield of product=98% (E/Z=11/1)

Reaction conditions:
[1] [silane]:[ester]:[RuH(Cl)(CO)(PPh_3)_3]=1:3:10^{-2}, benzene, sealed ampules, 110°C, 24 h
[2] [silane]:[ester]:[Cl_2(PCy_3)_2Ru(=CHPh)]=5:1:2x10^{-2}, dichlorometane (reflux), 3 h

The two reactions discussed can also be used for functionalization of multivinylsilicon compounds. While well-defined tungsten alkylidene complexes were found not to convert metathetically to vinyltrimethylsilane, the molybdenum complex. Fig. (4) was revealed to be a very effective catalyst in cross-metathesis of vinylsubstituted silsesquioxanes with a variety of alkenes [26] (100% conversion and high yields of cross-metathesis products). For all reactions 1.5-12 equivalents of alkene per $H_2C=CHSi$ group were used and no competitive alkene homo-metathesis was reported. No self-metathesis of vinylsilsesquioxane was observed, presumably due to steric demands of the Si-O framework. Exemplary results are presented below Fig. (4).

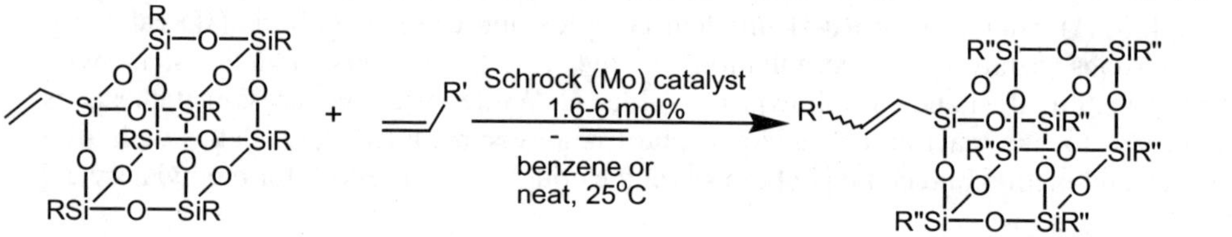

R
Si—O—SiR
Si—O-I-SiR
RSi—O—SiR
RSi—O—SiR
+
R'
Schrock (Mo) catalyst
1.6-6 mol%
benzene or
neat, 25°C
R'
R"Si—O—SiR"
Si—O-I-SiR"
R"Si—O—SiR"
R"Si—O—SiR"

where R = $HC=CH_2$; R' = Bu, Ph, $CH_2Si(OMe)_3$, $CH_2CH_2CH_2Br$, $CH_2(CH_2)_7COOEt$; R" = $HC=CHR'$

Fig. 4

Reactions of vinylsilsesquioxanes with pent-4-en-1-on and 5-brompent-1-ene were also checked in the presence of Grubbs catalyst, but much lower conversions were found (10 and 25%, respectively). However, a recent experiment performed in our laboratory with the application of Grubbs catalyst in the cross-metathesis of octavinylsilsesquioxanes has successfully given the product with high yield and 100% stereoselectivity. The X-ray structure of this compound was also solved [27]. The reaction offers a possibility of synthesis of highly functionalized silsequioxane and spherosilicate frameworks as well as silsesquioxane-based core for dendrimers. The Table 2 gives the successful results of the silylative coupling (**I**) and cross-metathesis **II** of octavinylsilsesquioxane with alkenes and with vinyl and allyl susbtituted organic and organosilicon compounds obtained recently in our laboratory.

Table 2. Silylative coupling vs. cross-metathesis of alkenes with octavinyl silsesquioxane catalyzed by Grubbs and [RuHCl(CO)(PCy$_3$)$_2$], respectively

Olefin	SILYLATIVE COUPLING [RuHCl(CO)(PCy$_3$)$_2$] conversion* [*trans-* / *cis-**]	CROSS-METATHESIS Grubbs catalyst conversion* [*trans-* / *cis-**]
1-hexene	39 % [-**]	71 (>99)*** % [91 / 9]
t-BuCH=CH$_2$	33 % [*trans-* major]	33 % [*trans-* major]
TMSCH$_2$CH=CH$_2$	49 % [-**]	78 (>99) *** % [93 / 7]
HOCHCH=CH$_2$	0 % [-]	3 % [*trans-* major]
PhCH=CH$_2$	>99 % [*trans-* only]	98 % [*trans-* only]
t-BuOCH=CH$_2$	>99 % [79 / 21]	0 % [-] 72% [68/32]
TMSOCH=CH$_2$	>99 % [45 / 55]	0 % [-] 61% [37/63]

* Based on vinylsilyl group calculated from NMR
** Isomerization of Si-CH=CH-CH$_2$-R was observed
*** 1 % [Ru=C] (8% to vinylsilyl group)

The reactions of vinylcyclosiloxanes and vinylcyclosilazanes with styrene have given products in the presence of [RuHCl(CO)(PCy$_3$)$_2$] (the yield 83-95%) and open a new route for functionalized monomers to ring-opening polymerization of cyclosiloxanes and cyclosilazanes [28].

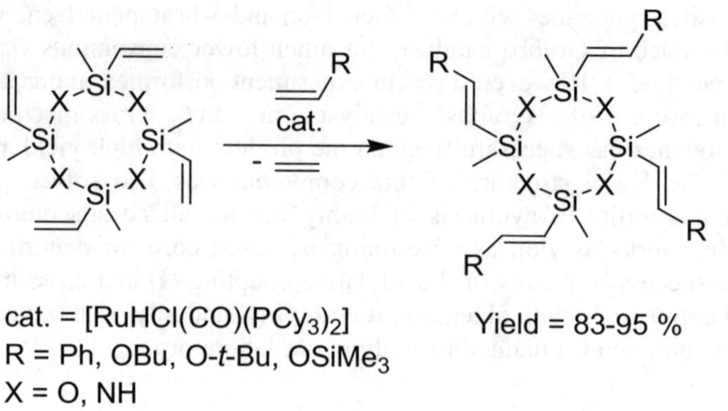
R
cat.
Yield = 83-95 %
cat. = [RuHCl(CO)(PCy3)2]
R = Ph, OBu, O-t-Bu, OSiMe3
X = O, NH

Scheme 7

The silylative coupling process can be used for syntheses of other types of unsaturated organosilicon compounds. Novel organosilicon dendrimers of the silicon-bridged-conjugated structure (which has potential optoelectronic properties) have been synthesized by the respective reactions of trivinylsubstituted silane (1,3,5-tris(dimethylvinylsilyl)benzene with conjugated diene (1,4-divinylbenzene) [29].

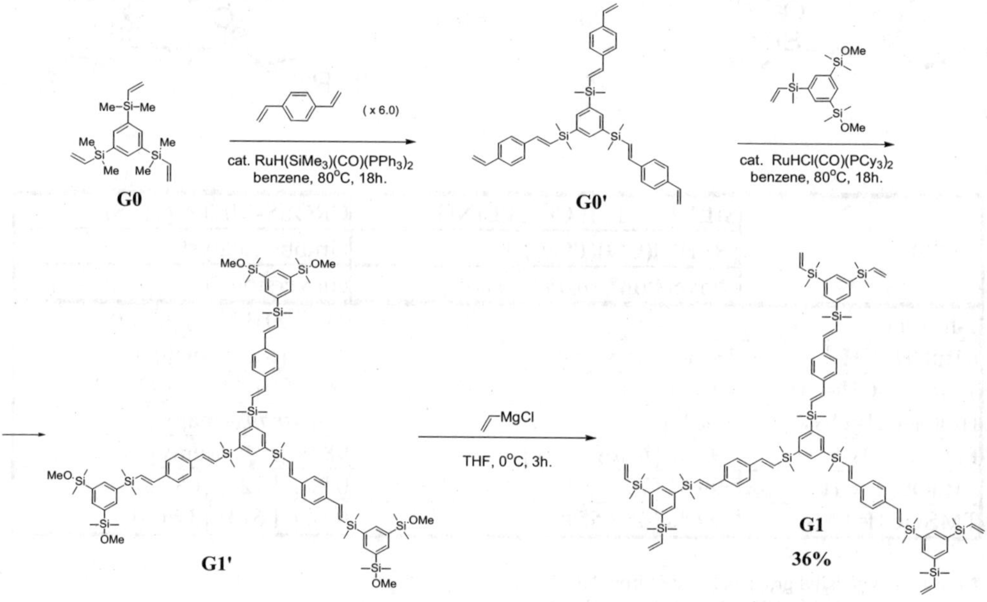
G0
(x 6.0)
cat. RuH(SiMe3)(CO)(PPh3)2
benzene, 80°C, 18h.
G0'
cat. RuHCl(CO)(PCy3)2
benzene, 80°C, 18h.
G1'
MgCl
THF, 0°C, 3h.
G1
36%

Scheme 8

In order to synthesize the first generation of dendrimer G1 direct silylation of G0' with tris(vinyldimethylsilyl)benzene was performed but the presence of terminal vinylsilyl groups caused silylative homo-coupling of G-1 itself or with another amount of tris(vinylsilyl)benzene (G0) to give a polymeric compound. We have therefore finally succeeded to synthesize G1 via a preliminary formation of iso-propoxydimethylsilyl terminated precursor (G1'). The silylation of (G0') with vinylsilyl-3,5-bis(iso-propoxysilyl)benzene was examined to synthesise G1, followed by vinylation with Grignard reagent (Scheme 8).

Conclusions

1. The two new catalytic reactions occurring between the same parent substances i.e. silylative coupling (also called dehydrogenative or *trans*-silylation, silyl group transfer) and cross-metathesis of alkenes with commercially available vinyl-substituted silicon compounds provide a universal route toward synthesis of well-defined molecular compounds with vinylsilicon functionality.

2. Stereo-, regio- and chemo-selective syntheses of p-substituted styrylsilanes and siloxanes occur in the presence of ruthenium, (but also rhodium and iridium) complexes containing or generating the M-H and/or M-Si bonds according to non-metallacarbene mechanism and, if catalyzed by Grubbs complexes (in the case of trialkoxy- and trisiloxy- substituents at silicon), according to the metallacarbene mechanism.

3. While vinylsilanes undergo productive cross-metathesis (Mo and Ru carbenes) with allyl-substituted functionalized alkenes, their effective transformation with derivatives containing a functionalized group attached directly to carbon-carbon double bond can be achieved via silylative coupling catalyzed by metal complexes containing (or generating) M-H and/or M-Si bonds (M = Ru, Rh, Ir).

4. The reactions of polyvinyl-substituted organosilicon compounds (e.g. 1,3,5-trivinylbenzene, vinylcyclosiloxanes and octavinylsilsesquioxane) with alkenes, e.g. styrene lead to synthesis of respective vinylsubstituted derivatives according to the non-metallacarbene process (all examples) and/or the metallacarbene process (substituted octavinylsilsesquioxanes).

Acknowledgement

My warmest thanks are due to the co-workers whose names appear in the references. Our recent research was partly supported by State Committee for Scientific research in Poland, projects No. K012/T09/2000 and K026/T09/2001.

References

1. Fehlner ed, *Inorganometallic Chemistry*, Plenum Press, New York, 1992, Chapter 1.1.
2. (a) Colvin, E.W. (1988) *Silicon Reagents in Organic Synthesis*. Acad. Press, Chapter 3; (b) Marciniec, B. (ed.) (1992) *Comprehensive Handbook on Hydrosilylation*. Pergamon Press, Oxford, Chapter 2 and other reviews cited there; (c) Seki, Y., Takeshita, K., Kawamoto, K., Murai, T. and Sonoda, N. (1986) *J. Org. Chem.* **51**, 3890; (d) Milan, A., Femandez, M.J., Bentz, P. and Maitlis, P.M. (1984) *J. Mol. Catal.* **26**, 89.
3. (a) Marciniec, B. (1997) *New J. Chem.* **21**, 815; (b) Marciniec, B. and Gulinski, J. (1984) *J. Organometal. Chem.* **266**, C19; (c) Marciniec, B., Maciejewski, H., Gulinski, J. and Rzejak, L. (1989) *J. Organometal. Chem.* **362**, 273; (d) Marciniec, B. and Pietraszuk, C. (1991) *J. Organometal. Chem.* **412**, Cl; (e) Marciniec, B., Pietraszuk, C. and Foltynowicz, Z. (1994) *J. Organometal. Chem.* **83**, 474;
4. (a) Marciniec, B., Rzejak, L., Gulinski J., Foltynowicz, Z. and Urbaniak, W. (1988) *J. Mol. Catal.* **46**, 329; (b) Marciniec, B., Foltynowicz, Z., Pietraszuk, C., Gulinski J. and Maciejewski, H. (1994) *J. Mol. Catal.* **90**, 213; (c) Foltynowicz, Z. and Marciniec, B. (1989) *J. Organometal. Chem.* **15**, 376; (d) Marciniec, B. and Pietraszuk, C. (1993) *J. Organometal. Chem.* **163**, 447; (e) Foltynowicz, Z. and Marciniec, B. (1991) *J. Mol. Catal.* **65**, 113; (f) Marciniec, B., Pietraszuk, C. and Foltynowicz, Z.

(1992) *J. Mol. Catal.* **76**, 307; (f) Foltynowicz, Z., Marciniec, B. and Pietraszuk, C. (1993) *Appl. Organometal. Chem.* **7**, 539; (g) Foltynowicz, Z. and Marciniec, B. (1997) *Appl. Organometal. Chem.* **11**, 667.

5.   Seki, Y., Takeshita, K. and Kawamoto, K. (1989) *J. Organometal. Chem.* **369**, 17.

6.   Wakatsuki, Y., Yamazaki, H., Nakano, M. and Yamamoto, Y. (1991) *J. Chem. Soc. Chem. Commun.* 703.

7.   Marciniec, B. and Pietraszuk, C. (1995) *J. Chem. Soc. Chem. Commun.* 2003.

8.   Marciniec, B. and Pietraszuk, C. (1997) *Organometallics* **16**, 4320.

9.   Marciniec, B., Walczuk-Gusciora, E. and Pietraszuk, C. (1998) *Catalysis Lett.* **55**, 125.

10. Marciniec, B., Kownacki, I. and Chadyniak, D. (1999) *Inorganic Chemistry Commun.* **2**, 581.

11. Marciniec, B., Kujawa, M. and Pietraszuk, C. (2000) *Organometallics* **19**, 1677.

12. Marciniec, B. In *Ring Opening Metathesis and Related Chemistry.* Khosravi, E. and Szymanska-Buzar, T. Eds; Kluwer Acad. Publ.: Dordrecht, (2002) p. 391-405.

13. Pietraszuk, C., Marciniec, B. and Fischer, H. (2000) *Organometallics.* **29**, 913.

14. Marciniec, B., Walczuk-Gusciora, E. and Pietraszuk, C. (2001) *Organometallics* **20**, 3423;

15. Marciniec, B., Walczuk-Gusciora, E. and Blazejewska-Chadyniak, P. (2000) *J. Mol. Catal. A.* **160**, 165.

16. Marciniec, B., Kownacki, I. and Kubicki, M. (2002) *Organometallics.* **21**, 3263.

17. Pol.Pat. P-355 875.

18. Pietraszuk, C., Fischer, H., Kujawa, M. and Marciniec, B. (2001) *Tetrahedron Letters.* **42**, 1175.

19. Marciniec, B., Kownacki, I., Kubicki, M., Krzyzanowski, P., Walczuk, E. and Blazejewska-Chadyniak, P. in *Perspective in Organometallic Chemistry.* RCS, Cambridge, UK (accepted for publication).

20. Marciniec, B., Kujawa, M. and Pietraszuk, C. (2000) *New J. Chem.* **24**, 671.

21. Kujawa-Welten, M. and Marciniec, B. (2002) *J. Mol. Catal.* **190**, 79.

22. Chadyniak, D. and Marciniec, B., to be published.

23. Jankowska, M. and Marciniec, B., to be published.

24. Kujawa-Welten, M., Pietraszuk, C. and Marciniec, B. (2002) *Organometallics.* **21**, 840.

25. Chadyniak, D. and Marciniec, B. (unpublished results).

26. Feher, F.J., Soulivong, D., Eklund, A.G. and Wydham, K.D. (1997) *Chem. Commun.* 1185.

27. Itami, Y., Marciniec, B. and Kubicki, M. (to be published).

28. Itami, Y. and Marciniec, B. *Organic Letters* (submitted for publication).

29. Itami, Y., Marciniec, B. and Kubicki, M. *Organometallics* (accepted for publication).

CATALYTIC PROPERTIES AND CHEMICAL TRANSFORMATIONS OF *CIS*-W(CO)$_4$(C$_5$H$_5$N)$_2$ INITIATOR IN RING OPENING METATHESIS POLYMERIZATION

L. BENCZE AND L. MIHICHUK[*]
Department of Organic Chemistry, Müller Laboratory,
University of Veszprém, H-8201 Veszprém, POB 158, Hungary

Abstract: The six-coordinate W^0 complex *cis*- W(CO)$_4$(C$_5$H$_5$N)$_2$ has been found to be active in the *in situ* formation of a carbene species from norbornene generating a typical ring opening metathesis product of the substrate. A mechanism of initiation is proposed illustrating the reaction follows the course of a 1,2-hydrogen shift in the coordinated norbornene ($\eta^2 \rightarrow \eta^1$). The initiating carbenoid group is identified from the products of the spontaneous carbene-CO coupling and Wittig reactions test. Formation of W(CO)$_5$(C$_5$H$_5$N) and W(CO)$_3$(η^6-C$_6$H$_5$CH$_3$) is blamed for catalyst deactivation.

Keywords: *bis*-pyridine-tetracarbonyl-tungsten; ROMP; Olefin metathesis; Norbornene

1. Introduction

Olefin metathesis is of great value in organic and polymer synthesis. The reactions do not occur spontaneously but all require the presence of a transition metal carbene complex ("*well defined catalyst*") or a *catalyst system* containing a transition metal complex (precursor) in conjunction with a second compound, generally an organo main group metal compound (co-catalyst), and sometimes a third (promoter) [1].

The transition metal carbenes formed in these systems trigger the transalkylidenation (i.e. metathesis) of olefins [2,3]. Catalyst systems having neither a preformed metallacarbene nor an organometallic hydrocarbon group in any component represent a relatively minor class of metathesis initiators [4, 5].

We wish to report here the catalytic properties of *cis* -W(CO)$_4$(C$_5$H$_5$N)$_2$ (**1**) in the ring opening metathesis polymerization (ROMP) of norbornene (NBE). This is the first six-coordinate W^0 non-carbenoid unimolecular catalyst that does not require a co-catalyst, a carrier or light for activation. The aim of this work is to study the initiation reactions of a system in which the catalyst precursor involves the minimum number of ligands whose reactions might interfere with a olefin $\rightarrow$ carbene transformation or side reactions other than metathesis.

---

[*]Permanent address: Department of Chemistry and Biochemistry, University of Regina, Regina, Saskatchewan, S4S 0A2, Canada

Y. Imamoglu and L. Bencze (eds.), Novel Metathesis Chemistry: Well-Defined Initiator Systems for Specialty Chemical Synthesis, Tailored Polymers and Advanced Material Applications, 65–71.

2. Experimental

cis -W(CO)$_4$(C$_5$H$_5$N)$_2$ **(1)** was prepared according to literature methods [6] with CO evolution being monitored until complete conversion of W(CO)$_6$ and formation of only the tetracarbonyl complex as seen by infrared analysis. All experiments were conducted using standard Schlenk techniques under deoxygenated argon. Toluene was distilled from sodium and benzophenone and NBE (Aldrich) was distilled from sodium before use. Tungsten hexacarbonyl (Aldrich) was used without further purification.

ROMP experiments were carried out in a reaction mixture composed of toluene (10 mL), olefin (0.52 mol/dm^3) and *cis* -W(CO)$_4$(C$_5$H$_5$N)$_2$ (4.4 mmol/dm^3). The reaction mixture was immersed in an oil bath heated to 116 °C and stirred by a magnetic stirrer.

The NBE reaction mixture changed in time from yellow to orange, to brown-orange and then to a transparent golden brown color with a gradual increasing of viscosity. The reaction was quenched by the addition of ethanol (20 mL). The precipitated polymer was soaked in ethanol, dried and analyzed by ^{13}C {^{1}H} NMR spectroscopy. The spectra of the polymers were recorded in CDCl$_3$ solution at 125 MHz using a GE GN Omega 500 spectrometer. The assignments of ^{13}C NMR shifts were based on previous analysis and the microstructural details of the polymers were calculated from the NMR integrals [7-10]. GC-MS analyses were carried out on a Shimadzu QP5000, 70eV instrument equipped with a 10 m×0.20 mm HPI column at a flow rate of 2 mL/min. IR spectra were obtained using a SPECORD M80 spectrophotometer in CH$_2$Cl$_2$ solution using KBr (0.065 mm) cells.

3. Results and discussion

The rate of polymer formation is rather slow, but nearly the same throughout the active period. The catalyst is poisoned or inactivated in some fashion within approximately 220 min. The maximum yield for the polymer is 25% attainable after 220 min of reaction as seen in Figure 1.

Figure 1. Norbornene polymer yield versus reaction time. Each point represents the average of 3 experimental runs.

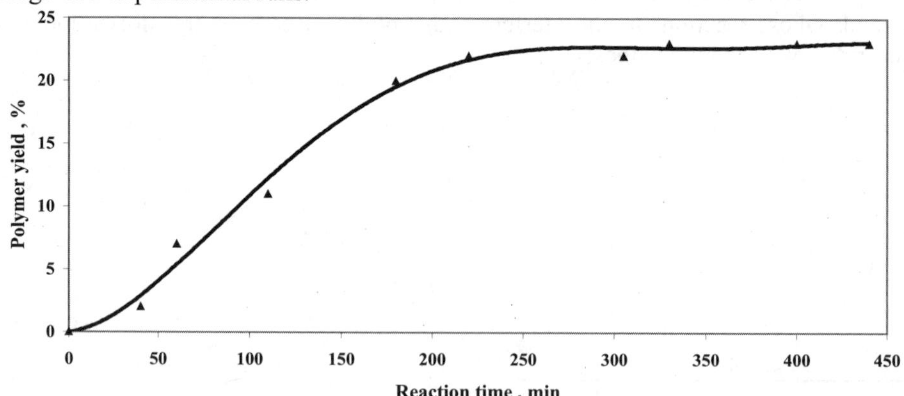

The solubility of the polymer in CDCl$_3$ is satisfactory for NMR examination but not free from some swollen gel. The ^{13}C NMR analysis indicates a typical ROMP product of NBE. The assignments of ^{13}C NMR shifts based on previous analysis [7-10] were made as shown in Figure 2 where a comparison of line intensity illustrates the high *cis* content (σ_c = 0.75-0.80) of the polymer. The distribution of the *cis* and *trans* double bonds in all cases is blocky ($r_c.r_t$~9-14) consistent with the high *cis*-content of the polymers as previously found [9,11].

Figure 2. Assignments of the ^{13}C NMR chemical shifts illustrating the high *cis* content of the NBE polymer.

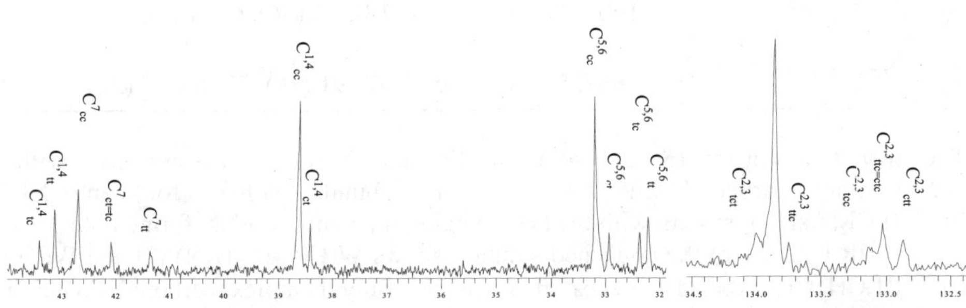

Adding benzaldehyde to the NBE—*cis* W(CO)$_4$(C$_5$H$_5$N)$_2$ toluene system prior to thermal activation inhibits the ROMP and leads to 2-benzylidenenorbornane (m/z 184(25% ;M$^+$), 155(27), 130 (100), 115 (38), 77 (25), the product of a Wittig type carbene - aldehyde reaction as illustrated in Scheme 1.

Scheme 1.

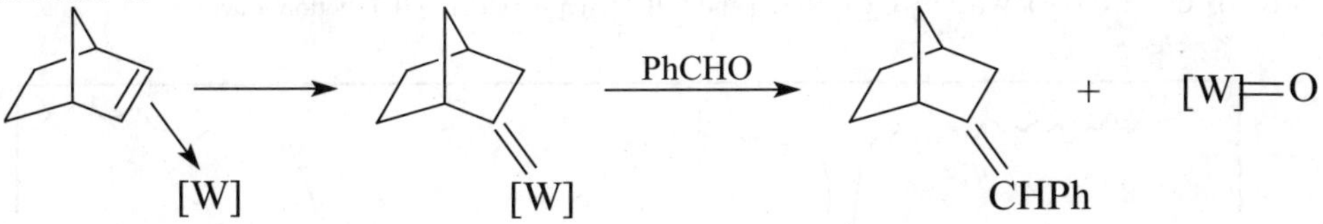

In order to elucidate the mechanism of carbene initiation a GC-MS analysis was carried out on the low molecular weight fraction of the ethanol quenched solution. The major products identified and mass spectral data are listed in Table 1.

TABLE 1. Low molecular weight products from the NBE - *cis* -W(CO)$_4$(C$_5$H$_5$N)$_2$ reaction

Product	Mass spectral data (m/z, % relative intensity)
—COOC$_2$H$_5$	168 (2%; M+),139 (2), 123 (4), 95 (100)
(ketone) O	218 (23%; M+), 151 (69), 123 (70), 95 (100), 67 (88)
—OH	112 (2%; M+), 94 (60), 79 (100), 67 (70)
(dimer)	190 (18%; M+), 161 (18), 95 (96), 67 (100)
—CH$_2$C$_6$H$_5$	186 (32%;M+), 95 (100), 91 (36), 77 (8), 67 (28)

The first two entries (**5** and **6** as in Scheme 2) are consistent also with a norbornylidene-carbene-ketene pathway as found earlier for an NBE-W(CO)$_3$Cl$_2$(AsPh$_3$)$_2$ system with the ketone being the major product formed [4c].

IR analysis of the quenched solution shows W(CO)$_5$(C$_5$H$_5$N) (**2**) and W(CO)$_3$ (η^6-C$_6$H$_5$CH$_3$) (**3**) (besides traces of **1**) as the only complex carbonyl containing products identified by comparison of the ν_{CO} bands with literature values [6, 12] (Figure 3). The presence of the toluene-derived product, which was the second major product, seemed to be perplexing in a metathesis catalyst system but it also forms from *cis*-W(CO)$_4$(C$_5$H$_5$N)$_2$ in refluxing toluene after a few hours, the result of a transformation pathway.

Figure 3. Infrared spectra in the CO stretching region for W(CO)$_4$(C$_5$H$_5$N)$_2$ (upper trace) and a mixture of W(CO)$_4$(C$_5$H$_5$N)$_2$ (A), W(CO)$_5$(C$_5$H$_5$N) (B) and W(CO)$_3$(η^6-C$_6$H$_5$CH$_3$) (C) (bottom trace).

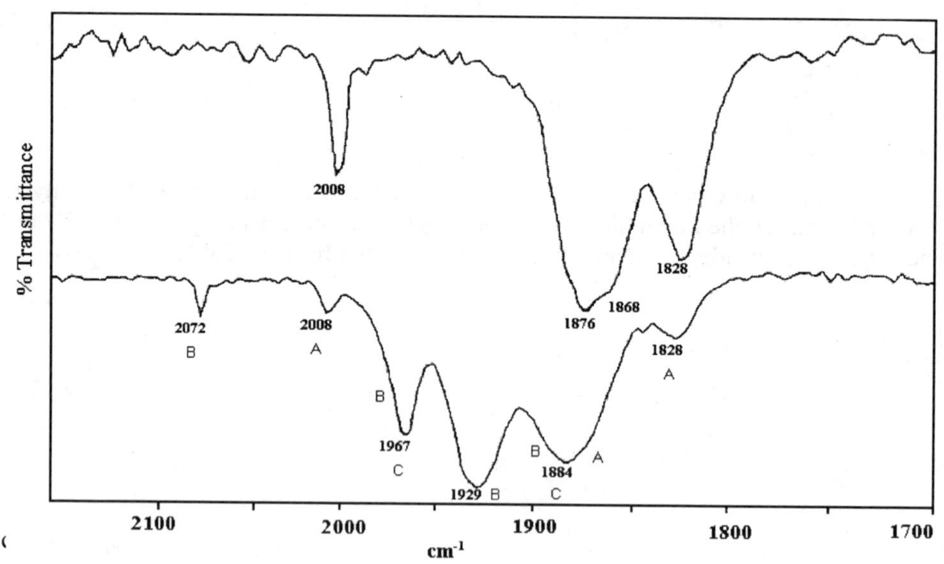

competition of CO, NBE and toluene for the vacant coordination site (Scheme 2). The basis for this deduction is provided by the fact that *if* the above reaction is carried out *in the presence of CO*, the only product found is $W(CO)_5(C_5H_5N)$. The presence of the penta- (**2**) and tricarbonyl (**3**) species at the beginning and by the end of the reaction in increasing quantity helps us to rule out their involvement as precursors in the formation of the active carbene species. Their presence does not maintain the catalytic activity. Due to the rapid disappearance of $W(CO)_4(C_5H_5N)_2$ and the quantity of the trapped primary alkylidene groups it is easily envisaged that the tetracarbonyl species is the intermediate of the initiating carbene and the latter form soon after the reaction commences.

Scheme 2.

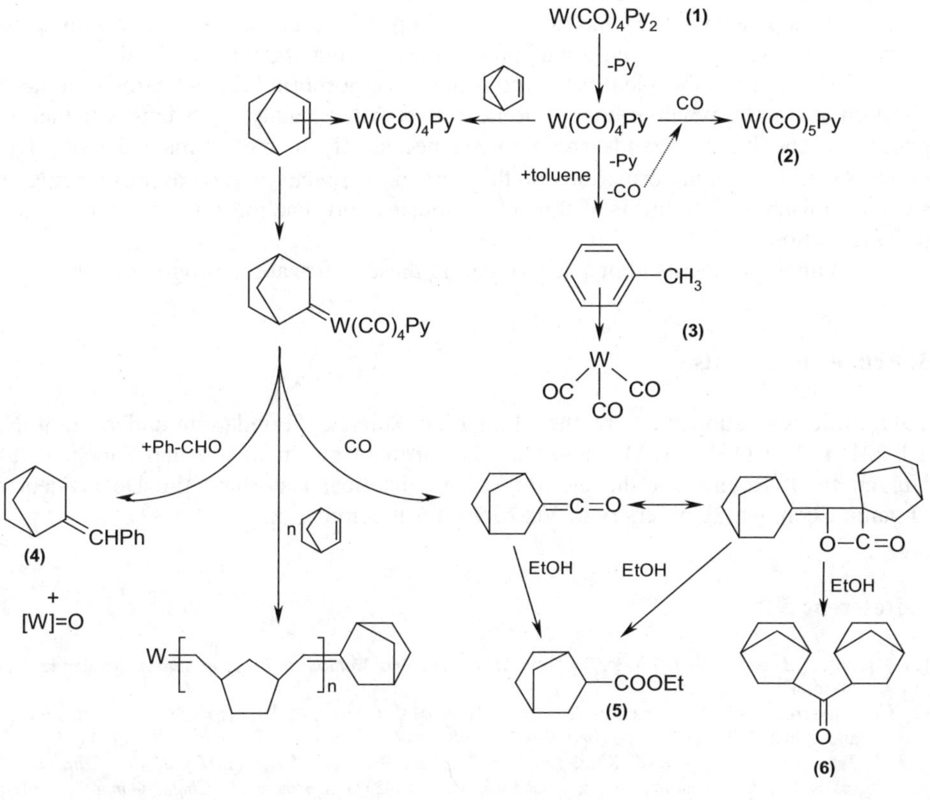

Thus we believe that *cis*-$W(CO)_4(C_5H_5N)_2$ initiates the ROMP of NBE at 116 °C in toluene through loss of one pyridine ligand followed by coordination of NBE forming a π-complex and rearrangement by a 1,2-hydrogen shift giving the initiating carbene which then propagates the ROMP. Considering the slow accumulation and the high molecular weight of the polymer product, the primary carbene complex is relatively stable [13] and the initiation is slow. Having no low molecular weight polymer fraction formed, living polymerization is assumed with little termination. The CO-carbene coupling reaction seems to be restricted also for the primary norbornylidene carbene that is exposed to the intact carbonyl ligands in the early stage of the reaction. The abrupt break down of the polymerization at about 25% polymer yield could be hardly rationalized at this stage of the experimental work. This might be attributed to the fall in concentration below a critical value and/or to the gel effect associated with an extended crosslinking of the polymer.

4. Conclusions

cis -$W(CO)_4(C_5H_5N)_2$ was found to be active in the *in situ* formation of a metathesis active carbene species from NBE substrate. The reactions follow the course of a 1,2-hydrogen shift in the coordinated olefin and the initiating carbene could be identified from the products of the carbene-CO coupling reactions and Wittig reactions as seen also in several seven-coordinate initiator systems.

This precatalyst (**1**) makes a unique simplistic catalytic system involving a six-coordinate W^0 species with no organophosphine or halide ligands involved.

Concerning the identity of the primary (2-norbornylidene-) carbene unit, the question is raised whether the pyridine or for other catalyst systems whether the phosphine and halide ligands are involved necessarily in the intimate details of the olefin $\rightarrow$ carbene transformation, or they are only spectator ligands influencing the stereochemistry and stability of the active species and reaction rate of the metathesis polymerization.

Our experiments aimed at elucidating these issues are in progress [14].

5. Acknowledgments

This work was supported by the Hungarian Science Foundation under grant No. T35221 and T016326. L.M. thanks the Department of Chemistry and Biochemistry, University of Regina for the granting of a sabbatical leave and the Department of Organic Chemistry, University of Veszprém for the invitation.

6. References

[1]   Ivin, K.J. and Mol, J.C. (1997) *Olefin Metathesis and Metathesis Polymerization*, Academic Press, London

[2]   (a)Harrison, J.L. and Chauvin, Y. (1970) *Makromol. Chem.* **141** 161; (b) Katz, T.J., McGinnis, J., and Altus, C.J. (1976) *J. Am.Chem.Soc.* **98**, 606; (c) Katz, T.J. and . Hersh, W.H. (1977) *Tetrahedron Lett.* 585; (d) Katz, T.J., Lee, S.J., and Shippey, M.A.J. (1980) *J. Mol. Catal.* **8**, 219; (e) Kress, J., Wesolek, M., and Osborn, J.A. (1982) *J. Chem. Soc., Chem. Commun.* 514; (f) Schrock, R.R., Rocklage, S., Wengrovius, J., Rupprecht, G., and Fellman, J. (1980) *J. Mol. Catal.* **8**, 73.

[3]   Grubbs, R.H., and Chang, S. (1998) *Tetrahedron* **54**, 4413.

[4]   (a) Laverty, D.T., McKervey, M.A., Rooney, J.J., and Stewart, A. (1976) *J. Chem. Soc., Chem. Commun.* 193; (b) Bencze, L., Ivin, K.J., and Rooney, J.J. (1980) *J. Chem. Soc., Chem. Commun.* 834; (c) Bencze, L., Kraut-Vass, A., and Prokai, L. (1985) *J. Chem. Soc., Chem Commun.* 911.

[5]   (a) Szymanska-Buzar, T., and T. Golowiak, T. (1995) *J. Organometal. Chem.* **489**, 207; (b) Szymanska-Buzar, T. Golowiak and I. Czelusniak, *Inorg. Chem.* 3 (2000) 102; (c) T. Szymanska-Buzar, T. Golowiak, T., and Czelusniak, I. (2001) *J. Organometal. Chem.* **640**, 72.

[6]   Kraihanzel, C.S., and Cotton, F.A. (1963) *Inorg. Chem.* **2**, 533.

[7]   Ivin, K.J., Laverty, D.T., and Katz, T.J. (1977) *Makromol. Chem.* **178**, 1545.

[8]   Ivin, K.J., Laverty, D.T., and Rooney, J.J. (1978) *Makromol. Chem.* **179**, 253.

[9]   Ivin, K.J., Laverty, D.T., O'Donnel, J.H., Rooney J.J., and Stewart, C.D. (1979) *Makromol. Chem.* **180,** 1989.

[10]   Greene, R.M.E., Hamilton, J.G., Ivin, K.J. and Rooney, J.J. (1986) *Makromol. Chem.* **187**, 619.

[11]   Bencze, L. Szalai, G., Hamilton, J.G., and Rooney, J.J. (1997) *J. Mol. Catal. A:Chemical* **115,** 193.

[12]   Zanotti, V., Rutar, V. and Angelici, R.J. (1991) *J. Organomet. Chem.* **414,** 177.

[13]   Bencze L. and Szilágyi, R.K. (1995) *J. Organometal. Chem.* **505**, 81.

[14]   Mihichuk, L., Bencze L., and Bíró, N. (2002) Manuscript in preparation.

THE IMPACT OF WELL-DEFINED TRANSITION METAL INITIATORS ON ROMP

E. KHOSRAVI
Interdisciplinary Research Centre in Polymer Science and Technology,
University of Durham,
Durham DH1 3LE. UK.

1. Introduction

Shortly after Ziegler's disclosures, Anderson and Merkling reported the polymerisation of norbornene using titanium tetrachloride and ethylmagnesium bromide at 50°C. Subsequently it was established that the polymer was the result of a ring opening process rather than the expected vinyl addition polymerisation. Using isotopic labelling (Dall'Asta and Motroni) it was established that in the polymerisation of cyclic olefins the reaction proceeded via complete cleavage of the carbon-carbon double bond.

The earlier initiator systems for ring opening metathesis polymerisation (ROMP), known as classical initiators, were based on transition metal chlorides, WCl_6, $MoCl_5$, $OsCl_3$, $RuCl_3$, $IrCl_3$, $ReCl_5$ activated with alkylating/activating agents, Ph_4Sn, $n\text{-}Bu_4Sn$, Me_4Sn, $n\text{-}Bu_3Al$, $iso\text{-}Bu_3Al$, Et_3Al, Et_2AlCl. The classical initiators suffer from many disadvantages. Of primary significance is the fact that such initiation system is ill-defined; in other words the precise nature of the active site at the metal centre is not known. The metal carbene must be generated before initiation and subsequent propagation can commence and this process usually proceeds with low yield resulting in poor control over molecular weight and molecular weight distributions (PDI). The polymers produced using classical systems had broad molecular weight distributions, generally showing values of Mw/Mn much greater than the most probable value for a well behaved chain growth polymerisation, i.e., Mw/Mn=2; an observation which is consistent with the presence of several kinds of initiator and/or propagating chain ends. The probability that most of these classical initiator systems contain several different kinds of initiating species is also consistent with the observation of complex chain microstructures for most of the product polymers although, of course, this does not prove the case. The polymerisation reactions were irreproducible and were not living. Moreover, with most of these "classical" systems complete removal of catalyst residues was difficult. The preparation of heteroatom-containing materials using classical metathesis catalysts proved difficult due to the sensitivity of these electrophilic metal complexes towards the heteroatom functionality. Also, as in the Ziegler-Natta field, most experimental effort in ROMP was concentrated on hydrocarbon monomers, cycloalkenes and bicycloalkenes.

Y. Imamoglu and L. Bencze (eds.), Novel Metathesis Chemistry: Well-Defined Initiator Systems for Specialty Chemical Synthesis, Tailored Polymers and Advanced Material Applications, 73–85.
© 2003 *Kluwer Academic Publishers. Printed in the Netherlands.*

2. Well-defined initiators

The development of the synthesis of well-defined single component transition metal complexes containing metal to carbon double bonds offered an attractive source of potential initiators for ROMP. Casey, and Fischer and Osborn carbenes were example of this type but they offered no advantages over the more readily available classical transition metal chloride systems.

The olefin metathesis has made remarkable developments with an incredible speed in various directions over the last decades. New catalyst systems have been developed which have resulted in the synthesis of novel materials. These new catalysts are now powerful tools for ROMP, Ring Closing Metathesis (RCM), and Acyclic Diene Metathesis (ADMET) and have found many applications in the synthesis of natural products. The success of olefin metathesis is attributed to outstanding advances in the field of catalysis and organometallic chemistry.

It has been established that ROMP using well-defined initiators allow control over several very important parameters such as molecular weight and molecular weight distribution, sequence and distribution of cis/trans vinylenes, tacticity, and functionalities. These controls therefore allow tuning gross polymer properties, e.g., mechanical, thermal, conductivity, dielectric, etc. We and others have demonstrated over the past few years the utility of ROMP in producing well defined polymeric materials such as functional polymers, block copolymers, stereoregular polymers, stereoblock copolymers, well defined graft copolymers, and well defined crosslinked materials [1-16].

2.1. WELL-DEFINED TITANIUM INITIATOR

The first truly well-defined initiator was reported by Grubbs and coworkers [17]. They isolated and determined the structure of a stable titanacyclobutane produced in the reaction of Tebbe reagent with norbornene in the presence of base and showed that it could initiate the ROMP, of norbornene, Scheme 1.

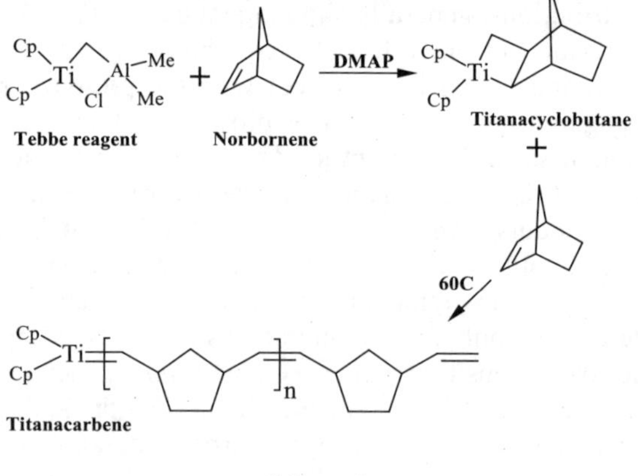

Scheme 1

The titanacyclobutane initiator gave well characterised living polymerisation with hydrocarbon monomers. However, the initiator appeared to be destroyed by fluorinated monomers, presumably the titanium centres are so electrophilic that the monomers can be regarded as a source of fluoride.

2.2. WELL-DEFINED TUNGSTEN AND MOLYBDENUM INITIATORS

Well-defined molybdenum [18] and tungsten [19] alkylidene complexes of the type $M(CH\text{-}t\text{-}Bu)(NAr)(OR)_2$ (Ar=2,6-diisopropylphenyl) have been synthesised whose reactivities toward carbon-carbon double bonds can be varied dramatically by varying the nature of the alkoxides. Electron withdrawing alokoxides produce the most active catalysts. Molybdenum complexes are more tolerant of functionalities compared to tungsten analogues. Coordinating solvents alter the activity of molybdenum initiators dramatically if the metal is relatively electron-deficient and/or solvent is a good ligand. Active protons (in water, acids, etc) and oxygen are not tolerated by these well defined initiatotrs. Molybdenum complexes containing tert-butoxides do not react readily with internal olefins but will react readily with the double bond in norbornenes. This results in a process where no chain termination or chain transfer occurs on the time scale of initiation and propagation, i.e., one that has all the characteristics of a living polymerisation [20,21].

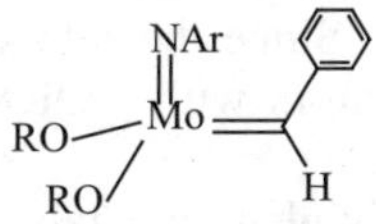

$Ar = C_6H_3\text{-}2,6\text{-}i\text{-}Pr_2$

I R = -CMe$_3$
II R = -CMe(CF$_3$)$_2$

Figure 1

The polymerisation reactions can be conveniently followed by ^{1}H NMR and the initiator alkylidene and propagating alkylidene protons can be observed. The reactions are conducted in solvents such as benzene, toluene or THF and the living polymerisation reactions can be terminated by the addition of aldehydes (e.g., benzaldehyde). The living polymer reacts with aldehydes readily to give metal oxide

Figure 2

in a Wittig-like capping reaction to give very narrow polydispersity values and Mn values determined by monomer:initiator ratios. ROMP of bis(trifluoromethyl)norbornadiene (BTFMND), initiated by Schrock alkylidenes, $Mo(CH\text{-}t\text{-}Bu)(NAr)(OR)_2$, gives all *trans* poly(BTFMND) when R=t-butyl (**I**, Figure 1) [1,2,22] and all *cis* poly(BTFMND) when R=hexafluoro-t-butyl (**II**, Figure 1) [23]. Detailed ^{13}C NMR analysis showed that the all *trans* polymer is 92% *tactic* and the all *cis* polymer is 75% *tactic*. The measurement of relaxed dielectric constant ε_R allowed an assignment of *syndiotacticity* unambiguously to all *trans* and, with a degree of uncertainty, to all *cis* polymers.

Two isomers are possible for the molybdenum initiators in any given circumstance, those in which the alkylidene ligand substituent is oriented toward the imido ligand (syn) or away from the imido ligand (anti), Figure 2. At room temperature the ^{1}H NMR spectrum of either of the two molybdenum initiators (**I & II**, Figure1) shows only one alkylidene resonance, assigned to the syn alkylidene isomer. Low temperature photolysis of the hexafluoro-t-butoxide initiator yields approximately 35% of the anti rotamer, which can be observed via ^{1}H NMR downfield of the syn alkylidene by approximately 1 ppm. By studying the relaxation of this mixture as it approaches 100% syn, the rate of rotamer interconversion has been determined [24]. The data for this study suggests that syn/anti rotamer conversion is fastest for the molybdenum initiator when OR=O-t-Bu, about five orders of magnitude greater than for when OR=OCMe(CF$_3$)$_2$. It has also been established that the anti alkylidene rotamer is much more reactive than the syn rotamer. A detailed proton NMR study by Schrock et al. [24] revealed that the streoregularity depends upon syn/anti rotamer interconversion rates. Since the rate of syn/anti rotamer interconversion is much slower for the molybdenum initiators with OR=OCMe(CF$_3$)$_2$ the ROMP is via the syn alkilydene to give all cis polymer. Since the rate syn/anti rotamer interconversion is very fast for the molybdenum initiators with OR=O-t-Bu the ROMP is via the anti alkilydene to give all trans polymer.

It has been shown that molybdenum catalysts can be prepared that contain C$_2$-symmetric chiral diolate ligands (Figure 3) which can produce polymers from BTFMND and 2,3-(dicarbomethoxy)norbornadiene that are >99% cis and >99% tactic [25].

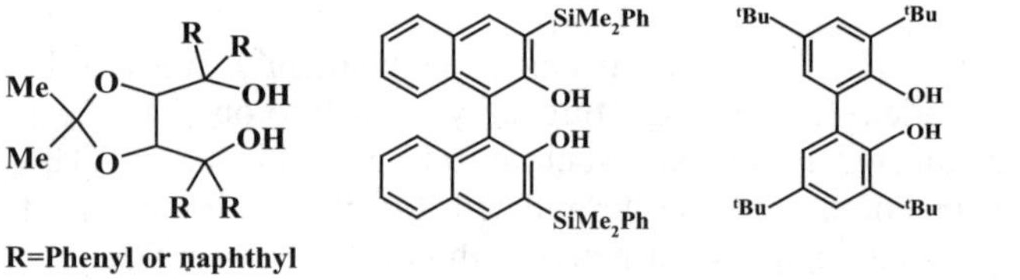

R=Phenyl or naphthyl

R$_4$tartH$_2$ **BINO(SiMe$_2$Ph)$_2$H$_2$**   **Biphenol(t-Bu)$_4$H$_2$**

Figure 3

The synthesis of functional polymers through the polymerisation of functionalised monomers is ideal, as it enables the direct incorporation of functionality into the polymer backbone chain and thus avoids the potential difficulty of chemical transformation on a polymeric substrate. The initial efforts in the search for a catalyst system effective for ROMP of functionalised substrate were met with limited success. The preparation of such heteroatom-containing materials using classical catalysts proved difficult in the past, due to the sensitivity of these electrophilic metal complexes towards the heteroatom functionality. As a result, poisoning of the catalyst and polymerisation become competitive processes. However, well-defined and living ROMP initiators based on molybdenum are deactivated to an extend that they do not react with the functionality, but still react with strained carbon-carbon double bond of the monomer on the time scale of a polymerisation.

It has been demonstrated that Schrock's molybdenum initiators allow the synthesis of completely functionalised polymers. By using the combination of chain transfer reaction on the initiator, ROMP of functionalised norbornenes and termination reaction with functionalised benzaldehyde, polymers with functional groups (X, Y and Z) at both chain ends and along the backbone chain can be produced, Scheme 2. Using this combination methodology, functional groups such as NH_2, OH, OMe, CF_3, NO_2, NMe_2, CN, CO_2Me, Cl can be incorporated into the polymer chains.

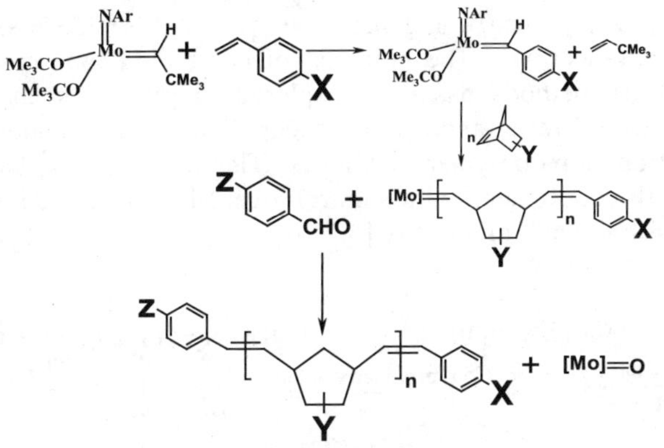

Scheme 2

Living chain growth polymerisation allows the possibility of making block copolymers, which in turn can allow control of morphology via the phase separation of incompatible blocks. Blocks derived from the same monomer but having different microstructures may be incompatible leading to the possibility of morphology control and hence bulk property control. Such stereoblock copolymer have been prepared via anionic and metallocene methods and it has been demonstrated that this is also possible using ROMP [6]. Poly(BTFMND) was obtained as a stereoblock copolymer containing *cis* and *trans* vinylene blocks via ligand exchange in living stereoselective ROMP initiated by a well defined Schrock type initiators.

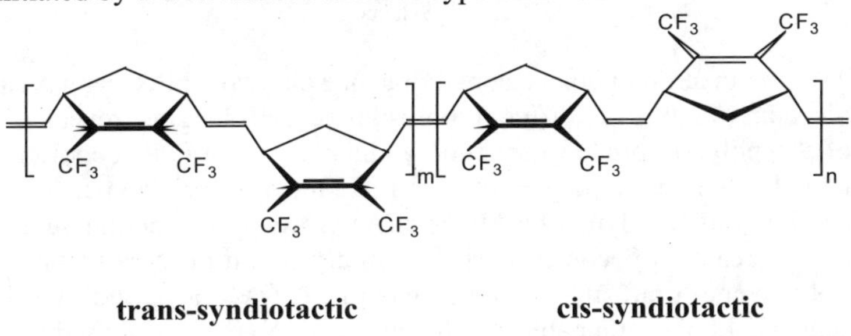

trans-syndiotactic   **cis-syndiotactic**

Figure 4

78

The ^{13}C nmr of the stereoblock copolymer displayed all the signals associated with *cis* and *trans* vinylene sequences in blocks of poly(BTFMND). The *cis/trans* blocks in the stereoblock copolymer are expected to have the same *tacticity* as their corresponding homopolymers, Figure 4. Differential Scanning Calorimetry (DSC) revealed two transitions at ca 95°C and 145°C as expected for the *trans* and *cis* blocks respectively.

There is considerable interest in developing synthetic methodologies capable of giving polymers with well-defined structures and with unusual topologies. Anionic polymerisation was for a long time the only viable process to engineer sophisticated polymeric structures. Recent advances in synthesis of well-defined initiators have shown that along with other living processes, ROMP is also a powerful tool for macromolecular engineering. The potential of combining the capabilities of living anionic and ROMP methods has been explored to prepare such materials. This approach offers access to a range of graft copolymers that cannot be prepared by grafting onto or from homopolymer backbones. The method allows rational design and synthesis of graft copolymers with control over the main chain and graft-chain molecular weights and the graft density [4,7,10].

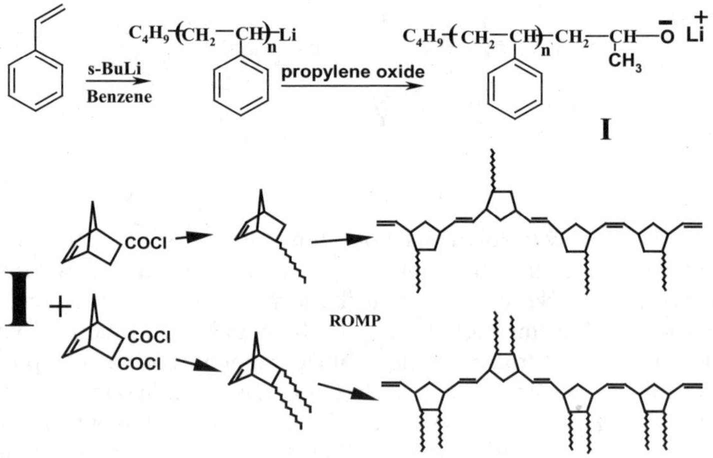

Scheme 3

ROMP of macromonomers containing one or two grafts per repeat unit of different molecular weights, i.e. different polystyrene graft lengths, revealed a limit to the length of the polynorbornene backbone attainable in the graft copolymer, which was related to the length of polystyrene graft in the macromonomer. The results suggested that the graft copolymer backbone chain grows up to a certain length beyond which the ROMP reaction becomes sterically hindered and eventually stops. As the length of polystyrene graft in the macromonomer was increased the length of polynorbornene backbone attainable in the graft copolymer decreased. The graft copolymers having one graft per repeat unit have both longer polynorbornene backbones and polystyrene grafts than was possible for the di-substituted macromonomers investigated previously. This is most probably because steric hindrance at the growing chain end during the ROMP step is significantly reduced.

The ^{1}H NMR spectrum of the living ROMP reaction mixture shows two propagating alkylidene signals due to *head or tail* insertion of macromonomer to the active site probably indicating that by adopting head-tail insertion the steric hindrance at the growing chain end is significantly reduced.

Homo-, block- and random-ROMP of norbornenyl macromonomers based on polystyrene, poly(ethylene oxide) and polybutadiene have been synthesised giving polymacromonomers which can adopt different kind of conformation and shape depending upon the degree of polymerisation of their backbone and the length of their side chain [26].

2.3. WELL-DEFINED RUTHENIUM INITIATORS

Schrock's molybdenum initiators suffer from high sensitivity to air and moisture and they are difficult to make. Among the first well-defined catalysts coming from Grubbs laboratory was the ruthenium(II) carbene, $(PPh_3)_2Cl_2Ru=CHCH=CPh_2$, active only for the living polymerisation of highly strained cyclic olefins [27]. The replacement of the PPh_3 ligands with bulkier and more basic PCy_3 ligands extended the activity to the metathesis of less strained cyclic olefins and acyclic olefins [28]. The next generation catalysts developed was $(PCy_3)_2Cl_2Ru=CHPh$, which exhibited faster initiation than its predecessors for ROMP, RCM, and ADMET [29]. These initiators tolerate a range of protic and polar functional groups including alcohols, acids and aldehydes. This was a major break through in ROMP and RCM, it allowed living ROMP in the presence of water and alcohol and using solvents and monomers without rigorous drying.

Molybdenum initiators particularly the hexafluoro-t-butoxy exhibit higher ROMP and RCM activity in comparison with ruthenium benzylidene. To increase the utility of the ruthenium initiator by increasing the activity, a number of derivatives of the ruthenium initiator have been prepared. Replacement of one of the phosphine ligands with a more electron-donating N-heterocyclic carbene ligand produced ruthenium initiators (Figure 5), which displayed dramatically improved metathesis activity, thermal stability and inertness towards oxygen and moisture when compared to the ruthenium benzylidene [30].

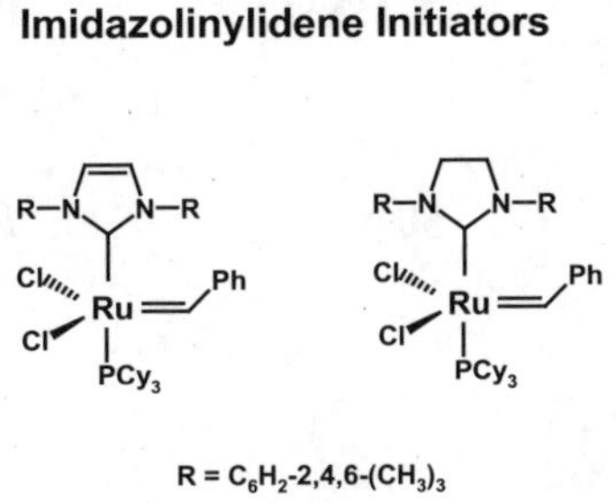

Imidazolinylidene Initiators

Figure 5

Grubbs et al. [31] have proposed a mechanism for the olefin metathesis reactions involving ruthenium initiators, shown in Scheme 4. It is proposed that the dissociative reactions of ruthenium complexes with olefinic substrates are governed by two important factors. The first factor is the rate of phosphine dissociation k_1 to produce 14-electron intermediate B, and a second consideration is the reactivity of this intermediate. Complex B can be trapped by free PCy_3 to regenerate the 16-electron starting material (at a rate proportional to k_{-1}) or it can bind olefinic substrate and undergo productive olefin metathesis reaction (at a rate proportional to k_2).

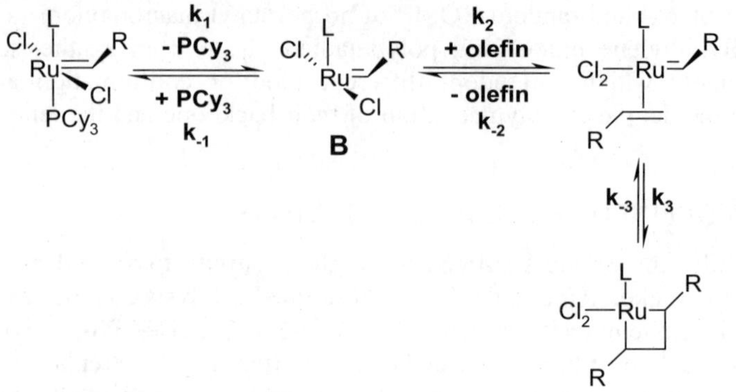

Scheme 4

The ratio of k_{-1}/k_2 is important and can be used to understand the activity of the ruthenium initiator. A comparison of complexes $(PCy_3)_2Cl_2Ru=CHPh$ and $(PCy_3)(DHIMes)Cl_2Ru=CHPh$ is indicative of the dramatic differences between these two catalysts. In the $(PCy_3)_2Cl_2Ru=CHPh$ systems, the intermediate B is formed frequently (k_1 is large). However, the recoordination of free PCy_3 is competitive with substrate binding ($k_{-1}/k_2 >> 1$). In contrast, the $(PCy_3)(DHIMes)Cl_2Ru=CHPh$ complexes dissociate phosphine relatively inefficiently (k_1 is small). However, once the phosphine comes off, coordination of olefin is facile compared to re-binding pf PCy_3 ($k_{-1}/k_2 \sim 1$). The decrease of 4 orders of magnitude in k_{-1}/k_2 between $(PCy_3)_2Cl_2Ru=CHPh$ and $(PCy_3)(DHIMes)Cl_2Ru=CHPh$ reflects a large increased selectivity for the latter to bind olefinic substrate in preference to PCy_3, Figure 6.

Initiators	T (°C)	k_{-1}/k_2
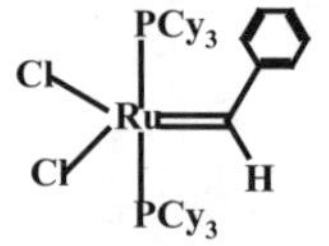	50	1.3×10^4
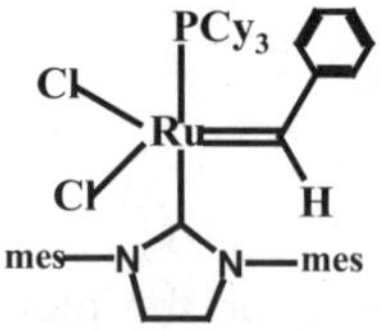	50	1.25

Figure 6

It has been observed that the ruthenium benzylidene catalyst reacted with terminal acyclic olefins to produce the new substituted alkylidenes in high yield [32].

The relative rates of metathesis of various olefinic substrates by the ruthenium benzylidene catalyst have been reported which have provided guidelines for the utilisation of the catalyst for organic syntheses in terms of the relative reactivities of the double bonds in a molecule, Figure 7. It has been shown that the kinetically preferred metallacyccle has the substituent on the olefin placed adjacent to the metal. However, if the final carbene complex is not stable due to steric interactions between a bulky carbene and bulky phosphines, the complex readily undergoes further metathesis until a sterically stable carbene is formed, which is the methylidene when the substrate is a bulky terminal olefin. It has been observed that the reaction with cis alkene (cis-hex-3-ene) is twice faster than the reaction with trans analogueand that the reaction with bulky terminal olefin (3,3-dimethyl butene) led directly to the formation of methylidene. No reaction was observed with the substituent on the double bond (2-methyl pentene).

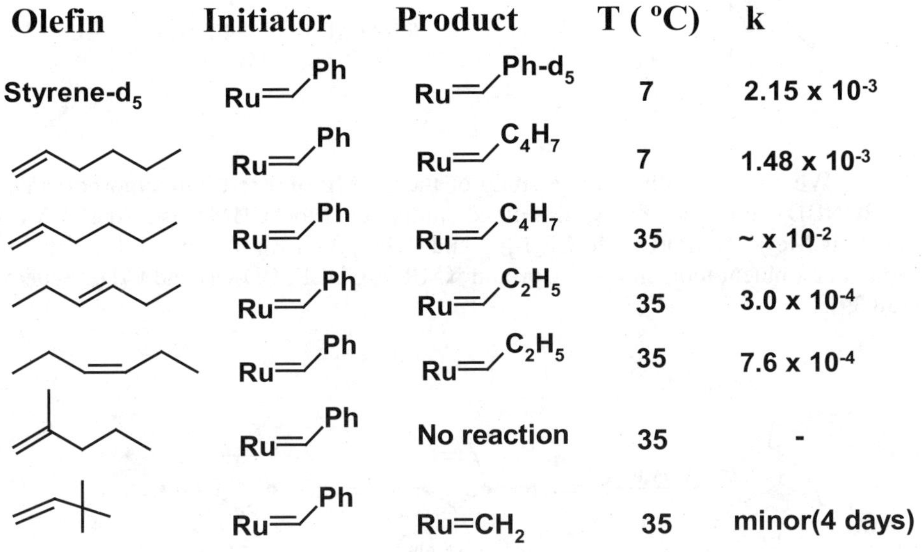

Olefin	Initiator	Product	T ($^{\circ}$C)	k
Styrene-d_5	Ru=CH-Ph	Ru=CH-Ph-d_5	7	2.15×10^{-3}
	Ru=CH-Ph	Ru=CH-C_4H_7	7	1.48×10^{-3}
	Ru=CH-Ph	Ru=CH-C_4H_7	35	$\sim \times 10^{-2}$
	Ru=CH-Ph	Ru=CH-C_2H_5	35	3.0×10^{-4}
	Ru=CH-Ph	Ru=CH-C_2H_5	35	7.6×10^{-4}
	Ru=CH-Ph	No reaction	35	-
	Ru=CH-Ph	Ru=CH_2	35	minor(4 days)

Figure 7

In recent years, the use of ruthenium carbene-based olefin metathesis initiators has gained wide acceptance in organic [12,33] and polymer syntheses [12,21,34]. Ruthenium-based catalysts exhibit greater functional group tolerance, as well as greatly enhanced air and water stability, relative to other polar single component catalyst systems based on molybdenum and tungsten [35,36]. However, thermolytic decomposition limits the usefulness of the ruthenium systems in many challenging reactions [37]. It has been reported that under standard decomposition conditions (0.023 M in C6D6 at 55°C), the propylidene has a half-life of 8 h while the methylidene has a half-life of approximately 40 minutes, Figure 8. Furthermore, it has been shown that a solution of benzylidene has a half-life of about 8 days.

Initiators	T(°C)	Conc.	Half-Life
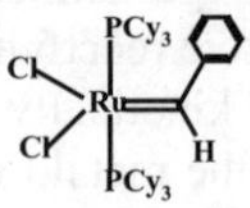 (PCy$_3$)$_2$Cl$_2$Ru=CHPh	55	0.023 M	8 days
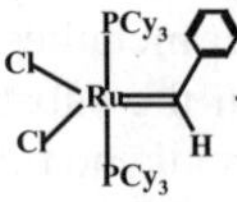 (PCy$_3$)$_2$Cl$_2$Ru=CHPh + CuCl	55	0.023 M	10 min
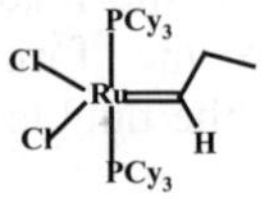 (PCy$_3$)$_2$Cl$_2$Ru=CHEt	55	0.023 M	8 hrs
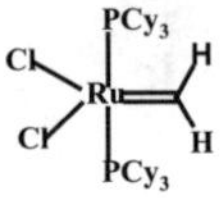 (PCy$_3$)$_2$Cl$_2$Ru=CH$_2$	55	0.023 M	40 min

Figure 8

We have recently made a study of the ROMP of 7-tert-butoxynorbornadiene (7-TBONBD) initiated by well-defined initiators Mo(=CHMe$_3$)(=NAr)(OCMe$_3$)$_2$, Mo(=CHCMe$_3$)(=NAR)[OCMe(CF$_3$)$_2$]$_2$ and (PCy$_3$)$_2$Cl$_2$Ru=CHPh following the reaction at ambient temperature by proton NMR in CDCl$_3$, CD$_2$Cl$_2$ and C$_6$D$_6$, Schemes 5 [38,39].

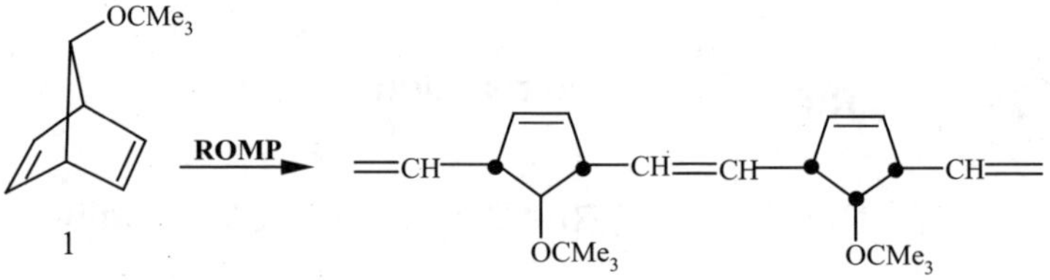

Scheme 5

The reaction with Mo(=CHCMe$_3$)(=NAR)[OCMe(CF$_3$)$_2$]$_2$ as initiator is extremely fast, but with Mo(=CHMe$_3$)(=NAr)(OCMe$_3$)$_2$ and (PCy$_3$)$_2$Cl$_2$Ru=CHPh the reaction proceeds at conveniently measurable speed. The remarkable observation has been made that with (PCy$_3$)$_2$Cl$_2$Ru=CHPh as initiator, after polymerisation is complete, the initiator, largely consumed (97%) in the initial reaction, is partially regenerated (50%) at the expense of the living propagating species Figure 9.

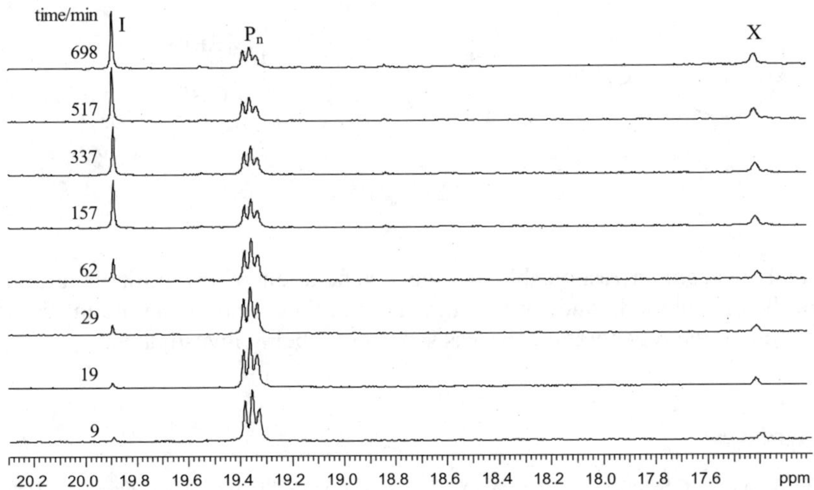

Figure 9

This is believed to be the result of secondary metathesis reactions. There may be two types of secondary metathesis reaction involved in this system; the intermolecular metathesis resulting in an increase in the average molecular weight of the polymer (Scheme 6) and ring-closing metathesis leading to the formation of cyclic species (Scheme 7). The regeneration of initiator has so far only been observed for the ROMP of 7-TBONBD initiated by $(PCy_3)_2Cl_2Ru=CHPh$ and has not been observed with this monomer using initiator $Mo(=CHCMe_3)(=NAR)[OCMe(CF_3)_2]_2$ or $Mo(=CHMe_3)(=NAr)(OCMe_3)_2$, nor using initiator $(PCy_3)_2Cl_2Ru=CHPh$ with other monomers, e.g. norbornene.

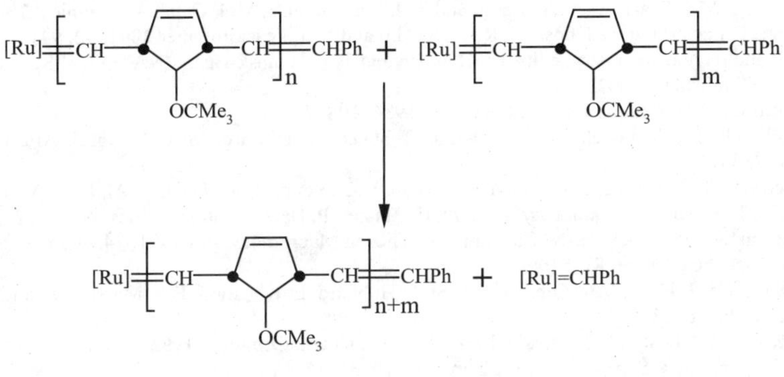

Scheme 6

84

CH=[Ru]
CH=CHPh
→
CH
CH
+
[Ru]
CHPh

Scheme 7

The actual mechanism responsible for the regeneration of the initiator and also the effect of the phosphine ligands of the ruthenium initiator and the nature the substituent in 7 position on the regeneration process is currently being investigated.

3. References

1. G. C. Bazan, R. R. Schrock, E., Khosravi, W. J. Feast and V. C. Gibson Polymer Commun. **1989**, 30, 258.

2. G. C. Bazan, E. Khosravi, R. R. Schrock, W. J. Feast, V. C. Gibson, M. B. Oregan, J. K. Thomas, W. M. Davis J. Am. Chem. Soc. **1990**, 112, 8378.

3. W.J. Feast, V.C. Gibson, L.M. Hamilton, E. Khosravi, and E.L. Marshall, J. Chem. Soc. Chem. Commun., 9, **1994**.

4. W. J. Feast, V. C. Gibson, A. F. Johnson, E. Khosravi and M. A. Mohsin Polymer **1994**, 35, 3542.

5. W. J. Feast and E. Khosravi, "Recent Development in ROMP", Chapter 3 in "New Methods of Polymer Synthesis", Vol 2, edits J. R. Ebdon and G. C. Eastmond, Blackie Academic and Professional, **1995**.

6. J. Broeders, W. J. Feast, V. C. Gibson and E. Khosravi J. Chem. Soc. Chem. Commun. **1996**, 343.

7. W. J. Feast, V. C. Gibson, A. F. Johnson, E. Khosravi and M. A. Mohsin J. Mol. Catal. A: Chemical **1997**, 115, 37.

8. E. Khosravi, In Modern Fluoropolymers, High Performance Polymers for Diverse Applications; Scheirs, J., Ed.; Wiley & Sons. **1997**, Chapter 8.

9. E. Khosravi, In Modern Fluoropolymers, High Performance Polymers for Diverse Applications; Scheirs, J., Ed.; Wiley & Sons. **1997**, Chapter 8.

10. A. C. M. Rizmi, E. Khosravi, W. J. Feast, M. A. Mohsin and A. F. Johnson Polymer **1998**, 39, 6605.

11. E. Khosravi, A. A. Al-Hajaji Polymer, **1998**, 39, 5619.

12. R. H. Grubbs, E. Khosravi, In Synthesis of Polymers - a Volume of Materials Science and Technology Series; Schluter, A. D., Ed.; Wiley-VCH, **1998**, p 63-104.

13. W. J. Feast, E. Khosravi J. Fluor. Chem. **1999**, 100, 117.

14. E. Khosravi, W. J. Feast, A. A. Al-Hajaji and T. J. Leejarkpai, J. Mol. Catal. A: Chemical **2000**, 160, 1.

15. P.J. Hine, T. Leejarkpai, E. Khosravi, R.A. Duckett and W.J. Feast Polymer **2001**, 42, 9413.

16. "ROMP and related chemistry, edits. E. Khosravi and T. Szymansk-Buzar, NATO ASI Series, Kluwer Academic Publishers, **2002**.

17. L.R. Golliom, R.H. Grubbs J. Am. Chem. Soc. **1986**, 108, 733.

18. R.R. Schrock, J.S. Murdzek, G.C. Bazan, J. Robbins, M. DiMare, M. O'Regan J. Am. Chem. Soc. **1990**, 112, 3875.

19. R.R. Schrock, R.T. DePue, J. Feldman, K.B. Yap, D.C. Yang, W.M. Davis, L.Y. Park, M. DiMare, M. Schofield, J. Anhaus, E. Walborsky, E. Evitt, C. Kruger, P. Betz Organometallics **1990**, 9, 2262.

20. G.C. Bazan, R.R. Schrock, H.-N. Cho and V.C. Gibson Macromolecules 1991, 24, 4495.

21. R.R. Schrock Acc. Chem. Res. **1990**, 23, 158.

22. G.R. Davies, W.J. Feast, V.C. Gibson, H.V.St.A. Hubbard, E. Khosravi, E.L. Marshall and I.M. Ward Polymer **1995**, 36, 235.

23. W.J. Feast, V.C. Gibson, E.L. Marshall J. Chem. Soc. Chem. Commun. **1992**, 1157.

24. J.H. Oskam and R.R. Schrock J. Am. Chem. Soc. **1993**, 115, 11831.

25. D.H. McConville, J.R. Wolf, and R.R. Schrock J. Am. Chem. Soc. **1993**, 115, 4413.

26. V. Heroguez and Y. Gnanou "ROMP and related chemistry, edits. E. Khosravi and T. Szymansk-Buzar, NATO ASI Series, Kluwer Academic Publishers, **2002**.

27. T. Nguyen, L.K. Johnson, R.H. Grubbs J. Am. Chem. Soc.**1992**, 114, 3974.

28. S.T. Nguyen, L.K. Johnson, R.H. Grubbs J. Am. Chem. Soc.**1993**, 115, 9858.
29. P. Schwab, R.H. Grubbs, J.W. Ziller J. Am. Chem. Soc.**1996**, 118, 100.
30. J.P. Morgan, R.H. Grubbs Org. Lett. 2000, 2, 3153.
31. M.S. Sanford, J.A. Love, and R.H. Grubbs J. Am. Chem. Soc. **2001**, 123, 6543
32. M. Ulman and R.H. Grubbs, Organometallics **1998**, 17, 2484.
33. R.H. Grubbs, S. Chang Tetrahedron **1998**, 54, 4413.
34. R.H. Grubbs, W. Tumas Science **1989**, 243, 907.
35. R.H. Grubbs J. Macromol. Sci., Pure Appl. Chem. **1994**, A31, 1829.
36. G.C. Fu, S.T. Nguyen, R.H. Grubbs J. Am. Chem. Soc.**1993**, 115, 9856.
37. M. Ulman and R.H. Grubbs, J. Org. Chem. **1999**, 64, 7202.
38. K. J. Ivin, A. M. Kenwright, E. Khosravi J. Chem. Soc., Chemical Commun. **1999**, 1209.
39. K. J. Ivin, A. M. Kenwright, E. Khosravi, J. G. Hamilton Macromol. Chem. Phys. **2001**, 202, 3624

DUAL ACTIVITY OF RUTHENIUM CATALYSTS IN CONTROLLED RADICAL REACTIONS AND OLEFIN METATHESIS

L. DELAUDE, S. DELFOSSE, A. DEMONCEAU, A. RICHEL, AND A. F. NOELS*
Center for Education and Research on Macromolecules (CERM), Institut de Chimie (B6a), University of Liège, Sart-Tilman par B-4000 Liège, Belgium

1. Introduction

The formation of carbon-carbon bonds using free radicals is of utmost importance both in synthetic organic chemistry and in polymer chemistry [1]. The developments that took place during the last decade have considerably modified the view that free radical reactions are commonly uncontrollable. Catalytic systems are now available, that allow radical reactions to be carried out in a precise and controlled manner. In particular, the past few years have witnessed a rapid growth in the development and understanding of controlled radical reactions based on the combination of suitable radical initiators and of transition-metal complexes. For instance, the addition of a polyhalogenated alkane to an olefin, also known as the Kharasch reaction [2], has largely benefited from the replacement of classical radical initiators such as peroxides or UV light by transition-metal complexes that promote a single-electron transfer or a redox-based chain reaction. The latter process is usually referred to as an Atom Transfer Radical Addition (ATRA). In the presence of a high ratio of olefin compared to the halogen derivative, successive insertions of the unsaturated monomer lead to a macromolecular chain, and the net process is known as an Atom Transfer Radical Polymerization (ATRP) (Scheme 1). Among the metals used for promoting ATRP, copper, nickel, iron, and ruthenium tend to display the highest activities, but complexes of rhenium, rhodium, and palladium have also been employed [3,4].

Scheme 1.

Thanks to the development of well-defined catalysts based on transition metals, the end of the twentieth century has also witnessed the emergence of olefin metathesis as a major tool for exchanging substituents across carbon-carbon double bonds, a task that was previously very difficult to accomplish efficiently [5]. Early catalytic systems were generated *in situ* from a transition-metal halide and a main group metal-alkyl cocatalyst. Thus, complex mixtures of titanium, molybdenum, tungsten, or rhenium salts, associated with lithium, aluminum, or tin organyl compounds were commonly employed. Nowadays, recourse to well-defined complexes

Y. Imamoglu and L. Bencze (eds.), Novel Metathesis Chemistry: Well-Defined Initiator Systems for Specialty Chemical Synthesis, Tailored Polymers and Advanced Material Applications, 87–100.
© 2003 *Kluwer Academic Publishers. Printed in the Netherlands.*

of transition metals bearing alkylidene fragments -some of which are readily available from commercial suppliers- allows the selective transformation of olefinic bonds in the presence of more polar functional groups and has opened the door to numerous applications, in both industry and academia [6-8]. Indeed, reactions such as the Ring-Closing Metathesis (RCM) or the Cross-Metathesis (CM) in organic synthesis, and the Ring-Opening Metathesis Polymerization (ROMP) in polymer synthesis are now routinely used in laboratories world-wide (Scheme 2).

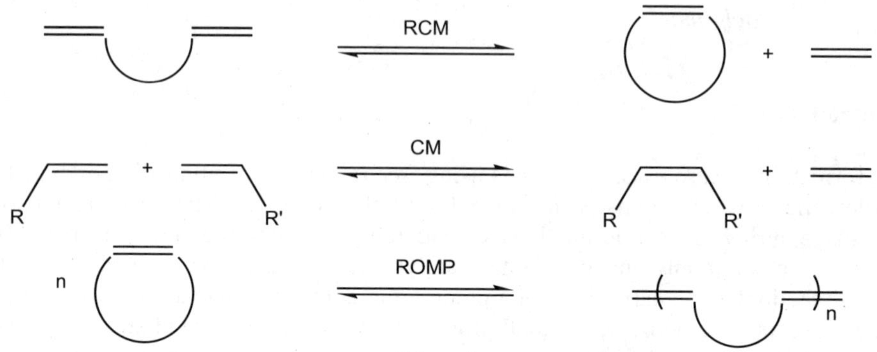

Scheme 2.

Ruthenium complexes hold a prominent position among the new catalyst precursors developed in both atom transfer radical reactions [9] and olefin metathesis polymerization [10]. A major breakthrough was achieved in the mid-nineties by Grubbs and coworkers with the discovery of the ruthenium-benzylidene complex **1**, a very efficient and highly tolerant pre-catalyst for all kinds of metathesis reactions [11,12]. At approximately the same time, we reported that the 18-electron ruthenium-arene complex **2** was a versatile and efficient promoter for the ROMP of both strained and low-strain cyclic olefins when activated by a suitable carbene precursor such as trimethylsilyldiazomethane (TMSD) [13,14]. We discovered also that both complexes **1** and **2** displayed an exceptional activity at promoting the ATRP of vinyl monomers in the absence of any cocatalyst [15].

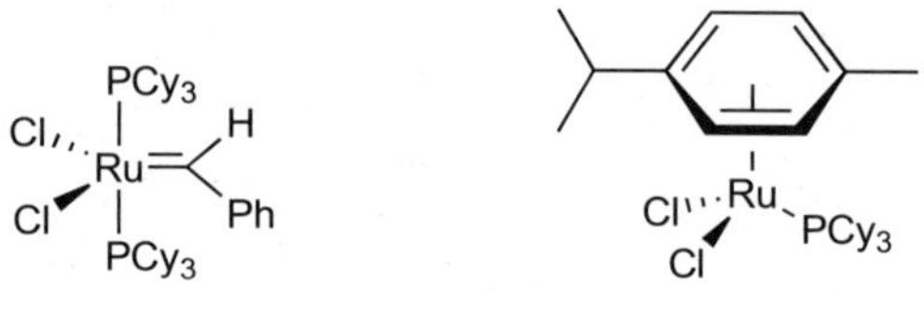

1

2

Over the past few years, a new generation of ruthenium catalysts bearing stable *N*-heterocyclic carbene ligands (NHCs) has almost completely superseded the phosphine-containing complexes of type **1** or **2** in terms of activity and stability [16]. Indeed, NHCs behave as phosphine mimics, yet they are better σ-donors and they form stronger bonds to metal centers than most phosphines [17]. Furthermore, they constitute a promising new class of ligands available for catalyst engineering and fine tuning, since their electronic and steric properties are liable to ample modification simply by varying the substituents on the nitrogen atoms. We have investigated the controlled radical polymerization of vinyl monomers with Herrmann-Grubbs complexes **3** and **4** bearing one or two NHC ligands instead of tricyclohexylphosphine. Good to excellent yields of polymers were obtained and the reactions proceeded in a controlled way with styrene and methyl methacrylate (MMA) [18]. We have also launched a detailed investigation on the role of the NHC ligand in ruthenium-(p-cymene) catalyst precursors of type **5** for the atom transfer radical addition [19] or polymerization [20] of vinyl monomers, and for the ring-opening metathesis polymerization of cyclic olefins [21,22]. In this chapter we focus only on a specific series of complexes **5** and we examine how structural variations in the R_1 and R_2 substituents of the NHC ligand affect the ATRA, ATRP, and ROMP processes.

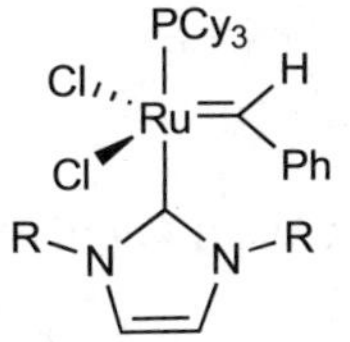

3 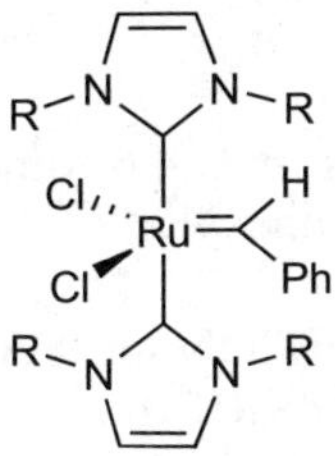   **4** 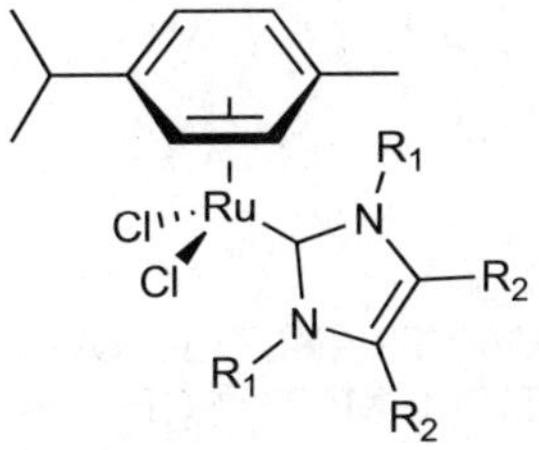   **5**

2. Results and Discussion

2.1. DEFINITION OF THE CATALYTIC SYSTEMS

The catalyst precursors used in this study were prepared by reacting the $[RuCl_2(p\text{-cymene})]_2$ dimer with two equivalents of a free carbene in THF [23,24]. The substituents on the nitrogen atoms of the NHC were either the mesityl or the cyclohexyl group. The choice of the former aryl group was motivated by the numerous successful applications of the corresponding imidazolylidene or imidazolinylidene ligands in organometallic catalysis [16]. The latter cycloalkyl chain was elected as a typical bulky alkyl group.

The steric and electronic properties of the NHC ligands were further modulated by varying the nature of the substituents on the carbon-carbon double bond of the imidazole ring. Thus, a series of complexes bearing N-mesityl groups (R_1) and either methyl, hydrogen, or chloro substituents on the C=C ring moiety (R_2) were synthesized (structures **5a-c**). The free carbene ligands were obtained by reduction with potassium of the parent imidazole-2(3H)-thione (in the case of **5a**) [25] or by deprotonation with potassium t-butoxide or hydride of the corresponding imidazolium chloride (in the case of **5b**) [26]. The dichlorocarbene in **5c** resulted from the substitution reaction between 1,3-dimesitylimidazol-2-ylidene and carbon tetrachloride [27].

Complexes **5d** and **5e** bearing cyclohexyl groups as R_1 units, and methyl groups or hydrogen atoms as R_2 substituents were synthesized following the same experimental procedures that afforded species **5a** and **5b**, respectively [25,26]. Attempts to chlorinate the double bond of the free carbene bearing N-cyclohexyl groups using the same method [27] that afforded the NHC with R_1 = Mes and R_2 = Cl failed. Preliminary investigations showed that the reaction of 1,3-dicyclohexylimidazol-2-ylidene with 2 equivalents of CCl_4 in THF afforded a mixture of 1,3-dicyclohexyl-2-chloroimidazolium chloride and another yet unidentified product instead of the desired 4,5-dichloro-1,3-dicyclohexylimidazol-2-ylidene. Thus, complex **5f** was not included in the series of catalyst precursors tested in this study.

5a	R_1 = Mes	R_2 = Me
5b	R_1 = Mes	R_2 = H
5c	R_1 = Mes	R_2 = Cl
5d	R_1 = Cy	R_2 = Me
5e	R_1 = Cy	R_2 = H
5f	R_1 = Cy	R_2 = Cl

Mes = 2,4,6-trimethylphenyl , Cy = cyclohexyl

2.2. ATOM TRANSFER RADICAL ADDITION OF CARBON TETRACHLORIDE TO METHYL METHACRYLATE AND STYRENE

The Kharasch addition of carbon tetrachloride across the double bonds of MMA and styrene was investigated using the ruthenium complexes **5a-e** as catalysts. The reactions were carried out at 90 °C in toluene under inert atmosphere. The halogenated derivative was introduced in small excess compared to the olefin and the catalyst loading was 0.3 mol%. Under these conditions, the methacrylic ester displayed a disappointingly low reactivity, whichever NHC ligand was present on the transition-metal redox center (Table I). Conversion remained below 20% and only minute amounts of the addition product (methyl 2,4,4,4-tetrachloro-2-methyl butanoate) were detected by gas chromatography (GC). Only when the dicyclohexyl complex **5e** served as catalyst, was the dimeric product resulting from the insertion of two olefinic units within the activated carbon-halogen bond visible in the chromatogram. In all cases higher oligomers also formed and accounted for the mass balance, but their high molecular weight prevented a satisfactory GC analysis.

Table I. Kharasch addition of carbon tetrachloride to MMA catalyzed by ruthenium complexes **5a-e**[a]

Catalyst	MMA conversion (%)[b]	Addition product (%)[b]
5a	10	<1
5b	7	<1
5c	7	<1
5d	19	1.5
5e	20	4

[a] Experimental conditions: MMA (9 mmol), carbon tetrachloride (13 mmol), ruthenium complex (0.03 mmol), toluene (4 mL), dodecane (0.25 mL), 24 h reaction at 90 °C under nitrogen
[b] Determined by gas chromatography (GC) using dodecane as internal standard

Styrene was more prone to undergo the atom transfer radical addition of carbon tetrachloride than MMA using the series of ruthenium-NHC complexes under study. A quantitative conversion of the unsaturated hydrocarbon into (1,3,3,3-tetrachloropropyl)-benzene was obtained in the presence of complex **5c** (R_1 = Mes, R_2 = Cl). Complex **5e** bearing a rather different carbene ligand (R_1 = Cy, R_2 = H) was the second best choice and afforded the Kharasch monoadduct in 62% yield after 24 h of reaction at 90 °C (Table II). Other substituent combinations on the imidazolylidene ring led to catalysts that were less efficient and selective. It should be pointed out, however, that stilbene formation did not contribute to more than 1 or 2% of the styrene consumption, indicative that ATRA was strongly favored over olefin metathesis under the experimental conditions adopted.

Table II. Kharasch addition of carbon tetrachloride to styrene catalyzed by ruthenium complexes **5a-e**[a]

Catalyst	Styrene conversion (%)[b]	Addition product (%)[b]
5a	47	27
5b	70	48
5c	100	97
5d	38	8
5e	88	62

[a] Experimental conditions: styrene (9 mmol), carbon tetrachloride (13 mmol), ruthenium complex (0.03 mmol), toluene (4 mL), dodecane (0.25 mL), 24 h reaction at 90 °C under nitrogen
[b] Determined by gas chromatography (GC) using dodecane as internal standard

In order to get a better picture of the reaction course, the Kharasch addition of carbon tetrachloride onto styrene mediated by complexes **5c** and **5e** was monitored by GC at regular time intervals. The results are illustrated in Figures 1 and 2, respectively. At 90 °C, the difference of activity between the mesityl- and cyclohexyl-substituted derivatives, although clearly noticeable, was not striking (cf. Table II). When the temperature was decreased to 60 °C, the substrate and product evolution curves recorded with complex **5c** were inflected, but not markedly affected (Fig. 1). Indeed, this catalyst still afforded a 95% styrene conversion and a 89% yield of monoinsertion adduct after 24 h. The activity of complex **5e**, on the other hand, almost completely collapsed when the reaction temperature was lowered (Fig. 2). After 24 h at 60 °C, the conversion remained limited to 22%, and only half of the monomer consumed led to the Kharasch reaction (11% yield from styrene).

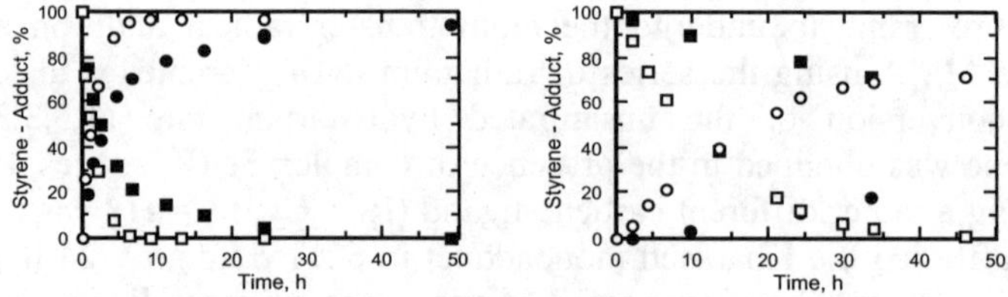

Figure 1 (left). Time course of the Kharasch addition of CCl$_4$ to styrene catalyzed by complex **5c** at 60 °C (styrene: □, adduct: ◻) or at 90 °C (styrene: ■, adduct: ●) (see Table II for experimental conditions).
Figure 2 (right). Time course of the Kharasch addition of CCl$_4$ to styrene catalyzed by complex **5e** at 60 °C (styrene: □, adduct: ◻) or at 90 °C (styrene: ■, adduct: ●) (see Table II for experimental conditions).

2.3. ATOM TRANSFER RADICAL POLYMERIZATION OF METHYL METHACRYLATE AND STYRENE

The homogeneous ATRP of methyl methacrylate (MMA) initiated by ethyl 2-bromo-2-methylpropionate and ruthenium-NHC complexes **5a-e** was investigated at 85 °C under inert atmosphere. Complexes **5b** and **5c** with R$_1$ = mesityl and R$_2$ = H or Cl, respectively, were the most efficient catalysts for this reaction (Table III). The semilogarithmic plots of ln([M]$_0$/[M]) *vs.* time were linear in both cases with a pseudo-first order rate constant (k_{app}) of 10.6 x10^{-6} s^{-1} for **5b** and 3.85 x10^{-6} s^{-1} for **5c**, indicating that the radical concentration remained constant throughout the polymerization run (Fig. 3). With these two complexes, the molecular weights increased linearly with conversion, indicative of a good control over M_n. (Fig. 5). Polydispersities (M_w/M_n) were quite low (typically *ca.* 1.3) and decreased with monomer conversion (Fig. 6). By contrast, complex **5a** displayed an induction period after which the semilogarithmic plot was almost linear (k_{app} = 16.4 x10^{-6} s^{-1}) (Fig. 4). Furthermore, M_n did not increase linearly with conversion and did not agree with theoretical values (Fig. 5) [20].

Table III. Polymerization of MMA catalyzed by ruthenium complexes **5a-e**[a]

Catalyst	Polymer yield (%)	M_n[b]	M_w/M_n[b]	f[c]
5a	28	52 000	1.6	0.2
5b	49	28 000	1.35	0.7
5c	24	13 000	1.33	0.75
5d	94	160 000	2.45	0.25
5e	51	36 000	1.75	0.55

[a] Experimental conditions: 16 h reaction at 85 °C under nitrogen with ethyl 2-bromo-2-methylpropionate as initiator, $[MMA]_0/[initiator]_0/[Ru]_0 = 800/2/1$

[b] Determined by size-exclusion chromatography (SEC) in THF with PMMA calibration

[c] Initiation efficiency $f = M_{n,theor.}/M_{n,exp.}$ with $M_{n,theor.} = ([monomer]_0/[initiator]_0) \times M_w(monomer) \times conversion$

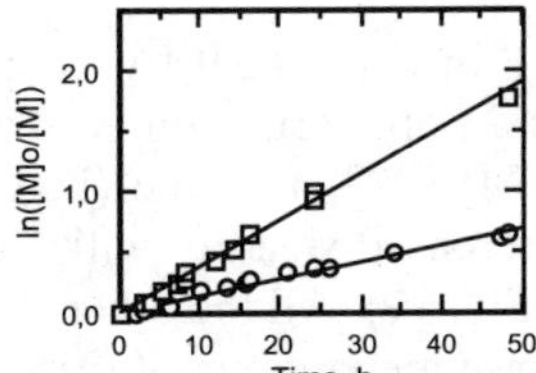
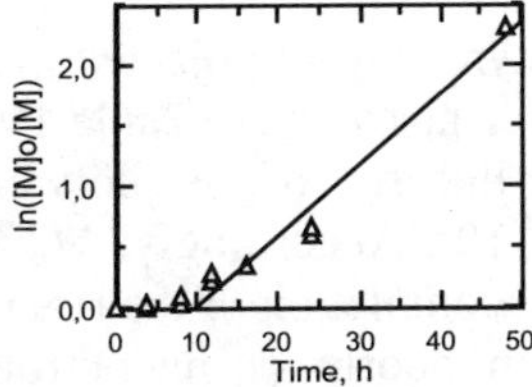

Figures 3 (left) and 4 (right). Time dependence of $ln([M]_0/[M])$ for the polymerization of MMA catalyzed by complex 5a ($\triangle$), 5b ($\square$), and 5c ($\square$) (see Table III for experimental conditions).

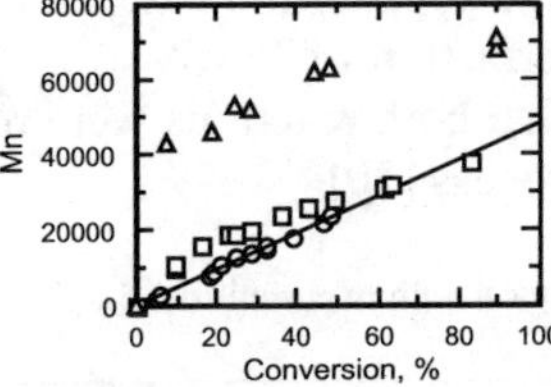
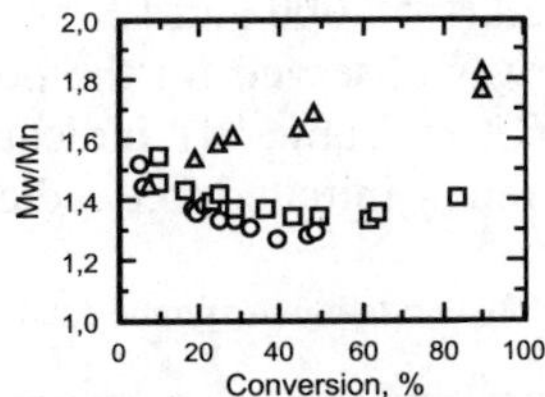

Figures 5 (left) and 6 (right). Dependence of the PMMA molecular weight M_n and molecular weight distribution M_w/M_n on monomer conversion for the polymerization of MMA catalyzed by complex 5a ($\triangle$), 5b ($\square$), and 5c ($\square$) (see Table III for experimental conditions).

Like ATRA, ATRP is believed to proceed via a radical mechanism. To test the validity of this hypothesis, the polymerization of MMA catalyzed by complex **5b** was carried out in the presence of galvinoxyl, a widely used free radical scavenger. Addition of a five-fold excess of this inhibitor with respect to ruthenium was necessary to effectively block the reaction, whenever it was introduced together with the catalyst at the beginning of the reaction (Fig. 7) or later during the run (Fig. 8).

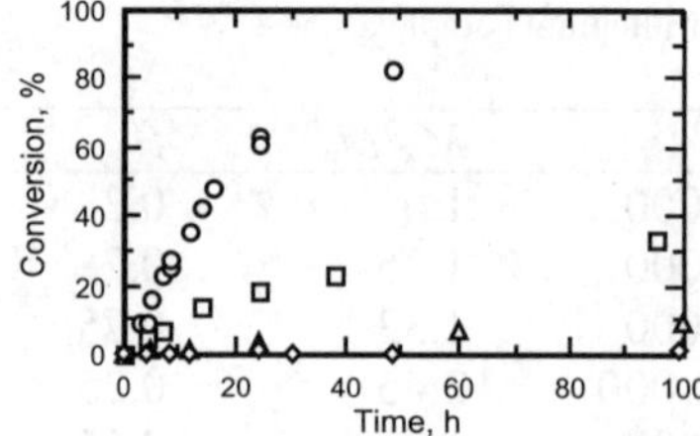

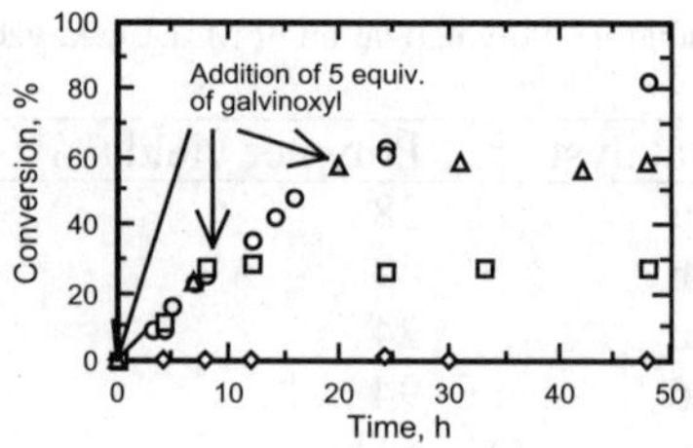

Figure 7 (left). Time dependence of conversion for the polymerization of MMA catalyzed by complex **5b** without (▪) and with 1 equiv. (□), 2.5 equiv. (△), and 5 equiv. (◊) of galvinoxyl (see Table III for experimental conditions).

Figure 8 (right). Time dependence of conversion for the polymerization of MMA catalyzed by complex **5b** in the absence (▪) or in the presence of 5 equiv. of galvinoxyl added at the beginning of the reaction (◊), after 8 h (□), or after 20 h (△) (see Table III for experimental conditions).

When styrene was subjected to ATRP instead of MMA, complex **5e** proved to be the most efficient catalyst (Table IV). The polymerization was well-controlled as indicated by the linearity of the plots of $\ln([M]_0/[M])$ *vs.* time (Fig. 9) and of M_n *vs.* conversion (Fig. 10). Accordingly, M_w/M_n decreased steadily with time and reached a low of 1.25 upon completion of the reaction (Fig. 11). All the other R_1-R_2 combinations tested resulted in poorly or uncontrolled polymerizations. Indeed, replacement of hydrogen atoms by methyl groups on both olefinic carbons of the NHC ligand dramatically altered the course of the reaction. With complex **5d**, the semilogarithmic plot of $\ln([M]_0/[M])$ *vs.* time was no longer linear (Fig. 12), and M_n as well as M_w/M_n remained constant ($\approx 55\ 000$ and 1.9, respectively) throughout the run (Figs 13 and 14). A similar trend was observed for the polymerization of MMA using **5e** ($M_n \approx 35\ 000$ and $M_w/M_n \approx 1.75$, *cf.* Table III), indicating that both reactions were most likely taking place through a redox-initiated free radical process [20].

Table IV. Polymerization of styrene catalyzed by ruthenium complexes **5a-e**[a]

Catalyst	Polymer yield (%)	M_n[b]	M_w/M_n[b]	f[c]
5a	66	53 000	1.75	0.5
5b	51	28 000	1.8	0.7
5c	10	10 200	1.9	0.35
5d	77	53 000	1.9	0.55
5e	86	47 000	1.25	0.7

[a] Experimental conditions: 16 h reaction at 110 °C under nitrogen with (1-bromoethyl)benzene as initiator, [styrene]$_0$/[initiator]$_0$/[Ru]$_0$ = 750/2/1

[b] Determined by size-exclusion chromatography (SEC) in THF with polystyrene calibration

[c] Initiation efficiency $f = M_{n,\text{theor.}}/M_{n,\text{exp.}}$ with $M_{n,\text{theor.}} = ([\text{monomer}]_0/[\text{initiator}]_0) \times M_w(\text{monomer}) \times \text{conversion}$

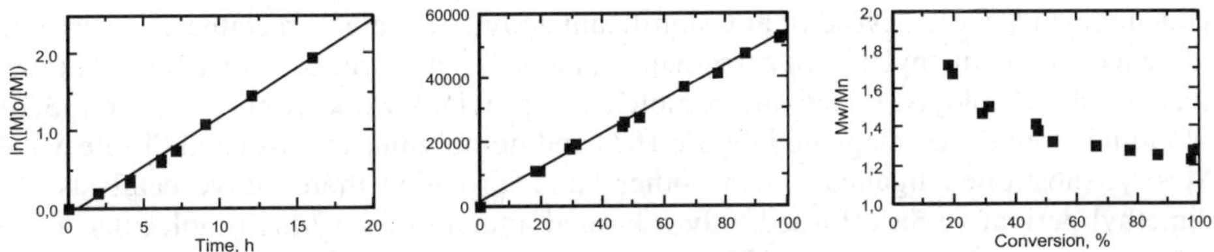

Figure 9 (left). Time dependence of $\ln([M]_0/[M])$ for the polymerization of styrene catalyzed by complex **5e**
Figure 10 (middle) and 11 (right). Dependence of the polystyrene molecular weight M_n and molecular weight distribution M_w/M_n on monomer conversion for the polymerization of styrene catalyzed by complex **5e** (see Table IV for experimental conditions).

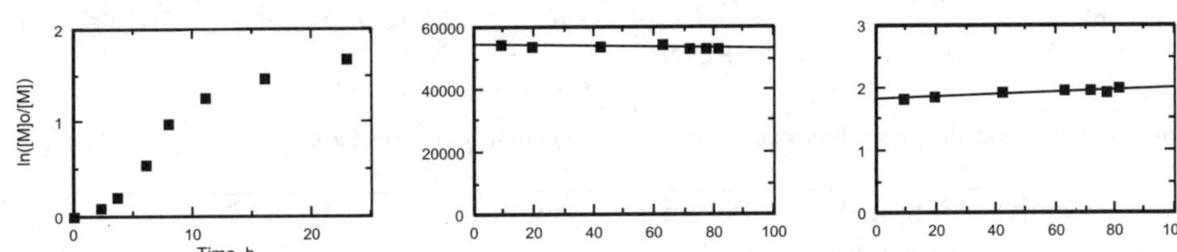

Figure 12 (left). Time dependence of $\ln([M]_0/[M])$ for the polymerization of styrene catalyzed by complex **5d**.
Figure 13 (middle) and 14 (right). Dependence of the polystyrene molecular weight M_n and molecular weight distribution M_w/M_n on monomer conversion for the polymerization of styrene catalyzed by complex **5d** (see Table IV for experimental conditions).

An important result in the frame of this discussion was that styrene underwent metathesis in addition to radical polymerization when ruthenium-NHC catalyst precursors bearing R_1 mesityl groups were present in the reaction medium. Thus, *cis*- and *trans*-stilbene (the latter isomer being largely predominant) were obtained in varying proportions according to the R_2 substituents. With complexes **5a** (R_2 = Me) and **5b** (R_2 = H), stilbene formation accounted for 25 and 5% of the monomer conversion, respectively, whereas ATRP gave rise to polystyrene in 66 and 51% yields, respectively. With complex **5c** (R_2 = Cl) olefin metathesis took the precedence over the chain reaction and afforded a 70% yield of stilbene after 16 h at 110 °C. Within the same period, a mere 10% polymer yield was obtained. Interestingly, complexes **5d** and **5e** bearing *N*-cyclohexyl-substituted NHC ligands were devoid of any significant activity for the metathesis of styrene and initiated only the radical polymerization process, with a very good control in the case of **5e**. Therefore, with both methyl methacrylate and styrene, subtle modifications of the R_1 and R_2 substituents led to dramatic changes in the ability of the resulting ruthenium complexes to favor the occurrence of a well-behaved ATRP [20].

2.4. RING-OPENING METATHESIS POLYMERIZATION OF CYCLOOCTENE

Polymerization of cyclooctene, a typical low-strain cycloolefin, served as a standard test reaction to evaluate the ability of complexes **5a-e** at promoting ROMP. The reactions were carried out under inert atmosphere in chlorobenzene at 60 °C for 4 h. They led to dichotomous results that confirmed the trend already observed for the cross-coupling of styrene. Complexes **5d** and **5e** bearing *N*-cyclohexyl imidazol-2-

ylidene ligands were devoid of any significant activity for olefin metathesis, even in the presence of trimethylsilyldiazomethane (TMSD), a carbene precursor that was successfully employed to activate complex **2** in previous work from our group [13,14]. Monomer conversion stagnated below 10% and no polymer was isolated (Table V). *N*-Mesityl-substituted ligands, on the other hand, provided more active catalysts. The dimethyl derivative **5a** afforded only a limited amount of very high molecular weight polymer, but with complexes **5b** and **5c** an almost complete consumption of the monomer was attained after 4 h at 60 °C and, accordingly, near quantitative yields of mostly *trans* polyoctenamer were isolated.

Table V. Polymerization of cyclooctene catalyzed by ruthenium complexes **5a-e**[a]

Catalyst	Monomer conversion (%)	Polymer yield (%)	M_n[b]	M_w/M_n[b]	σ_c[c]
5a	15	12	n. d.[d]	n. d.[d]	0.48
5b	>99	84	269 000	1.97	0.17
5c	>99	91	625 000	1.29	0.22
5d	4	0	-	-	-
5e	6	0	-	-	-

[a] Experimental conditions: 2 h reaction in PhCl at 60 °C under argon, [cyclooctene]$_0$/[Ru]$_0$ = 250
[b] Determined by size-exclusion chromatography (SEC) in THF with polystyrene calibration
[c] Fraction of *cis* double bonds within the polyoctenamer determined by ^{13}C NMR
[d] Not determined (insoluble polymer)

When the polymerizations were carried out at room temperature (*ca.* 20 °C), the influence of the various R$_2$ substituents within ruthenium-NHC catalyst precursors bearing R$_1$ mesityl group became more blatant (Fig. 15). Thus, the time course of the reactions carried out in the presence of complexes **5a-c** was followed by gas chromatography. It confirmed that the presence of methyl groups on the 4,5 positions of the heterocyclic imidazole ring had a detrimental effect on the catalytic activity of the corresponding ruthenium-NHC species. It also helped discriminate between the hydrogen and chloro substituents. In the presence of complex **5a** (R$_2$ = Me), conversion quickly leveled off and remained under the 10% threshold. The dichloro derivative **5b** displayed a steady albeit moderate activity. With this species, conversion reached 69% after the two hours imparted for the reaction but did not seem to have reached its ceiling yet. Complex **5b** bearing the parent unsubstituted 1,3-imidazol-2-ylidene ligand (R$_2$ = H) outperformed its two congeners. Under the experimental conditions adopted, an almost quantitative conversion of cyclooctene into polyoctenamer occurred in less than one hour [21].

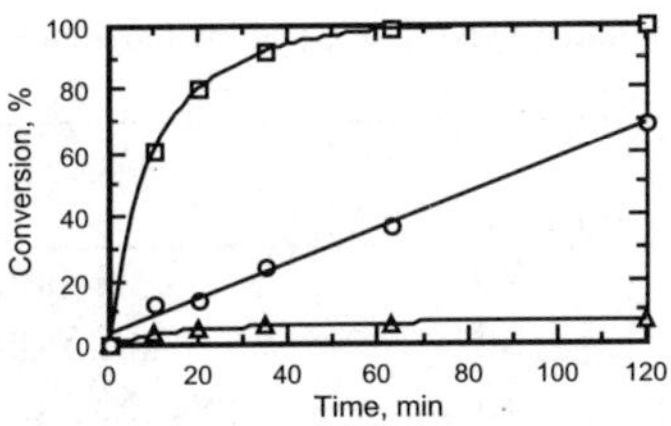

Figure 15. Time dependence of conversion for the polymerization of cyclooctene catalyzed by complex **5a** (△), **5b** (□), and **5c** (□) at room temperature (see Table VI for experimental conditions).

Most interestingly, we found that the polymerization of cyclooctene with catalysts **5b** and **5c** performed equally well in the absence of TMSD, whereas the addition of a diazo compound was a requisite with complex **2** [13,14]. Lighting, as a counterpart, played a key-role in the nucleophilic carbene-based metathesis system. The intervention of a photochemical effect was first discovered by serendipity during kinetics measurements on the ROMP of cyclooctene catalyzed by complex **5b**. Indeed, recourse to NMR spectroscopy for monitoring the olefin disappearance gave unexpected results. The reaction mixture did not show any sign of evolution until the NMR tube was eventually removed from the magnet cavity for visual inspection. When it was placed again into the spectrometer probe, some polyoctenamer had formed. Subsequent insertion/removal sequences confirmed that the progress of the polymerization was conditioned by the exposure of the reaction medium to visible light (Fig. 16).

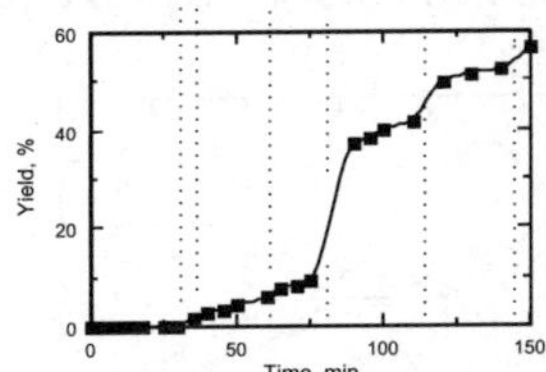

Figure 16. Time dependence of conversion for the polymerization of cyclooctene catalyzed by complex **5b** monitored by ^{1}H NMR spectroscopy in chlorobenzene-d_5 at room temperature (the dashed lines indicate the periods of exposure to daylight).

The crucial influence of a photochemical activation in the ROMP process catalyzed by ruthenium-arene complex **5b** was further supported by the results gathered in Table VI. In the darkness, a modest 20% yield of polymer was obtained after 2 h at room temperature. Normal lighting in the laboratory, a combination of daylight and of fluorescent light, was sufficient to raise the conversion to 93% within the same period. More intense visible light sources brought the reaction to completion while ensuring reproducible conditions. Thus, recourse to an ordinary 40 W "cold white" fluorescent tube or to a 250 W incandescent light bulb standing 10 cm away from the standard Pyrex reaction flasks afforded quantitative yields of polyoctenamer. The differences in the emission spectra of the two light sources did not have any incidence on the polymer microstructure. In both cases, the polyoctenamer obtained was mainly *trans* and had a relatively narrow molecular weight distribution [21].

Table VI. Effect of light on the polymerization of cyclooctene catalyzed by complex **5b**[a]

Lighting conditions	Monomer conversion (%)	Polymer yield (%)	M_n^b	M_w/M_n^b	σ_c^c
Darkness	22	20	21 000	1.53	0.36
Normal	93	84	625 000	2.00	0.27
Neon tube	99	93	553 000	1.33	0.17
Light bulb	>99	91	537 000	1.33	0.18

[a] Experimental conditions: 2 h reaction in PhCl at room temperature under argon, [cyclooctene]$_0$/[Ru]$_0$ = 250

[b] Determined by size-exclusion chromatography (SEC) in THF with polystyrene calibration

[c] Fraction of *cis* double bonds within the polyoctenamer determined by ^{13}C NMR

The UV/Vis spectra of complexes **2** and **5b** freshly dissolved in chlorobenzene were recorded under the exclusion of air and moisture (Fig. 17). The phosphine derivative had an absorption maximum at 369 nm while the carbene compound displayed 2 less intense bands centered at *ca.* 350 and 450 nm. The former corresponds to the absorption of the free carbene ligand (not shown). The latter was more visible when [RuCl$_2$(*p*-cymene)]$_2$ and 1,3-dimesitylimidazol-2-ylidene (2 equivalents) were mixed in the UV cell immediately prior to the analysis. Upon exposure to intense visible light for 30 min, the band around 450 nm completely disappeared. Hence, we tentatively assign this visible absorption to the *p*-cymene moiety in **5b**. A light-induced decomplexation of the η^6-arene ligand would liberate a highly reactive coordinatively unsaturated ruthenium species that is believed to trigger the catalytic process, although the exact nature of the active species in solution remains elusive so far.

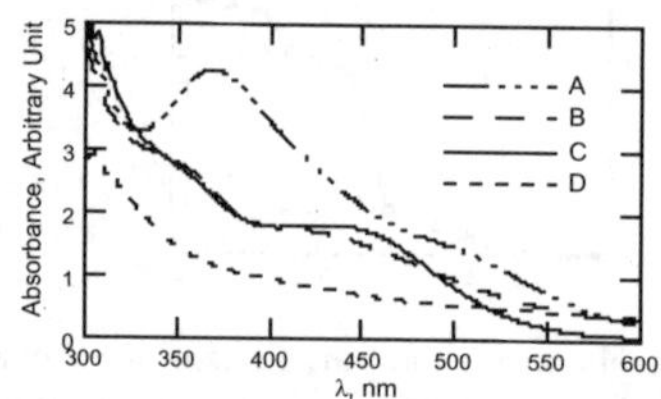

Figure 17. UV/Vis spectra of 1 mM solutions in PhCl of complexes **2** (A), **5b** (B), [RuCl$_2$(*p*-cymene)]$_2$ + 1,3-dimesitylimidazol-2-ylidene (2 eq.) freshly mixed (C) and irradiated 30 min with intense visible light (D).

3. Conclusion

The ruthenium-arene complexes **5a-e** described in this study are fine examples of readily accessible ruthenium *N*-heterocyclic carbene catalysts. Depending on the substituents of the carbene ligand, they can be tuned to promote atom transfer radical reactions or olefin metathesis of carbon-carbon double bonds. The quantitative ATRA of carbon tetrachloride onto styrene was achieved in the presence of complex **5c** (R$_1$ = Mes, R$_2$ = Cl), but complex **5e** bearing two different carbene substituents (R$_1$ = Cy, R$_2$ = H) was also a rather efficient catalyst at 90 °C. Complexes **5b** and **5c** with R$_1$ = mesityl and R$_2$ = H or Cl, respectively, were most suitable for promoting the ATRP of methyl methacrylate. Their use resulted in well-behaved polymerizations and afforded macromolecular products with narrow molecular weight distributions, M_w/M_n, and high initiation efficiencies, f. In the case of styrene, complex **5e** (R_1 = Cy, R_2 = H) was the most efficient catalyst precursor for initiating a controlled radical polymerization. A switch in the reaction pathway could be induced, however, by replacing the cycloalkyl group with a mesityl substituent on the nitrogen atoms. Thus, complexes **5a-c** displayed a dual activity and afforded both the cross-metathesis and the ATRP products. Among the three species tested, the chloro derivative **5c** led to the highest proportion of stilbene compared to polystyrene. When the ROMP of cyclooctene was investigated at room temperature, complex **5c** lost its superiority as a metathesis catalyst precursor to the benefit of its parent **5b** (R_1 = Mes, R_2 = H).

At the present time, it remains difficult to put forward general guidelines to rationalize the choice of a specific ruthenium-NHC catalyst precursor for a given reaction. Depending on the substrate and the experimental conditions adopted (temperature, solvent, presence of an initiator or a cocatalyst,...) the coordination sphere around the metal center must be specifically tailored to afford the most efficient catalytic system. Fine tuning of the electronic, steric, and solubility parameters of the carbene ligand, undoubtedly contributes to these adjustments, but any correlation between well-defined 18-electron ruthenium-arene complexes and the active coordinatively unsaturated species generated *in situ* is blurred by the elusive nature of the actual catalytic system.

4. References

1. Renaud, P. and Sibi, M. P. (eds.) (2001) *Radicals in Organic Synthesis*, Wiley-VCH, Weinheim.
2. Kharasch, M.S., Jensen, E.V., and Urry, W.H. (1945) Addition of carbon tetrachloride and chloroform to olefins, *Science* **102**, 128.
3. Matyjaszewski, K. and Xia, J. (2001) Atom transfer radical polymerization, *Chem. Rev.* **101**, 2921-2990.
4. Kamigaito, M., Ando, T., and Sawamoto, M. (2001) Metal-catalyzed living radical polymerization, *Chem. Rev.* **101**, 3689-3746.
5. Ivin, K.J. and Mol, J.C. (1997) *Olefin metathesis and metathesis polymerization*, Academic Press, San Diego.
6. Schuster, M. and Blechert, S. (1997) Olefin metathesis in organic chemistry, *Angew. Chem., Int. Ed.* **36**, 2036-2056.
7. Fürster, A. (2000) Olefin metathesis and beyond, *Angew. Chem., Int. Ed.* **39**, 3012-3043.
8. Rouhi, A.M. (2002) Olefin metathesis: big-deal reaction, *Chem. Eng. News* December 23, 29-33.
9. Simal, F., Demonceau, A., and Noels, A.F. (1999) Atom transfer radical addition (ATRA) versus atom transfer radical polymerization (ATRP) catalysed by ruthenium complexes, *Recent Res. Devel. Org. Chem.* **3**, 455-464.
10. Frenzel, U. and Nuyken, O. (2002) Ruthenium-based metathesis initiators: development and use in ring-opening metathesis polymerization, *J. Polym. Sci., Part A: Polym. Chem.* **40**, 2895-2916.
11. Schwab, P., France, M.B., Ziller, J.W., and Grubbs, R.H. (1995) A series of well-defined metathesis catalysts – Synthesis of [$RuCl_2$(=CHR')(PR_3)$_2$] and its reactions, *Angew. Chem., Int. Ed. Engl.* **34**, 2039-2041.
12. Trnka, T.M. and Grubbs, R.H. (2001) The development of L_2X_2Ru=CHR olefin metathesis catalysts: an organometallic success story, *Acc. Chem. Res.* **34**, 18-29.
13. Stumpf, A.W., Saive, E., Demonceau, A., and Noels, A.F. (1995) Ruthenium-based catalysts for the ring-opening metathesis polymerisation of low-strain cyclic olefins and of functionalised derivatives of norbornene and cyclooctene, *Chem. Commun.* 1127-1128.
14. Demonceau, A., Stumpf, A.W., Saive, E., and Noels, A.F. (1997) Novel ruthenium-based catalyst systems for the ring-opening metathesis polymerization of low-strain cyclic olefins, *Macromolecules* **30**, 3127-3136.
15. Simal, F., Demonceau, A., and Noels, A.F. (1999) Highly efficient ruthenium-based catalytic systems for the controlled free-radical polymerization of vinyl monomers, *Angew. Chem., Int. Ed. Engl.* **38**, 538-540.
16. Herrmann, W.A. (2002) *N*-heterocyclic carbenes: a new concept in organometallic catalysis, *Angew. Chem., Int. Ed.* **41**, 1290-1309.

17. Bourissou, D., Guerret, O., Gabbaï, F.P., and Bertrand, G. (2000) Stable carbenes, *Chem. Rev.* **100**, 39-91.

18. Simal, F., Delfosse, S., Demonceau, A., Noels, A.F., Denk, K., Kohl, F.J., Weskamp, T., and Herrmann, W.A. (2002) Ruthenium alkylidenes: modulation of a new class of catalysts for controlled radical polymerization of vinyl monomers, *Chem. Eur. J.* **8**, 3047-3052.

19. Richel, A., Delfosse, S., Cremasco, C., Delaude, L., Demonceau, A., and Noels, A.F. (2003) Ruthenium catalysts bearing *N*-heterocyclic carbene ligands in Kharasch chemistry, manuscript in preparation.

20. Delaude, L., Delfosse, S., Richel, A., Demonceau, A., and Noels, A.F. (2003) Tuning of ruthenium *N*-heterocyclic carbene catalysts for ATRP, manuscript submitted for publication.

21. Delaude, L., Demonceau, A., and Noels, A.F. (2001) Visible light induced ring-opening metathesis polymerisation of cyclooctene, *Chem. Commun.* 986-987.

22. Delaude, L., Szypa, M., Demonceau, A., and Noels, A.F. (2002) New in situ generated ruthenium catalysts bearing *N*-heterocyclic carbene ligands for the ring-opening metathesis polymerization of cyclooctene, *Adv. Synth. Catal.* **344**, 749-756.

23. Herrmann, W.A., Köcher, C., Goossen, L.J., and Artus, G.R.J. (1996) Heterocyclic carbenes: a high-yielding synthesis of novel, functionalized *N*-heterocyclic carbenes in liquid ammonia, *Chem. Eur. J.* **2**, 1626-1636.

24. Jafarpour, L., Huang, J., Stevens, D., and Nolan, S.P. (1999) (*p*-cymene)RuLCl$_2$ (L = 1,3-bis(2,4,6-trimethylphenyl)imidazol-2-ylidene and 1,3-bis(2,6-diisopropylphenyl)imidazol-2-ylidene) and related complexes as ring closing metathesis catalysts, *Organometallics* **18**, 3760-3763.

25. Kuhn, N. and Kratz, T. (1993) Synthesis of imidazol-2-ylidenes by reduction of imidazole-2(3*H*)-thiones, *Synthesis* 561-562.

26. Arduengo, A.J., III, Krafczyk, R., Schmutzler, R., Craig, H.A., Goerlich, J.R., Marshall, W.J., and Unverzagt, M. (1999) Imidazolylidenes, imidazolinylidenes and imidazolidines, *Tetrahedron* **55**, 14523-14534.

27. Arduengo, A.J., III, Davidson, F., Dias, H.V.R., Goerlich, J.R., Khasnis, D., Marshall, J., and Prakasha, T.K. (1997) An air stable carbene and mixed carbene "dimers", *J. Am. Chem. Soc.* **119**, 12742-12749.

O,N- BIDENTATE LIGANDS CO-ORDINATED ON Ru-BASED OLEFIN METATHESIS CATALYSTS.

B. D. CLERCQ, AND F. VERPOORT[*]
Division of Organometallic Chemistry and Catalysis, Department of Inorganic and Physical Chemistry, Ghent University, Krijgslaan 281 (S-3), 9000 Gent, Belgium

Abstract

A new class of Schiff Base Ruthenium arene complexes has been prepared. These complexes has been tested in ring-opening metathesis polymerisation (ROMP) and ring-closing metathesis (RCM) reactions. The results point out that without trimethylsilyldiazomethane (TMSD) very low activities are obtained. In case the complexes are activated with TMSD good yields are obtained for RCM. Furthermore, a mechaniscm has been suggested.

The Schiff Base Ruthenium arene complexes has been tested on their ROMP and RCM behaviour. The obtained results suggest that the catalytic activity strongly depends on the steric and electronic environment of the Schiff Base. The different behaviour of the Schiff Base Ruthenium carbenes for ROMP and RCM has been explained.

1. Introduction

Over the years Schiff Base ligands has been used extensively in catalysis due to some advantages:

i)   They are easy to synthesize by simple condensation of an aldehyde compound with an amine derivative. These ligands also provide chiral or *cis/trans* selective catalysts [1].

ii)   Furthermore, N and O ligands are more resistant to oxidation than phosphanes. This is one of the reasons why the Jacobsen Catalyst prove so successful in stereoselective epoxidation [2,3].

High optical inductions are also seen for Cu^{II}- or Co^{II}- attached semicorrine and Schiff base ligands in olefin cyclopropanation reactions with diazoalkanes [4].

High enantioselectivity was obtained using Ti-complexes having a salicylal-type Schiff base ligand for the asymmetric addition of hydrogen cyanide to aldehydes [1a].

iii)   Variation in steric and/or electronic effects are reflected on the resulting complexes [5].

Y. Imamoglu and L. Bencze (eds.), Novel Metathesis Chemistry: Well-Defined Initiator Systems for Specialty Chemical Synthesis, Tailored Polymers and Advanced Material Applications, 101–119.
© 2003 *Kluwer Academic Publishers. Printed in the Netherlands.*

Recent years well-defined single-component Ru-carbene complexes have been prepared and extensively utilized in olefin metathesis reactions. Of particulary utility among the various well-defined catalyst systems has been the Ru-carbene systems **1-5**, figure 1.

All these Ru-systems exhibits high reactivity for a variety of metathesis processes under mild conditions and are remarkably tolerant of many organic functional groups [6].

Olefin metathesis is a catalytic reactions in which alkenes are converted into new products via rupture and reformation of C-C double bonds. Depending on the starting material (cyclic or acyclic olefines) and the reaction conditions, ring closing metathesis (RCM), acyclic diene metathesis (ADMET) or ring opening metathesis polymerisation (ROMP) proceed (scheme 1).

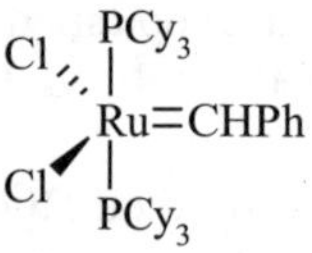

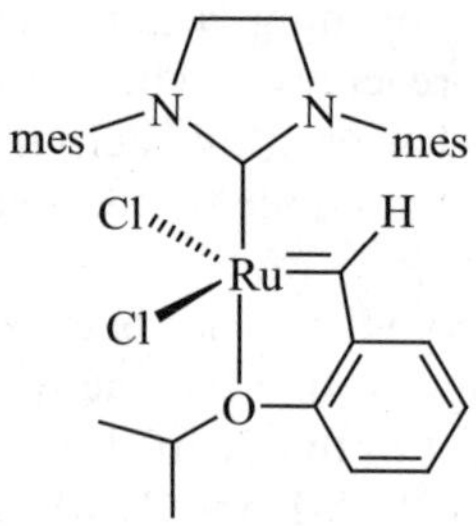

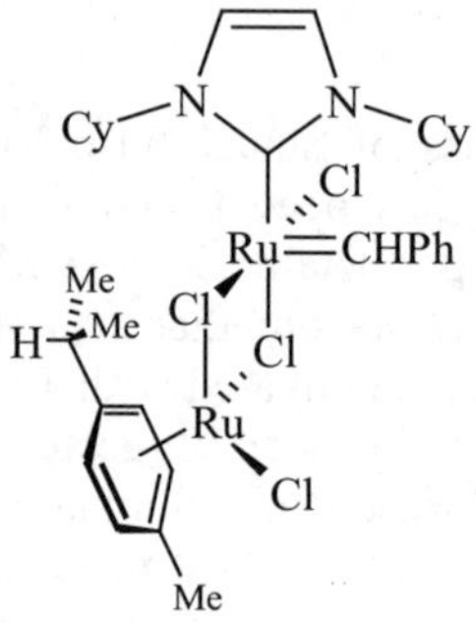

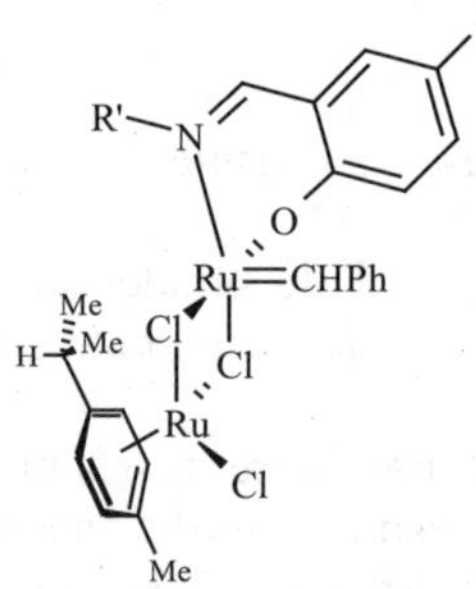

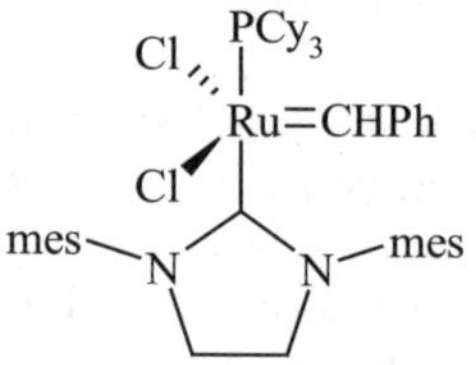

Figure 1.

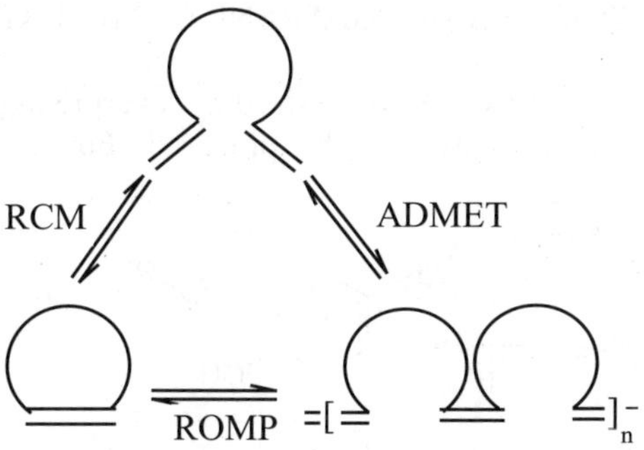

Scheme 1.

The first implementation of Schiff Base ligands in Ru-catalysts for olefin metathesis was in 1998 by Grubbs et al. [7]. Although these complexes **6**, figure 2, are less reactive at room temperature then the carbene complex **1** for metathesis reactions, the reactivity increases dramatically at higher temperatures.

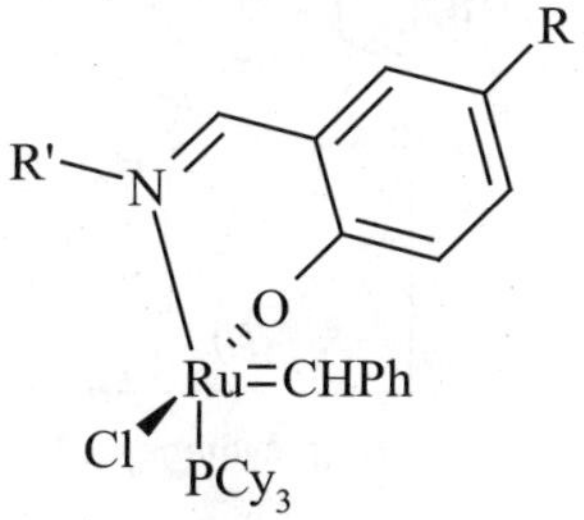

6

Figure 2

However, no further details concerning the ability of these complexes are available in literature. This prompted us to further explore the ability of Schiff base Ru-carbenes for RCM and ROMP of several substrates and monomers.

In addition, there exist an extensive chemistry of [(arene)-RuX_2PR_3] complexes and their activity in RCM, ROMP, Atomic Transfer Radical Polymerization (ATRP) and Atomic Transfer Radical Addition (ATRA) has been truly investigated [8]. In contrast analogue half sandwich Schiff base Ru-complexes are virtually unknown. In this study some new arene Schiff Base Ru-complexes were developed and screened on their activity for C-C double bond formation.

2. Results

2.1. SYNTHESIS OF SCHIFF BASE LIGANDS.

The salicylaldimine ligands **9** were synthesized by condensation of salicylaldehydes **7** with aliphatic or aromatic amine derivatives **8** in high yields (Scheme 2).

2.2. SYNTHESIS OF SCHIFF BASE Ru-ARENE COMPLEXES.

The Ru-arene compounds **11a-c** are synthesized by a surprisingly general reaction that allows for a variety of derivatives to be prepared. For example, the reaction of phosphines

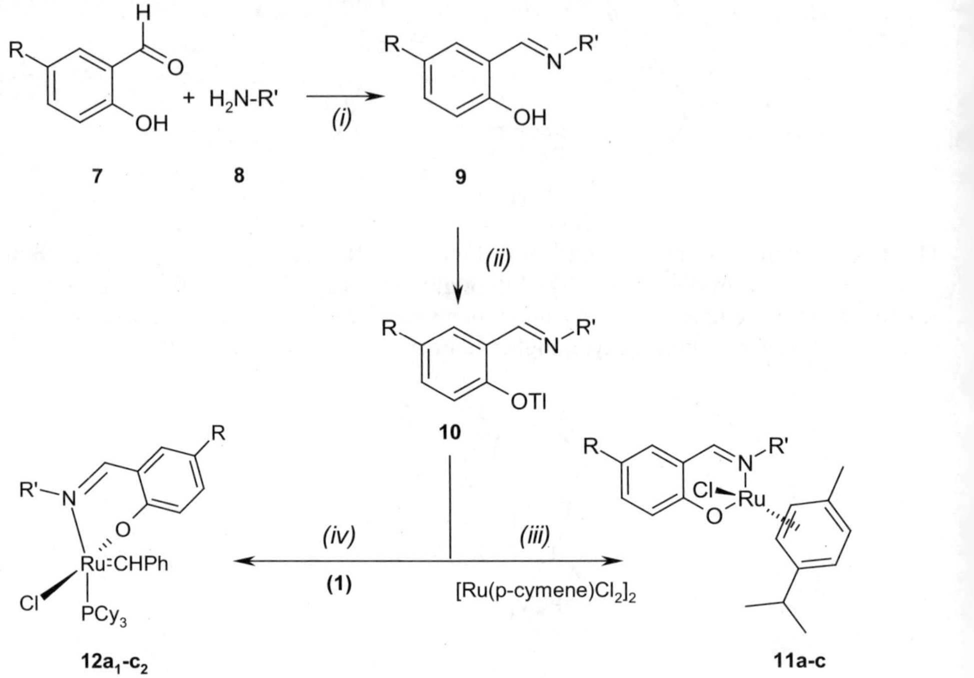

7   **8**   **9**

10

(1) $[Ru(p\text{-cymene})Cl_2]_2$

12a$_1$-c$_2$ **11a-c**

a1   R = H,   R' = Me
a2   R = NO_2, R' = Me
b1   R = H,   R' = 2,6-Me-4-BrC_6H_2
b2   R = NO_2, R' = 2,6-Me-4-BrC_6H_2
c1   R = H,   R' = 2,6-$iPrC_6H_3$
c2   R = NO_2, R' = 2,6-$iPrC_6H_3$

a   R = H,   R' = Me
b   R = H,   R' = t-Bu
c   R = H,   R' = 2,6-Me-4-BrC_6

Scheme 2: Reaction sequence for the synthesis of the different Schiff base substituted Ru-complexes. (i) THF, Δ, 2h for H_2NR' (R'= Me, t-Bu) and EtOH, 80°C, 2h for H_2NR' (R' = 2,6-Me-4-BrC_6H_2, 2,6-$iPrC_6H_3$); (ii)TlOEt, THF, RT, 2h; (iii) [(*p-cymene*)-RuCl₂]₂, THF, RT, 6h; (iv) **1**, THF, RT, 4h.

with transition-metal dimmers containing bridging chloride ligands has been well established [9]. The reaction of immobilized diphenyl phosphino ethyltriethoxysilane (DIPHES) on silica with [(*p-cymene*)-RuCl₂]₂ is depicted in figure 3 [10].

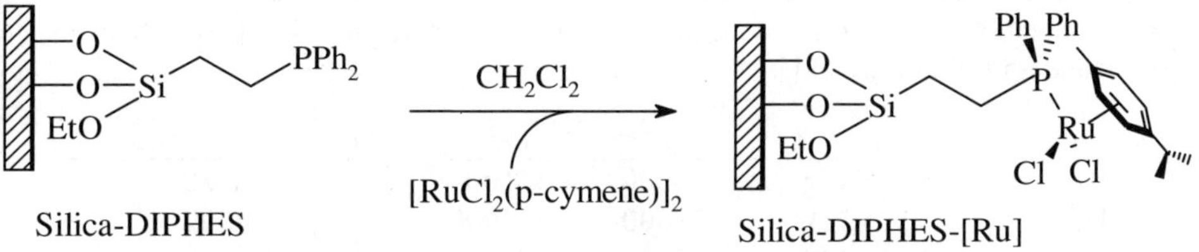

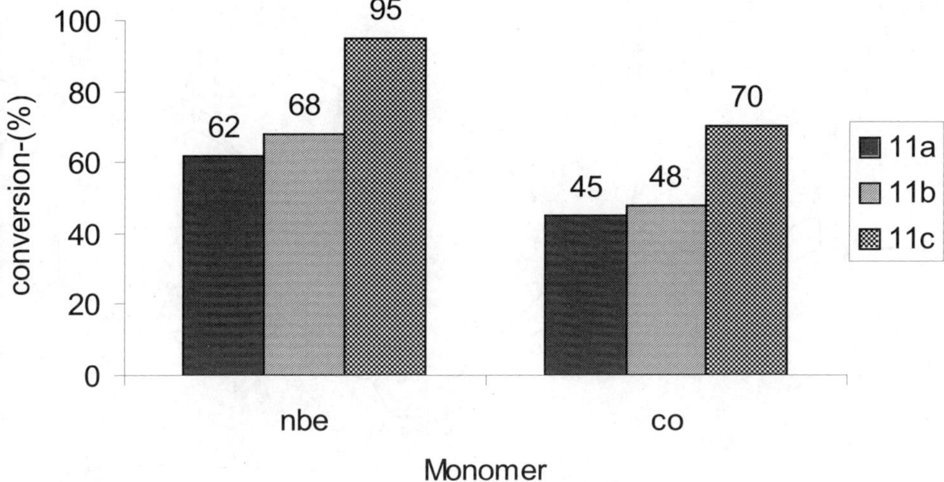

Fig.3: Preparation of MCM-41-DIPHES-[Ru].

The reaction (**9 → 11**) of the Ru-dimer with Schiff-bases **9a-c** were performed after converting the ligands to the corresponding Tl-salts. The thallium salts **10a-c** were quantitatively obtained upon treatment of Schiff base ligands **9a-c** with thallium ethoxide. Further treatment of **10a-c** with [(*p-cymene*)-RuCl$_2$]$_2$ at room temperature afforded the desired complexes **11a-c** as red-brown solids. The complexes are stable to air and moisture.

2.3. ROMP-ACTIVITY STUDIES OF SCHIFF BASE Ru-ARENE COMPLEXES.

The obtained conversions after activating the complexes **11a-c** with trimethylsilyl diazomethane (TMSD) for norbornene (NBE) and cyclo-octene (CO) are depicted in figure 4.

Fig. 4: %-conversion for ROMP of NBE and CO using catalysts 11a, 11b and 11c. Reaction conditions see Experimental Section.

The properties of the poly-NBE and poly-CO are presented in table 1.
A maximum conversion for both NBE and CO polymerisation is reached at 80 min. reaction time and this for all three systems. Catalytic system having the aromatic substituted imine ligand **11c** is clearly the most active.

The polydispersities are in all cases broad indicating the occurance of backbiting and transfer reactions. A Trans configuration is preferred in all cases which is in accordance with literature [11].

		M_n	M_w	PDI	cis/trans
	11a	74000	471000	6.34	0.38
NBE	**11b**	78000	462000	5.95	0.32
	11c	98000	682000	6.95	0.39
	11a	55000	181000	3.26	0.36
CO	**11b**	59000	196000	3.35	0.34
	11c	57000	262000	4.62	0.38

Table 1: Properties of polyNBE and polyoctene using using catalysts 11a, 11b and 11c activated with TMSD. Reaction conditions see Experimental Section

In a next experiment no activator (TMSD) was used. Only NBE and catalyst **11c** were mixed (ratio of Ru/NBE : 1/800) in 1 ml toluene at 85°C. After 6h a conversion of 6% was observed. In this case the initiating Ru-carbene must result from a reaction between the substrate olefin and the Ru-arene Schiff base complex.

To elucidate the mechanism of carbene formation, 0.5 mmol of **11c** solution in C_6D_6 was added to 0.5 mmol of NBE solution in C_6D_6. The reaction mixture was then heated for 4h at 85°C. Thereafter the reaction mixture was quenched with ethyl vinylether.

Investigating the resulting solution using ^{1}H-NMR the presence of an alkoxy substituted carbene ([Ru]=C*H*OEt, 14.95 ppm (s)) was confirmed (compound **15**, scheme 3).

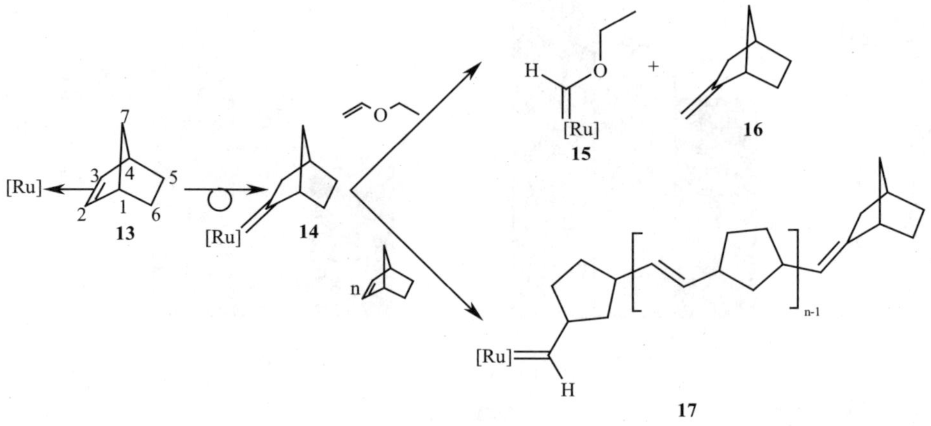

Scheme 3: Possible mechanism for the formation of the initial metal-carbene in the absence of TMSD.

Moreover, the presence of compound **16** was unambiougously identified using ^{1}H-and 13C-NMR after purification and concentration of the reaction mixture.

Furthermore, the reaction between 0.5 mmol of complex **11c** in C_6D_6 and 5 equivalents of NBE solution in C_6D_6 at 85°C for 4h, a propagating species appeared as a doublet at 18.85 ppm (compound **17**).

In addition, ^{1}H-NMR measurements reveal that the propagating Ru-carbene peak could be integrated for ± 2.5 % of the total ruthenium present in the solution. The amount of free p-cymene released in the solution is in agreement with the amount of the propagating Ru-carbene.

From these data it is reasonable to assume that the catalyst (an 18 electron species) polymerises norbornene through the loss of p-cymene followed by the coordination of NBE and rearrangement to an initiating ruthenium carbene complex which then propagates ROMP. The initiating species (compound **14** , scheme 3) itself is produced by a 2,3 H-shift in the ruthenium-olefin complex (compound **13**).

To elucidate the mechanism of carbene formation in the presence of TMSD the following experiment was set up. Two equivalents of TMSD were added to a solution containing 0.5 mmol of catalyst **11c** in C_6D_6. Generation of N_2-gas took place and a signal at 23.66 ppm (singlet) was observed and assigned to the proton of [Ru]=C*H*SiMe$_3$. The methyl groups of

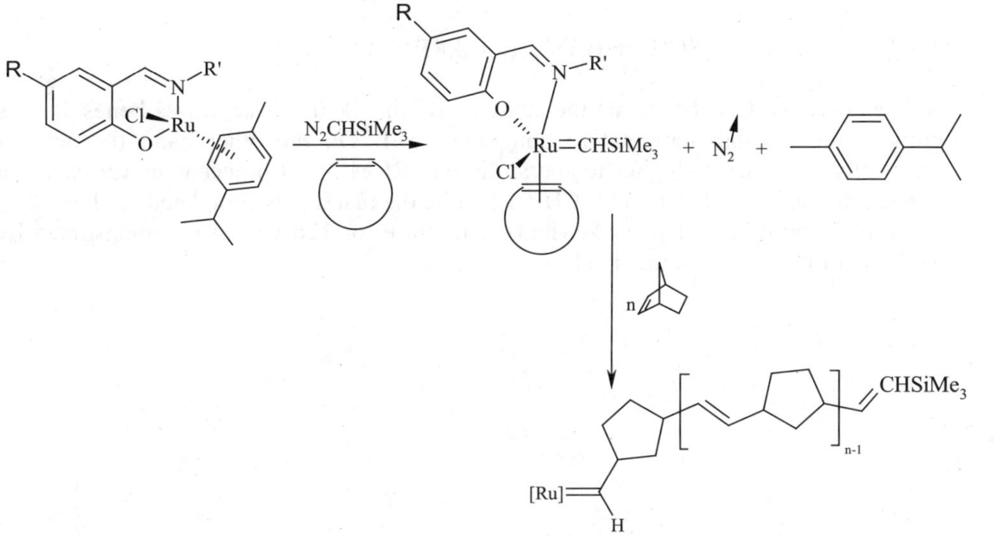

Scheme 4: Possible mechanism for the formation of the initial metal-carbene using TMSD.

the carbene species appeared as a singlet at 0.46 ppm. By addition of 5 equivalents of NBE to the reaction mixture, the signal at 23.66 ppm disappeared and a new signal at 18.85 ppm appeared. The latter being the propagating carbene of the growing polymer chain. Integration of the propagating Ru-carbene signal amounts in 12% of the total ruthenium amount present in the solution. These results are again in good agreement with the free p-cymene amount released after addition of TMSD. Studies by other teams dealing with metathesis reactions mediated by ruthenium arene complexes have shown that the release of the arene ligand is crucial and is responsible for the generation of the active catalyst [12].

Therefore, we suggest that the mechanism depicted in scheme 4 is responsible for the formation of the catalytic active species in olefin metathesis.

2.4. RCM ACTIVITY STUDIES OF SCHIFF BASE Ru-ARENE COMPLEXES.

The ring-closing metathesis activities of complexes **11a-c** are depicted in table 2. Catalytic RCM of dienes 1,7 octadien (entry 4) and diallyl ether (entry 5) was performed quantitatively by all three catalysts. Diallyl phatalate (entry 6a) and linalool (entry 7a) are converted smoothly. In harsh conditions (85°C and 17h) the yields are drastically improved (entry 6b and 7b). With the best catalytic system **11c** conversions of 94 and 76 % respectively are obtained. The RCM of diallylmalonate (the standard test for catalytic activity in olefin metathesis) was performed quantitatively by all three systems (entry 1).
Moreover, the reactivity of the complexes is sufficiently high to allow the preparation of trisubstituted cyclo-alkenes in moderated yields (entry 2a). Increasing temperature and reaction time results in dramatical higher conversions. Under these conditions, the most performing system **11c** reaches conversions 71 and 23 % for the tri- and tetrasubstituted malonate derivates, respectively (entry 2b and 3b).

2.5. SYNTHESIS OF SCHIFF BASE Ru-CARBENES.

As reported by Grubbs et al. the activity of the Schiff base Ru-carbenes is less pronounced at room temperature compared to **1.** On the other hand the activity drastically increases at higher temperature, e.g. RCM of **13** proceeds in 12h at room temperature using **12d** (8 mol %, CH_2Cl_2) while the reaction is completed in 1h at 70°C using **12d** (3mol %, C_6D_6, 96 % yield). The utility of **12d** was also demonstrated by RCM of **14** in MeOH, figure 4 [7].

Entry	Substrate[b]	Product	Yield,%[a]		
			11a	**11b**	**11c**
1			100	100	100
2a[c]			22	26	32
2b[d]			46	53	71
3a[c]			<5	<5	<5
3b[d]			8	15	23
4	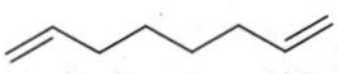		100	100	100
5	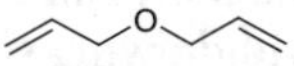		100	100	100
6a[c]			48	55	60
6b[d]			62	78	94
7a[c]			16	21	28
7b[d]			33	65	76

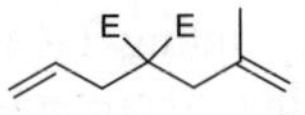

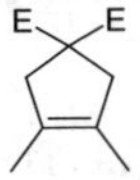

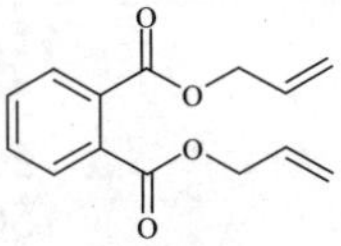

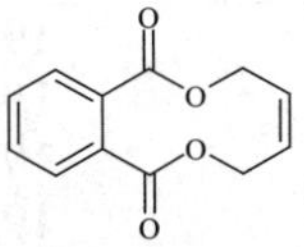

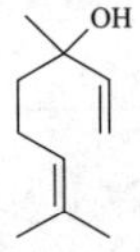

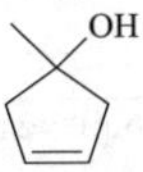

Table 2: RCM results of representative diolefins.
[a]: Yield as determined with H-NMR analysis and confirmed by GC using 1,3,5-mesitylene as internal standar. The formation of cycloisomers, oligomers or telomers was ruled out by GC analysis of the reaction mixture.
[b]: E = COOEt.
[c]: Reaction conditions: 70°C; 5 mol% **11a-c** in toluene; 2.2 equiv. TMSD.
[d]: Reaction conditions: 85°C; 5 mol% **11a-c** in toluene; 2.2 equiv. TMSD.

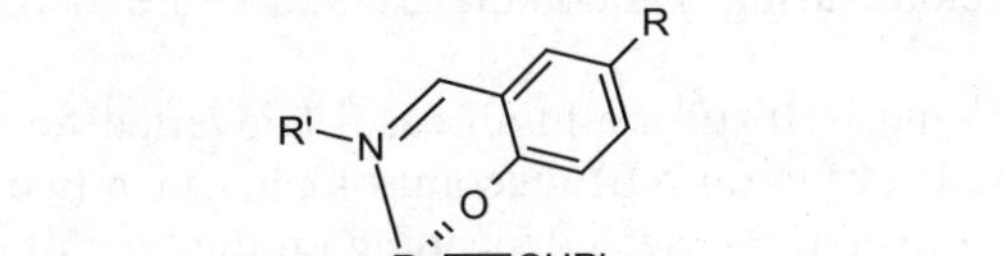

13 X = $C(CO_2Et)_2$
14 X = NH.HCl

12d : R = NO_2, R' = 2,6-iPr-4-NO_2-C_6H_3

Fig. 5: RCM of **13** and **14** using **12d** as catalyst.

Due to the fact that no more ROMP and RCM data using Schiff base substituted parent Grubbs' catalyst were available in literature we were intreged what the results would be.

Therefore, six different monometallic Schiff base Ru-alkylidenes were developed. To synthesize the Schiff base ligands the same procedure was followed as described earlier in this work. The substitution reactions ($10\rightarrow12$) of complex **1** with Schiff bases **9a$_1$-c$_2$** were performed after converting the ligands to the corresponding Tl-salts which were quantitatively obtained upon treatment with TlOEt. Further treatment of the Tl-salts with **1** at room temperature afforded the desired complexes **12a$_1$-c$_2$** in moderated to good yields.

The substitution of one chloride and one phosphine with the bidentate O-N-ligands were unambiguously indicated by characteristic ^{1}H and ^{31}P NMR spectral changes and this for all substitution reactions ($10\rightarrow12$), table 3.

The benzylidene protons in the compounds **12a$_1$-c$_2$** appear between 19.45 and 19.99 ppm as doublets (table 3). These carbene protons are shifted more downfield compared to complex **1** (singlet, 20.1 ppm in CD_2Cl_2). This downfield shift can be explained by the fact that the carbene complexes having more electron-withdrawing substituents.

Compound 12	$^1H_\alpha$	^{31}P
a$_1$ R = H, R' = Me	19.95	52.44
a$_2$ R = NO$_2$, R' = Me	19.99	52.47
b$_1$ R = H, R' = 2,6-Me-4-BrC$_6$H$_2$	19.45	50.56
b$_2$ R = NO$_2$, R' = 2,6-Me-4-BrC$_6$H$_2$	19.49	50.66
c$_1$ R = H, R' = 2,6-iPrC$_6$H$_3$	19.68	52.23
c$_2$ R = NO$_2$, R' = 2,6-iPrC$_6$H$_3$	19.77	52.40

Table 3: ^{1}H- (carbene proton) and ^{31}P-NMR of the complexes **12a$_1$-c$_2$**.

As found in the proton NMR spectroscopy, the ^{31}P shifts for the coordinated phosphine ligands in **12a-c** are also dependent on the electronic nature of the Schiff base ligands. The chemical shift of phosphorous is in the range of 50 – 53 ppm for all complexes **12a$_1$-c$_2$**.

2.6. ROMP ACTIVITY STUDIES OF SCHIFF BASE RU-CARBENES.

Ten different monomers were applied to perform ROMP with the six different Ru-complexes. The obtained conversions using Ru-complexes **12a-c** are depicted in figures 6 and 7.

From figure 6 it follows that NBE and NBE-Et are in all cases converted for the full 100%. When the alkyl chain of the R-group on NBE becomes longer than two carbon atoms a deviation in the obtained conversion is seen depending on the type of catalyst used. For the catalytic systems **12b-c** a lower conversion is reached. The lower conversion is a result of the increased steric demands in the Ru-Schiff base complexes. Also minor differences of the obtained conversions are seen if one compare the system **12b$_1$** with **12b$_2$** and **12c$_1$** with **12c$_2$**. This suggests that although the ligand coordinated around the ruthenium metal centre is similar the presence of an electron withdrawing group on the 4 position of the phenoxy fragment slightly increases the activity of the system. From figure 7 it is obvious that the monomers where R= phenyl, CH_2Cl, $Si(OEt)_3$ can be converted in high yields. Here the same explanation as given for figure 6 can be used for the observed trends.

Only in case of NBE-cyano and NBE-CH$_2$OH low conversions are obtained but still the same trends can be observed.

Summarizing, the following conversion sequence can be reported:

12a$_2$ > 12a$_1$ > 12b$_2$ > 12b$_1$ > 12c$_2$ > 12c$_1$

Also, the bulkiness of the Schiff base and the electron withdrawing properties exert a profound influence on the ROMP-activities.

For instance, to confirme the influence of the bulkiness of the Schiff base ligand on the ROMP-activities the results of NBE-phenyl are compared, see table 4.

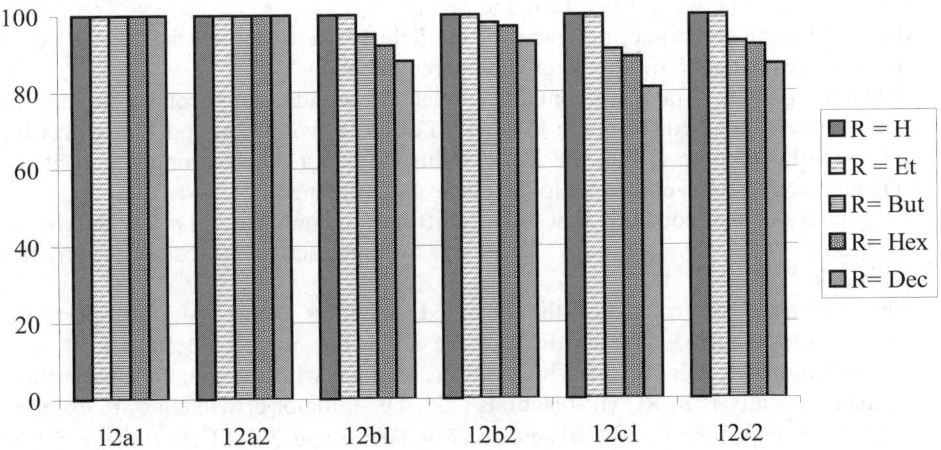

Fig.6: %-Conversion of alkylsubstituted NBE using catalysts 12a$_1$-12c$_2$.

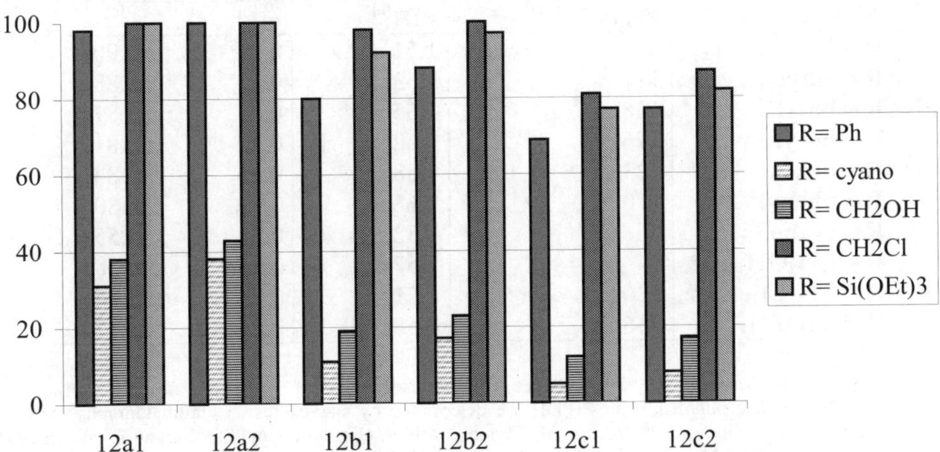

Fig. 7: %-Conversion of substituted NBE (substituent = Ph, CN, CH$_2$OH, CH$_2$Cl, Si(OEt)$_3$) using catalysts 12a$_1$-12c$_2$.

Catalyst	Yield, %	Yield, %	Catalyst	
$12a_1$ ↓	98	100	$12a_2$	
$12b_1$	80	88	$12b_2$	**BULKINESS**
$12c_1$	69	77	$12c_2$	

Table 4: ROMP yields (%) of NBE-phenyl using catalysts $12a_1$-$12c_2$.

When the imine fragment bearing a methyl group ($12a_1$) a conversion of 98 % is reached. Introduction of a 2,6-Me-4-BrC$_6$H$_2$ on the imine ($12b_1$) results in a drastic decrease of the conversion, 80 %. In case even more bulkiness is introduced on the imine ($12c_1$) the conversion decreases further. The same conclusion can be drawn for the comparison between $12a_2$, $12b_2$ and $12c_2$.

To confirm the electronic influences of the Schiff base ligand on the ROMP activities again the results of NBE-phenyl are compared, table 4.

When the phenoxy-fragment contains an electron withdrawing group a slightly higher conversion is obtained (compare $12a_2$ with $12a_1$, $12b_2$ with $12b_1$ and $12c_2$ with $12c_1$).

Also worthy of note is that the steric influence has a much stronger impact on the obtained conversions compared to the electronic influences.

In general one can conclude that O-N Schiff base ligands having a low bulkiness and electronic withdrawing groups give rise to high active Ru-catalysts for ROMP reactions.

The polydispersity values of the obtained polymers using the most performance system, $12a_2$, table 5, are narrow and are all in the range of 1.50 – 1.70. A trans configuration is predominant. This is in accordance with the general observation for ruthenium catalysts in ROMP-reactions [11]. The initiator efficiencies are excellent (f_i > 80 %) except in case of NBE-cyano ($f_i \approx 50$ %) and NBE-CH$_2$OH ($f_i \approx 50$ %). It seems that the OH group of NBE-CH$_2$OH and the CN group of NBE-cyano have a detrimental effect on the activity of the catalysts.

NBE-R	M_n (x 10^3)[a]	PDI[a]	σ_c[b]	f_i[c]
R = H	84	1.51	0.25	0.90
R = ethyl	111	1.63	0.22	0.88
R = butyl	138	1.55	0.24	0.87
R = hexyl	176	1.62	0.22	0.81
R = decyl	223	1.58	0.26	0.84
R = phenyl	170	1.68	0.29	0.80
R = cyano	66	1.52	0.22	0.55
R = CH$_2$OH	74	1.57	0.26	0.58
R = CH$_2$Cl	136	1.71	0.24	0.84
R = Si(OEt)$_3$	236	1.58	0.27	0.87

Table 5: Properties of polymers obtained with catalyst $12a_2$.
[a] M_n and the polydispersities (PDI) are determined by size-exclusion chromatography (SEC) with polystyrene calibration, [b] fraction of polymers with *cis* configuration, [c] f_i = initiation efficiency = $M_{n, theor.}$/ $M_{n, exp.}$ with $M_{n, theor.}$ = ([monomer]$_0$/[initiator]$_0$) * MW(monomer) * conversion.

2.7. RCM ACTIVITY STUDIES OF SCHIFF BASE Ru-CARBENES.

The RCM results are depicted in figure 8. RCM of dienes 1,7 octadieen (substrate 4) and diallylether (substrate 5) was performed quantitatively for all catalysts **12a-c**. Diallylphatalate (substrate 6) and linalool (substrate 7) are converted smoothly. Still the obtained conversions are depending on the type of catalyst used. Now, for the catalytic systems **12a** the lowest conversions are reached. The lower conversion is a result of the lack of steric demands in the imine fragment of the Schiff base. This observation is opposite to the findings for the ROMP activities. Also minor differences of the obtained conversions are seen if one compare the systems **12a₁** with **12a₂**, **12b₁** with **12b₂** and **12c₁** with **12c₂**. Again this suggests that the presence of an electron withdrawing groups on the 4-position of the phenoxy fragment slightly increases the activity of the system. The RCM of diethyl malonate was performed quantitatively by all systems (substrate 1). However, the reactivity of the complexes is too low to allow the ringclosing of substrate 2 and substrate 3 to generate respectively trisubstituted and tetrasubstituted cycloalkenes in moderated yields.

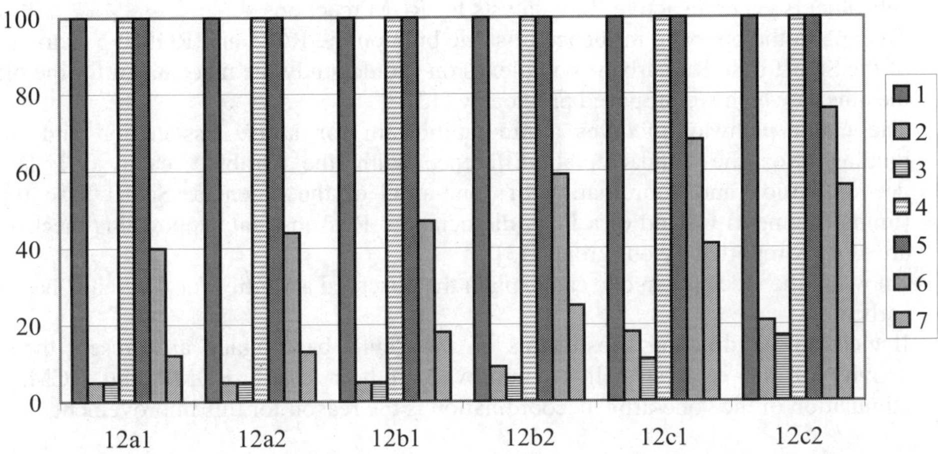

Fig. 8: RCM yields-(%) of some representative substrates. 1 = diethyl malonate, 2 = trisubstituted diethyl malonate, 3 = tetra substituted diethyl malonate, 4 = 1,7 octadieen, 5 = diallylether, 6 = diallylphatalate, 7 = linalool.

Moreover, the ability to produce tri- and tetrasubstituted cyclo-alkenes is much more pronounced for the in this work earlier described Ru-arene Schiff base catalysts which are easy to synthesize and unexpensive compared to the Ru-Schiff base carbene systems.

Summarizing, the following conversion sequence can be drawn:

12c₂ > 12c₁ > 12b₂ > 12b₁ > 12a₂ > 12a₁

As expected here again the bulkiness of the Schiff base and the electronic withdrawing properties exert a profound influence on the RCM-activities.

Confirmation of the steric influence of the Schiff base ligand on the RCM activities is best seen by comparing the results of linalool, table 6.

Catalyst	Yield, %	Yield, %	Catalyst	
12a$_1$	12	13	12a$_2$	
12b$_1$	18	25	12b$_2$	BULKINESS
12c$_1$	41	56	12c$_2$	

Table 6: RCM yields (%) of linalool using catalysts 12a$_1$-12c$_2$.

Applying a catalyst with an imine fragment bearing a methyl group (**12a$_1$**) a conversion of only 12 % is obtained. When more bulkiness is introduced the conversion can be drastically increased going from 18 % (**12b$_1$**) to 41 % (**12c$_1$**). The same is true for the comparison between **12a$_2$**, **12b$_2$** and **12c$_2$**.

The electronic influences of the Schiff base ligand on the RCM-activities can easily be confirmed using the data of table 6.

An electron withdrawing substituent on the phenoxy fragment results in a slightly higher conversion (compare **12a$_2$** with **12a$_1$**, **12b$_2$** with **12b$_1$** and **12c$_2$** with **12c$_1$**).

Noteworthy is that the bulkiness is more advantageous than the electronic parameters on the obtained conversions.

Generally, Schiff base ligands bearing highly bulky groups and electronic-withdrawing substituents generate active Ru-catalysts for RCM reactions.

To explain the opposite influence of steric bulk on the RCM and ROMP performances of the Schiff base Ru-carbene complexes on should study the mechanism for the olefin metathesis which we proposed previously [13].

The major pathway involves a pre-equilibrium for imine dissociation and olefin binding. So, the fundamental difference with the Grubbs' catalyst **1** is the decoordination and coordination of "one-arm" of the bidentate Schiff base ligand (imine fragment) instead of a PCy$_3$ dissociation. Reasons that support this mechanism are already reported by our group [13].

Knowing the mechanism one can explain the observed activities for the Schiff base Ru-carbenes.

If electron-withdrawing substituents on the Schiff base ligand are present then the activity of the catalyst will be improved in both cases, ROMP and RCM. The stimulation of the "one-arm" decoordination is the reason for this improvement.

The opposite influence of bulky substituents on the Schiff base ligand on ROMP and RCM can be explained as follows:

i) In case of ROMP, the produced polymer is connected to the ruthenium via the Ru=C bond. If a bulky group is present on the imine fragment of the Schiff base ligand then the presence of the polymer chain will produce even more steric hindering. In this way the monomer will receive much more retention to reach the vacant site on ruthenium. In the absence of a bulky group on the imine fragment, only steric hindering of the generated polymer chain is present. Now the monomer can reach the active site in an easier way.

ii) In case of RCM, no retention due to the polymer chain is possible. Moreover, more bulkiness on the imine fragment will stabilize the reactive intermediate because the strength of [Ru]-N bonds is strongly effected by steric effects [13].

3. Conclusion

We succeed in synthesizing a new class Ru-based catalysts, Ru-arene Schiff base complexes. The ROMP activities are poor without activation but increases dramatically when TMSD is added. The RCM-yields are good to moderate. Using harsh conditions (85°C) the conversions increases dramatically. Tri- and tetrasubstituted malonate derivates can be prepared.

Moreover, the activities are influenced by steric properties of the Schiff base ligand.

A mechanism for the formation of the initial ruthenium carbene with and without TMSD was proposed.

Furthermore, polymers with moderate PDI-values (1.5 – 1.7) were obtained with the Schiff base Ru-carbenes. The RCM-yields are excellent to moderate. The observed results are explained using an earlier proposed mechanism for the Schiff base Ru-carbene systems. The activities are influenced by electronic and steric properties of the Schiff base ligands.

4. Acknowledgements

BDC is indebted to the IWT (Vlaams instituut voor de bevordering van het wetenschappelijk-technologisch onderzoek in de industrie) for a research grant. FV is indebted to the FWO (Fonds voor Wetenschappelijk Onderzoek-Vlaanderen) for a research grant. Financial support by The Research funds of Ghent University is gratefully acknowledged.

5   Experimental Section

5.1. GENERAL

All reactions and manipulations were performed under an argon atmosphere by using conventional Schlenck-tube techniques. Argon gas was dried by passage through P_2O_5 (Aldrich 97%). ^{1}H-NMR (500 MHz), ^{13}C-NMR (75.41 MHz) and ^{31}P-NMR (121.40 MHz) spectra were recorded on a Varian Unity 300 spectrometer. Chemical shifts are reported in ppm downfield from tetramethylsilane (TMS) with TMS employed as the internal solvent for proton spectra and 85% phosphoric acid employed as the internal solvent. The number- and weight average molecular weights (M_n and M_w) and polydispersity (M_w/M_n) of the polymers were determined by gel permeation chromatography (GPC) (CHCl$_3$, 25°C) using polystyrene standards. The GPC instrument used is a Shimadzu CLASS-VP^{TM} system equipped with three serial placed columns.

The ruthenium dimer [RuCl$_2$(p-cymene)]$_2$ was prepared according to literature procedures [14] and the structure and purity was checked with IR and ^{1}H-NMR and ^{13}C-NMR spectroscopy. Cyclooctene and norbornene were purchased from Aldrich and distilled from CaH$_2$ under nitrogen prior to use. The other norbornene derivatives were purchased from INEOS and used as received. Commercial grade solvents were dried and deoxygenated for at least 24 h over appropriate drying agents under nitrogen atmosphere and distilled prior to use. Unless otherwise noted, all other compounds were purchased from Aldrich Chemical Co., and used as received.

5.2. GENERAL PROCEDURE FOR PREPARATION OF THE SCHIFF BASE LIGANDS.

The Schiff base ligands **9** were prepared and purified using well-established procedures [7, 13, 15].

The condensations of salicylaldehydes with aliphatic amine derivatives were carried out with stirring in THF at reflux temperature for 2 hours. After cooling to room temperature, the viscous yellow oily condensation products were purified by silica gel chromatography (silica gel 60, 0.063-0.200 mm, Merck, a 5:1 benzene-tetrahydrofuran mixture was used as an eluant) and the desired salicylaldimine ligands were obtained in excellent yields. The condensations of salicylaldehydes with aromatic amine derivatives were carried out with stirring in ethanol at 80 °C for 2 hours. Upon cooling to 0 °C, a yellow solid precipitated from the reaction mixture. The solids were filtered, washed with cold ethanol and then dried in vacuo to afford the desired salicylaldimine ligand in quantitative yields.

The spectroscopic properties of the Schiff bases can be found in references 7, 13 and 15.

5.3. GENERAL PROCEDURE FOR PREPARATION OF THE SCHIFF BASE Ru-CARBENES $12a_1$-$12c_2$.

The Schiff base ruthenium carbenes $12a_1$-$12c_2$ were prepared and purified using well-established procedures[7, 13].

To a solution of the appropriate Schiff base ($9a_1$-c_2) in THF (10 ml), a solution of thallium ethoxide in THF (5 ml) was added dropwise at room temperature. Immediately after the addition, a pale yellow solid formed and the reaction mixture was stirred for 2 hour at room temperature. Filtration of the solid under an argon atmosphere gave the respective thallium salt in quantitative yield. The salt was immediately used in the next step without further purification. A solution of the appropriate thallium salt in THF (5 ml) was added to a solution of the first generation Grubbs catalyst [$RuCl_2(PCy_3)_2$=CHPh] in THF (5 ml). The reaction mixture was stirred at room temperaure for 4 hours. After evaporation of the solvent, the residue was dissolved in a minimal amount of benzene and cooled to 0 °C. The thalliumchloride was removed via filtration. After evaporation of the solvent, the solid residue was recrystallized from pentane (- 70 °C) to give the respective Schiff base ruthenium carbenes ($12a_1$-c_2) in good yield as a brown solid.

The spectroscopic properties of Schiff base Ruthenium carbenes $12a_1$-c_2 can be found in references 7 and 13.

5.4. GENERAL PROCEDURE FOR RING OPENING METATHESIS POLYMERIZATION USING 12a₁-c₂.

5.4. GENERAL PROCEDURE FOR RING OPENING METATHESIS POLYMERIZATION USING $12a_1$-c_2.

In a typical ROMP experiment 0.005 mmol of the catalyst solution in chlorobenzene (0.1004 M) was transferred into a 15 ml vessel followed by the addition the appropriate amount of monomer solution (800 equivalents for the norbornene derivatives) in chlorobenzene. The reaction mixture was then kept stirring at 70 °C for 4 hours. To stop the polymerization reaction, 2-3 ml of an ethylvinylether/BHT solution was added and the solution was stirred for 0.5 hour to make sure that the deactivation of the active species was completed. The solution was poured into methanol (50 ml containing 0.1% BHT) and the polymers were precipitated and dried in vacuum overnight.

5.5. GENERAL PROCEDURE FOR RING CLOSING METATHESIS REACTIONS USING $12a_1$-c_2.

All reactions were performed on the benchtop in air by weighing 5 mol% of the catalyst into a dry NMR tube and dissolving the solid in 1 ml C_6D_5Cl. A solution of the appropriate substrate (0.1 mmol) in C_6D_5Cl (1 ml) was added. The tube was then capped, wrapped with parafilm, and shaken for 4 hours at 55 °C. Product formation (all reaction products were unambiguously identified previously [13] and diene disappearance were monitored by integrating the allylic methylene peaks.

5.6. GENERAL PROCEDURE FOR PREPARATION OF THE SCHIFF BASE Ru-ARENES 11a-c.

To a solution of Schiff bases **9a-c** in THF (15 ml) was added dropwise a solution of thalliumethoxide in THF (5 ml) at room temperature. Immediately after the addition, a yellow solid formed and the reaction mixture was stirred for 2 h at room temperature. Filtration of the solid under a nitrogen atmosphere gave the thallium salts in quantitative yields. The salts were immediately used in the next step without further purification.

To a solution of $[RuCl_2(p\text{-cymene})]_2$ in THF (5 ml) was added a solution of the corresponding thalliumsalts in THF (5 ml). The reaction mixture was stirred at room temperature for 6 h. The thalliumchloride was removed via filtration. After evaporation of the solvent, the residu was dissolved in a minimal amount of toluene and cooled to 0 °C. The obtained crystals were then washed with cold toluene (3x10ml) and dried. The Schiff base ruthenium arene complexes **9a-c** appeared as red-brown solids. The spectroscopic properties of Schiff base Ruthenium arenes **9a-c** can be found in reference 15.

5.7. GENERAL PROCEDURE FOR RING OPENING METATHESIS POLYMERIZATION USING 11a-c.

In a typical ROMP experiment 0.005 mmol of the catalyst solution in toluene was transferred into a 15 ml vessel followed by the addition of a catalytic amount (2 equiv.) of trimethylsilyldiazomethane (TMSD) diluted in 1 ml toluene via a precision syringe over 0.5 h to allow the formation of the initiating metal carbene species. Then the right amount of monomer solution in toluene (800 equivalents for norbornene) was added and the reaction mixture was then kept stirring at 85 °C. To stop the polymerisation reaction, 2-3 ml of an ethylvinylether/BHT solution is added and the solution is stirred till the deactivation of the active species is completed. The solution is poured into 50 ml methanol (containing 0.1% BHT) and the polymers are precipitated and dried in vacuum overnight.

5.8. GENERAL PROCEDURE FOR RING CLOSING METATHESIS REACTIONS USING **11a-c.**

All reactions were performed on the bench top in air by weighing 5 mol% of the catalyst into a dry 10 ml vessel and suspending the solid in 2 ml toluene. Then addition of a catalytic amount (2 equiv.) of trimethylsilyldiazomethane (TMSD) diluted in 1 ml toluene via a precision syringe over 0.5 h to allow the formation of the initiating metal carbene species. A solution of the appropriate substrate (0.1 mmol) in toluene (2 ml) was added, together with the internal standard dodecane. The reaction mixture was stirred for 1 hour at 70 °C. or 17 hours at 85 °C (harsh conditions). Product formation (all reaction products were unambiguously identified previously [15] and diene disappearance were monitored by GC analysis and confirmed in reproducibility experiments by ^{1}H-NMR spectroscopy through integration of the allylic methylene peaks (here the solvent was deuterated toluene and the internal standard 1,3,5-mesitylene). GC analysis of the reaction mixture also ruled out the formation of cycloisomers, oligomers or telomers.

6. References

[1]   (a) Nitta, H., Yu, D., Kudo, M., Mori, A. and Inoue, S. (1992) *J. Am. Chem. Soc,* **114**, 7969; (b) De Clercq, B. and Verpoort, F. (2002) *Macromolecules,* **35**, 8943.
(c) De Clercq B, Verpoort F. (2002) *Catal Lett* **83**, 9;(d) Opstal T, Verpoort F. (2002) *Synlett* (6) 935; (e) El-Hendawy, A.M., Alkubaisi, A.H., El-Ghany El-Kourashy, A., Shanab, M.M. (1993) *Polyhedron,* **12**, 2343.

[2]   (a) Platz, A.(1989) *Mod. Synth. Methods* 5, 199. (b) Aratani, T. (1985) *Pure Appl. Chem.,* 57, 1839.

[3]   (a) Jacobsen, E.N., Deng, L., Furukawa, Y., Martinez, E. (1994) *Pure Appl. Chem.,* **50**, 4323.(b) Brandes, B.D., Jacobsen, E.N. (1995) *Tet. Lett.,* **36**, 5123.(c) Brandes, B.D., Jacobsen, E.N. (1994) *J. Org. Chem.* **59**, 4378.

[4]   (a) Kitamura, M., Suga, S., Noyori, R. (1986) *J. Am. Chem. Soc.,* **108**, 6071.
(b) Kitamura, M., Okada, S., Suga, S., Noyori, R. (1989) *J. Am. Chem. Soc.,* **111**, 4028.

[5]   (a) De Clercq B, Verpoort F (2001) *Tet. Lett.* **42**, 8959.(b) De Clercq B, Verpoort (2002) *Tet. Lett.* **43**, 4687. (c) Opstal T., Verpoort F. (2002) *Tet. Lett.* **43**, 9259.

[6]   (a) Fu, G.C., Nguyen, S.T., Grubbs, R.H. (1993) *J. Am. Chem. Soc.,* **115**, 9856; Clark, T.D., Ghadiri, M.R. (1995) *J. Am. Chem. Soc.,* **117**, 12364; Diver, S.T., Schreiber, S.L. (1997) *J. Am. Chem. Soc.,* **119**, 5106.
(b) Dias, E.L., Nguyen, S.T., Grubbs, R.H. (1993) *J. Am. Chem. Soc.* **119**, 3887   (c) Weskamp, T., Kohl, F. J., Hieringer, W., Gleich, D., Herrmann, W.A. (1999) *Angew. Chem. Int. Ed.,* **38**, 2416.
(d) De Clercq B, Verpoort F (2002) *Adv Synth Catal.,* **34**, 639.
(e) Heck, M.-P., Baylon, C., Nolan, S.P., Mioskowski, C. (2001) *Org. Lett.,* **3**, 1989; Fürstner, A., Thiel, O.R., Ackermann, L., Schanz, H.-J., Nolan, S.P. (2000) *J. Org. Chem.,* 65, 2204.

[7]   Chang, S., Jones II, L., Wang, C., Henling, L. M. and Grubbs, R. H. (1998) *Organometallics*, **17**, 3460.

[8]   (a) Demonceau, A., Noels, A.F., Saive, A.J. (1992) *J. Mol. Cat.* **76**, 123; (b) Demonceau, A., Stumpf, A.W., Saive, A.J., Noels, A.F. (1997) *Macromolecules*, **30**, 3127; (c) Hansen, H.D., Nelson, J.H. (2000) *Organometallics*, **19**, 4740.

[9]   Bennett M.A., Smith, K.J. (1974) *J. Chem. Soc., Dalton Trans.*, 233.

[10]   Opstal T, Melis K, Verpoort F. (2001) *Catal. Lett.* **74**, 155.

[11]   (a) Ivin, K.J., Mol, J.C.(1996) *Olefin metathesis and metathesis polymerisation*, Academic Press, Cornwall. 2416. (b) Scholl, M., Trnka, T.M., Morgan, J.P., Grubbs, R.H. (1999) *Tetrahedron Lett.*, **40**, 2247.

[12]   Snapper, M.L. (1999) *J. Org. Chem.*, **64** (N° 2), 344.

[13]   De Clercq B, Verpoort F (2002) *Adv Synth Catal* **34**, 639.

[14]   Bennet, M.A., Smith, A.K. (1974) *J. Chem. Soc. Dalton Trans.*, 233.

[15]   De Clercq B, Verpoort F (2002) *J. Mol. Catal. A: Chemical*, **180**, 67.

CATALYTIC ACTIVITY OF W–Sn AND Mo–Sn BIMETALLIC COMPOUNDS IN METATHESIS AND RELATED REACTIONS

T. SZYMAŃSKA-BUZAR
Faculty of Chemistry, University of Wrocław,
ul. F. Joliot-Curie 14, 50-383 Wrocław, Poland

Abstract: This review summarises the most interesting results concerning the application of W–Sn and Mo–Sn heterobimetallic compounds as initiators for the metathesis polymerisation of alkynes and the ring-opening metathesis polymerisation of cyclic olefins (ROMP). Special attention is given to compounds formed in the reaction of tungsten or molybdenum complexes and an organic substrate, *viz.* alkyne or cyclic olefin.

1. Introduction

The development of soluble organometallic compounds as catalysts for the metathesis and related reactions of olefins remains an area of intense research activity [1–3]. Over the past 20 years, a diverse array of tungsten, molybdenum and ruthenium complexes have been found to catalyse the metathesis of olefins [1]. To date, the most widely used are the well-defined alkylidene complexes based on molybdenum (Schrock's catalysts [2]) and ruthenium (Grubbs' catalysts [3]). However, many others transition metal compounds can be used as initiators for metathesis reactions. There are multi-component and one-component systems. In a multi-component system a transition metal complex usually gives an alkylidene species in reaction with an organometallic compound of a main group metal. One-component initiators transform to alkylidene species in reaction with organic substrates. Such initiators include tungsten(II) and molybdenum(II) seven-coordinate compounds developed as metathesis initiators by Bencze *et al.* in 1980 [4–7]. In connection with research in this area our group has also been developing tungsten(II) and molybdenum(II) seven-coordinate compounds suitable for such applications. During the past few years, we have been investigating a new and rare kind of heterobimetallic compounds of the type $[MCl(SnCl_3)(CO)_3(NCR)_2]$, M = W, Mo; R = Me, Et, which under mild conditions initiate the metathesis polymerisation of terminal alkynes and the ring opening metathesis polymerisation of cyclic olefins such as norbornene and norbornadiene [8–14].

Seven-coordinate bis(nitrile) W–Sn and Mo–Sn compounds are stable enough in an inert atmosphere and can be easily isolated in pure crystalline form. The crystal structure of the bis(propinitrile) Mo-Sn complex [15] is similar to that observed for the analogous tungsten compound, with a direct metal-metal bond and one chlorine atom occupying a bridging position between the two metal atoms (Fig. 1, [9]).

Y. Imamoglu and L. Bencze (eds.), Novel Metathesis Chemistry: Well-Defined Initiator Systems for Specialty Chemical Synthesis, Tailored Polymers and Advanced Material Applications, 121–129.
© 2003 *Kluwer Academic Publishers. Printed in the Netherlands.*

122

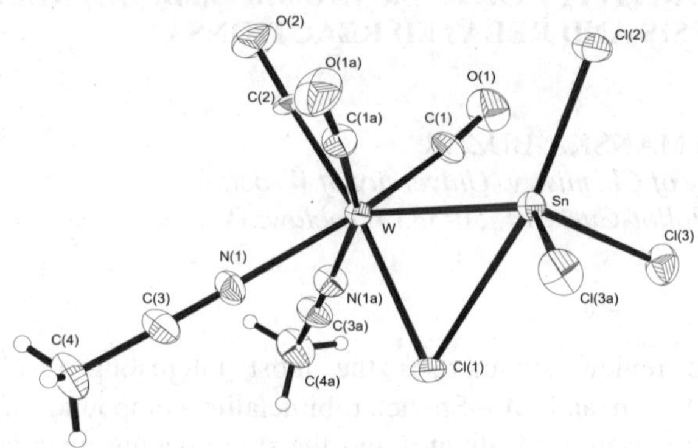

Figure 1. Crystal structure of [WCl(SnCl₃)(CO)₃(NCMe)₂] [9]

All of the investigated heterobimetallic W–Sn and Mo–Sn compounds initiate the metathesis polymerisation of terminal alkynes and the ring-opening metathesis polymerisation of cyclic olefins. The first step in all catalytic reactions is the coordination of the organic substrate to the transition metal and the formation of a complex. Sometimes it was possible to isolate the products in crystalline form but frequently such adducts were observed only *in situ* by NMR investigations.

2. Coordination of Unsaturated Hydrocarbons to the Metal Atom

2.1. REACTION OF [MCl(SnCl₃)(CO)₃(NCR)₂], M = W, Mo; R = Me, Et, COMPOUNDS WITH ALKYNES

The most important role of nitriles in coordination chemistry stems from their ability to act as labile ligands. This reactivity has been widely applied in numerous ligand exchange processes. As far as the chemistry of tungsten(II) and molybdenum(II) is concerned, the ready accessibility of the complexes [MCl(SnCl₃)(CO)₃(NCR)₂], M = W [9], Mo [10,15]; R = Me, Et, has allowed them to be used as useful precursors of a wide series of derivatives, which have prompted the development of a chemistry of both stoichiometric and catalytic transformations.

In the reaction of bis(nitrile) compounds of the type [MXX'(CO)₃(NCR)₂], M = W, Mo; X = X' = Cl, Br, I; X = Cl, X' = SnCl₃, GeCl₃; R = Me, Et, with alkynes, the substitution of two carbonyl group and one nitrile ligand by two molecules of alkyne has been observed [9–11,16–24]. As was first shown by Templeton *et al.* and then by Baker *et al.*, all bis(alkyne) complexes of tungsten(II) and molybdenum(II) contain two alkyne ligands in mutually *cis* positions [16–23]. However, heterobimetallic W–Sn and Mo–Sn alkyne compounds are very unstable and have been observed only *in situ* by NMR investigations [9–11,24].

The reaction of terminal alkynes with a metal complex can be easily followed using ¹H-NMR due to characteristic acetylenic proton signals of coordinated alkyne at about 10 ppm. In the reaction of the W–Sn bis(nitrile) compound and 2 equiv of *tert*-butylacetylene we have been able to observe by ¹H-NMR the formation of a bis(alkyne) complex due to two equal-intensity signals of acetylenic protons at $\delta_H = 11.27$ and 11.07 ($^2J_{WH} = 9.3$ Hz) [11,24]. The analogous bis(alkyne) complex of molybdenum(II) has a signal for an acetylenic proton at $\delta_H = 10.37$ and 10.24 [11]. After the addition of an excess of *tert*-butylacetylene to the solution of the initially formed W–Sn bis(alkyne) compounds in the NMR tube, other proton signals appear. Two signals at $\delta_H = 12.73$ and 12.13 ($^2J_{WH} = 12$ Hz) can be attributed to the alkylidene ligands coordinated to tungsten. In a similar region we also observed two signals ($\delta_H = 12.98$ and 11.89) assigned to the alkylidene ligands coordinated to molybdenum. The intensity of these proton signals never achieved a high level. In this type of reaction we also observed other signals: at $\delta_H = 6.19$ and 5.98 due to poly-(*tert*-butylacetylene) containing respectively *cis* and *trans* configurations of the olefinic protons and at $\delta_H = 7.3$ ppm due to the alkyne cyclotrimerisation product 1,3,5-tri(*tert*-butyl)benzene. The intensity of the latter signal increases considerably only in reactions initiated by the molybdenum compound [11].

The heterobimetallic W–Sn and Mo–Sn compounds react rapidly at room temperature with an excess of the terminal alkyne, such as phenylacetylene or *tert*-butylacetylene, to give an unsaturated polymer in high yield (Scheme 1) [8–11].

$$n \ RC\!\!\equiv\!\!CH \ \xrightarrow{\ [M\text{-}Sn]\ } \ -\!\!(RC\!=\!CH)\!\!\overline{)_n}$$

Scheme 1. Polymerisation of terminal alkynes by W–Sn and Mo–Sn compounds

The poly-*tert*-butylacetylene is air-stable and totally soluble in halogeneted hydrocarbons. The weight-average molecular weights of polyalkynes obtained in the presence of W–Sn and Mo–Sn compounds are high, up to 8×10^5 gmol^{-1} [11].

In the residue obtained after the separation of the polymer, dimers and cyclotrimers of the alkyne have also been detected, which indicates that the formation of an alkylidene species initiating the polymerisation involves the transformation of the two mutually *cis* alkynes ligands to a metallacycle. The insertion of an alkyne into the metal-carbon bond of the initially formed metallacyclopentadiene moiety leads to the formation of metallacycloheptatriene. The metallacycle formed with three or four molecules of alkyne can then rearrange to an alkylidene ligand initiating the increase of the polymer chain. There is competition between cyclotrimer and alkylidene ligand formation. Molybdenum compounds lead to the formation of arenes in higher yields than tungsten compounds [11].

2.2. REACTION OF [MCl(SnCl$_3$)(CO)$_3$(NCR)$_2$], M = W, Mo; R = Me, Et, COMPOUNDS WITH CYCLIC OLEFINS

Heterobimetallic W–Sn and Mo–Sn compounds are also one-component initiators in the ring-opening metathesis polymerisation of cyclic olefins such as norbornene and norbornadiene. The first step in this catalytic reaction is the coordination of olefin to metal. The substitution of two acetonitrile ligands by a norbornadiene molecule at the tungsten atom can be followed by means of proton NMR spectra due to the appearance of proton signals of the coordinated norbornadiene at $\delta_H = 5.04$ (=CH), 3.97 (CH), 3.71(=CH), 1.50, 1.48 (CH$_2$) (Fig. 2).

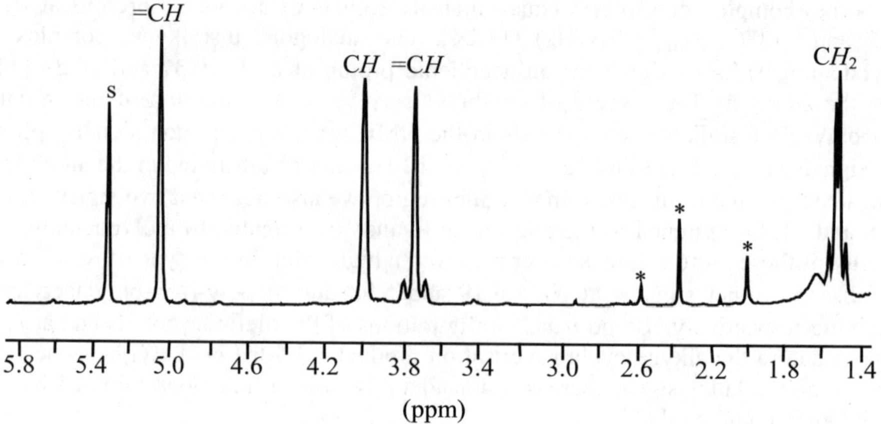

Figure 2. ^{1}H-NMR spectrum of [WCl(SnCl$_3$)(CO)$_3$(η^4-C$_7$H$_8$)] [25] in CD$_2$Cl$_2$ (s) solution (asterisks denote impurities)

The coordination of the norbornadiene ligand in the tungsten complex [25] and in the analogous molybdenum complex [12] is arranged in two types. One of the olefinic bonds is *trans* to CO and the other *trans* to the chloride ligand, with different olefinic carbon-molybdenum distances: one, longer by 10 pm, *trans* to the CO ligand and a shorter c

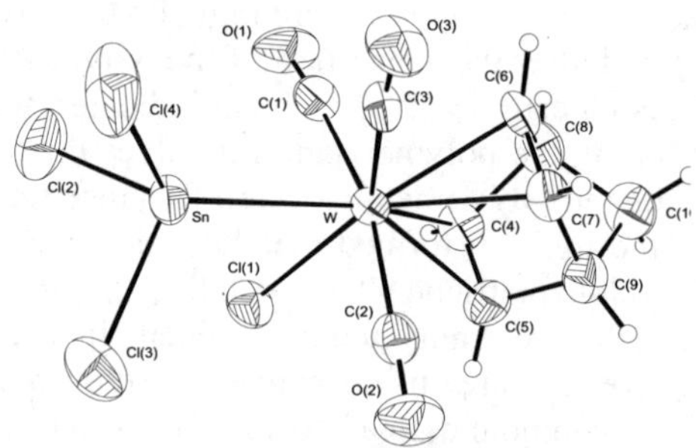

Figure 3. Structure of [WCl(SnCl$_3$)(CO)$_3$(η^4-C$_7$H$_8$)] [25]

However, during the reaction of the bis(nitrile) Mo–Sn compound with norbornadiene, the formation of at least four complexes containing the norbornadiene ligand has been observed [26]. The first complex to be formed contains three carbonyl ligands, the second two, and the third one carbonyl group. In acetonitrile solution, a redistribution of the anionic ligands between the two molybdenum atoms occurs to give a bis(trichlorostannyl) and a bis(chloride) complex. X-ray structure investigations revealed an unusual coordination of a norbornadiene ligand to a molybdenum centre in a complex containing two mutually *cis* carbonyl ligands. Four different olefinic carbon–molybdenum distances were observed: two longer ones, differing from each other by 5 pm, and two shorter by 10 pm, also differing from each other by 5 pm (Fig. 4 [26]).

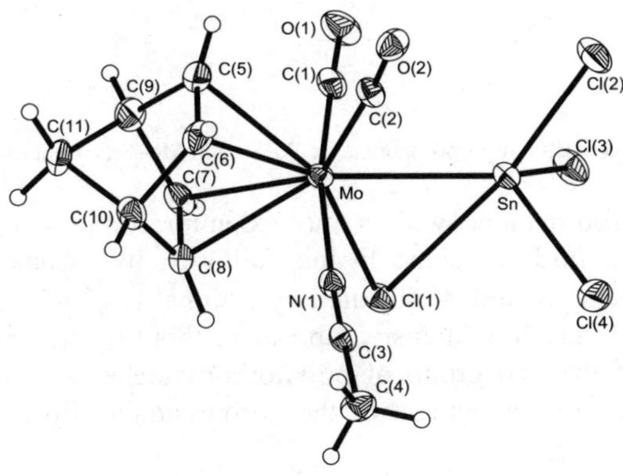

Figure 4. Structure of $[MoCl(SnCl_3)(CO)_2(\eta^4\text{-}C_7H_8)(NCMe)]$ [26]

In the NMR spectra of the latter compound four olefinic proton signals: δ_H = 5.43, 4.84, 4.00 and 3.91, and four olefinic carbon signals: δ_C = 96.00, 94.07, 53.86, 52.31, have been detected (Fig. 5 [26]).

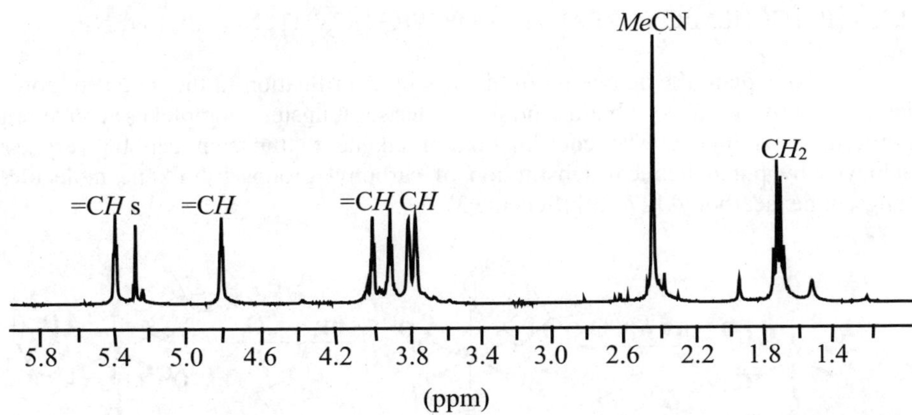

Figure 5. ^{1}H-NMR spectrum of $[MoCl(SnCl_3)(CO)_2(\eta^4\text{-}C_7H_8)(NCMe)]$ [26] in CD_2Cl_2 (s) solution

The NMR and X-ray data suggest different lability of the two olefinic bonds of the norbornadiene ligand in heterobimetallic W–Sn and Mo–Sn compounds.

The coordination of olefin to the tungsten or molybdenum atom is accompanied by ring-opening polymerisation and the formation of the unsaturated polymer with olefinic proton signals in the region of 5.6–5.2 ppm.

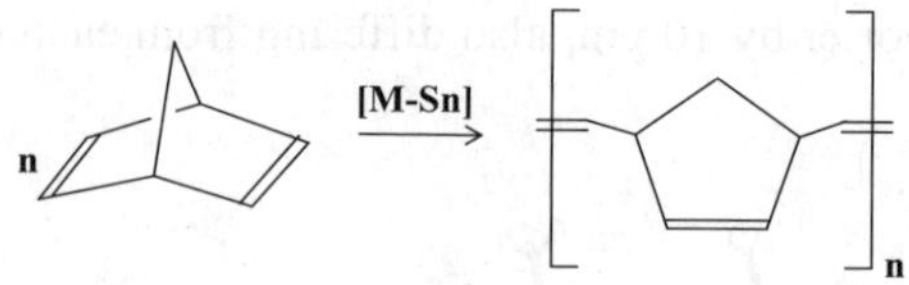

Scheme 2. Polymerisation of norbornadiene by W–Sn and Mo–Sn compounds

In the reaction of norbornadiene with a molybdenum complex containing two *cis* carbonyl groups and a norbornadiene ligand followed by means of ^{1}H-NMR, the formation of an alkylidene ligand is indicated by a signal at $\delta_H = 10.5$. Along with that signal, other signals, of very low intensity, appear at about 6 and 3 ppm. These can be assigned to protons of the end group of polynorbornadiene, *e.g.* a cyclopentadienyl moiety formed after the rearrangement of the norbornadiene ligand to an alkylidene [26].

During the reaction of norbornadiene with a tungsten compound containing a norbornadiene ligand monitored by means of ^{1}H-NMR we observed that the decay of signals due to the norbornadiene coordinated to tungsten is accompanied by the increase of the signals due to polynorbornadiene. The simultaneously increasing signals at about 6 and 3 ppm were assigned to protons of the end group of polynorbornadiene, formed after the rearrangement of the coordinated norbornadiene.

2.3. PHOTOCHEMICAL REACTION OF W(CO)$_6$ WITH NORBORNENE

To explain the activation of alkenes by coordination to the tungsten atom, we have recently tried to obtain and characterise tungsten complexes containing a norbornene as ligand. The coordination of alkene to tungsten can be very easily achieved by photochemical substitution of carbonyl groups by alkene molecules in tungsten hexacarbonyl [27–29] (Scheme 3).

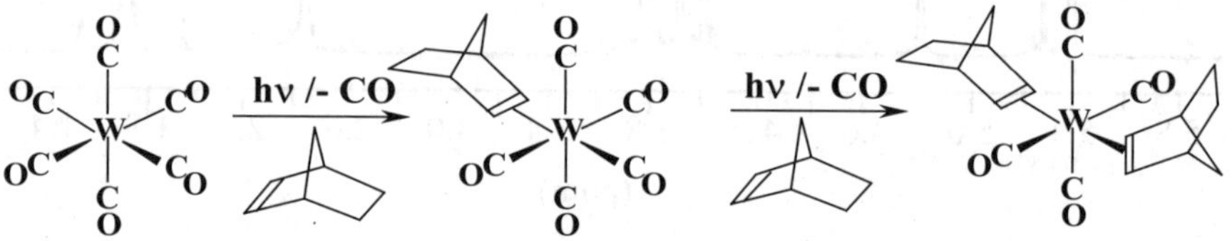

Scheme 3. Photochemical substitution of carbonyl groups by norbornene in W(CO)$_6$

So far we have been able to obtain and characterise a very stable compound of tungsten(0) with two mutually *trans* norbornene molecules, namely *trans*-[W(CO)$_4$((η^2-C$_7$H$_{10}$)$_2$]. The unstable pentacarbonylnorbornene complex [W(CO)$_5$(η^2-C$_7$H$_{10}$)] decomposes to give quantitatively polynorbornene and W(CO)$_6$. The oxidative addition of tin tetrachloride to a *trans*-bis(norbornene) complex leads to the formation of another very labile norbornene tungsten(II) complex. Its characterisation has been possible only by means of low-temperature NMR studies (Table 1).

TABLE 1. NMR data for free nornborene and norbornene complexes of tungsten in $CDCl_3$ solution

Assign of carbon atom			$[W(CO)_5(\eta^2\text{-}C_7H_{10})]$ at 298 K		$trans\text{-}[W(CO)_4(\eta^2\text{-}C_7H_{10})_2]$ at 298 K		"$[W^{(2+)}\text{-}(\eta^2\text{-}C_7H_{10})]$" at 258 K	
	δ_H	δ_C	δ_H	δ_C	δ_H	δ_C	δ_H	δ_C
2,3	5.97	135.2	4.81	88.03	3.27	60.93	5.05	89.68
1,4	2.89	41.8	2.92	42.00	2.71	41.34	3.11	42.76
5,6	1.57	24.6	1.77	26.12	1.77	28.88	2.10	25.24
	0.94		1.09		1.25		1.47	
7	1.32	48.5	0.98	37.06	0.99	34.28	0.88	38.12
	1.07		0.71		0.59		-0.72	

The coordination shift of the olefinic carbon and the olefinic proton depends strongly on the oxidation state and the electronic configuration of the transition metal. For a very stable complex of tungsten(0) with a d^6 configuration and mutually *trans* alkene ligands, the coordination shift is much higher than for a complex of tungsten(II) with a d^4 configuration. It should be noted that only labile complexes containing loosely coordinated olefinic ligands rearrange to species initiating the catalytic reaction.

3.   Conclusions

The heterobimetallic tungsten-tin and molybdenum-tin complexes which we have developed are easy to synthesise and isolate in pure crystalline form. These complexes readily react with alkynes or cyclic olefins giving the well-characterised transition metal compounds under mild conditions. In the presence of an excess of an organic substrate such compounds initiate the catalytic process.

4.   Acknowledgements

The author's thanks go to Prof. Tadeusz Głowiak, who resolved the crystal structure of all the compounds presented in the paper, to Izabela Czeluśniak, who studied the catalytic activity of seven-coordinate compounds, and to Marcin Górski, who synthesised and characterised the norbornene complexes of tungsten.

5.   References

1.   Ivin, K.J., Mol, J.C. (1997) Olefin Metathesis and Metathesis Polymerization, Academic Press, San Diego CA, p.12-49.

2.   Schrock, R.R., (1990) Living Ring-Opening Metathesis Polymerization Catalyzed by Well-Characterized Transition-Metal Alkylidene Complexes, *Acc. Chem. Res.* **23**, 158-165.

3.   Trnka, T.M., Grubbs, R.H. (2001) The Development of $L_2X_2Ru=CHR$ Olefin Metathesis Catalysts: An Organometallic Success Story, *Acc. Chem. Res.* **34**, 18-29.

4.   Bencze, L., Ivin, K.J., Rooney, J. (1980) The nature of the preferred chain-carrying metallacarbene intermediate in metathesis reactions involving alk-1-enes, *J. C. S. Chem. Comm.* 834–835.

5.   Bencze, L., Kraut-Vass, A. (1984) Ring-opening polymerization of norbornene catalysed by $W(CO)_3Cl_2(AsPh_3)_2$ in the presence of other olefins, *J. Organomet. Chem.* **270**, 211–220.

6. Bencze, L., Kraut-Vass, A. Prókai, L. (1985) Mechanism of Initiation of the Metathesis of Norbornene using $W(CO)_3Cl_2(AsPh_3)_2$ as Catalyst, *J. Chem. Soc. Chem. Comm.*, **13**, 911–912.

7. Bencze, L., Szalai, G., Hamilton, J.G., Rooney, J.J. (1997) Composition vs catalytic properties of $M(CO)_3X_2(L)_2$ (M. = Mo or W; X = Cl, Br or I; L = PPh_3, $AsPh_3$ or $SbPh_3$) initiators in romp of norbornene, *J .Mol. Catal. A: Chem.*, **115**, 193–197.

8. Szymańska-Buzar, T. (1997) Tungsten(II) chlorocarbonyls as alkene metathesis, arene alkylation and alkyne polymerization catalysts, *J. Mol. Catal., A: Chem.*, **123**, 113–122.

9. Szymańska-Buzar, T., Głowiak, T. (1998) Spectroscopic properties, structure and reactivity of $[WCl(SnCl_3)(CO)_3(NCMe)_2]$, *Polyhedron*, **17**, 3419–3426.

10. Szymańska-Buzar, T., Głowiak, T. (1999) Synthesis and reactivity of Mo-Sn compounds. X-ray crystal structure of a novel $[Mo(SnCl_3)_2(CO)_2(NCEt)_3]$, *J. Organomet. Chem.*, **575**, 98–107.

11. Szymańska-Buzar, T., Czeluśniak, I. (2000) Polymerization of *tert*-butylacetylene by seven-coordinate heterobimetallic tungsten(II) and molybdenum(II) compounds, *J. Mol. Catal. A: Chem.*, **160**, 133–143.

12. Szymańska-Buzar, T., Czeluśniak, I., Głowiak, T. (2001), The initiation of ring-opening metathesis polymerization of norbornadiene by seven-coordinate molybdenum(II) compounds. X-ray crystal structure of $[Mo(\mu-Cl)(SnCl_3(CO)_3(\eta^4-NBD)]$, *J. Organomet. Chem.*, **640**, 72–78.

13. Czeluśniak, I., Szymańska-Buzar, T., Kenwright, A., and Khosravi, E., (2002) Ring-opening metathesis polymerization of 5,6-bis(chloromethyl)-norbornene by tungsten(II) and molybdenum(II) complexes, *Catal. Lett.*, **81**, 157-161.

14. Czeluśniak, I., Szymańska-Buzar, T., (2002) Ring-opening metathesis polymerization of nornornene and norbornadiene by tungsten(II) and molybdenum(II) complexes, *J. Mol. Catal. A: Chem.*, **190**, 131–143.

15. Szymańska-Buzar, T., Głowiak, T., Czeluśniak, I. (2000) A novel Mo-Sn chlorocarbonyl compound. X-ray crystal structure of $[(CO)_4Mo(\mu-Cl)_3Mo(SnCl_3)(CO)_3]$ and $[MoCl(SnCl_3)(CO)_3(NCEt)_2]$, *Inorg. Chem. Comm.* **3**, 102–104.

16. Templeton, J.L and Herrick, R.S. (1982) Synthesis, Spectral Properties, and Dynamic Solution Behavior of Bis(alkyne)bis(dithiocarbamato)molybdenum(II) Complexes, *Organometallics*, **1**, 842-851.

17. Amstrong, E.M., Baker, P.K., and Drew, M.G.B. (1987) Reaction of $[WI_2(CO)(NCMe)(\eta^2-RC_2R)_2]$ (R = Me, Ph) with carbon monoxide to give either $[WI_2(CO)_2(\eta^2-MeC_2Me)_2]$ or $[W(\mu-I)_2(CO)_2(\eta^2-PhC_2Ph)_2]_2$. Crystal structure of $[WI_2(CO)_2(\eta^2-MeC_2Me)_2]$ *J. Organomet. Chem.*, **336**, 377–383.

18. Baker, P.K., Amstrong, E.M., and Drew, M.G.B. (1988) Reaction of $[WI_2(CO)(NCMe)(\eta^2-RC_2R)_2]$ (R = Me, Ph) with Pyridines and Related ligands. X-ray Crystal Structure of $[WI(CO)(bpy)(\eta^2-MeC_2Me)_2][BPh_4]$, *Inorg. Chem.*, **27**, 2287-2291.

19. Baker, P.K., (1996) The Organometallic Chemistry of Halocarbonyl Complexes of Molybdenum(II) and Tungsten(II), *Adv. Organomet. Chem.*, **40**, 45-115.

20. Baker, P.K., (1998) Seven-coordinate halocarbonyl complexes of the type $[MXY(CO)_3(NCMe)_2]$ (M = Mo, W; X,Y = halide, pseudo halide) as highly versatile starting materials, *Chem. Soc. Rev.*, **27**, 125-131.

21. Baker, P.K., Drew, M.G.B., Meehan, M.M. (2001) Synthesis, molecular structure and catalytic activity of the bis(phenylacetylene) complex [$WI_2(CO)(NCMe)(\eta^2-HC_2Ph)_2]$, *Inorg. Chem. Commun.*, **2**, 124-125.

22. Al-Jahdali, M., Baker, P.K., (2001) Synthesis and reaction of the bis(3-hexyne) complex $[WCl_2(CO)(NCMe)(\eta^2-EtC_2Et)_2]$, *J. Organomet. Chem.*, **628**, 91–98.

23. Baker, P.K., Drew, M.G.B., Moore, D.S. (2002) 3,3-Dimethyl-1-butyne complexes of tungsten(II). Crystal structure of $[WI_2(CO)(NCPh)(\eta^2-HC_2Bu^t)_2]$ and $[WI_2(CO)\{P(OR)_3\}_2(\eta^2-HC_2Bu^t)_2]$ (R = Me and Et), *J. Organomet. Chem.*, **658**, 77–87.

24. Czeluśniak, I., Głowiak, T., Szymańska-Buzar, T. (2000) Reactivity of $[WCl(SnCl_3)(CO)_3(NCMe)_2]$ towards $^tBuC\equiv CH$. X-ray crystal structure of $[WCl(CO)(^tBuC\equiv CH)(NCMe)(PPh_3)_2]_2[SnCl_6]$, *Inorg.Chem.Comm.* **3**, 285–288.

25. Szymańska-Buzar, T., Głowiak, T. (1997) Solid state structure and spectroscopic properties of $[WCl(SnCl_3)(CO)_3(\eta^4-NBD)]$, NBD = norbornadiene, *Polyhedron* **16**, 1599–1603.

26. Szymańska-Buzar, T., Głowiak, T., Czeluśniak, I. (2002) Reactivity of $[Mo(\mu-Cl(SnCl_3)(CO)_3(NCMe)_2]$ towards norbornadiene. X-ray crystal structure of $[Mo(\mu-Cl)(SnCl_3(CO)_2(\eta^4-C_7H_8)(NCMe)]$, *Polyhedron* **21**, 2505–2513.

27. Szymańska-Buzar, T., Górski, M. (manuscript in preparation).

28. Grevels, F.-W., Jacke, J., Klotzbücher, W.E., Mark, F. Skibbe, V., and Schaffner, Angermund, K., Krüger, C., and Lehmann, C.W., Özkar, S., (1999) Sequential Photosubstitution of Carbon Monoxide

by (*E*)-Cyclooctene in Hexacarbonyltungsten: Structural Aspects, Multistep Photokinetics, and Quantum Yields, *Organometallics*, **18**, 3278-3293.

29. Szymańska-Buzar, T., Kern, K., (1999) NMR studies of the structure, isomerism and photochemical reactions of *trans*-$[W(CO)_4((\eta^2\text{-alkene})_2]$ complexes, *J. Organomet. Chem.*, **592**, 214-224.

POLYMERIZATIONS CATALYZED WITH RHODIUM COMPLEXES

J. VOHLÍDAL[a], M. PACOVSKÁ[a], J. SEDLÁČEK[a], J. SVOBODA[a], J. ZEDNÍK[a], H. BALCAR[b]

[a] Department of Physical and Macromolecular Chemistry, Charles University, Laboratory of Specialty Polymers*, Albertov 2030, CZ-128 40, Prague 2, Czech Republic

[b] J. Heyrovský Institute of Physical Chemistry, Academy of Sciences of the Czech Republic, CZ-182 23 Prague 8, Czech Republic

Abstract

In the last decade, rhodium complexes are increasingly used as catalysts for a preparation of specialty polymers of diverse functionality, since a use of them brings about important advantages. Rh-catalysts (i) show unusually high tolerance to the reaction surroundings as well as to functional groups of reactants and products, (ii) they often show a precise control of the configurational structure of formed macromolecules (particularly those of polyvinylenes), (iii) they can be transformed to the living polymerization systems, (iv) they can be anchored on various inorganic and organic supports to give effective heterogeneous catalysts, and (v) they can catalyze reactions in the ionic liquid systems. Rhodium complexes are prevailingly used for a preparation of stereoregular (head-to-tail, cis-transoid) polymers of monosubstituted acetylenes, molecules of which easily adopt the helical conformation in the solid state, some of them even in solutions (e.g., molecules of poly(propiolate)s). In addition to it, Rh-complexes are nowadays used as catalysts of (i) atom transfer radical polymerization, (ii) polymerization of arylallenes taking place exclusively via 2,3-addition mode and copolymerization of allenes with carbon monooxide to give alternating copolymers, (iii) cross-dehydrocoupling polymerization of dihydrosilanes and bis(hydrosilane)s with diols, disilanols and dithiols, (iv) silylative coupling polymerization of bis(vinylsilane)s, (v) hydrosilylative addition copolymerization of bis(silane)s and diethynyl monomers, and (vi) ring-opening polymerization of silaferrocenophanes and 1,3-disilacyclobutanes. In spite of a high synthesis potential, a practical application of these expensive catalysts in a medium-to-large scale production of polymers depends on successful solving of questions related to their effective and reliable recycling.

1. Homogeneous polymerization of acetylenes with Rh-based catalysts

Rhodium complexes are mainly known as effective catalysts for hydrocarbonylation of olefins (industry-scale process) and hydrogenation of unsaturated carbon-to-carbon bonds under mild conditions [1]. Besides, they also catalyze reactions such as isomerization, hydrosilylation and silylation [1-3], Kharash reaction [4], methanation

* Supported by the Ministry of Education, Youth and Sport of Czech Republic, project MSMT 113100001.

Y. Imamoglu and L. Bencze (eds.), Novel Metathesis Chemistry: Well-Defined Initiator Systems for Specialty Chemical Synthesis, Tailored Polymers and Advanced Material Applications, 131–154.
© 2003 Kluwer Academic Publishers. Printed in the Netherlands.

132

of carbon oxides [5] and transformations of diazocompounds [6]. A use of Rh complexes in polymer chemistry had been for a long time limited mainly to π-allyl and carbonyl Rh complexes which were used as catalysts for polymerization of dienes and allenes [7]. A breakthrough in this field was achieved in 1986 by Furlani et al. [8-10] who discovered that Rh^I complexes such as $[\{Rh(cod)\}_2(\mu\text{-}Cl)_2]$, $[Rh(cod)(bipy)]PF_6$ and $[Rh(nbd)(bipy)]PF_6$ (where cod stands for $\eta^2{:}\eta^2$-cycloocta-1,5-diene, nbd for $\eta^2{:}\eta^2$-norborna-1,4-diene and bipy for 2,2'-bipyridyl) polymerize phenylacetylene (PA) to a highly stereoregular, cis-transoid poly(PA). Various Rh(diene) complexes, mainly dinuclear Rh(nbd) and Rh(cod) complexes with μ-chloro and μ-methoxo ligands, have been used to polymerize substituted acetylenes to stereoregular cis-transoid polyvinylenes in next few years [10-15].

1.1. PRODUCT-SELECTIVITY OF RHODIUM CATALYSTS

The ability of Rh(diene) complexes to control the molecular structure of formed polyacetylenes is unique. The overall isomerism of polymers derived from mono-substituted (or non-symmetrically substituted) acetylenes is namely very complex involving the configurational and head-tail (H-T) isomerism. Because of conjugation,

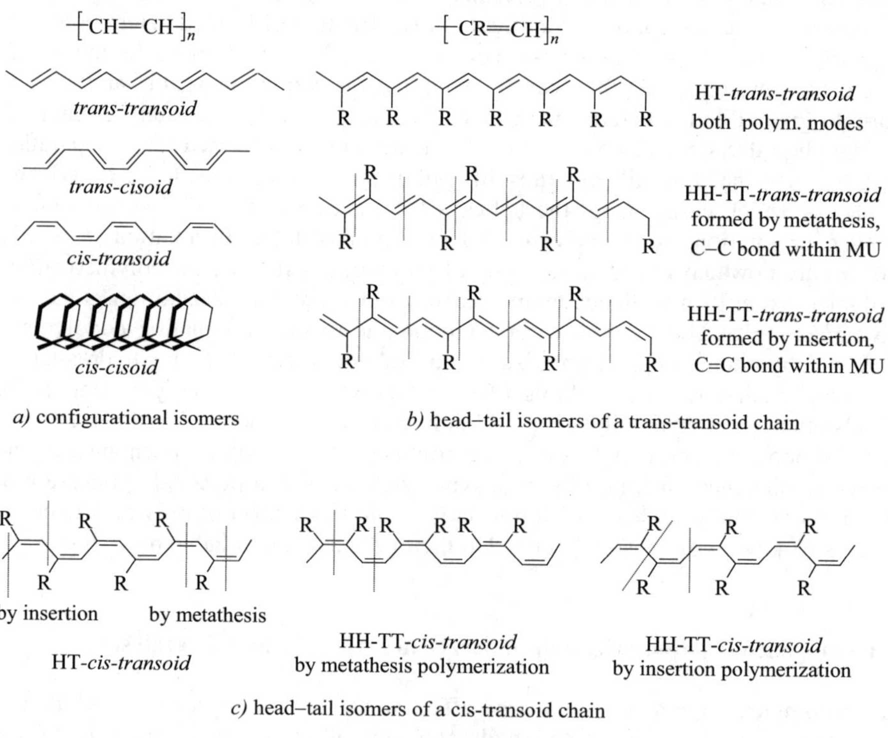

Fig. 1. Isomerism of polyvinylene chains

main-chain C–C single bonds have restricted free rotation, thus exhibiting cisoid–transoid isomerism (i.e., cis-like–trans-like isomerism). As a result, there are four different basal configurations that a polyvinylene chain can acquire (Fig. 1*a*). In addition to it, three different H-T isomers can be derived from each of these configurations (see isomers of a trans-transoid chain, Fig. 1*b*, and cis-transoid chain, Fig. 1*c*). This means that there are, on the whole, twelve possible basal structures of a polyvinylene chain composed of asymmetric monomeric units (MU) of one kind. From these twelve possible isomeric forms and virtually infinite number of their combinations, Rh(diene) catalysts provide the only one, HT-cis-transoid form, which clearly shows a unique product selectivity of them. Other known catalysts fairly do not show so high stereoselectivity [16].

The HT-cis-transoid structure of poly(PA) prepared with Rh(diene) catalysts has already been suggested in the first papers of Furlani et al. [8-10] and it is well established today [17]. This structure is characterized by well resolved ^{1}H NMR signals at 5.84 ppm (s, olefinic proton), 6.63 ppm (d, 2o-H aromatic) and 6.94 ppm (m, 3H aromatic) and ^{13}C NMR signals at 126.7, 127.5 and 127.8 ppm (aromatic), 131.8 ppm (olefinic), 139.3 ppm (quaternary, olefinic) and 142.8 ppm (ipso). The HT-cis-transoid polymers of substituted PA also show a sharp singlet of olefinic hydrogen in the region from 5 to 6 ppm while signals of their aromatic protons are displaced according to the substituent structure (see, e.g., [17] and Refs in Table I). The observed cis-units' content of a prepared polyvinylene is usually, lower than 100 % owing to the cis-to-trans isomerization of macromolecules in solution [18,19], which takes place during a polymerization as well as during NMR analysis (*Fig. 2*).

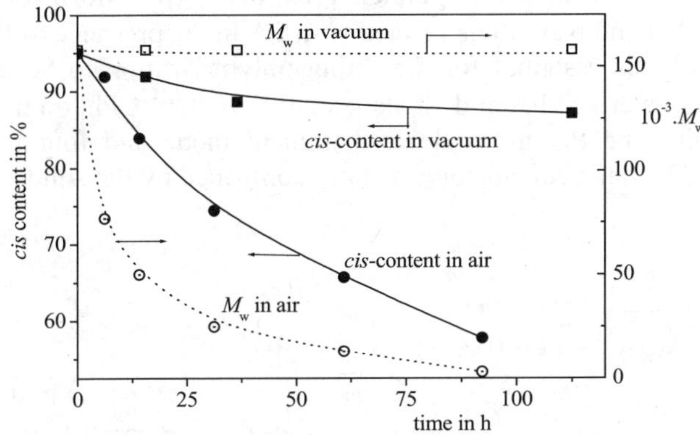

Figure 2. Time-dependence of the cis-units content in poly(4-fluoroPA) in THF solution kept in vacuum and in air. Polymer prepared with [{Rh(nbd)}₂(μ-OMe)₂], *cis*-contents calculated from ^{1}H NMR spectra (unpublished data of authors)

As to occurrence of HH-TT polymers, it comes into consideration mainly in the case of polymers prepared with metathesis catalysts, because head-to-head reactions of substrate molecules are typical of metathesis. Recently, the presence of HH-TT sequences was observed in poly(o-CF$_3$PA) prepared with MoOCl$_4$/Bu$_4$Sn/EtOH catalyst [20]. Indirect evidences on the presence of HH-TT sequences in substituted polyvinylenes were obtained earlier by pyrolysis gas chromatography [21]. This method is based on the GC/MS analysis of products formed in a very fast, actually "flesh" decomposition of a polymer deposited on a resistance wire that is suddenly

134

heated up to 500 °C for ca 0.5 s only. 1,3,5-Trisubstituted benzenes are the main products of the decomposition of HT polymers while those containing HH-TT sequences also provide 1,2,4-trisubstituted benzenes. The flesh pyrolysis proceeds as the intramolecular process, as it has been shown by experiments performed with mixture of two homopolymers [poly(PA) and poly(hex-1-yne)] and a corresponding random copolymer poly(PA-*co*-hex-1-yne) [21]. Therefore, this method is useful for an investigation of the H-T isomerism of polyvinylene chains.

1.2 LIVING POLYMERIZATION WITH RHODIUM CATALYSTS

First paper on polymerization of an acetylene monomer (*m*-chlorophenylacetylene) exhibiting partly living polymerization behavior presented Tabata et al. in 1990 [12]. They combined [{Rh(nbd)}$_2$(μ-Cl)$_2$] and triethylamine (Et$_3$N) in the mole ratio from 1:2 to 1:100 and found that Et$_3$N increases the catalyst efficiency (up to value of 0.2) as well as lifetime of growing species. In 1994, Kishimoto et al. have published first example of living polymerization of PA which was supported with a successful preparation of the poly(PA-*block*-4-methoxyPA) block copolymer [22]. They used mononuclear complex Rh(nbd)(C≡CPh)(PPh$_3$)$_2$ as a catalyst and observed that the living polymerization takes only place in the presence of 4-(dimethylamino)pyridine (DMAP). The catalyst prepared by reacting [{Rh(nbd)}$_2$(μ-Cl)$_2$] with LiC≡CC$_6$H$_5$ and PPh$_3$ (mole ratios 1 : 2,5 : 4,5) in diethyl ether was characterized by the single-crystal X-ray analysis. Later on [23], *in situ* formation of this complex has been proposed in polymerization systems in which either [{Rh(nbd)}$_2$(μ-OMe)$_2$] or a mixture of [{Rh(nbd)}$_2$(μ-Cl)$_2$] and NaOMe is reacted with PA in the presence of PPh$_3$. Again, the presence of DMAP is essential for the living polymerization to be achieved. These authors also copolymerized PA and ^{13}C-labeled PA, H^{13}C≡^{13}CPh, on this Rh catalyst in order to determine the PA molecule enchainment mode and found that this is the insertion mode [22]. This conclusion was later confirmed by the solid-

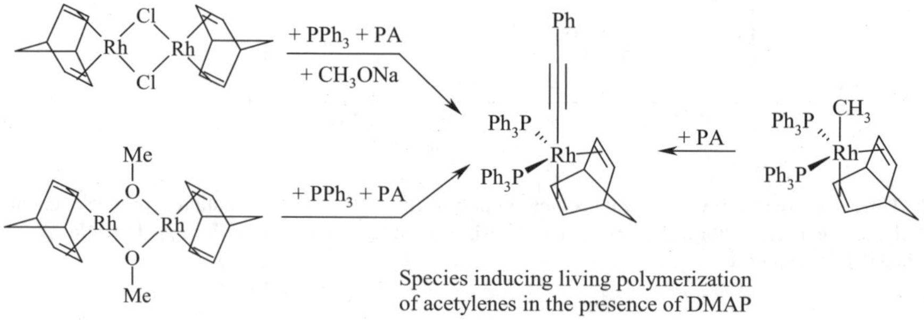

Species inducing living polymerization of acetylenes in the presence of DMAP

Figure 3. Catalyst for living polymerization developed by Kishimoto, Noyori et al. [17,21,22].

state NMR study of this copolymer [24], in which also slow π-flip motion of phenyl rings and conformational inhomogeneity in the cis-transoidal poly(PA) was revealed.

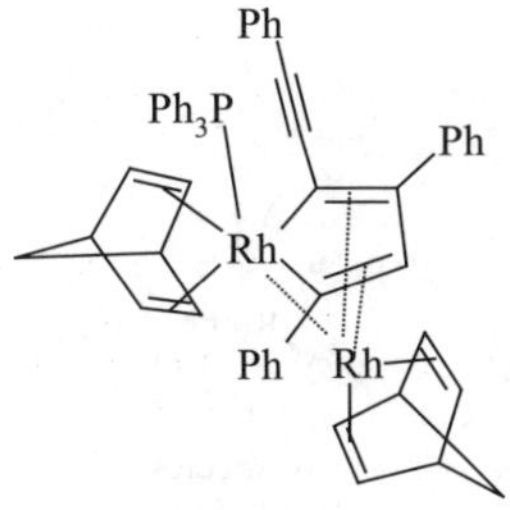

A role of DMAP in the above living polymerization systems was remaining unclear till 1999, when Kishimoto et al. [17] discovered that DMAP prevents Rh-species from their transformation into dinuclear rhodacyclopentadiene species, which proceeds in the presence of PA. The rhodacyclopentadiene complex (Fig. 4) also polymerizes PA, but to a high-molecular-weight polymer of $M_n > 10^5$ and polydispersity index $M_w/M_n > 2$, so as it cannot be a precursor of living polymerization species. DMAP inhibits this complex formation thus preferring a pathway to living polymerization species. Similar background can stay behind the effect of Et_3N observed by Tabata et al. [12], however, no evidence for it has been given.

Figure 4.
Rhodacyclopentadiene complex formed in absence of DMAP

Further progress in the living polymerization induced by Rh complexes has been achieved by Masuda et al. [25-27] who developed new ternary catalyst composed of $[\{Rh(nbd)\}_2(\mu\text{-Cl})_2]$, $Ph_2C=CPhLi$, and PPh_3. The catalyst was shown to induce living polymerization of PA as well as ring-substituted PAs carrying chemically divers groups: CH_3, OCH_3, Cl and $CO\text{-}OCH_3$, in the para position. Quantitative transformation of the Rh complex into growing species and easy synthesis of block copolymers of all the above monomers in any block-sequence order are reported. Reacting $[\{Rh(nbd)\}_2(\mu\text{-Cl})_2]$, and $Ph_2C=CPhLi$ with $(4\text{-}ClC_6H_4)_3P$ and $(4\text{-}FC_6H_4)_3P$ the same authors isolated corresponding Rh complexes [28] which were also shown to form living polymerization systems of this class. They concluded that the presence of a bulky ligand such as phenyl or terc.butyl group on the α-carbon of vinyl ligand is a key factor for the Rh-complex transformation into living polymerization species, whereas a role of anionic μ-ligand (chloro, methoxo, acetoxo) seems unimportant in this respect.

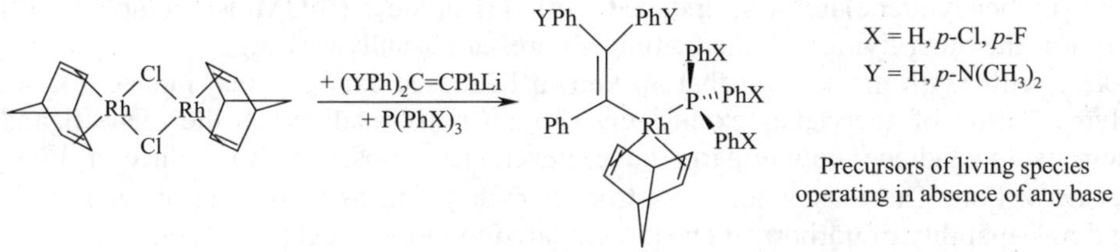

Figure 5. Living polymerization catalysts developed by Masuda and Misumi [25-28].

Recently, Farnetti et al. [29] reported a living polymerization of PA with $[\{Rh(nbd)\}_2(\mu\text{-OMe})_2]$ complex in the presence of bidentate phosphines of the general structure $Ph_2P(CH_2)_xPPh_2$ where x is 2, 3 and 4. They observed low polydispersity of poly(PA) formed and increase in M_n as the PA conversion increases. However, their results also show very low value of the initiator efficiency. They obtained polymers of M_n, from 10^5 to above 10^6 at the monomer/Rh mole ratio of 50, for which the ideal, "living-polymerization" value of M_n is ca 5 000. This means that less than 1 % of Rh atoms only gives rise to growing species. For a comparison, catalyst systems introduced by Kishimoto et al. and Masuda et al. show the efficiency from 25 to 70 % [17,22-24] and 100 % [25-28], respectively.

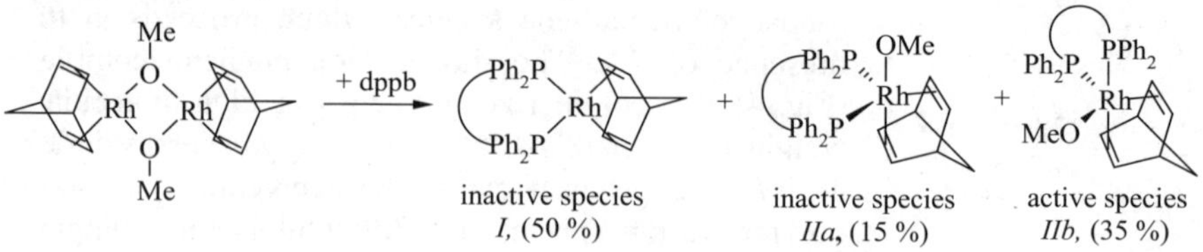

Figure 6. Living polymerization catalyst described by Farnetti [29].

Kishimoto et al. [17] report that a replacement of both PPh_3 ligands of with two PPh_2Me ligands or one $Ph_2P(CH_2)_4PPh_2$ (dppb) bidentate ligand makes their catalyst almost or totally inactive. However, Farnetti et al. has detected two isomeric species Rh(nbd)(OMe)(dppb), **II**, formed, together with Rh(nbd)(dppb), **I**, as products of the reaction of Rh(nbd)(OMe) with dppb (Fig. 6). Upon addition of PA, species **IIb** with MeO group in the equatorial position disappear, whereas species **I** and **IIa** remain intact. It is worth noting that species **IIa** with MeO group in axial position correspond to the Rh(nbd)(C≡CPh)(dppb) species which were found inactive by Kishimoto et al. Therefrom, Farnetti et al. concluded that species **IIb** are precursors of the observed living polymerization centers. However, it means that also species **IIb** show very low catalyst efficiency up to 3 % only. Thus, it is clear that the system with bidentate phosphine ligands needs further development to become an effective living polymerization system. Partial attributes of the living-like behavior, a consumption of additional portion of the monomer, is occasionally observed, e.g., in the case of polymerization of acetylenes with aromatic Shiff-base pendent groups, 4-substituted *N*-(4-ethynylbenzylidene)anilines, induced with [{Rh(cod)}$_2$(μ-OMe)$_2$] complex [30], however, the other living polymerization features are usually missing.

It is still worth mentioning that all known Rh-based catalysts that induce a living polymerization of acetylenes exclusively contain norbornadiene as the diene ligand. Their cyclooctadiene counterparts have never been observed to induce a living polymerization. The reason for it is stronger σ-donating as well as stronger π-back-bonding capability of norbornadiene as compared to those of cyclooctadiene.

1.3 POLYMERIZATION POTENTIAL OF RHODIUM CATALYSTS

As far as polymerization of acetylenes is concerned, Rh(I)-based catalysts show favorable properties that make them attractive despite their high prize. They show:

(i)   precise control of the propagation stereochemistry (part 1.1.),
(ii) ability to create living polymerization systems (part 1.2.),
(iii) unprecedented tolerance to the substrate functional groups,
(iv) stability in a high variety of chemically diverse solvents.

It is almost sure that Rh(I) catalysts indeed operate as stereospecific catalysts as the observed small configurational defects can be fully ascribed to the additional isomerization of formed macromolecules. Although Rh(diene) catalysts are not sensitive to oxygen or moisture, a protection against air is desirable during the preparation and additional treatment of a polymer, because monosubstituted polyvinylenes are mostly sensitive to oxygen [18,19,31-33]. When dissolved and exposed to air, they undergo quite rapid autoxidative degradation, which can even affect the results of the polymer SEC analysis [34-37]. It is worthy to note here that this degradation is faster for cis-transoid polymers than for the irregular or trans-rich ones [33,34] and that, in addition to it, the degradation speeds up the cis-to-trans isomerization of high-cis polymers, as can be seen from Fig.1.

Rh(I) catalysts effectively operate under mild conditions in solvents typically used in coordination polymerization, such as in hydrocarbons, chlorinated hydrocarbons and ethers, as well as in solvents that are rather rare in this field, such as amines, alcohols and water. This opens up a wide scope of possibilities of tailoring the polymerization conditions for a given monomer. The Rh(I) complexes also show unusually high tolerance to functional groups of monomers and so the scope of substituted acetylene monomers which were polymerized with Rh(I)-based catalysts is very wide (see *Table I*). Disturbing effects on polymerization show mainly groups that can reduce Rh(I) to Rh(0), such as aldehyde group. Acetylene monomers bearing a carboxylic group should be polymerized under basic conditions (i.e., transformed into salts), as carboxylic acids easily form Rh(I) carboxo-complexes which show too low or no polymerization activity toward acetylenes [17,60].

An inspection of *Table I* shows that only monosubstituted acetylenes have been polymerized with Rh(I) complexes. There is no example of successful polymerization of a disubstituted acetylene published in the literature, only negative results are reported, e.g. [38]. This clearly indicates that only ethynyl monomers can be polymerized by Rh(I) catalysts. Further, it is seen that from monosubstituted acetylenes only those containing an electron-withdrawing substituent have been polymerized by neutral Rh(I) complexes (in cyclohexen-1-yl acetylene, the electron density in the triple bond is decreased due to conjugation). Alkylacetylenes have been polymerized by the ionic Rh complex **6** only while zwitterionic complex **9**[*] was found to be inactive. These observations suggest that the polymerizability of an ethynyl monomer is a function of the acetylenic hydrogen acidity, besides other factors. Existence of such dependence has been demonstrated on a systematic series of 4-substituted *N*-(4-ethynylbenzylidene)anilines, for which a close correlation between the Hammett constant σ and the ^{1}H-NMR shift of acetylenic proton and polymer yield has been observed [30,66]. Qualitatively similar result were obtained with a series of ring-4-substituted PAs, for which the polymerization rate increases in the following order of substituents: OMe < H < COOCH$_3$ [39]. The last observation as well as solvent effects observed in the polymerization of PA, has led Ogawa et al. [39] to a suggestion of the propagation mechanism in which a proton transfer from a

TABLE I. Substituted acetylenes polymerized with various Rh(diene) complexes:

1 [{Rh(nbd)}$_2$(μ-Cl)$_2$],   1* [{Rh(cod)}$_2$(μ-Cl)$_2$],
2 [{Rh(nbd)}$_2$(μ-OMe)$_2$],   2* [{Rh(cod)}$_2$(μ-OMe)$_2$],
3 [Rh(nbd)(PPh$_3$)$_2$(C≡CPh)],   4 [Rh(nbd)P(Ph)$_3$(CPh=CPh$_2$)],
5* [Rh$^+$(cod){CH$_3$CO$^-$=CHCOO(CH$_2$)$_2$-OCOC(CH$_3$)=CH$_2$}],
6 [Rh$^+$(nbd)(η^6-Ph-B$^-$Ph$_3$)],   7* [{Rh(cod)}$_2$(μ-S-C$_6$F$_5$)$_2$],
8 [Rh$^+$(nbd)(Tosyl$^-$) (H$_2$O)]   8* [Rh$^+$(cod)(Tosyl$^-$)(H$_2$O)]
9* [Rh$^+$(cod){tris(pyrazolyl)borate}$^-$]   10* [Rh(cod)$_2$]BF$_4^-$

Monomer	Catalyst	References
SUBSTITUTED PHENYLACETYLENES		
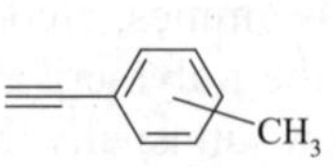 ethynylbenzene–CH$_3$	**1,1*,2,2*,3,** **4,5*,6,7,8,** **8*,9***	o-[17,38] p-[19,27,39,55,58]
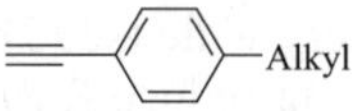 ethynylbenzene–Alkyl	**1,2***	n-Bu, t-Bu [33] n-C$_5$H$_{11}$ [40,41]
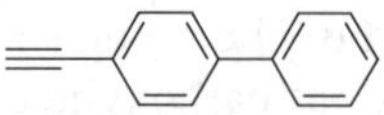 ethynyl-biphenyl	**1,3**	[17,42]
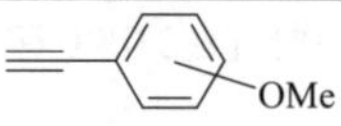 ethynylbenzene–OMe	**1,2,2*,3,4,** **6,7***	o-[17,38] m-[17,38] p-[17,22,27,33,38,39, 43]
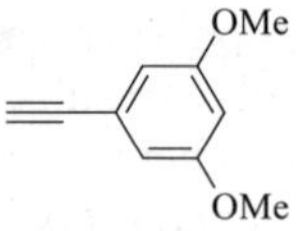 ethynylbenzene (3,5-di-OMe)	**3**	[17]
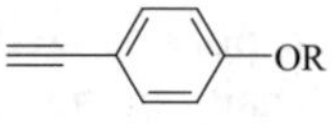 ethynylbenzene–OR R = Et, ^{i}Bu	**1**	[43,68]
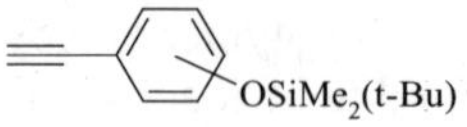 ethynylbenzene–OSiMe$_2$(t-Bu)	**1**	m-, p- [42]
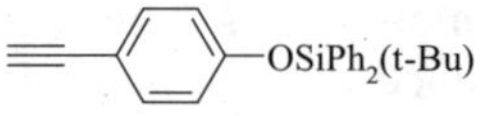 ethynylbenzene–OSiPh$_2$(t-Bu)	**1**	[42]
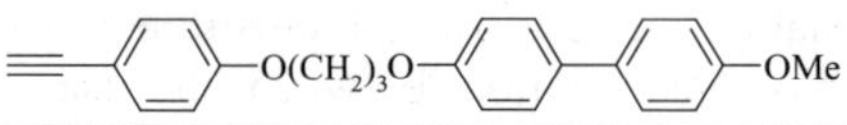 ethynylbenzene–O(CH$_2$)$_3$O-biphenyl-OMe	**3**	[23]
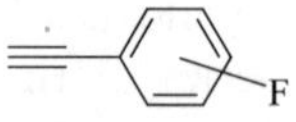 ethynylbenzene–F	**2***	o-[40] p-[40,41]
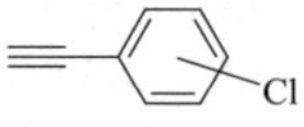 ethynylbenzene–Cl	**1,4, 9***	m-[11,12] p-[11,15, 27,58]
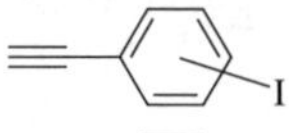 ethynylbenzene–I	**2,2***	m-, p-[44,45]
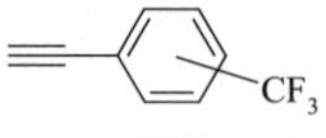 ethynylbenzene–CF$_3$	**1,2,2*,3,6**	o-[17,20,38]
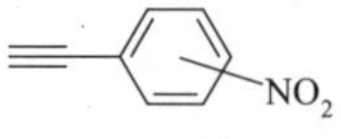 ethynylbenzene–NO$_2$	**1,2, 9***	p-[11,46,58]
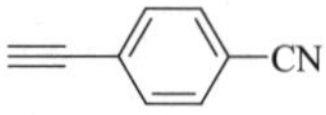 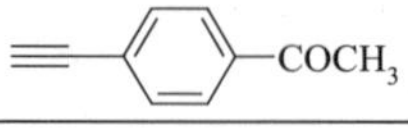 ethynylbenzene–CN ethynylbenzene–COCH$_3$	**9***	[58]

TABLE I. *continued*

Structure		
≡—⟨⟩—NMe₂	**1**	*p*-[11]
≡—⟨⟩—COOCH₃	**1,2,3,4,6, 7*,9***	*m*-[17,38] *p*-[17,22,27,33,38,39,58]
≡—⟨⟩—COO(-)menthyl	**3**	*p*-[17]
≡—⟨⟩—COO(CH₂)ₙO—⟨⟩—⟨⟩—CN	**1**	n = 6, 12 [47]
≡—⟨⟩—≡—Si^{i}Pr₃	**2,2*,3**	[45,48,49]
≡—⟨⟩—≡—⟨⟩—X, X = H, I, CN	**2*,3**	[45]
≡—⟨⟩—≡—⟨⟩—≡—Si^{i}Pr₃	**2*,3**	[45,49]
≡—⟨⟩(R)(R), R is: ≡—⟨⟩(t-Bu)(t-Bu) or ≡—⟨⟩—SiMe₃	**1**	[50]
≡—⟨⟩—N=CH—⟨⟩—X, X = H, Me, t-Bu, F, Br, CN, NO₂, NMe₂, OH, C≡CSiMe₃	**1*,2***	[30,45,51]
≡—⟨⟩—N=CH—⟨⟩(HO)(t-Bu (or H))	**1**	*m-, p-* [59]
≡—⟨⟩—CH=N—⟨⟩—X, X = Me, t-Bu, F, Br, I, CN, NO₂, C≡CSiMe₃	**2***	[30,45]
≡—⟨⟩—N=N—⟨⟩	**1**	*m-, p-* [52]
≡—⟨⟩—≡—(ferrocenyl)	**2**	[45,43]
≡—⟨⟩—N=CH—(ferrocenyl)	**2***	[45]
≡—⟨⟩—O-CO-NH-CH(Me)—⟨⟩ (or —(naphthyl))	**1**	*m-, p-* [42]
≡—⟨⟩—Ge(CH₃)(Ph)—(naphthyl) ≡—⟨⟩—Si(CH₃)(Ph)—(naphthyl)	**1**	[56,57]

TABLE I. *continued*

OTHER SUBSTITUTED ACETYLENES

Structure		Catalyst	Ref.
(ethynylnaphthalene)		**1**	[54] [42]
≡—t-Bu		**6**	[38]
(1-phenyl... Me/Ph), (isobutyl Me/Me), (ethynylcyclohexane)		**6**	[38]
(ethynylcyclohexene)		**1**	[60]
(propargyl ether with NEt₂), (propargyl ether with NO₂)		**2*,3**	[45]
(propargyl ether coumarin with CH₃)		**3**	[45]
≡—COO (menthyl), ≡—COO (bornyl)		**1**	[61,64]
≡—COOH ≡—COONa		**10***	[62]
≡—COO(CH₂)₆OCO—⟨⟩—⟨⟩—O(CH₂)₆CH₃		**1**	[63]
≡—COOR R = (CH₂)ₙCH₃ (n = 2 - 14), (CH₂)₄Cl, (+) or (−)-(CH₂)ₙCH(CH₃)C₂H₅ (where n = 0 to 5)		**1**	[64,69,70]
≡—CH₂—NH—CO—R R = Me, Et, Pr, iPr, Bu, iBu, pentyl, heptyl		**6**	[65]

*The list of monomers as well as catalysts and papers is not comprehensive.

coordinated monomer molecule to the last monomeric unit of growing chain is assumed. However, a solid evidence for such mechanism has not been given yet.

As it is obvious, Rh(I)-based catalysts have opened a feasible path to substituted polyvinylenes carrying various functional groups, through which their physical properties, such as non-linear optical properties and charge carrier photogeneration and transport in these materials, can be tuned (see, e.g. [11,47,49,52]). Other, probably more promising applications of these polymers can be based on the high stereoregularity of their molecules. In 1991, Tabata et al. [67] observed a solvent-induced crystallization (toluen was found as preferable solvent) of an amorphous high-molecular-weight cis-transoid poly(PA) that was prepared with [{Rh(nbd)}$_2$(μ-Cl)$_2$]. This transition was later observed for other cis-transoid polyvinylenes, too. It was recognized that, in these crystallites, macromolecules adopting the helical conformation are hexagonally packed forming so-called columnar structures [68-70]. Under high pressure, the macromolecules assembled in columnar crystallites undergo cis-to-trans isomerization [68,69]. Hexagonal packing was interestingly found to make polyvinylenes, mainly poly(propiolic ester)s, more resistant to autoxidative degradation [70].

Research of columnar polyvinelenes has revealed that the unprecedented HT-cis-transoid stereospecifity of Rh(I) catalysts affords a synthesis of highly stereoregular polymers that can spontaneously adopt a helical conformation. It was found that cis-transoid macromolecules of substituted polyvinylenes partly adopt this conformation also in solutions. The more bulky side groups the longer are helical sequences of a macromolecule [57,65,72]. Also presence of chiral substituents increases a tendency of polyvinylene molecule to adopt helical conformation [57,61,71]. Surprisingly, also molecules of poly(propiolic ester)s and propargylamides smoothly adopt the helix conformation regardless bulkiness of their substituents. Thus, a stabilization of their helical conformations must be associated with the presence of polar carboxy and carboxamide groups [61-65,72]. Research of π-conjugated helical polymers is undoubtedly important, since such polymers are potentially useful as polarization-sensitive electrooptical materials for asymmetric electrodes and sensors.

2. Other polymerization reactions catalyzed by Rh complexes

Nowadays scope of a use of Rh catalysts in polymerization reactions is fairly not limited to a preparation of polyvinylenes only. Rh complexes have recently been introduced as catalysts of the atom transfer radical polymerization (ATRP), they have underwent a renascence in the polymerization of allenes, and they are increasingly used in the polycondensation and ring-opening polymerization involving organo-silicon compounds.

The first Rh-complex-induced ATRP polymerization was accomplished with the Wilkinson catalyst RhCl(PPh$_3$)$_3$, styrene as monomer and p-(CH$_3$O)C$_6$H$_4$SO$_2$Cl as initiator [73]. Rather poor control of the polymerization was observed, $M_w/M_n \geq 1.8$. Better results were achieved for polymerization of methyl (MMA), butyl (BuMA) and 2-hydroxyethyl (HEMA) methacrylates with the same catalyst and p-ClC$_6$H$_4$-CO-C$_6$H$_4$-p-Cl as initiator [74]. Polymers with M_w/M_n values down to 1.3 were prepared and shown to act as a macroinitiator in the consecutive polymerization of MMA and BuMA. Besides, a random copolymer of MMA and HEMA was prepared, and a positive effect of added water on behavior of these ATRP systems was found. Another

142

ATRP-active Rh catalyst, [RhCl(CO)(PPh$_3$)(Et$_2$NH)], was prepared via the bridge-splitting reaction of [{Rh(CO)(PPh$_3$)}$_2$(μ-Cl)$_2$] with Et$_2$NH [75]. Also with this catalyst, the ATRP of MMA yielded better results than that of styrene (M_w/M_n values of 1.43 and 2.08, respectively, are reported). Rh(I)(diene) complexes [{Rh(cod)}$_2$(μ-O-C$_6$H$_4$-p-CH$_3$)] and [{Rh(cod)}$_2${μ-OCO(CH$_2$)$_{20}$CH$_3$}] have been reported as ATRP catalysts recently [76]. By itself, they promote a controlled polymerization of MMA (CH$_3$CHBr-COOMe initiator) and styrene (CH$_3$CHBrPh initiator), affording a medium to good yield of high-molecular-weight (M_w up to 400 000) polymers with M_w/M_n values from 1.45 to 1.65 but too low initiator efficiency (0.04 to 0.20). The addition of Bu$_2$NH was found to improve the initiator efficiency (up to 0.85) as well as the polymer molecular weight control (M_w/M_n down to 1.27) in both systems, but more distinctly for the μ-docosanoato catalyst. The bridge splitting of dinuclear Rh-complex by Bu$_2$NH is supposed to be responsible for this improvement, but corresponding mononuclear complexes have not been isolated.

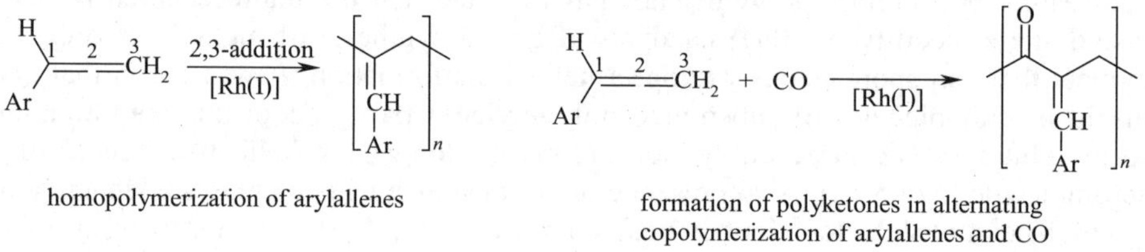

homopolymerization of arylallenes

formation of polyketones in alternating copolymerization of arylallenes and CO

Figure 7. Homopolymerization and copolymerization of substituted arylallenes via 2,3-addition.

Advances in allene polymerization have been achieved with a use of nonplanar Rh(I) complexes such as RhH(PPh$_3$)$_4$, RhH(CO)(PPh$_3$)$_3$, RhH(dppe)$_4$ (where dppe is Ph$_3$PCH$_2$CH$_2$PPh$_3$), and Rh(π-allyl) complexes formed by single or multiple insertion of an arylallene into the Rh-H bond of the above complexes [77-80]. These catalysts polymerize arylallenes exclusively via 2,3-addition (*Fig.* 7) so as regular unsaturated polymers with saturated main chain and arylmethylene (typically benzylidene) pendant groups directly linked by double bonds to quaternary main-chain atoms are formed. In addition, the above mentioned Rh(π-allyl) complexes with tetrameric π-allyl ligands catalyze alternating copolymerization of both alkyl- and arylallenes and CO giving regular polyketones (*Fig.* 7) [80,81], which corresponds with well known ability of Rh(I) complexes to catalyze the CO insertion into organic molecules. The CO molecule insertion to the Rh(π-allyl) growing species was revealed as the rate-determining step of these copolymerizations, and low polydisperzity of so prepared polyketones indicates a living character of this reaction [81].

In the last few years, some Rh-complex-catalyzed polymerizations of silanes have been introduced as new effective tools for a preparation of silicon polymers from relatively easily available monomers. One class of these reactions are cross-dehydrocoupling copolymerizations of bis(hydrosilanes) and/or dihydrosilanes with comonomers containing labilized hydrogen, such as with diols [82], disilanols [83] and dithiols [84,85], which easily proceed in the presence of Wilkinson catalyst (*Fig.* 8). These reactions proceed easily at ambient room or little elevated temperature and

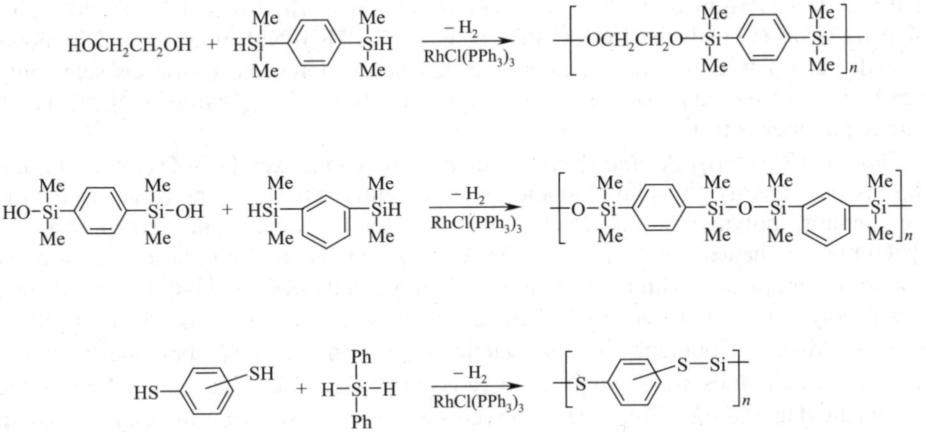

Figure 8. Examples of the cross-dehydrocoupling polymerization of bis(hydrosilanes) and dihydrosilane with a diol [82], disilanol [83] and dithiol [84].

yield high-molecular-weight polymers including those that are at this quality hardly available by other polymerization reactions (e.g., high-MW silane/diol copolymers). Also polymers with main-chain organometallic moieties can be prepared via these polycondensations [85].

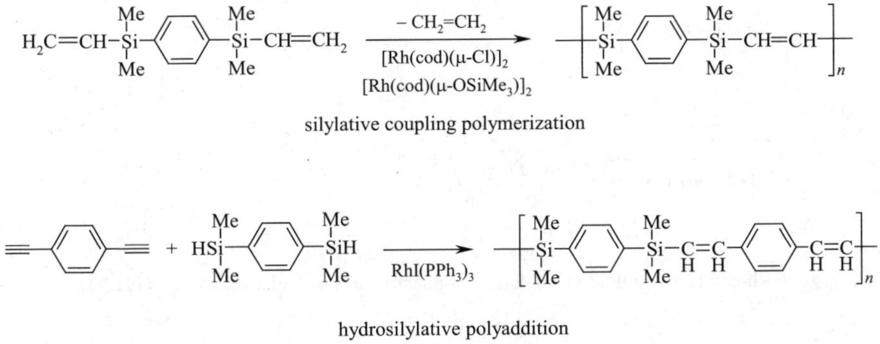

silylative coupling polymerization

hydrosilylative polyaddition

Figure 10. Rh-catalyzed silylative and hydrosilylative non-chain polymerization reactions.

Next two newly introduced processes open up a way to poly(silylene-vinylene)s and related polymers which meet demands of practice for new field-responsive organic materials. The first of these processes is referred to as silylative coupling polycondensation of divinylsilanes and bis(vinylsilane)s, which from the formal stoichiometry point of view proceeds as metathesis, but their mechanism consists of the alternating sequence of Si–C and C–H bond cleavages [86-88]. This reaction is the subject of the B. Marciniec's contribution to this issue. The alternative process is based on the hydrosilylation of diethynyl compounds with bis(hydrosilane)s [89,90]. This polymerization is catalyzed by $RhI(PPh_3)_3$ or $RhCl(PPh_3)_3/NaI$ mixture, and it shows unusually distinctive and high temperature-dependent stereoselectivity: cis-selectivity of ca 95 % is observed at low reaction temperatures (0 to 20 °C) while, at 60 °C, ca 95 % trans-polymer is formed.

Cationic $[Rh(cod)_2]^+A^-$ and $[Rh(cod)(dmpe)]^+A^-$ complexes {A = OTf and PF_6 and dmpe is bis(dimethylphosphino)ethane} were recently introduced as catalysts for the ring-opening polymerization (ROP) of disilacyclobutanes and silicon-bridged [1]silaferrocenophanes [90,91]. The former catalyst gives 100 % conversion of [1]silaferrocenophanes within 2 min, however, it paralelly cleaves Si–C(Cp) bonds in a formed polymer inducing a rapid degradation of it. On the other hand, the ROP of tetramethyldisilacyclobutane is also facile with $[Rh(cod)_2]^+A^-$ but the resulting polymer, which does not contain Si–C(unsaturated) bonds, does not undergo the subsequent degradation. Thus the observed cleavage of Si–C(unsaturated) bonds in poly(ferrocene-1,1′-diyldialkylsilane)s can also be regarded as an additional prove supporting the mechanism suggested for the silylative coupling polymerization (see above). Complex $[Rh(cod)(dmpe)]^+A^-$ shows lower activity in polymerization of [1]silaferrocenophanes but no activity in the polymer degradation. The presence of superfluous cod or dmpe inhibits catalytic activity of Rh complexes, thus indicating the presence of one cod ligand a growth center.

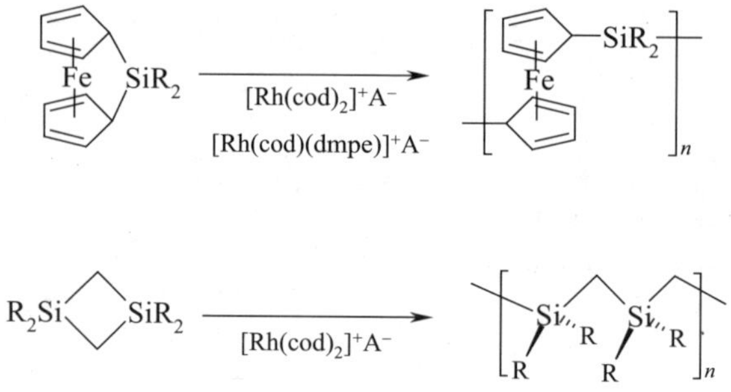

Figure 9. Rh-catalyzed ROP of [1]silaferrocenophanes and 1,3-disilacyclobutanes [91,92].

3. Polymerization of acetylenes in heterogeneous reaction systems

3.1. POLYMERIZATION WITH IMMOBILIZED RHODIUM COMPLEXES

Purity is of increasing importance in the case of specialty polymers with potential application in areas such as electronics, medicine and pharmacy. If a polymer is prepared by homogeneous coordination polymerization, transition-metal-catalyst residues buried in the polymer can disable it from a practical use. The residues can be poisonous, which is inadmissible for any application of the polymer in medicine or pharmacy. In conjugated polymers, the catalyst residues can act as dopants or, in the opposite way, as quenchers of excited states or traps of charge carriers. A removal of the residues is a time consuming process in which undesirable waste solvents are produced. Therefore, a development of heterogeneous polymerization catalysts is desirable, since these catalysts can easy be separated from the reaction mixture, which opens a way to pure polymers that are free of catalyst residues.

Acetylene and its methyl and ethyl derivatives were polymerized with H-ZSM5 zeolites and Co^{II} and Ni^{II} exchanged zeolites more than ten years ago [93-96]. These experiments gave inorganic matrix – conjugated polymer composites and attempts to isolate polymers formed via dissolving the matrix in diluted hydrofluoric acid were unsuccessful, yielding products of hydrofluorination of the polymers. Nevertheless, it should be mentioned here that just a preparation of composites containing organic nanowires was the main objective of authors of these papers, since such composites are believed to find practical applications in energy storage and related fields [97-99]. In the last two years, similar composites based on mesoporous silica and alumina sieves and polyacetylenes have been prepared and studied as to their functional properties [99-102].

A preparation of isolable polymers with microporous catalysts is practically impossible because of transport limitations existing for macromolecules in narrow pores. Macromolecules can hardly escape from narrow pores of typical diameter d from 0.4 to 1 nm. As a result, they remain jailed in pores, where they block active species and, in addition to it, a formed polymer cannot be isolated without the catalyst decomposition. Transport limitations can be overcome by using mesoporous ($d = 2 - 50$ nm) polymerization catalysts instead of the microporous ones. Such catalysts can be prepared via immobilization of species of active polymerization catalysts on inner walls of a mesoporous support of either inorganic or organic class.

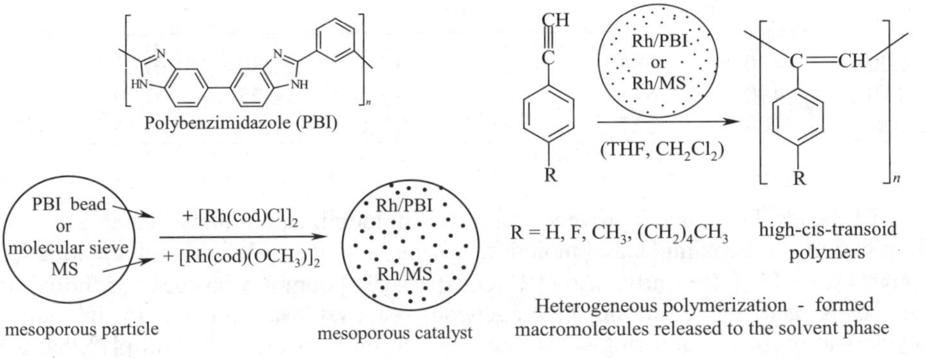

Figure 11. Polymerization of ring-substituted PAs with Rh catalysts immobilized in pores of mesoporous supports.

Among inorganic supports, mesoporous molecular sieves are perhaps the most promising materials, since they typically have narrow pore-size distribution and their pore size and pore architecture can be tuned [103]. They are increasingly used as support to immobilize transition-metal catalysts for transformations of hydrocarbons

[103,104] and also for polymerization of alkenes [105] and ring-opening metathesis polymerization of norbornene [106]. Recently, also Rh(diene) complexes such as [{Rh(cod)}$_2$(μ-Cl)$_2$] and [{Rh(cod)}$_2$(μ-OMe)$_2$] have been anchored on all-siliceous mesoporous sieves MCM-41 (d = 3.7 nm), MCM-48 (d = 3.7 nm) and SBA-15 (d = 7.4 nm) and resulting mesoporous catalysts have been successfully used in the polymerization of PA and ring-substituted PAs [40,107,108] yielding high yields of isolable polymers (see paper [108] in this issue).

Another support used for immobilization of homogeneous catalysts are commercial (Hoechst-Celanese Corporation) porous polybenzimidazole (PBI) beads. PBI is known as a thermally very stable polymer that is resistant to oxidation, and which lacks solubility in common solvents mainly due to the rigidity of polymer chains and interchain hydrogen bonding. Dry PBI beads do not appear to be a good support for a catalyst. They have a pore volume of 0.11 cm^3/g and the overall BET surface area about 20 m^2/g, of which 10 m^2/g represents the surface of macropores and 10 m^2/g the surface of meso- and micropores in walls between macropores. However, upon swelling in a solvent such as THF, water and toluene, the inner walls swell and high number of mesopores is created in them [109]. Such high porosity is possible only thanks to the foam-like morphology of PBI beads, as revealed by SEM [109]. Even if the smallest mesopores would not be accessible, there are many wider mesopores of d from 10 to 20 nm in the PBI support, in which catalyst species can be easily immobilized.

TABLE 2. Characteristics of PBI beads of the outer diameter 250-500 μm according to [105].

d, nm	specific surface area, m^2/g			pore volume, cm^3/g		
	3	10 – 20	40 (80[*])	3	10 – 20	40 (80[*])
toluene	410	384	8	0.31	1.30	0.08
THF[*]	490	643	14.5	0.37	2.25	0.29
water	1040	227	15	0.78	1.02	0.15

PBI beads have been proved useful for immobilization metal complexes via coordination to benzimidazole moieties, giving active oxidation and epoxidation catalysts [110-112]. Recently, also [{Rh(cod)}$_2$(μ-Cl)$_2$] complex has been anchored on PBI beads and the resulting heterogeneous catalyst was proved to be active polymerization of PA and ring-substituted PAs to cis-transoid polymers [41] that are continuously released to the solvent phase. In this Rh/PBI catalyst, Rh(I) species are assumed to be anchored via direct interaction with nitrogen atoms of the PBI imidazole moieties, in analogy with the structure of Rh(I) complexes with the modified purine bases [113]. Another polymer-anchored Rh(I) catalyst has been prepared by the copolymerization of Rh(cod) complex containing 2-(acetoacetoxy)ethyl methacrylate ligand with acrylates and cross-linkers. Also this catalyst induce polymerization of PA and p-Me-PA to cis-transoid polymers [114].

Rh complexes have also been anchored on polymers such as poly(styrene-*co*-divinylbenzene) or polyfluoroacrylate copolymers, which, by themselves, have not binding sites for Rh species, to which, however, such sites of phosphine class have been introduced via copolymerization with a functionalized monomer [115,116] or via modification of pristine polymer [117]. However, these complexes were tested as catalysts in reactions such as hydrogenolysis and hydrogenation but not in polymerization.

3.2. POLYMERIZATION IN THE IONIC LIQUID SYSTEMS

In the last few years, the polymerization in ionic liquids is increasingly used as the method alternative to heterogeneous catalysis, which also enables a preparation of polymers that are free of catalyst residues as well as effective catalyst recycling. This method uses two-phase systems composed of a molten salt such as 1-butyl-3-methylimidazolium tetrafluoroborate, [bmim]BF$_4$, or butylpyridinium tetrafluoroborate, [bupy]BF$_4$, as a solvent for an ionic catalyst, and an immiscible organic liquid as a solvent for a polymer formed. Upon polymerization, a formed polymer as well as unreacted monomer can be easily removed and the catalyst solution in ionic liquid can be reused.

This polymerization method is perhaps the most frequently used in ATRP. However, recently it was also applied in the ring-opening metathesis polymerization of norbornene with cationic ruthenium allenylidene precatalyst [118] and in the polymerization of PA with complexes [{Rh(cod)}$_2$(μ-Cl)$_2$], [{Rh(nbd)}$_2$(μ-Cl)$_2$], [Rh(cod)(acac)] and [Rh(nbd)(acac)] [119] in the presence of Et$_3$N. High yields of poly(PA) of M.W. from 55,000 to 200,000 were obtained with both [bmim]BF$_4$ and [bupy]BF$_4$ ionic liquids and the last catalyst has been recycled without a significant loss in activity.

4. Concluding remarks

Until recently, rhodium complexes have been prevailingly used as catalysts for the polymerization of monosubstituted acetylenes to highly stereoregular, HT-cis-transoid polymers. However, nowadays scope of their use is substantially wider, involving also (i) atom transfer radical polymerization, (ii) polymerization of arylallenes taking place exclusively via 2,3-addition mode and copolymerization of arylallenes with carbon monooxide to give alternating copolymers, (iii) cross-dehydrocoupling polymerization of dihydrosilanes and bis(hydrosilane)s with diols, disilanols and dithiols, (iv) silylative coupling polymerization of bis(vinylsilane)s, (v) hydrosilylative addition copolymerization of bis(silane)s and diethynyl monomers, and (vi) ring-opening polymerization of sila-ferrocenophanes and 1,3-disilacyclobutanes. It means that the polymerization reactions catalyzed with these complexes involve not only a cleavage of the C≡CH triple bonds and a formation of C–C bonds, but also a cleavage of bonds Si–C, Si–H, O–H and S–H, and a formation of bonds Si–C, Si–O, Si–S and C–H.

As can be seen, Rh-based catalysts have become a special class of polymerization catalysts in the last decade. They are increasingly used in a synthesis of the specialty polymers of various chemical structure and functionality, many of which are not available via other methods. Besides, they also afford a control over some polymerization processes leading to commodity polymers such as polystyrene and acrylic polymers.

There is no doubt that the wide scope of chemical transformations that can be accomplished with Rh complexes is the main reason for steadily increasing number of applications of Rh catalysts in polymer chemistry. However, there are still other important factors supporting this increase, which can be summarized to the points given below.

(i) Rh-based catalysts often exhibit the unique product selectivity. A precise control of the configurational structure of formed macromolecules of polyvinylenes, exclusive 2,3-addition in polymerization of allenes and highly selective transformations of silicon-containing monomers can be mentioned here as typical examples. The preparation of unique stereoregular poly(propiolate)s, molecules of which easily adopt helical conformations not only in the solid state but also in solutions, is worth mentioning in this connection.

(ii) Rh-based catalysts exhibit unusually high tolerance to functional groups of monomers and polymers, which makes tailoring as well as synthesis of specialty polymers carrying various functional groups easier.

(iii) Rh catalysts show unusually high tolerance to the reaction surroundings. They can be used in solvents such as hydrocarbons, chlorinated hydrocarbons, ethers, alcohols, water, carbon dioxide and ionic liquids. Since they are mostly stable in air, they need not be handled in a dry box or in vacuum. These properties offer a wide scope of tuning the reaction conditions for a particular system monomer(s) – polymer and, in addition, they are important for a potential use of these complexes in a medium-to-large scale production of polymers.

(iv) Some Rh complexes can be transformed to living polymerization systems allowing a precise control of the distribution of molecular weights of the polymer under formation.

In spite of very high chemical versatility, Rh-catalysts applied under classical homogeneous reaction conditions can hardly find a place in the medium-to-large scale production of polymers. They are very expensive and if they remain more or less buried in the resulting polymer, the price of lost catalyst overbalance its advantages. Therefore, a development of heterogeneous polymerization processes enabling easy and cheep recycling of Rh catalysts is highly desirable. The first steps in this direction have been done, already: (i) first immobilized heterogeneous Rh catalysts have been prepared and examined, and (ii) some Rh catalysts were shown to be active in the ionic-liquid polymerization systems. The most important conclusion arising from these experiments is that the Rh complexes retain their selectivity even under heterogeneous conditions. Nevertheless, further extensive research of their stability under heterogeneous conditions and the recycling effectiveness is needed.

5. Acknowledgements

Financial support of the Grant Agency of the Czech Republic (projects No. 203/03/0281, 203/01/PO14 and 203/02/0976) and Grant Agency of Charles University (projects No. 228/2000/B-CH/PrF and 241/2002/B-CH/PrF) is gratefully acknowledged.

6. References

1. Cornils B., Herrmann W.A., Schögl R., Wong C.H. (2000) *Catalysis from A to Z. A Concise Encyclopedia*, Wiley-VCH: Weinheim.
2. Crivello J.V., Fan M. (1992) Regioselective rhodium-containing catalysts for ring-opening polymerizations and hydrosilylations, *J. Polym. Sci. Polym. Chem.* **30**, 1-11.
3. Biffis A., Castello E., Zecca M., Basato M. (2001) Fluorous biphasic catalysis with dirhodium(II) perfluorocarboxylates: selective silylation of alcohols under fluorous biphasic conditions, *Tetrahedron* **57**, 10391-10394.
4. Murai S., Sugise R., Sonoda N. (1981) [(-)-Diop]RhCl-catalyzed asymmetric addition of bromotrichloromethane to styrene, *Angew. Chem., Int. Ed. Engl.* **20**, 475-476.
5. Solymosi F., Erdöhelyi A. (1980) Methanation of CO_2 on supported rhodium catalyst, in T. Seiyama and K. Tanabe (eds.) *New Horizons in Catalysis, Studies in Surface Science and Catalysis* **7B**, Elsevier, Amsterdam, pp. 1448-1449.
6. Bertani R., Michelin R.A., Mozzon M., Sassi A., Basato M., Biffis A., Martinati G., Zecca M. (2001) Catalytic activity of dicationic platinum(II) and rhodium (II) complexes towards 9-diazafluorene, *Inorg. Chem. Commun.* **4**, 281-284.
7. Cooper W. (1976) Kinetics of polymerization initiated by Ziegler-Natta and related catalysts, Chapter 3 in C.H. Bamford and C.F.H. Tipper (eds.), *Comprihensive Chemical Kinetics, Vol. 15, Non-Radical Polymerization,* Elsevier, Amsterdam, pp. 133-257.
8. Furlani A, Napoletano C, Russo M.V., Feast W.J. (1986) Stereoregular polyphenylacetylene, *Polym. Bull.* **16**, 311-317.
9. Furlani A., Licoccia S., Russo M.V., Camus A., Marsich N. (1986) Rhodium and platinum complexes as catalysts for the polymerization of phenylacetylene, *J. Polym. Sci. A - Pol. Chem.* **24**, 991-1005.
10. Furlani A., Napoletano C., Russo M.V., Camus A., Marsich N. (1989) The influence of the ligands on the catalytic activity of a series of Rh(I) complexes in reactions with phenylacetylene - synthesis of stereoregular poly(phenyl) acetylene, *J. Polym. Sci. Polym. Chem.* **27**, 75-86.
11. Lindgren M., Lee H.S., Yang W., Tabata M., Yokota K. (1991) Synthesis of soluble polyphenylacetylenes containing a strong donor function, *Polymer* **32**, 1531-1534.
12. Tabata M., Yang W., Yokota K. (1990) Polymerization of meta-chlorophenylacetylene initiated by [Rh(norbornadiene)Cl]₂-triethylamine catalyst containing long-lived propagation species, *Polym. J.* **22**, 1105-1107.
13. Furlani A., Paolesse R., Russo M.V., Camus A., Marsich N. (1987) Polymerization of *N*-benzyl-propargylamine in the presence of ionic rhodium(I) complexes - A new functionalized polyacetylene - investigation of its conducting properties, *Polymer* **28** 1221-1226.
14. Goldberg Y., Alper H. (1994) Polymerization of phenylacetylene catalyzed by a zwitterionic rhodium(I) complex under hydrosilylation conditions, *J. Chem. Soc., Chem. Commun.* 1209-1210.
15. Tabata M., Yang W., Yokota K. (1994) ^{1}H-NMR and UV studies of Rh complexes as a stereoregular polymerization catalysts for phenylacetylenes - effects of ligands and solvents on its catalyst activity *J. Polym. Sci. Polym. Chem.* **32**, 1113-1120.
16. Shirakawa H., Masuda T., Takeda K. (1994) Synthesis and properties of acetylenic polymers, in S. Patai, ed., *The Chemistry of Tripple-Bonded Functional Groups*; Supplement C2, Wiley New York, Chapter 17, pp. 945-1016.
17. Kishimoto Y., Eckerle P., Miyatake T., Kainosho M., Ono A., Ikariya T., Noyori R. (1999) Well-controlled polymerization of phenylacetylenes with organorhodium(I) complexes: Mechanism and structure of the polyenes, *J. Am. Chem. Soc.* **121**, 12035-12044.
18. Percec V., Rudick J.G., Nombel P., Buchowicz W. (2002) Dramatic decrease of the cis content and molecular weight of cis-transoidal polyphenylacetylene at 23 °C in solutions prepared in air, *J. Polym. Sci. Pol. Chem.* **40**, 3212-3220.
19. Mastrorilli P., Nobile C.F., Rizzuti A., Suranna G.P., Acierno D., Amendola E . (2002) Polymerization of phenylacetylene and of p-tolylacetylene catalyzed by µ-dioxygenato rhodium(I) complexes in homogeneous and heterogeneous phase, *J. Mol. Cat. A-Chem.* **178**, 35-42.
20. Sone T., Asako R., Masuda T., Tabata M., Wada T., Sasabe H. (2001) Polymerization of *o*-trifluoromethyl(phenylacetylene) initiated by [Rh(norbornadiene)Cl]₂ and MoOCl₄–*n*-Bu₄Sn–EtOH catalysts. Formation of order and disorder trans sequences, *Macromolecules* **34**, 1586-1592.
21. Duc S., Petit A. (1997) Microstructure characterization of acetylenic polymers by Curie-point pyrolysis capillary gas chromatography mass spectrometry, *J. Anal. Appl. Pyrol.* **40-41**, 55-68.

150

22. Kishimoto Y., Eckerle P., Miyatake T., Ikariya T., Noyori R. (1994) Living polymerization of phenylacetylenes initiated by $Rh(C\equiv CC_6H_5)(2,5$-norbornadiene) $[P(C_6H_5)_3]_2$, *J. Am. Chem. Soc.* **119**, 12131-12132.

23. Kishimoto Y., Miyatake T., Ikariya T., Noyori R. (1996) An efficient rhodium(I) initiator for stereospecific living polymerization of phenylacetylenes, *Macromolecules* **29**, 5054-5055.

24. Hirao K., Ishii Y., Terao T., Kishimoto Y., Miyatake T., Ikariya T., Noyori R. (1996) Solid state NMR study of poly(phenylacetylene) synthetized with a rhodium complex initiator, *Macromolecules* **31**, 3405-3408.

25. Misumi Y., Masuda T. (1998) Living polymerization of phenylacetylene by novel rhodium catalysts. Quantitative initiation and introduction of functional groups at the initiating chain end, *Macromolecules* **31**, 7572-7573.

26. Misumi Y., Kanki K., Miyake M., Masuda T. (2000) Living polymerization of phenylacetylene by rhodium-based ternary catalysts, (diene)Rh(I) complex/vinyllithium/phosphorus ligand. Effects of catalyst components, *Macromol. Chem. Phys.* **201**, 2239-2244.

27. Isomura M., Misumi Y., Masuda T. (2000) Living polymerization and block copolymerization of various ring-substituted phenylacetylenes by rhodium-based ternary catalyst, *Polym. Bull.* **45**, 335-339.

28. Miyake M., Misumi Y., Masuda T. (2000) Living polymerization of phenylacetylene by isolated rhodium complexes, $Rh[C(C_6H_5)=C(C_6H_5)_2](nbd)(4$-X-$C_6H_4)_3P$ (X = F, Cl), *Macromolecules* **31**, 6636-6639.

29. Falcon M., Farnetti E., Marsich N. (2001) Stereoselective living polymerization of phenylacetylene promoted by rhodium catalysts with bidentate phosphines, *J. Organomet. Chem.* **629**, 187-193.

30. Balcar H., Sedláček J., Zedník J., Blechta V., Kubát P., Vohlídal J. (2001) Polymerization of isomeric N-(4-substituted benzylidene)-4-ethynylanilines and 4-substituted N-(ethynylbenzylideneanilines) by transition metal catalysts. Preparation and characterization of new substituted polyacetylenes with aromatic Schiff-base pendant groups, *Polymer* **42**, 6709-6721.

31. Vohlídal J., Rédrová D., Pacovská M., Sedláček J. (1993) Autooxidative Degradation of Poly(phenylacetylene), *Collect. Czech. Chem. Commun.* **58**, 2651-2662.

32. Sedláček J., Vohlídal J., Grubušic-Gallot Z., (1993) Molecular-weight determination of poly(phenyl-acetylene) by size exclusion chromatography/low-angle laser light scattering. Influence of polymer degradation, *Makromol. Chem. Rapid Commun.* **14**, 51-53.

33. Karim A.S.M., Nomura R., Masuda T. (2001) Degradation behavior of stereoregular cis-transodial poly(phenylacetylene), *J. Polym. Sci. Polym. Chem.* **39**, 3130-3136.

34. Vohlídal J., Sedláček J. (1999) Degradation of substituted polyacetylenes and effect of this process on SEC analysis of these polymers, in T. Provder ed., *Chromatography of Polymers: Hyphenated and Multidimensional Techniques*, ACS Symposium Series, Volume 731, Washington 1999, Chapter 19, pp. 263-287.

35. Vohlídal J., Kabátek Z., Pacovská M., Sedláček J., Grubišic-Gallot Z. (1996) Size exclusion chromatography of substituted acetylene polymers. Effect of autoxidative degradation of the polymer during analysis, *Collect. Czech. Chem. Commun.* **61**, 120-125.

36. Sedláček J., Pacovská M., Etrych T., Dlouhý M., Patev N., Cabioch S., Lavastre O., Balcar H., Žigon M., Vohlídal J. (1997) SEC study of autoxidative degradation of substituted acetylene polymers, *Polym. Mater. Sci. Eng.* **77**, 52-53.

37. Kabátek Z., Gaš B., Vohlídal J. (1997) Size exclusion chromatography of polymers degrading randomly in SEC column. theoretical treatment, *J. Chromatogr. A*, **786**, 209-218.

38. Kishimoto Y., Itou M., Miyatake T., Ikariya T., Noyori R. (1995) Polymerization of monosubstituted acetylenes with a zwitterionic rhodium(I) complex, $Rh^+(2,5$-norbornadiene)$[(\eta^6$-$C_6H_5)$-$B^-(C_6H_5)_3]$, *Macromolecules* **28**, 6662-6666.

39. Escudero A., Vilar R., Salcedo R., Ogawa T. (1995) Effects of substituent groups and substituted benzenes in the polymerization of phenylacetylenes initiated by di-μ-pentafluorothiophenolate bis(1,5-cyclooctadiene) rhodium(I), *Eur. Polym. J.* **31**, 1135-1138.

40. Balcar H., Sedláček J., Čejka J., Vohlídal J. (2002) MCM-41 Immobilized Rh(cod)(OCH_3)]_2 complex – a hybrid catalyst for polymerization of phenylacetylene and its ring-substituted derivatives, *Macromol. Rapid Commun.* **23**, 32-37.

41. Sedláček J., Pacovská M., Rédrová D., Balcar H., Biffis A., Corain B., Vohlídal J., (2002) Polybenzimidazole-supported $[Rh(cod)Cl]_2$ complex: Effective catalyst for the polymerization of substituted acetylenes, *Chem. Eur. J.* **8**, 366-371.

42. Yashima E., Huang S., Matsushima T., Okamoto Y. (1995) Synthesis and conformational study of optically active poly(phenylacetylene) derivatives bearing a bulky substituent, *Macromolecules*, **28**, 4184-4193.

43. Tabata M., Sone T., Sadahiro Y., Yokota K., Nozaki Y. (1998) Pressure-induced cis to trans isomerization of aromatic polyacetylenes prepared using a Rh complex catalyst: A control of π-conjugation length, *J. Polym. Sci. Polym. Chem.* **36,** 217-223.

44. Vohlídal J., Sedláček J., Patev N., Pacovská M., Lavastre O., Cabioch S., Dixneuf P.H., Blechta V., Matějka P., Balcar H. (1998) Comparative study of polymerization of 2-, 3- and 4-iodophenylacetylenes with Rh-, Mo- and W-based catalysts, *Collect. Czech. Chem. Commun.* **63,** 1815-1838.

45. Balcar H., Sedláček J., Zedník J., Vohlídal J., Blechta V. (2002) Polymerization of unconventional monosubstituted acetylenes with metathesis and insertion catalysts, in E. Khosravi and T. Szymanska-Buzar eds., *Ring Opening Metathesis Polymerization and Related Chemistry,* Kluwer Academic Publishers, Amsterdam, pp. 417-424.

46. Russo M.V., Furlani A., D'Amato R. (1998) Synthesis and properties of *p*-nitrophenylacetylene-phenylacetylene copolymers, *J. Polym. Sci. Polym. Chem.* **36,** 93-102.

47. Tang B.Z., Kong X., Wan X., Feng X.-D. (1997) Synthesis and properties of stereoregular polyacetylenes containing cyano groups, poly[[4-[[[*n*-[(4′cyano-4-biphenylyl)oxy]alkyl]oxy]carbonyl]-phenyl]acetylenes], *Macromolecules,* **1997,** 5620-5628.

48. Vohlídal J., Sedláček J., Patev N., Lavastre O., Dixneuf P.H., Cabioch S., Balcar H., Pfleger J., Blechta V. (1999) New substituted polyacetylenes with phenyleneethynylene side groups [–(C$_6$H$_4$–C≡C)$_n$–Si^{i}Pr$_3$; n = 1, 2]: Synthesis, characterization, spectroscopic and photoelectric properties; *Macromolecules, 32,* 6439-6449.

49. Lavastre O., Cabioch S., Dixneuf P.H., Sedláček J., Vohlídal J. (1999) New route to conjugated polymer networks: Synthesis of poly(4-ethynylphenylacetylene) and its transformation into conjugated network, *Macromolecules, 32,* 4477-4481.

50. Kaneko T., Horie T., Asano M., Aoki T., Oikawa E. (1997) Polydendron: Polymerization of dendritic phenylacetylene monomers, *Macromolecules, 30,* 3118-3121.

51. Balcar H., Sedláček J, Vohlídal J, Zednik J., Blechta V. (1999) New polyacetylenes with aromatic Schiff's base pendant groups by polymerization of benzylidene-ring-substituted N-benzylidene-4-ethynylanilines with Rh-based catalysts, *Macromol. Chem. Phys.* **200,** 2591-2596.

52. Teraguchi M., Masuda T. (2000) Synthesis and properties of polyacetylenes having azobenzene pendant groups, *Macromolecules, 33,* 240-242.

53. Sedláček J., Vohlídal J., Patev N., Pacovská M., Cabioch S., Lavastre O., Dixneuf P.H., Balcar H., Matějka P. (1999) Polymerization of 4-(ferrocenylethynyl)phenylacetylene with transition metal catalysts, *Macromol. Chem. Phys.* **200,** 972-976.

54. Tabata M., Yokota K., Namioka M. (1995) An electron-spin-resonance study of poly(alpha-ethynylnaphthalene) polymerized with [Rh(norbornadiene)Cl]$_2$ and WCl$_6$ as catalysts, *Macromol. Chem. Phys.* **196,** 2969-2977.

55. Tang B.Z., Poon W.H., Leung S.M., Leung W.H., Peng H. (1997) Synthesis of stereoregular poly(phenylacetylene)s by organorhodium complexes in aqueous media, *Macromolecules, 30,* 2209-2212.

56. Kwak G., Masuda T. (2002) Poly(phenylacetylene) with bulky chiral germyl groups: synthesis and effects of measuring solvents and temperature on chiroptical properties, *Polymer,* **43,** 665-669.

57. Kwak G., Masuda T. (2000) Synthesis, chiroptical properties, and high gas permeability of poly(phenylacetylene) with bulky chiral silyl groups, *Macromolecules,* **33,** 6633-6635.

58. Katayama H., Yamamura K., Miyaki Y., Ozawa F. (1997) Stereoregular polymerization of phenylacetylenes catalyzed by [hydridotris(pyrazolyl)borato]rhodium(I) complexes, *Organometallics,* **16,** 4497-4500.

59. Karim S.M.A., Nomura R., Masuda T. (2002) Synthesis and properties of polyacetylenes with salicylideneaniline groups, *J. Polym. Sci. Polym. Chem.* **40,** 2458-2463.

60. Ochiai B., Tomita I., Endo T. (2001) Coordination polymerization of a conjugated enyne: Synthesis of a novel polyacetylene derivative bearing conjugated double bond moieties, *Macromol. Rapid Commun.,* **22,** 1485-1487.

61. Nakako H., Nomura R., Tabata M., Masuda T. (1999) Synthesis and structure in solution of poly[(–)-menthylpropiolate] as a new class of helical polyacetylene, *Macromolecules, 32,* 2861-2864.

62. Maeda K., Goto H., Yashima E. (2001) Stereospecific polymerization of propiolic acid with rhodium complexes in the presence of bases and helix induction on the polymer in water, *Macromolecules* **34,** 1160-1164.

63. Lam J.W.Y., Luo J.D., Dong Y.P., Cheuk K.K.L., Tang B.Z. (2002) Functional polyacetylenes: Synthesis, thermal stability, liquid crystallinity, and light emission of polypropiolates, *Macromolecules,* **35,** 8288-8299.

64. Nomura R., Fukushima Y., Nakako H., Masuda T. (2000) Conformational study of helical poly(propiolic ester)s in solution, *J. Am. Chem. Soc.* **122,** 8830-8836.

65. Nomura R., Tabei J., Masuda T. (2002) Effect of side chain structure on the conformation of poly(*N*-propargylamide), *Macromolecules* **35**, 2955-2961.
66. Balcar H., Čejka J., Kubišta J., Petrusová L., Kubát P., Blechta V. (2000) Preparation and properties of isomeric N-(4-substituted benzylidene)-4-ethynylanilines and 4-substituted N-(4-ethynylbenzylidene)-anilines, *Collect. Czech. Chem. Commun.* **65**, 203-215.
67. Yang W., Tabata M., Kobayashi S., Yokota K., Shimizu A. (1991) Synthesis of ultra-high-molecular-weight aromatic polyacetylenes with [Rh(norbornadiene)Cl]$_2$ – triethylamine and solvent-induced crystallization of the obtained amorphous polyacetylenes, *Polym. J.* **23**, 1135-1138.
68. Tabata M., Takamura H., Yokota K., Nozaki Y., Hoshina T., Minakawa H., Kodaira K. (1994) Pressure-induced cis to trans isomerization of poly(o-methoxyphenylacetylene) polymerized by Rh complex catalyst - a Raman, X-ray, and ESR study, *Macromolecules,* **27**, 6234-6236.
69. Mawatari Y., Tabata M., Sone T., Ito K., Sadahiro Y. (2001) Origin of color of π-conjugated columnar polymers. I. Poly(*p*-3-methylbutoxy)phenylacetylene prepared using a [Rh(norbornadiene)Cl]$_2$ catalyst, *Macromolecules,* **34**, 3776-3782.
70. Tabata M., Sadahiro Y., Nozaki Y., Inaba Y., Yokota K. (1996) Hexagonal columns of poly(*n*-alkylpropiolate) produced with rhodium complex catalyst. X-ray analysis and oxygen permeability, *Macromolecules,* **29**, 6673-6675.
71. Nakako H., Mayahara Y., Nomura R., Tabata M., Masuda T. (2000) Effect of chiral substituents on the helical conformation of poly(propiolic ester)s, *Macromolecules,* **33**, 3978-3982.
72. Tabei J., Nomura R., Masuda T. (2002) Conformational study of poly(*N*-propargylamide)s having bulky pendant groups, *Macromolecules,* **35**, 5405-5409.
73. Percec V., Barboiu B., Neumann A., Ronda J. C., Zhao M. (1996) Metal-catalyzed "living" radical polymerization of styrene initiated with arenesulfonyl chlorides. From heterogeneous to homogeneous catalysis *Macromolecules*, **29**, 3665-3668.
74. Moineau G., Granel C., Dubois P., Jérôme R., Teyssié P. (1998) Controlled radical polymerization of methyl methacrylate initiated by an alkyl halide in the presence of the Wilkinson catalyst *Macromolecules*, **31**, 542-544.
75. Petrucci M.G.L., Lebuis A.M., Kakkar A.K. (1998) Rhodium(I) mixed CO/phosphine/amine complexes: Synthesis, structure, and reactivity, *Organometallics*, **17**, 4966-4975.
76. Opstal T., Zedník J., Sedláček J., Svoboda J., Vohlídal J., Verpoort F. (2002) Atom Transfer Radical Polymerization of Styrene and Methyl Methacrylate Induced by Rh^I(Cycloocta-1,5-diene) Complexes, *Collect. Czech. Chem. Commun.* **67**, 1858–1871.
77. Choi J.C., Osakada K., Yamaguchi I., Yamamoto T. (1997) Polymerization of arylallenes catalyzed by organo-rhodium(I) and -cobalt(I) complexes to give structurally regulated high-mass polymers, *Appl. Organomet. Chem.* **11**, 957-961.
78. Choi J.C., Osakada K., Yamamoto T. (1998) Single and multiple insertion of arylallene into the Rh-H bond to give (π-allyl)rhodium complexes, *Organometallics* **17**, 3044-3050.
79. Osakada K., Takenaka Y., Choi J.C., Yamaguchi I., Yamamoto T. (2000) Synthesis of linear and branched polyketones from the Rh complex catalyzed living alternating copolymerization of (4-alkylphenyl)allene with CO, *J. Polym. Sci. Polym. Chem.* **38**, 1505-1511.
80. Takenaka Y., Osakada K. (2001) Rh complex catalyzed alternating copolymerization of alkylallene or aryloxoallene with carbon monoxide: Influence of monomer structures on the reaction rate, *Macromol. Chem. Phys.* **202**, 3571-3578.
81. Takeuchi D., Choi J.C., Takenaka Y., Kim S., Osakada K. (2002) Polymerization of high potential monomers by transition metal complex catalysts, *Kobunshi Ronbunshu*, **59**, 342-355.
82. Li Y.N., Kawakami Y. (1999) Efficient synthesis of poly(silyl ether)s by Pd/C and RhCl(PPh3)(3)-catalyzed cross-dehydrocoupling polymerization of bis(hydrosilane)s with diols, *Macromolecules,* **32**, 6871-6873.
83. Zhang R.Z., Mark J.E., Pinhas A.R. (2000) Dehydrocoupling polymerization of bis-silanes and disilanols to poly(silphenylenesiloxane) as catalyzed by rhodium complexes, *Macromolecules,* **33**, 3508-3510.
84. Osakada K. (2000) Structure and chemical properties of mononuclear and dinuclear silylrhodium complexes. Activation of the Si-C bond and formation of Si-Cl and Si-SR bonds promoted by Rh complexes, J. Organomet. Chem., **611**, 323-331.
85. Yamaguchi I., Ishii H., Sakano T., Osakada K., Yamamoto T. (2001) Rhodium- and ruthenium-complex-catalyzed condensation of ferrocene-containing dithiols and diols with diarylsilanes to give silaferrocenophanes and ferrocene polymers, *Appl. Organomet. Chem.* **15**, 197-203.
86. Marciniec B., Malecka E. (1999) Synthesis of silazanylene-vinylene oligomers via catalytic polycondensation of divinyltetramethyldisilazane, *Macromol. Rapid Commun.* **20**, 475-479.

87. Marciniec B. (2000) New unsaturated organosilicon oligomers via catalytic polycondensation of divinylsubstituted silicon compounds, *Mol. Cryst. Liquid Cryst.* **354**, 761-770.
88. Marciniec B. (2000) Silicometallics and catalysis, *Appl. Organomet. Chem.* **14**, 527-538.
89. Chen R.M., Chien K.M., Wong K.T., Jin B.Y., Luh T.Y., Hsu J.H., Fann W.S. (1997) Synthesis and photophysical studies of silylene-spaced divinylarene copolymers. Molecular weight dependent fluorescence of alternating silylene-divinylbenzene copolymers, *J. Am. Chem. Soc.* **119**, 11321-11322.
90. Mori A., Takahisa E., Kajiro H., Nishihara Y., Hiyama T. (2000) Regio- and stereocontrolled hydrosilylation polyaddition catalyzed by RhI(PPh$_3$)$_3$. Syntheses of polymers containing (*E*)- or (*Z*)-alkenylsilane moieties, *Macromolecules,* **33**, 1115-1116.
91. Ogawa T., Murakami M. (1996) Synthesis, thermal and mechanical properties of poly(methylphenylsilmethylene)s, *Chem. Mater.* **8**, 1260-1267.
92. Temple K., Dziadek S., Manners I. (2002) Highly active cationic rhodium(I) precatalysts for the ambient temperature ring-opening polymerization of [1]silaferrocenophanes and tetramethyldisilacyclobutane, *Organometallics*, **21**, 4377-4384.
93. Dutta P.K., Puri M. (1988) Formation of trans-polyacetylene on transition-metal zeolites – A resonance Raman study, *J. Catal.* **111**, 453-456.
94. Cox S.D., Stucky G.D. (1991) Polymerization of methylacetylene in hydrogen zeolites, *J. Phys. Chem.* **95**, 710-720.
95. Pereira C,. Kokotailo G.T., Gorte R.J. (1991) Acetylene polymerization in a H-ZSM-5 zeolite, *J. Phys. Chem.* **95**, 705-709.
96. Bordiga S., Ricchiardi G., Spoto G., Scarano D., Carnelli L., Zecchina A., Arean C.O. (1993) Acetylene, methylacetylene and ethylacetylene polymerization on H-ZSM5 - a spectroscopic study, *J. Chem. Soc. - Faraday Trans.* **89**, 1843-1855.
97. RuizHitzky E., Aranda P. (1997) Confinement of conducting polymers into inorganic solids, *Anales Quim.* **93**, 197-212.
98. Cardin J. (2002) Encapsulated conducting polymers, *Adv. Mater.* **14**, 553-563.
99. Alvaro M., Ferrer B., Garcia H., Lay B., Trinidad F., Valenciano J. (2002) Remarkably high charge uptake for modified electrodes of polyacetylene molecular wires encapsulated within zeolites and mesoporous MCM-41 aluminosilicate, *Chem. Phys. Lett.* **356**, 577-584.
100. Cardin D.J., Constantine S.P., Gilbert A., Lay A.K., Alvaro M., Galletero M.S., Garcia H., Marquez F. (2001) Polymerization of alkynes in the channels of mesoporous materials containing Ni and Zn cations: Almost complete filling of the voids, *J. Am. Chem. Soc.* **123**, 3141-3142.
101. Galletero M.S., Alvaro M., Garcia H., Gomez-Garcia C.J., Lay A.K. (2002) Spontaneous doping and magnetic properties of polyacetylene and polypropyne synthesized in situ in Ni-exchanged mordenite and mesoporous MCM-41, *Phys. Chem. Chem. Phys.* **4**, 115-120.
102. Lin V.S.Y., Radu D.R., Han M.K., Deng W.H., Kuroki S., Shanks B.H., Pruski M. (2002) Oxidative polymerization of 1,4-diethynylbenzene into highly conjugated poly(phenylene butadiynylene) within the channels of surface-functionalized mesoporous silica and alumina materials, *J. Am. Chem. Soc.* **124**, 9040-9041.
103. Corma A. (1997) From microporous to mesoporous molecular sieve materials and their use in catalysis, *Chem. Rev.* 97, 2373-2419.
104. Ying J.Y., Mehnert C.P., Wong M.S. (1999) Synthesis and applications of supramolecular-templated mesoporous materials, *Angew. Chem. Int. Edit.* **38**, 56-77.
105. Rahiala H., Beurroies I., Eklund T., Hakala K., Gougeon R., Trens P., Rosenholm J.B. (1999) Preparation and characterization of MCM-41 supported metallocene catalysts for olefin polymerization, *J. Catal.* **188**, 14-23.
106. Melis K, De Vos D, Jacobs P, Verpoort F. (2001) ROMP and RCM catalysed by (R$_3$P)$_2$Cl$_2$Ru=CHPh immobilised on a mesoporous support, *J. Mol. Catal. A-Chem.* **169**, 47-56.
107. Balcar, H., Čejka, J., Sedláček, J., Svoboda, J., Zedník, J., Bastl, Z., Bosáček, V., Vohlídal, J. (2003) [Rh(cod)Cl]$_2$ Complex Immobilized on Mesoporous Molecular Sieves MCM-41 – A New Hybrid Catalyst for Polymerization of Phenylacetylene, *J. Mol. Catal. A: Chem.* submitted.
108. Balcar H., Čejka J., Sedláček J., Svoboda J., Bastl Z., Pacovská M., Vohlídal J. (2003) Mesoporous molecular sieves immobilized catalysts for polymerization of phenylacetylene and its derivatives, *this issue*.
109. D'Archivio A.A., Galantini L., Biffis A., Jeřábek K., Corain B. (2000) Polybenzimidazole as a promising support for metal catalysis: Morphology and molecular accessibility in the dry and swollen state, *Chem. Eur. J.* **6**, 794-799.
110. Leinonen S., Sherrington D.C., Sneddon A., McLoughlin D., Corker J., Canevali C., Morazzoni F., Reedijk J., Spratt S.B.D. (1999) Molecular structural and morphological characterization of polymer-supported Mo(VI) alkene epoxidation catalysts, *J. Catal.* **183**, 251-266.

111. Olason C., Sherrington D.C. (1999) Oxidation of cyclohexene by t-butylhydroperoxide and dioxygen catalysed by polybenzimidazole-supported Cu, Mn, Fe, Ru and Ti complexes, *React. Funct. Polym.* **42**, 163-172.

112. Leadbeater N.E., Marco M. (2002) Preparation of polymer-supported ligands and metal complexes for use in catalysis, *Chem. Rev.* **102**, 3217-3273.

113. Sheldrick W.S., Gunther B. (1989) Synthesis and structure of the mixed bridged diene-rhodium(I) complex [(cod)Rh(μ-Cl)(μ-OAc)Rh(cod)] - reactions with the modified purine bases *N*-6,*N*-6-dimethyladenine and 8-aza-9-methyladenine, *J. Organomet. Chem.* **375**, 233-243.

114. Mastrorilli, P., Nobile, C.F., Rizzuti, A., Suranna, G.P., Acierno, D. and Amendola, E. (2002) Polymerization of phenylacetylene and p-tolyacetylene catalyzed by beta-dioxygenato rhodium(I) complexes in homogeneous and heterogeneous phase, *J. Mol. Catal. A: Chem.* **178**, 35-42.

115. Bianchini C., Frediani M., Mantovani G., Vizza F. (2001) Synthesis of polymer-supported rhodium(I)-1,3-bis(diphenylphosphino)propane moieties and their use in the heterogeneous hydrogenation of quinoline and benzylideneacetone, *Organometallics* **20**, 2660-2662.

116. Bianchini C., Frediani M., Vizza F. (2001) Synthesis of the first polymer-supported tripodal triphosphine ligand and its application in the heterogeneous hydrogenolysis of benzo[b]thiophene by rhodium catalysis, *Chem. Commun.* 479-480.

117. Lopez-Castillo Z.K., Flores R., Kani I., Fackler J.P., Akgerman A. (2002) Fluoroacrylate copolymer-supported rhodium catalysts for hydrogenation reactions in supercritical carbon dioxide, *Ind. Eng. Chem. Res.* **41**, 3075-3080.

118. Csihony S., Fischmeister C., Bruneau C., Horvath I.T., Dixneuf P.H. (2002) First ring-opening metathesis polymerization in an ionic liquid. Efficient recycling of a catalyst generated from cationic ruthenium allenylidene complex, *New J. Chem.* **26**, 1667-1670.

119. Mastrorilli P., Nobile C.F., Gallo V., Suranna G.P., Farinola G. (2002) Rhodium(I) catalyzed polymerization of phenylacetylene in ionic liquids, *J. Mol. Cat. A-Chem.* **184**, 73-78.

MESOPOROUS MOLECULAR SIEVES IMMOBILIZED CATALYSTS FOR POLYMERIZATION OF PHENYLACETYLENE AND ITS DERIVATIVES

H. BALCAR[a], J. ČEJKA[a], J. SEDLÁČEK[b], J. SVOBODA[b], Z. BASTL[a], M. PACOVSKÁ[b] AND J. VOHLÍDAL[b]

[a] J. Heyrovský Institute of Physical Chemistry, Academy of Sciences of the Czech Republic, CZ-182 23 Prague 8, Czech Republic

[b] Department of Physical and Macromolecular Chemistry, Laboratory of Specialty Polymers Faculty of Science, Charles University, Albertov 2030, CZ-128 40, Prague 2, Czech Republic

1. Introduction

Substituted polyvinylenes, polymers with conjugated polyene main chains, attract attention because of their unique properties implicating applications in electronics and optics [1] (electro- and photoconductivity, electro- and photoluminescence, optic non-linearity). So far, these polymers have been prepared almost exclusively by homogeneously catalyzed polymerization of corresponding substituted acetylenes with transition metal catalysts (most frequently compounds of W, Mo, Rh and Pd) [2-4]. However, catalyst residues in polymers may undesirably affect the polymer properties (especially those essential for their applications) and polymer purification (e.g. by repeated polymer precipitation) leads often to the decrease of molecular weight and changes in microstructure resulting from polymer degradation [5,6]. Therefore, new polymerization procedures minimizing the content of catalyst residues in the resulting polymers can be of important advantage.

Hybrid catalysts prepared by immobilization of catalytically active transition metal complexes on meso- and macroporous supports promise to combine high activity and selectivity with an easy separation of product from catalyst. The first hybrid catalysts for polymerization of phenylacetylene (PhA) using organic insoluble polymers as supports have been reported recently [7,8]. To improve thermal and mechanical stability of hybrid catalysts and to avoid effect of support swelling on catalyst activity, we decided to use inorganic support instead of polymeric ones, especially siliceous mesoporous molecular sieves, which represent a modern inorganic support with a high surface area (about 1000 m^2/g), regular architecture and controlled pore size (pore diameter from 3 to 10 nm) [9].

In this contribution we report new hybrid catalysts for acetylenes polymerization prepared by anchoring rhodium(I) complexes $[Rh(cod)X]_2$ (X = OCH_3, Cl), (cod = η^4-cycloocta-1,5-diene) on siliceous molecular sieves MCM-41 and SBA-15. The given Rh-complexes are widely used as homogeneous catalysts for polymerization of substituted acetylenes. They exhibit high activity, stereoselectivity in a sense of providing polymers of a high cis-transoidal microstructure, and high tolerance to a variety of functional groups in polymerization systems (monomer/solvent). An important advantage for their employment as a component of hybrid catalysts is also their low moisture- and oxygen-sensitivity.

155

Y. Imamoglu and L. Bencze (eds.), Novel Metathesis Chemistry: Well-Defined Initiator Systems for Specialty Chemical Synthesis, Tailored Polymers and Advanced Material Applications, 155–165.

2. Catalysts preparation

2.1. PREPARATION OF MOLECULAR SIEVES MCM-41 AND SBA-15

The synthesis of mesoporous molecular sieves was carried out using a new approach where the self-assembled molecular aggregates or supramolecular assemblies were applied as structure-directing agents. Mesoporous molecular sieves MCM-41 were synthesized using hexadecyltrimethylammonium bromide, sodium silicate and methyl acetate at 90 °C for 48 h. Mesoporous molecular sieves SBA-15 were prepared from tetraethyl orthosilicate using Pluronic 10400 as structure-directing agent at 80 °C for 48 h. After the synthesis, both materials were washed by distilled water, dried at 50 °C overnight and calcined at 550 °C for 8 h in a stream of air [9,10]. X-ray powder diffraction confirmed regular hexagonal structure of both MCM-41 and SBA-15 molecular sieves. Surface area (S_{BET}), pore volume (V_p) and pore diameter (d) of samples prepared were determined from N_2 adsorption isotherms (Fig. 1) as follows: S_{BET} = 1 032 m^2/g, V_p = 0.81 cm^3/g, d = 3.7 nm for MCM-41, and S_{BET} = 720 m^2/g, V_p = 0.95 cm^3/g, d = 7.4 nm for SBA-15. Particle size about 3µm was determined by SEM for both samples.

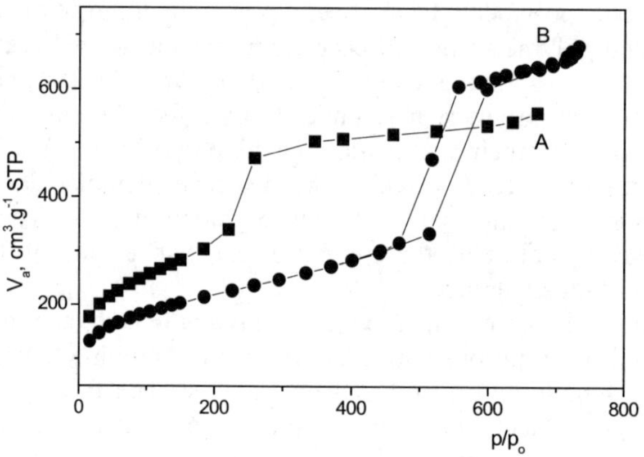

Figure 1. N_2 adsorption and desoption isotherms for MCM-41 (A) and SBA-15 (B) samples. Accusorb 2100E Micrometrics, -196°C

2.2. RHODIUM COMPLEXES ANCHORING

The ways of Rh-complexes anchoring are shown in Scheme 1. [Rh(cod)OCH$_3$]$_2$ anchoring was achieved by mixing dry MCM-41 with THF solution of [Rh(cod)OCH$_3$]$_2$ at room temperature [11]. [Rh(cod)OCH$_3$]$_2$ reacts with surface Si-OH groups under methanol elimination. After 24 h of stirring, [Rh(cod)OCH$_3$]$_2$ is consumed quantitatively and Rh-loading can be calculated directly from the amount of Rh-complex submitted for immobilization. Hybrid catalyst, labelled as MCM-41/Rh

(Rh-loading = 1 wt. %), was thus prepared. [Rh(cod)Cl]$_2$ cannot be immobilized by a direct reaction with sieves Si-OH groups. However, both MCM-41 and SBA-15 modified on the surface by the reaction with 3-aminopropyltrimethoxysilane (APTMS), react with [Rh(cod)Cl]$_2$ smoothly under quantitative [Rh(cod)Cl]$_2$ consumption, providing hybrid catalysts MCM-41/APTMS/Rh [12] and SBA-15/APTMS/Rh, respectively (Rh-loading = 1 wt.% for both catalysts).

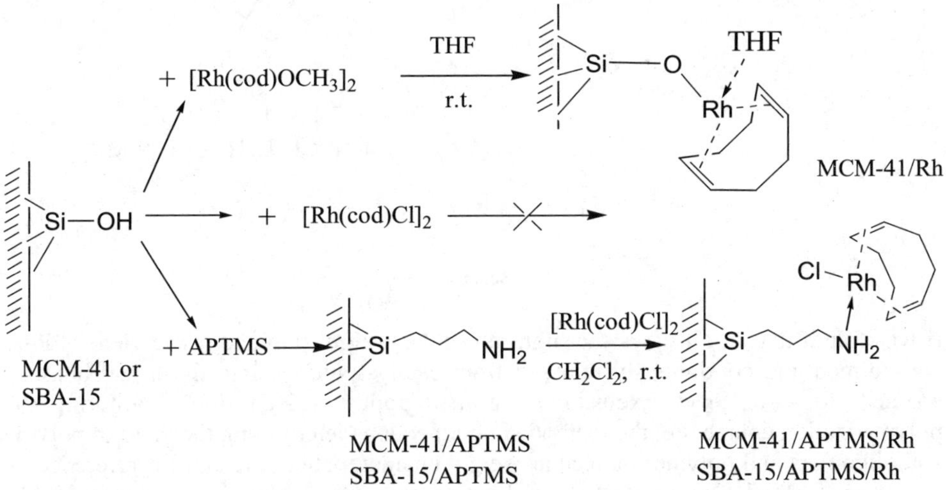

Scheme 1

3. Polymerization activity

The activity of hybrid catalysts was tested in polymerization of PhA (Scheme 2, X = H).

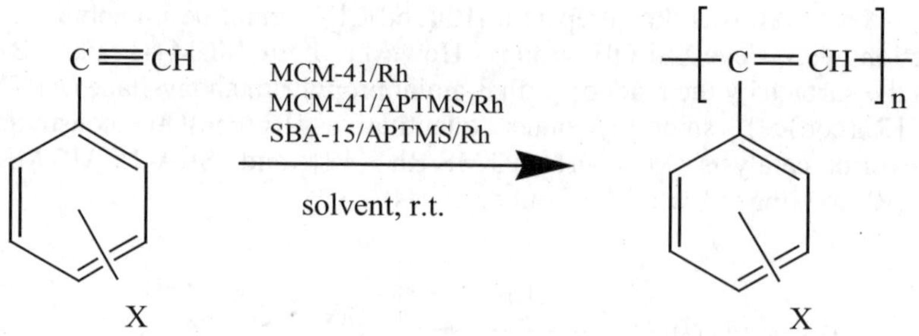

$$C \equiv CH \quad \xrightarrow[\text{solvent, r.t.}]{\substack{\text{MCM-41/Rh} \\ \text{MCM-41/APTMS/Rh} \\ \text{SBA-15/APTMS/Rh}}} \quad \left[C = CH \right]_n$$

$$X = H,\ 2\text{-F},\ 4\text{-F},\ 4\text{-CH}_2(\text{CH}_2)_3\text{CH}_3,$$
$$4\text{-N=CH-C}_6\text{H}_4\text{-C} \equiv \text{C-SiMe}_3$$

Scheme 2

It was found that in all cases the high-molecular-weight polyphenylacetylene (PPhA) was formed and continuously released from catalyst surface into the liquid phase. It enabled to use Size Exclusion Chromatography (SEC) for monitoring the polymerization course, i.e. the method simultaneously determining the yield of polymer and oligomers and polymer molecular-weight characteristics (M_w and M_n referenced to polystyrene standards) in a given polymerization time. The time courses of PhA polymerizations performed in tetrahydrofuran (THF) and catalyzed by MCM-41/Rh and MCM-41/APTMS/Rh are given in Fig. 2a together with the time courses of PhA polymerizations induced with [Rh(cod)Cl]$_2$ and [Rh(cod)OCH$_3$]$_2$ complexes applied as homogeneous catalysts. As it is evident the initial polymerization rate with hybrid catalysts is lower than that with homogeneous ones most probably due to slow diffusion processes; the final polymer yield achieved with hybrid catalysts is also somewhat lower compared to that gained in homogeneously performed polymerizations. On the other hand, molecular weight of PPhA prepared with hybrid catalysts is higher than that of PPhA prepared with their homogeneous counterparts (Fig. 2b). The slight decrease of polymer molecular weight in the course of polymerization may be due to the formation of shorter polymer chains at later stages of polymerization and/or to the slow polymer degradation [5,6]. Polydispersity index, M_w/M_n, in the range from 2.0 to 3.5 was determined for PPhA(s) prepared with both kinds of catalysts. Lower initial polymerization rate achieved with [Rh(cod)Cl]$_2$ in comparison with that for [Rh(cod)OCH$_3$]$_2$ catalyst results from a lower rate of [Rh(cod)Cl]$_2$ dissociation into mononuclear catalytically active species. Contrary to this, polymerization activity of MCM-41/Rh and MCM-41/APTMS/Rh is practically the same, because complex dissociation has already occurred during the anchoring process.

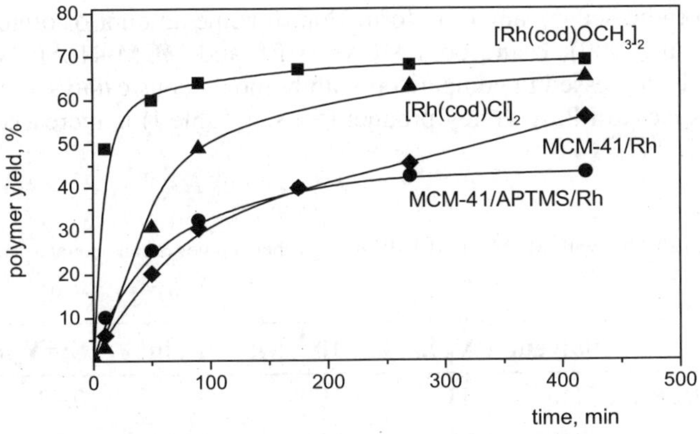

Figure 2a. PPhA yield vs. time for PhA polymerization with hybrid catalysts MCM-41/Rh and MCM-41/APTMS/Rh and parent complexes [Rh(cod)OCH₃]₂ and [Rh(cod)Cl]₂. Room temperature, THF, [Rh] = 1.5 mmol/l, [PhA]₀ = 0.6 mol/l. Data taken from [11,12].

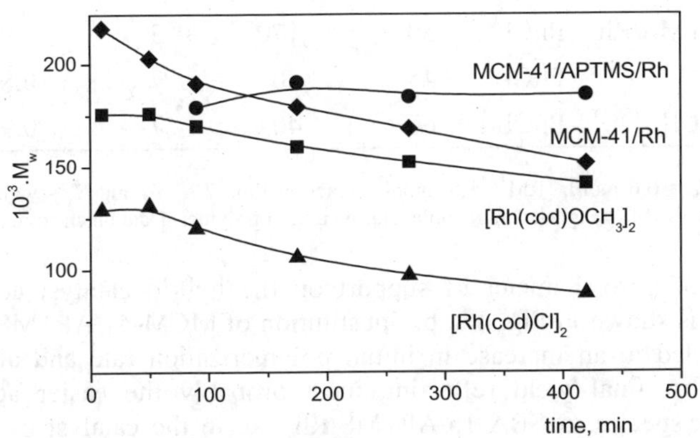

Figure 2b. M_w of PPhA vs. time for PhA polymerization with hybrid catalysts MCM-41/Rh and MCM-41/APTMS/Rh and parent complexes [Rh(cod)OCH₃]₂ and [Rh(cod)Cl]₂. Room temperature, THF, [Rh] = 1.5 mmol/l, [PhA]₀ = 0.6 mol/l. Data taken from [11,12].

Polymerization of PhA with MCM-41/APTMS/Rh was studied also in chlorobenzene (PhCl) and CH_2Cl_2, i.e. in solvents less polar than THF. Polymerization results (Table I) were compared with those achieved with two homogeneous catalysts: (i) parent [Rh(cod)Cl]₂ and (ii) (η^4-cycloocta-1,5-diene)(n-propylamino)chloro rhodium(I), [Rh(cod)(PA)Cl], structure of which is close to the structure of anchored Rh-species. It can be seen that, contrary to the homogeneous catalysts, polymerization activity of MCM-41/APTMS/Rh does not depend on the solvent used. In comparison with both homogeneous catalysts, MCM-41/APTMS/Rh provides (i) comparable (in THF, PhCl) or even higher (in CH_2Cl_2) PPhA yield, and (ii) considerably higher molecular weight of PPhA formed (in all solvent tested). Polymerization of PhA with Rh-homogeneous catalysts is known to be accompanied by formation of some amount of oligomeric by-products (MW about 2 000). Using both MCM-41/Rh and MCM-41/APTMS/Rh the yield of oligomers is depressed in comparison with homogeneous catalysts and catalyst selectivity with respect to PPhA as key product (S_p, see Table I) is increased (Table I, ref. [11]).

TABLE I. Polymerization of PhA with MCM-41/APTMS/Rh and their homogeneous counterparts in various solvents

Catalyst	Solvent	Y_p in %	$10^{-3} \cdot M_w$	Y_o in%	$S_p=Y_p/(Y_p+Y_o)$
MCM-41/APTMS/Rh	THF	43	190	3	0.93
[Rh(cod)Cl]$_2$	THF	65	90	6	0.92
[Rh(cod)(PA)Cl]	THF	31	61	6	0.84
MCM-41/APTMS/Rh	CH$_2$Cl$_2$	64	130	3	0.96
[Rh(cod)Cl]$_2$	CH$_2$Cl$_2$	15	15	12	0.56
[Rh(cod)(PA)Cl]	CH$_2$Cl$_2$	22	21	7	0.76
MCM-41/APTMS/Rh	PhCl	50	170	3	0.94
[Rh(cod)Cl]$_2$	PhCl	45	20	7	0.87
[Rh(cod)(PA)Cl]	PhCl	60	40	7	0.90

Room temp, $[PhA]_0$ = 0.6 mol/l, [Rh] = 1.5 mmol/l, reaction time 7 h, Y_p and Y_o - yield of polymer and oligomers, respectively, M_w weight-average molecular weight of polymer. Data taken from [11,12].

The influence of pore diameter of support on the hybrid catalyst activity in PhA polymerization is shown in Fig. 3 a, b. Substitution of MCM-41/APTMS/Rh for SBA-15/APTMS/Rh led to an increase in initial polymerization rate and also to a slight increase in PPhA final yield reflecting most probably the easier accessibility of anchored active species in SBA-15/APTMS/Rh, i.e. in the catalyst of a higher pore diameter. On the other hand, SBA-15/APTMS/Rh provides PPhA of lower molecular weight than MCM-41/APTMS/Rh. The effect of pore diameter on polymerization rate was more pronounced in polymerization of a more bulky monomer, N-(4-trimethysilylethynylbenzylidene)-4-ethynylaniline [TMSEBA, Scheme 2, X = 4-(N=CH-C$_6$H$_4$-C≡C-SiMe$_3$)] (Fig. 4 a, b). Similarly to PhA polymerization, SBA-15/APTMS/Rh provided P(TMSEBA) of a lower molecular weight as compared to MCM-41/APTMS/Rh. However, it is worth to notice, that molecular weight of P(TMSEBA) prepared with both hybrid catalysts is considerably higher than that achieved with [Rh(cod)OCH$_3$]$_2$ homogeneous catalyst (M_w = 170 000) which was applied for TMSEBA polymerization in the recent study [13].

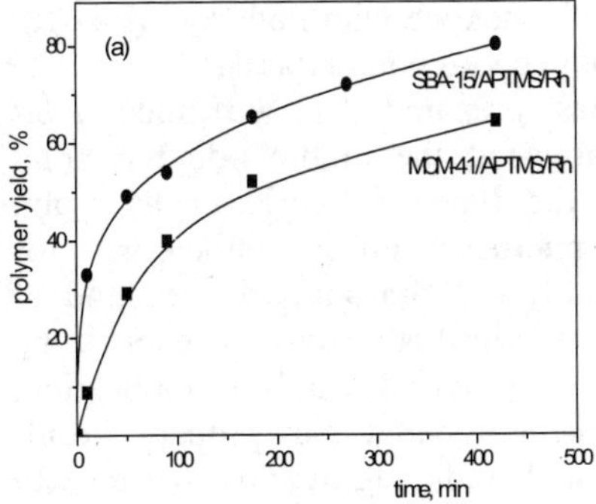

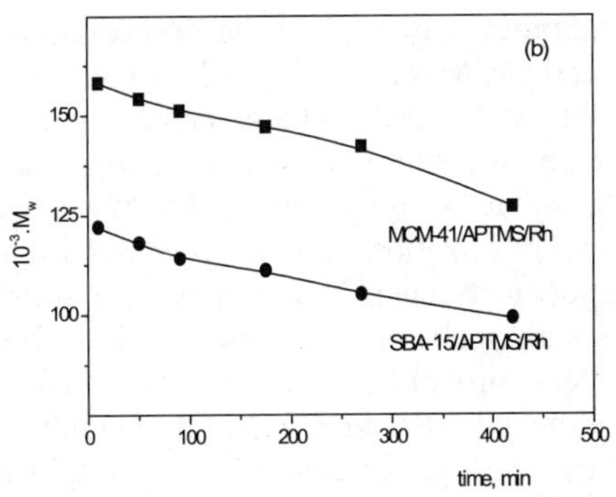

Figure 3 a, b. PPhA yield (a) and M_w (b) vs. time for PhA polymerization with MCM-41/APTMS/Rh and SBA-15/APTMS/Rh. Room temperature, CH_2Cl_2, [Rh] = 1.5 mmol/l, $[PhA]_0$ = 0.6 mol/l.

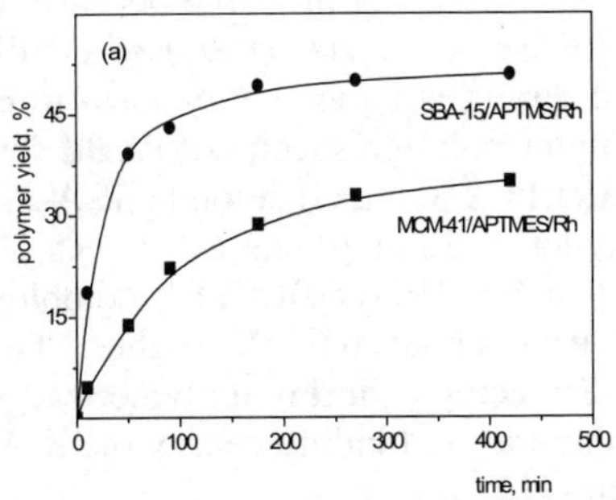

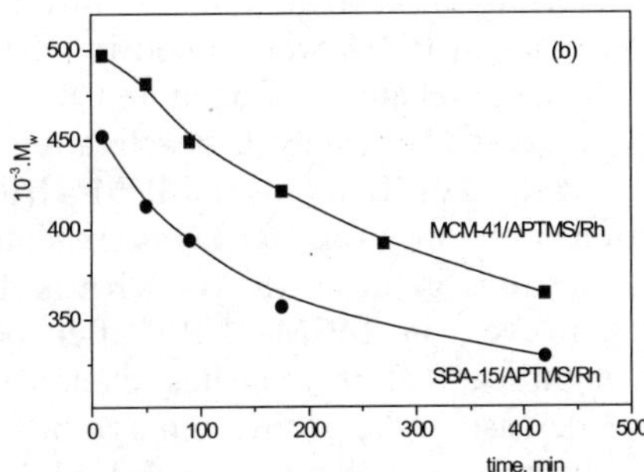

Figure 4 a, b. P(TMSEBA) yield (a) and M_w (b) vs. time for TMSEBA polymerization with MCM-41/APTMS/Rh and SBA-15/APTMS/Rh. Room temperature, THF, [Rh] = 0.3mmol/l, $[TMSEBA]_0$ = 0.15 mol/l.

The influence of PhA ring substituents X of a less bulky-type, i.e. 2-F, 4-F, 4-$CH_2(CH_2)_3CH_3$ (Scheme 2) on the yield and molecular weight of polymers prepared was studied using MCM-41/Rh. It was found [11] that both these quantities decrease in the order of decreasing acidity of acetylene hydrogen of monomer: 2-F > H > 4-F > 4-$CH_2(CH_2)_3CH_3$. The positive effect of acetylene hydrogen acidity on the monomer polymerizability has been already observed in polymerizations with $[Rh(cod)OCH_3]_2$ homogeneous catalyst [13]. The very low acidity of acetylene hydrogen of linear alkynes may explain the failure in polymerization of 1-hexyne with both MCM-41/Rh and MCM-41/APTMS/Rh as well as with their homogeneous counterparts.

IR and NMR characterization of polyacetylenes prepared (i) confirmed their polyvinylene structure (Scheme 2), and (ii) established the high-cis-double bond content in polymers (90-95%) irrespectively of the type of hybrid catalyst and polymerization solvent used. As high cis polymer microstructure is characteristic for polyacetylenes prepared with Rh-homogeneous catalysts, high similarity in kind of catalytically active centres in both homogeneous and hybrid Rh catalysts seems to be very probable. The observed effects connected with hybrid catalyst application, especially the increase in molecular weight of polymers prepared, may probably result from a specific environment in hybrid catalyst channels reducing the polymer growth termination.

4. Catalysts stability and re-usability

Both MCM-41/Rh and MCM-41/APTMS/Rh are stable and can be stored (under inert atmosphere) for several months. They are also resistant to Rh leaching during their applications. However, attempts for their re-using in PhA polymerization failed. In the second polymerization cycle, only about 2 % yield of PPhA was achieved after 7 h. A blocking of catalyst pores by insoluble polymer deposition is not probable because no signals of PPhA were found in DRIFT spectra of the used catalysts. Thus, irreversible chemical changes of anchored Rh species occurred during polymerization seem to be the reason for catalysts deactivation. Fig. 5 shows high resolution spectra of Rh 3d core electrons of Rh atoms in MCM-41/Rh before and after PhA polymerization (ESCA 310 electron spectrometer, Grammadata Scienta, Sweden). Binding energy of Rh 3d electrons in fresh MCM-41/Rh is the same as that in free $[Rh(cod)OCH_3]_2$ complex, however, in MCM-41/Rh after polymerization, it is about 0.6 eV higher. This significant shift indicates chemical changes of Rh centres during polymerization (increase in the positive charge of Rh atoms). Similar shift in binding energy of Rh 3d electrons was observed for MCM-41/APTMS/Rh, too [12].

In a special experiment, PhA polymerization with MCM-41/Rh was followed by ^{1}H NMR spectroscopy ([Rh] = 4 mmol/l, [PhA]$_0$ = 0.09 mol/l, room temperature, THF-d$_8$). Signals of free cycloocta-1,5-diene at 2.33 ppm and 5.52 ppm appeared in supernatant 60 min after the onset of polymerization. After PhA total consumption, volatiles were distilled off in vacuo and cycloocta-1,5-diene and also cycloocta-1,3-diene were found in this distillate by GC (total amount of cyclooctadienes corresponded to the amount of Rh species in MCM-41/Rh). Therefore, liberation of *cod* ligand from Rh species during polymerization (probably due to its replacement by PhA molecule(s) and/or polymer or oligomer segments) may be the main reason for the observed hybrid catalyst deactivation.

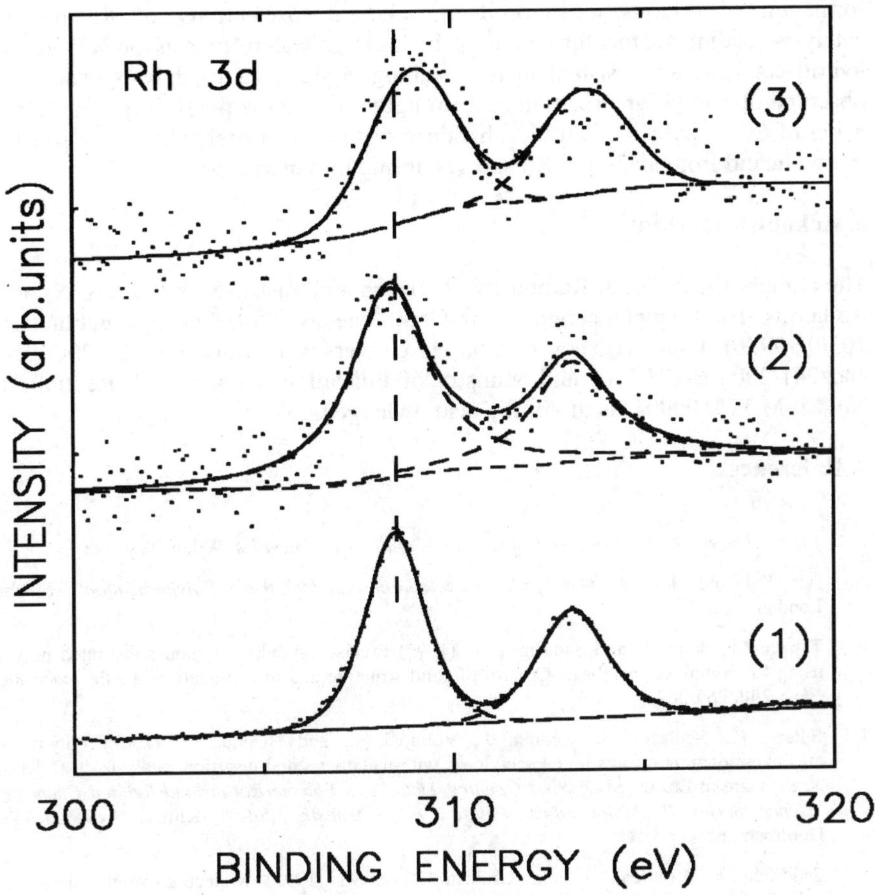

Figure 5. Spectra of Rh 3d electrons in [Rh(cod)OCH$_3$]$_2$ (1), MCM-41/Rh before (2) and after (3) PhA polymerization.

5. Conclusions

New hybrid catalysts for polymerization of phenylacetylene and its derivatives into high-molecular-weight polyphenylacetylenes were prepared by anchoring [Rh(cod)X]$_2$ (X = Cl, OCH$_3$) on siliceous mesoporous molecular sieves MCM-41 and SBA-15. The polymerization proceeds on the catalyst surface and the polymer formed is continuously released into the liquid phase from which it can be easily isolated. In comparison with homogeneous analogues, the hybrid catalysts exhibit comparable or even higher polymerization activity. Especially, the immobilization of Rh-complexes brings about increase in molecular weight of polymers formed and parallel reduction in formation of oligomeric by-products. The stereoselectivity of Rh homogeneous catalysts leading to the formation of high-cis-polyacetylenes is preserved. Although hybrid catalysts are resistant to Rh leaching, a significant catalysts deactivation was observed during polymerization that strongly reduces the possibility of an efficient re-using of tested hybrid catalysts. This deactivation most probably consists in liberation of *cod* ligand from anchored Rh species during polymerization.

6. Acknowledgement

The authors thank Dr. J. Rathouský (J. Heyrovský Inst.) for measuring N_2 adsorption isotherms. The financial support from Grant Agency of the Czech Republic (Grant No. 203/02/0976), Grant Agency of Charles University (Grants No. 228/2000/B-CH/PrF and 241/2002/B-CH/PrF) and Ministry of Education of the Czech Republic (Project No. MSM 113100001) is gratefully acknowledged.

7. References

1. Nalva, H.S. ed. (1996) *Handbook of Organic Conductive Molecules,* Wiley, New York.

2. Ivin, K.J. and Mol, J.C. (1997) *Olefin Metathesis and Metathesis Polymerization,* Academic Press, London.

3. Tabata, M., Sone, T. and Sadashiro, Y. (1999) Precise synthesis of monosubstituted polyacetylenes using Rh complexes catalysts. Control of solid structure and pi-conjugation length, *Macromol. Chem. Phys.* **200,** 265-282.

4. Balcar, H., Sedláček, J., Zedník, J., Vohlídal, J., and Blechta, V. (2002) Polymerization of unconventional monosubstituted acetylenes with metathesis and insertion catalysts, in E. Khosravi and T. Szymanska-Buzar (eds.), *Ring Opening Metathesis Polymerization and Related Chemistry, NATO Science Series II. Mathematics, Physics and Chemistry Vol.56,* Kluwer Academic Publishers, Dordrecht, pp 417-424.

5. Sedláček, J., Vohlídal, J. and Grubišic-Gallot, Z. (1993) Molecular-weight determination of polyphenylacetylene by size-exclusion chromatography/low-angle laser light scattering. Influence of polymer degradation, *Makromol. Chem. Rapid Commun.* **14,** 51-53.

6. Karim, S.M.A., Nomura, R. and Masuda, T. (1999) Synthesis and properties of poly(phenylacetylene)s with pendant imino groups, *Polym. Bull.* **43,** 305-310.

7. Mastrorilli, P., Nobile, C.F., Rizzuti, A., Suranna, G.P., Acierno, D. and Amendola, E. (2002) Polymerization of phenylacetylene and p-tolyacetylene catalyzed by beta-dioxygenato rhodium(I) complexes in homogeneous and heterogeneous phase, *J. Mol. Catal. A: Chem.* **178,** 35-42.

8. Sedláček, J., Pacovská, M., Rédrová, D., Balcar, H., Biffis, A., Corain, B. and Vohlídal, J. (2002) Polybenzimidazole-Supported [Rh(cod)Cl]$_2$ Complex: Effective Catalyst for the Polymerization of Substituted Acetylenes, *Chem. Eur. J.* **8,** 366-371.

9. Kooyman, P., Slabová, M., Bosáček, V., Čejka, J., Rathouský, J. and Zukal, A. (2001) The influence of pH on the structure of templated mesoporous silicas prepared from sodium metasilicate, *Collect. Czech. Chem. Commun.* **66,** 555-566.

10. Čejka, J., Žilková, N., Rathouský, J. and Zukal, A. (2001) Nitrogen adsorption study of organised mesoporous alumina, *Phys. Chem. Chem. Phys.* **3,** 5076-5081.

11. Balcar, H., Sedláček, J., Čejka, J. and Vohlídal, J. (2002) MCM-41-Immobilized [Rh(cod)OCH$_3$]$_2$ Complex – A Hybrid Catalyst for the Polymerization of Phenylacetylene and Its Ring-Substituted Derivatives, *Macromol. Rapid Commun.* **23**, 32-37.

12. Balcar, H., Čejka, J., Sedláček, J., Svoboda, J., Zedník, J., Bastl, Z., Bosáček, V. and Vohlídal, J. (2002, submitted) [Rh(cod)Cl]$_2$ Complex Immobilized on Mesoporous Molecular Sieves MCM-41 – A New Hybrid Catalyst for Polymerization of Phenylacetylene, *J. Mol. Catal. A: Chem.*

13. Balcar, H., Sedláček, J., Zedník, J., Blechta, V., Kubát, P. and Vohlídal, J. (2001) Polymerization of isomeric N-(4-substituted benzylidene)-4-ethynylanilines and 4-substituted N-(4-ethynylbenzylidene)anilines by transition metal catalysts: preparation and characterization of new substituted polyacetylenes with aromatic Schiff base type pendant groups, *Polymer* **42**, 6709-6721.

STUDY OF THE STABILITY AND ACTIVITY OF ELECTROCHEMICALLY PRODUCED TUNGSTEN-BASED METATHESIS CATALYST WITH SYMMETRICAL ALKENES

S. ÇETINKAYA, B. DÜZ AND Y. IMAMOĞLU*
Department of Chemistry, HacettepeUniversity,
06532 Beytepe, Ankara, Turkey

1. Introduction

The olefin metathesis reaction has been carried out with many olefins of different types over various kinds of both homogeneous and heteregeneous catalysts [1-3]. The activity of a particular catalyst system depends on a number of factors , such as proportions of the components, the order in which the components are mixed, pretreatment procedures and reaction time [4]. Optimization of a given catalyst system by adjustment of the various parameters inturn can be quite a lengthy procedure.

Electrochemical reduction of WCl_6 or $MoCl_5$ in methylene chloride, with an aluminum anode, gives in situ formation of a species that catalyzes olefin metathesis [5-10]. Good activity and high selectivity is maintained even after several charges of olefin, e.g. pent-2-ene, have undergo metathesis. The WCl_6-e^--Al-CH_2Cl_2 catalyst systems have the advantage of being selective and able to catalyse the reaction under mild conditions [7]. The application of the WCl_6-e^--Al-CH_2Cl_2 catalyst system to cross-metathesis reactions of non-functionalized, functionalized olefins and acylic diene metathesis (ADMET) polymerization is reported [8,10].

Time is important in determining the activity of a given catalyst system. This study deals with investigations concerning the stability and activity of electrochemically produced W-based catalyst in cross-metathesis reaction of 7-tetradecene and 4-octene. Optimum reaction time was found to be 24 hours for cross-metathesis of symmetrical alkenes.

2. Experimental

Dichloromethane (Merck) was washed with concentrated H_2SO_4, water and an aqueous solution of Na_2CO_3 (5 %,wt), then dried over $CaCl_2$ and distilled from P_2O_5 under nitrogen [11]. Olefins were obtained from Aldrich, purified by distillation over calcium hydride and kept under nitrogen. WCl_6 (Aldrich) was purified by sublimation at 220 °C under nitrogen to remove more volatile WO_2Cl_2 and $WClO_4$ impurities [12].

The electrochemical instrumentation consisted of a EGG-PAR Model 273 coupled with a PAR Model Universal Programmer. The measurements were carried out under a nitrogen atmosphere in a three-electrode cell having a jacket through which water from a constant temperature bath was circulated. In the electrochemical experiments, the reference electrode consisted of AgCl coated on an silver wire in

Y. Imamoglu and L. Bencze (eds.), Novel Metathesis Chemistry: Well-Defined Initiator Systems for Specialty Chemical Synthesis, Tailored Polymers and Advanced Material Applications, 167–171.
© 2003 *Kluwer Academic Publishers. Printed in the Netherlands.*

CH_2Cl_2/ 0.1 N tetra-n-butyl ammonium tetrafluoroborate (TBABF$_4$), which was separated from the electrolysis solution by a sintered glass disc. Experiments were carried out in an undivided cell with a macro working platinum foil electrode (2.0 cm^2) and aluminum foil (2.0 cm^2) counter electrode. Electrolysis was carried out without a supporting electrode due to its deleterious effect on the catalyst system. For this reason, the distance between the platinum working and aluminum counter electrode was kept constant and as small as possible (i.e. 2.0 mm) in order to keep the solution resistance to a minimum.

All electrochemical and catalytic work was done under nitrogen atmosphere. WCl_6 (0.2 g, 0.50 mmol) was introduced into the electrochemical cell containing CH_2Cl_2 (25ml) and a red solution was observed. The electrodes were introduced into the deep red solution and reductive electrolysis at +0.9 V was applied to the solution for 3 hours. The color of the solution darkened progressively. Aliquots from this catalytic solution were used in cross-metathesis reactions.

The metathesis experiments were carried out in a stirred glass vessel at room temperature. In a typical experiment, a certain amount of olefins was introduced into the reactor. 385 µl of the catalytic solution was taken with an automatic pipette from the electrochemical cell and added to the olefin in the glass vessel under nitrogen atmosphere. Metathesis reactions were performed under different reaction times. The product mixtures obtained from the cross-metathesis experiments were analysed by a Shimadzu GCMS-QP5050A.

3. Results and discussion

The reaction of 7-tetradecene with an equimolar amount of 4-octene, catalyzed by electrochemically produced fresh tungsten-based catalyst, resulted in cross-metathesis product, *4-undecene* (C_{11}). To investigate the stability of the catalyst with time, catalyst was stored under nitrogen atmosphere. It was used after 24 and 48 hours in the same catalytic reaction. Chromatograms of the products are given in Figure 1. C_8 and C_{14} are unreacted olefins.

TABLE 1. Effect of catalyst on the yield of cross-product, 4-undecene

Catalyst (WCl_6)	4-C_{11}, Yield (%)
Fresh	59
24 hours stored	31
48 hours stored	-

As it is seen clearly from Figure 1, the amount of cross-product is the highest when the fresh active catalyst is used. A yield of 59 % 4-C_{11} was obtained when the molar ratio of
WCl_6/7-C_{14}/4-C_8 was 1:40:40 [8]. The stored catalyst was used after 24 hours for the same catalytic reaction.

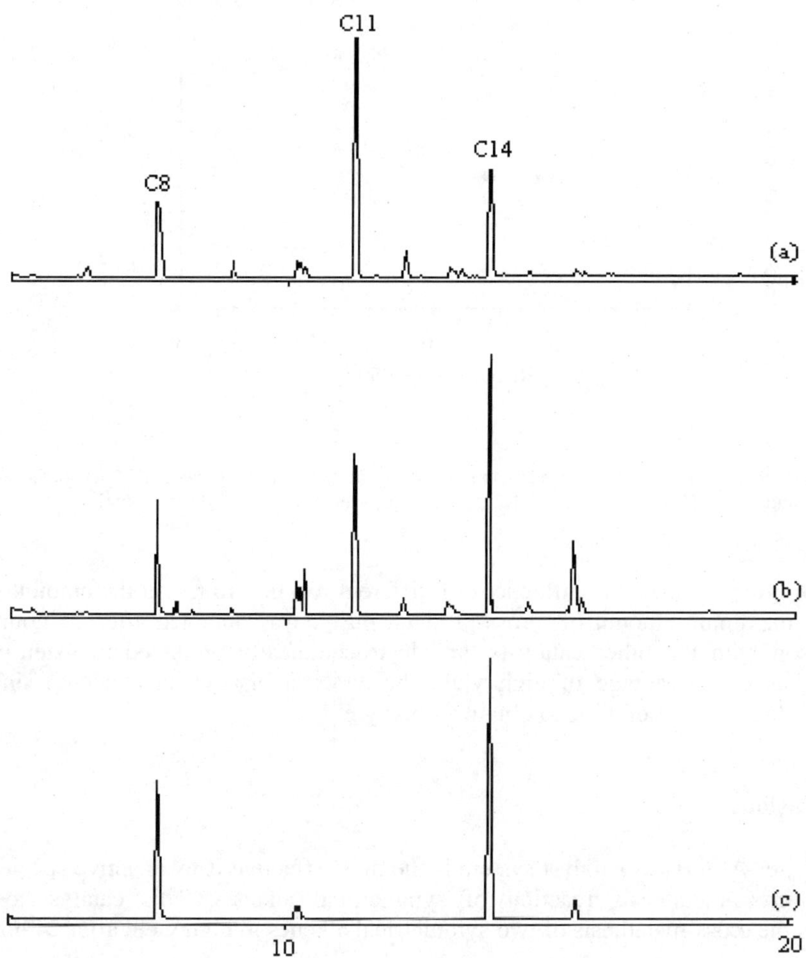

Figure 1. Chromatogram of the cross-metathesis products of 7-tetradecene and 4-octene with electrochemically produced; (a) fresh catalyst , (b) 24 hours stored catalyst, (c) 48 hours stored catalyst.

The activity of catalyst decreased and the yield of 4-C_{11} decreased to % 31. 48 hours later, W-based catalyst lost its activity and C_{11} did not form. It is better to use fresh active catalyst in the cross-metathesis of symmetrical olefins.

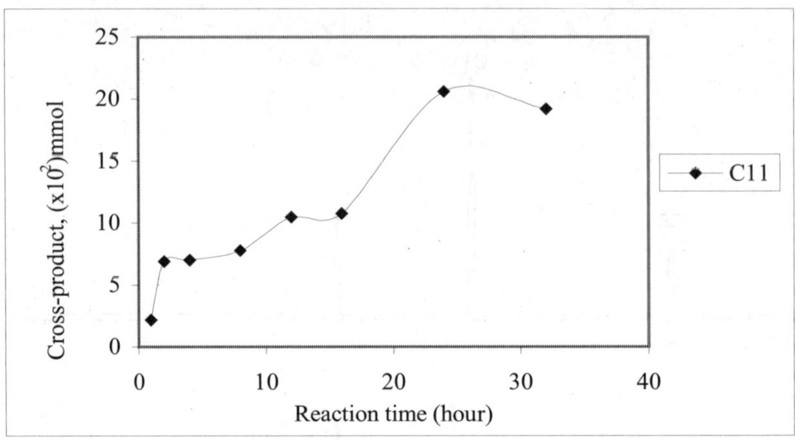

Figure 2. Effect of reaction time on the amount of 4-C_{11} in the cross metathesis of 7-tetradecene and 4-octene. Condition : (WCl_6/7-C_{14}/4-C_8 = 1:15:15, 4-C_8 = 0.12 mmol, WCl_6 = 8.0 x10^{-3} mmol).

Figure 2 shows the influences of different reaction times on the amount of 4-C_{11}. The maximum amount of cross-product, 4-C_{11}, was obtained after 24 hours. In comparison with the other catalysts, an electrochemically produced tungsten-based catalyst has the advantage of high yield; however, a disadvantage occurs since it requires a higher reaction time to obtain a good yield.

4. Conclusions

The WCl_6-e$^-$-Al-CH_2Cl_2 catalyst system is the most effective if fresh active species are used in cross-metathesis reaction of symmetrical alkenes. This catalyst system catalyzes the cross-metathesis of two symmetrical alkenes in high yield after 24 hours.

5. Acknowledgments

This work was supported by The Scientific and Technical Research Council of Turkey Project No TBAG-2148 (102T039).

6. References

[1]   Grubbs, R.H. and Chang, S. (1998) *Tetrahedron* **54**, 4413.
[2]   Spronk, R., Mol, J.C. (1991) *Appl. Catal.* **70**, 295.
[3]   Fox, H.H., Schrock, R.R., O'Dell, R. (1994) *Organometallics* **13**, 2804.
[4]   Ivin, K.J. and Mol, J.C. (1997) *Olefin Metathesis and Metathesis Polymerization*, Academic Press, London.

[5]   Gilet, M., Mortreux, A., Folest, J.C., Petit, F. J. Am. (1983) *Chem. Soc.* **105**, 3876.
[6]   Bages, S., Petit, M., Mortreux, A., Petit, F. (1990) *NATO SCI Ser.* **C326**, 89.
[7]   Gilet, M., Mortreux, A., Nicole, J., Petit, F. (1979) *J. C. S. Chem. Comm.* 521.
[8]   Çetinkaya, S., Düz, B. and Imamoğlu, Y. (2003) *Appl. Organometal. Chem.* **17**, 232-235.
[9]   Dereli, O., Düz, B., Zümreoğlu-Karan, B. and Imamoğlu, Y. (2003) Appl. *Organometal. Chem.* **17**, 23-27.
[10]   Çetinkaya, S., Düz, B., Imamoğlu, Y. (2002) *Appl. Organometal. Chem..* submitted.
[11]   Calderon, N., Ofstead, E.A., Ward, J.P., Judy, W.A. and Scott, K.W. (1968) *J. Am. Chem. Soc.* **90**, 4133.
[12]   Uchida, A., Hamano, Y., Mukai, Y. and Matsuda, S. (1971) *Ind. Eng. Chem. Prod. Res. Develop.* **10**, 372.

IMMOBILIZED TUNGSTEN-CONTAINING CATALYSTS FOR THE METATHESIS OF LINEAR AND FUNCTIONALIZED OLEFINS

O. V. SHOUVALOVA, N. B. BESPALOVA, P. NIECZYPOR* and J. C. MOL*

A.V. Topchiev Institute of Petrochemical Synthesis, Russian Academy of Sciences, 29, Leninsky prospect, Moscow,V-71, 11999, Russia.
**Institute of Molecular Chemistry, Universiteit van Amsterdam, Nieuwe Achtergracht 166, 1018 WV Amsterdam, The Netherlands.*

1. Introduction

Olefin metathesis is an important catalytic reaction. This is largely due to the opportunity of applying the reaction to the synthesis of a wide range of products, from industrial polymers and bulk chemicals to small amounts of biologically active compounds, such as nucleotides, peptides, sugars etc. Olefin metathesis can be used for petrochemical processes, oleochemical processes and in fine chemistry [1,2]. The introduction of the well-defined Ru- and Mo-carbene catalysts, developed in the laboratories of Grubbs and Schrock, respectively, greatly enhanced the application of metathesis in the field of functionalized olefins. In this field, less than 10% of the catalyst concentration is generally sufficient for an efficient reaction, although sometimes it is necessary to increase this to 25% and even 50% [3,4]. Because ruthenium complexes are expensive, it is very attractive to develop new catalysts of which small amounts can be used, or to try to use catalysts where the metal-carbene initiator is generated during the reaction. The latter catalysts can not only be effective, but also much cheaper.

The most popular classical catalyst systems for the metathesis of functionalized olefins where the metal-carbene initiators are generated *in situ*, are the homogeneous systems based on WCl_6 or $WOCl_4$ in combination with an organotin cocatalyst [5] and the heterogeneous systems Re_2O_7/Al_2O_3 and $Re_2O_7/Al_2O_3–SiO_2$, promoted by an organotin (or -lead) compound [2,5,6]. Another type of catalysts are the homogeneous systems WCl_6 or $WOCl_4$ in combination with an organosilicon compound, *viz.* 1,1,3,3-tetramethyl-1,3-disilacyclobutane or H_2SiPh_2 [7,8]. Silicon-containing catalysts have low toxicity compared to tin- or lead-containing systems. Moreover, they are tolerant to moisture and oxygen, and relatively easy to prepare. These catalysts can be used for a number of olefin metathesis reactions, such as self- and cross-metathesis of linear substrates with functional groups, ring-closing metathesis (RCM), and ring-opening polymerization (ROMP) of cyclic and bicyclic olefins [9].

Well-defined organometallic compounds (W, Mo) show catalytic properties in metathesis when they are supported on SiO_2 or Al_2O_3 [10]. Sometimes tungsten carbyne complexes become active for metathesis only after reaction with SiO_2 [11].

Heterogenisation of homogeneous precursors by immobilization onto solid support materials allows the catalyst to be quantitatively separated from the reaction products. In the past, the catalytic systems WCl_6- and $WOCl_4$ -Me_4Sn were immobilized on dry SiO_2 and the catalytic activity of the catalysts thus obtained was tested in the metathesis of linear olefins and unsaturated fatty acid esters [12]. The solid catalyst system WCl_6-SiO_2/Me_4Sn showed the same good activity as the homogeneous system $WOCl_4/Me_4Sn$ for the metathesis of 2-pentene, and could be reused several times.

Y. Imamoglu and L. Bencze (eds.), Novel Metathesis Chemistry: Well-Defined Initiator Systems for Specialty Chemical Synthesis, Tailored Polymers and Advanced Material Applications, 173–177.
© 2003 *Kluwer Academic Publishers. Printed in the Netherlands.*

174

These results open up new directions for creating stable solid catalysts for olefin metathesis as well as for RCM of non-conjugated dienic compounds. We anticipated that the homogeneous catalyst systems WCl_6- and $WOCl_4$-1,1,3,3-tetramethyl-1,3-disilacyclobutane after fixation on SiO_2 keep their catalytic properties and become more stable compared to the homogeneous analogues. We here report some preliminary results.

2. Results and Discussion

We investigated the system $WCl_6/SiO_2/1,1,3,3$-tetramethyl-1,3-disilacyclobutane as an immobilised catalyst system for the metathesis of linear and functionalised olefins. The formation of the organotungsten precursor of the catalyst was realized in a two-step process.

1. WCl_6 was grafted to high surface-area silica by wet impregnation [12] using a toluene solution of WCl_6 that was stirred in the presence of silica until all the WCl_6 was adsorbed on the silica surface:

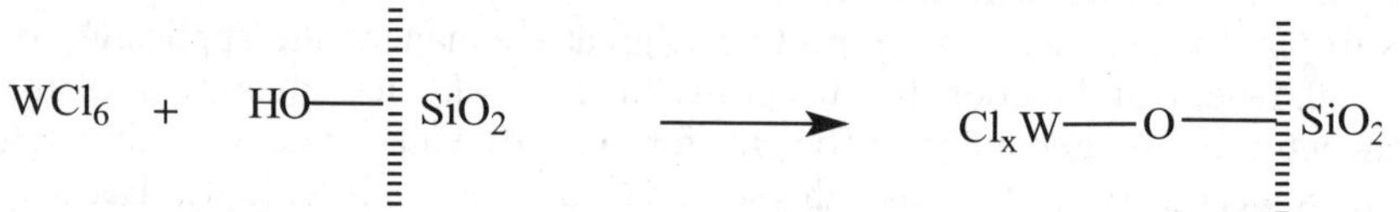

$$WCl_6 \ + \ HO-SiO_2 \ \longrightarrow \ Cl_xW-O-SiO_2$$

2. Interaction of the supported tungsten chloride with the organosilicon cocatalyst, followed by the formation of the organotungsten precursor of the active species:

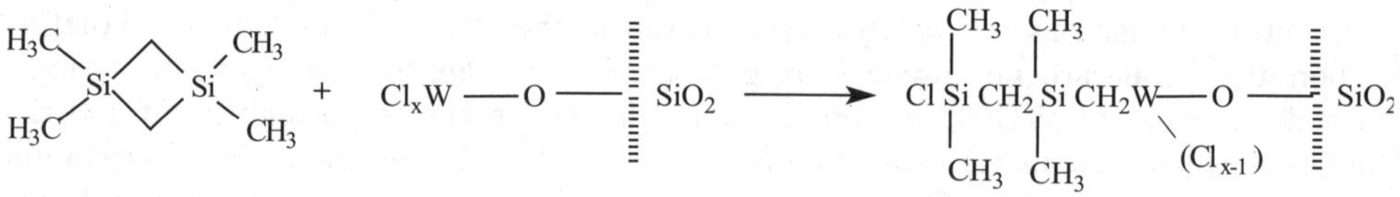

Formation of the active species of the catalyst can be realized via destruction of the organotungsten compounds with elimination of two kinds of tungsten-carbene initiators in the same way as for the corresponding homogeneous system (Scheme 1).

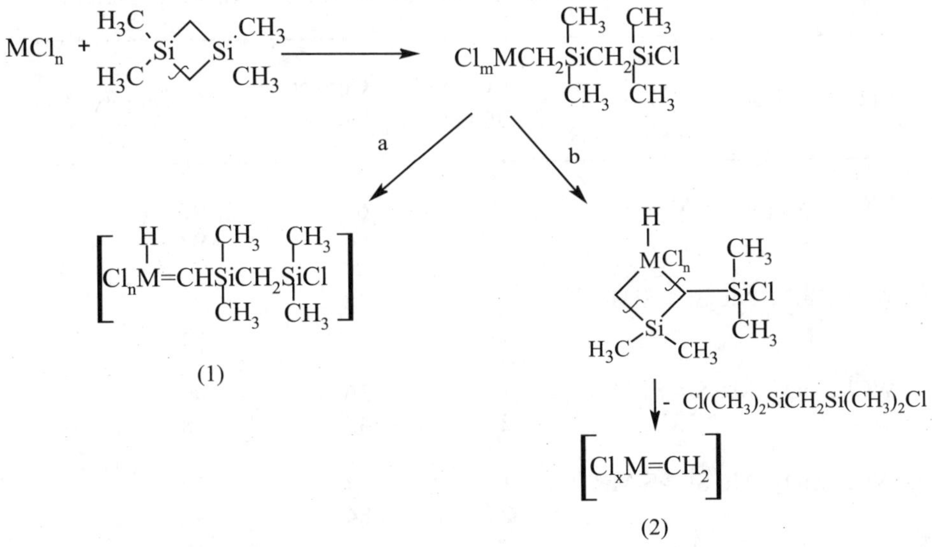

Scheme 1

We tested the catalytic system based on $WCl_6/SiO_2/$-1,1,3,3-tetramethyl-1,3-disilacyclobutane in the metathesis of 1-octene. Table 1 shows the results.

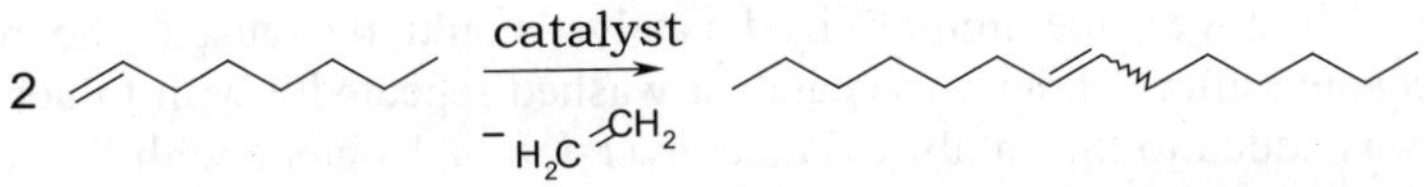

We observed that the catalytic activity depends on the method of preparation of the catalyst. The catalytic activity of the immobilized catalyst was less compared with the homogeneous analogue when the catalyst was prepared before use and kept 1 hour at room temperature or at 60°C. That means that the active species formed on the surface were destroyed. The best results were obtained when the catalyst was formed in the presence of the substrate and the reaction was carried out without solvent. When a mixture of 1-octene and 1,1,3,3-tetramethyl-1,3-disilacyclobutane was added to a calculated amount of WCl_6/SiO_2, the yield of the metathesis products was high and the selectivity was about 100%.

Table 1. Activity of the homogeneous and immobilized catalysts in the metathesis of 1-octene.[a]

Catalytic System	Reaction time, h	Conversion, %	Selectivity, %
$WCl_6 - Me_2Si\Diamond SiMe_2$	1	60	97
	2	66	97
$WCl_6/SiO_2 - Me_2Si\Diamond SiMe_2$ [b]	1	48	99
	2	58	98
$WCl_6/SiO_2 - Me_2Si\Diamond SiMe_2$ [c]	1	36	99
	2	45	98
$WCl_6/SiO_2 - Me_2Si\Diamond SiMe_2$ [d]	1	77	99
	2	84	99

[a] 60°C, toluene, [1-octene]: [W]: [Si] = 100:1:2.
[b] Catalyst was kept for 1 h at room temperature.
[c] Catalyst was kept for 1 h at 60°C.
[d] Reactions were carried out without preliminary preparation of the catalyst and without solvent.

It thus appears that the activity of the immobilized catalyst is higher than the activity of the corresponding homogeneous system when the reaction is performed without an organic solvent. Moreover, the immobilized catalyst could be reused: the reaction mixture was decanted after 2 hours, the catalyst washed repeatedly with toluene, then fresh 1-octene was added to the catalyst. The conversion of 1-octene with the recycled catalyst was 62% compared to 77% for the fresh catalyst at 60°C at a 1-octene:[W]:[Si] ratio of 100:1:2 in toluene.

We also investigated the efficiency of the immobilised catalyst in the metathesis of a functionalised olefin, *viz.,* the ring–closing metathesis of diethyl diallylmalonate:

Both the homogeneous and the immobilized catalyst showed a low activity (conversion 10-15%) at a molar ratio of [substrate]:[catalyst] = 100:1 and 10:1, but a 100% selectivity.

3. Conclusions

• The heterogeneous catalyst system consisting of WCl_6, immobilised on silica gel, in combination with 1,1,3,3-tetramethyl-1,3-disilacyclobutane shows good activity and high selectivity for the metathesis of 1-octene.

• The system is environmentally benign, because it does not require a solvent. Moreover, it does not contain toxic components, while the catalyst can also be reused.

4. References

1. Ivin, K.J. and Mol, J.C. (1997) *Olefin Metathesis and Metathesis Polymerization*, Academic Press, London.
2. Mol, J. C. (2002) *Green Chemistry* **4**, 5.
3. Ivin, K. J. (1998) *J. Mol. Catal.* **133,** 1.
4. Overkleeft, H. S., Bruggeman, P. and Pandit, U.K. (1998) *Tetrahedron Lett.* **39**, 3869.
5. Mol, J. C. (1994) *J. Mol. Catal.* **90**, 185.
6. Mol, J. C. (1983) *CHEMTECH*, **13**, 258.
7. Bespalova, N. B. and Bovina, M. A. (1992) *J. Mol. Catal.* **76**, 181.
8. Bespalova, N. B., Bovina, M .A., Sergeeva, M.B., Oppengeim, V.D. and Zaikin, V.G. (1994) *J. Mol. Catal.* **90**, 21.
9. Bespalova N. B., Bovina, M. A., Popov, A. V. and Mol, J. C. (2000) *J. Mol. Catal.* **160,** 157.
10. Startsev, N., Bogdanovich, B., Bonnemann, B., Rodin, V.N. and Yermakov, Y. V. (1986) *Chem. Commun.*, 381.
11. Weiss, K. and Lossen, G. (1989) *Angew. Chem. Int. Ed. Engl.* **28**, 62.
12. Van Roosmalen, A. J, Polder, K. and Mol, J.C. (1980) *J. Mol. Catal.* **8**, 185.

THE INCORPORATION OF AMINO ACIDS INTO POLYMERS ADMET

Timothy E. Hopkins, Joshua M. Priebe, Kenneth B. Wagener
Dept of Chemistry
University of Florida
P.O. Box 117200
Gainesville, Fl 32611-7200

Abstract:
Amino acid based polymers are of interest for a variety of biomaterial applications including drug delivery, proteomics, and tissue engineering. A new class of polymers bearing amino acids has been prepared using acyclic diene metathesis (ADMET) to create copolymers of polyethylene with linear amino alcohol, branched amino acid, or branched dipeptide substituents. The synthesis of the monomers and the polymerization using the second generation Grubbs' ruthenium catalyst is discussed. The resulting highly functionalized polymers are strong, film-forming materials (moduli up to 220 mPa with up to 260 % elongation) with molecular weights typical of polycondensation polymers, *i.e.* Nylon and PET.

1. Introduction

The synthesis of polymers containing amino acid and peptide moieties is an area of great interest due to the high amount of functionality possessed by the resulting polymers. This functionality can lead to secondary structure formation and enhanced solubility. There are two main ways of incorporating amino acids into a polymer: linearly in the main chain of the polymer or as pendant groups. The first has the potential to be biodegradable, and the latter could be used to make functional, nonbiodegradable materials. The resulting polymers could be used for a variety of biomedical applications such as proteonomics, membranes, artificial surfaces, and drug delivery.

The idea of incorporating amino acids into polymers originated with the work by Pino and coworkers in the 1960's,[1] and with the help of modern separation techniques have become quite common in the literature.[2] Specifically work by Morcellot,[3] Endo and Sanda,[4] and North[5] has focused on the synthesis of amino acid and peptide-branched polyolefins. Morcellot *et. al* have investigated the metal binding of such polymers, and they found that the polymers were much more efficient metal binders than similar model compounds. Endo and Sanda have investigated polyacrylamides with a variety of amino acid and peptide substituents.[4] In particular, they found that when a helix forming peptide was attached to the monomer the resulting polymer maintained the same CD spectrum as the monomer when a nonpolar solvent was used for the polymerization, but a CD spectrum correlating to a lack of secondary structure was observed when the polymerization was ran in a polar solvent. Thus, they concluded that the helix survived the polymerization only when nonpolar solvents were used.[6] North and coworkers have synthesized a variety of polyacrylates where the amino acids are attached to methyl methacrylic acid through the alcohol

179

Y. Imamoglu and L. Bencze (eds.), Novel Metathesis Chemistry: Well-Defined Initiator Systems for Specialty Chemical Synthesis, Tailored Polymers and Advanced Material Applications, 179–189.

moiety of serine.[5] This methodology was extended to include tripeptides by attaching an amino acid to the N and C terminus of the serine based monomer.[7]

Other work by Grubbs and Menard involves the use of ring opening metathesis (ROMP) chemistry to polymerize norbornene derivatives with amino acid and peptide branches.[8] Specifically, they have prepared a polymer bearing the biologically active peptide sequence RGD.[9] Further, binding studies were performed to see if the polymers would bind to fibronectin, and they found that the polymer did indeed bind to fibronectin, which could lend it useful for a variety of biomedical applications.[9] Endo and coworkers have examined the incorporation of amino acids and amino alcohols in the backbone of a polymer by condensation[10] and radical polymerizations with dithiols[11], respectively. They have synthesized polymers with amino alcohol moieties in the polymer backbone, and have demonstrated that the resulting polymers are enzymatically degraded.[11] Recently Brezinska *et. al* have used acyclic diene metathesis (ADMET) to make a telechelic amine terminated poly(ethylene)-based polymer followed by the metal catalyzed, living N-carboxyanhydride chemistry to make triblock[12] and most recently pentablock copolymers.[13]

Acyclic Diene Metathesis (ADMET) is a well defined polycondensation reaction that allows for the synthesis of unique polymer architectures by simple monomer design (Figure 1).[14] Recent advanced in catalyst development, specifically the Grubbs' 2nd generation Catalyst (1) (tricyclohexylphosphine[1,3-bis(2,4,6-trimethylphenyl)-4,5-dihydroimidazol-2-ylidene] [benzylidine] ruthenium (IV) di-

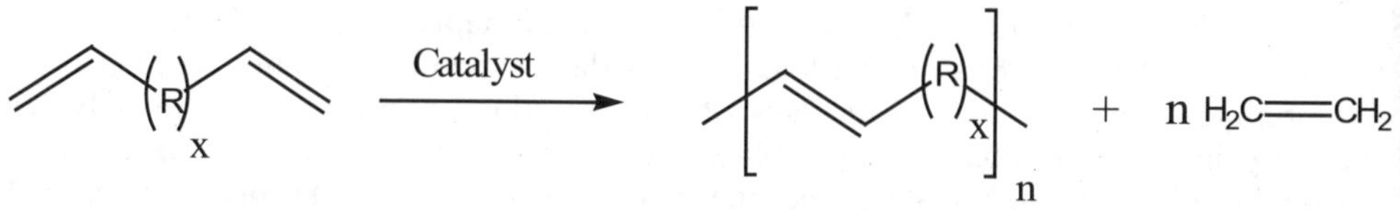

Figure 1. General ADMET reaction.

chloride) (Figure 2) have enabled the ADMET polymerization of monomers with a variety of functional groups.[15] We have recently reported the polymerization of various protected amino acid containing dienes as further examples of the functional group tolerance of the 2nd generation Grubbs' catalyst.[16,17] The molecular weights obtained by ADMET polymerizations resemble those obtained by typical polycondensation reactions *e.g.* Nylon and PET – molecular weights of only 10-20,000 are required for good physical properties of functionalized polyolefins, which is in the same molecular weight range commonly obtained by ADMET polymerizations.

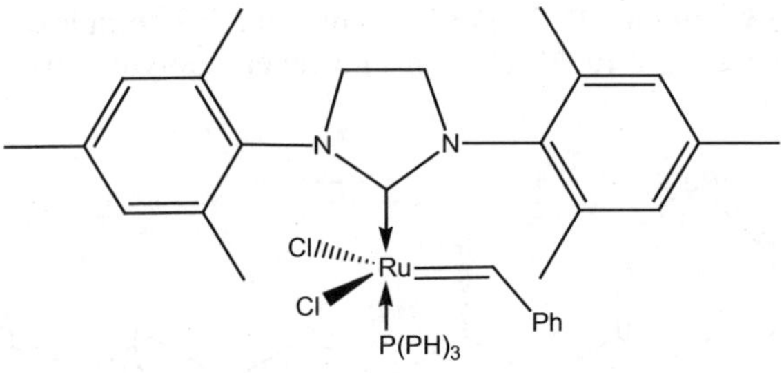

1

Figure 2. The 2nd Generation Grubbs' Ru catalyst.

Here we report a summary of our efforts to synthesize new biomaterials using ADMET chemistry. We have synthesized linear amino alcohol-containing polymers, that should be biodegradable based on the work by Endo *et. al*,[4] and a variety of amino acid and dipeptide branched polyolefins, which are designed to be non-water soluble, functionalized materials. These polymers are strong film-forming polymers with moduli of up to 220 MPa with up to 260 % elongation, which could be useful for a variety of biomaterial applications including biodegradable polymers, membranes, proteomics, and surfaces for artificial implants.

2. Results and Discussion
2.1 LINEAR AMINO ALCOHOL-CONTAINING POLYMERS

Initial attempts in the Wagener research group at the University of Florida to make chiral polymers focused on the preparation of chiral hydrocarbon monomers. The synthesis of such monomers proved to be futile, so we decided to use the abundant chiral pool of molecules, amino alcohols, which are widely available by the reduction of naturally occurring amino acids. These amino alcohols can be coupled to two equivalents of the corresponding acid using any of the common peptide coupling reagents, *e.g.* EDC·HCl/DMAP[11] or DIC/DMAP, to prepare the 2, 3, and 8-spacer amino alcohol containing dienes (2) (Figure 3) in high yields.[16]

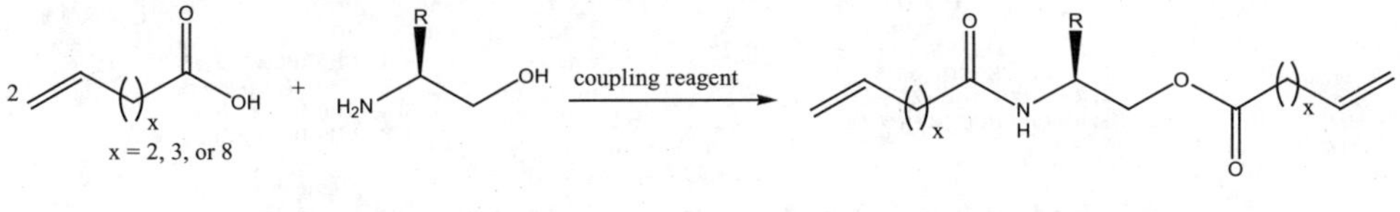

Figure 3. Synthesis of the linear amino alcohol containing monomers.

The catalyst chosen for the polymerizations was the 2nd Generation Grubbs' catalyst (1), due to its high tolerance of functional groups. The polymerization of monomer 2 with x = 2 (Figure 3) gave interesting and surprising results. All attempts for polymerization of these monomers yielded only oligomer formation. A review of the literature on ring closing metathesis (RCM) gave insight on what was preventing conversion to high polymer. Several groups have reported an intermolecular complexation that had occurred when the amide carbonyl was located two methylenes away from the olefin (Figure 4).[18] In order to prevent this complexation, Ti(O Pr)$_4$ was used to allow for successful RCM reactions; however, applying this methodology to the

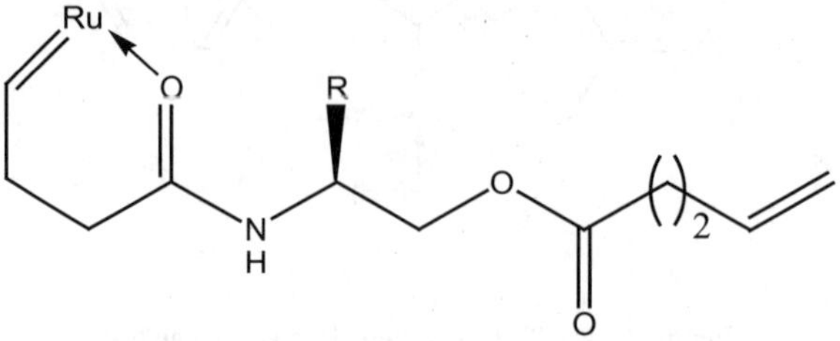

Figure 4. Suggested intermolecular complexationpreventing conversion to high polymer.

ADMET reaction of the above monomers yielded only the initial monomers with isomerized olefins. Isomerization of the external olefins to internal olefins is a common side effect of the second generation Grubbs' catalyst. An important difference between RCM and ADMET is the high conversion required to yield high molecular weight polymer, *i.e.* 50-60 % yield per coupling will only result in oligomer formation.

To successfully polymerize these monomers, the methylene spacer length was increased to three carbons, which is a method commonly employed in ADMET chemistry when polymerization is only yielding low molecular weight products. This allowed for successful ADMET polymerization; however, only low molecular weight polymers (~3,000 g/mol) were obtained. This monomer still has the potential to form a 7-membered ring complex through the amide oxygen and the Ru carbene, which must slow the kinetics of the reaction to the point where high conversion is not obtained. Once the spacer length was increased to eight methylene units, the monomers were polymerized rapidly to high molecular weight (up to 53,000 g/mol) (Figure 5); the 8-

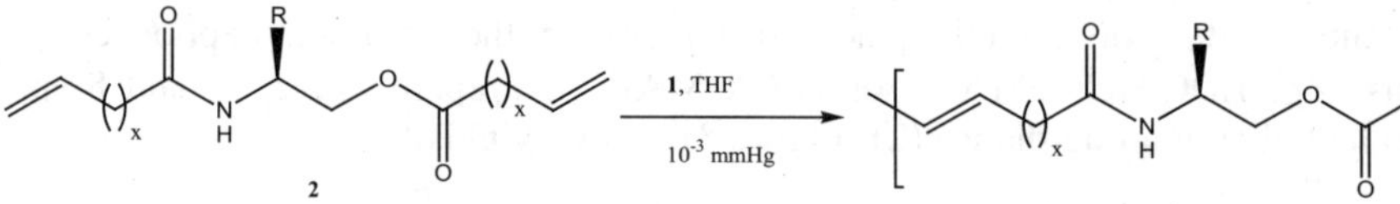

2

2) R = CH$_2$CH(CH$_3$)$_2$, x = 3
3) R = CH$_2$CH(CH$_3$)$_2$, x = 8
4) R = CH(CH$_3$)$_2$, x = 8
5) R = CH(CH$_3$)CH$_2$CH$_3$, x = 8

6) R = CH$_2$CH(CH$_3$)$_2$, x = 3
7) R = CH$_2$CH(CH$_3$)$_2$, x = 8
8) R = CH(CH$_3$)$_2$, x = 8
9) R = CH(CH$_3$)CH$_2$CH$_3$, x = 8

Figure 5. Polymerization of linear amino alcohol containing dienes (2).

spacer was synthesized due to the wide availability of 10-undecenoic acid. The molecular weight obtained for polymer **9** is approximately twice that of polymers **7** and **8**, because the reaction was allowed to stir for 6 days under vacuum rather than the 5 days that are normally used for ADMET polymerization. The resulting film-forming polymers are semicrystalline with T_m's around 39 °C from solvent crystallization (Table 1), but polymer **7** was totally amorphous when melt crystallized at 10 °C per minute (as determined by differential scanning calorimetry (DSC)). As of now only polymer **8** has been completely saturated by exhaustive hydrogenation using Pd(C), and the T_m of the polymer was increased to 70 °C with a large increase in delta H after hydrogenation. These polymers should be biodegradable based on the work by Endo *et al.*, who demonstrate that very similar polymers were enzymatically degraded through the ester and amide bonds. Indeed, we intend to investigate the biodegradability of these polymers as well.

2.2. AMINO ACID BRANCHED POLYOLEFINS

Polymers possessing a nonbiodegradable backbone that could be used as biologically active surfaces are of much interest, due to the potential applications of such materials. Therefore, we decided to synthesize amino acid branched polymers having a hydrocarbon backbone, and ADMET provides an excellent route to synthesizing these designed biomaterials by simple monomer design. Also, the functionality can be placed at specific intervals; a property not easily duplicated using copolymerizations. We have shown that the amine functionality of the amino acid can be pointed in towards the backbone (Figure 7) or away from the backbone (Figure 8), which ultimately will give rise to polymers with drastically different properties.[16,17] Once deprotected, the first will yield a polymer with carboxylic acid functionality located at the surface of the polymer, and the latter have primary amine functionality located at the surface of the polymer — the incorporation of primary amine branches into polyolefins is very difficult using common techniques.

The branched monomers can be prepared using simple peptide coupling chemistry to attach the amino acid or dipeptides to the corresponding amine or carboxylic acid branched dienes. The carboxylic acid branched dienes can be prepared by the reaction of the 5-hexenoic acid or 6-heptenoic acid with 2 equivalents of LDA, followed by the addition of the 5-pentenyl bromide (Figure 6.1). Whereas, the amine branched dienes can be prepared easily by two different methods. The first entails forming the Grignard reagent of 11-bromoundecene followed by the addition of ½ of an equivalent of ethyl formate to give the resulting alcohol, which can be treated with pyridinium chlorochromate (PCC) to yield the desired ketone (Figure 6.2). The amine can then be prepared via reductive amination using ammonium acetate and sodium cyanoborohydride. All three steps can be performed without stringent purification, . . only extractions are performed. The final method for preparation of the 8-spacer amine branched diene involves the formation of the ketene by the reaction of 10-undecenyl chloride with 1.8 equivalents of triethylamine, followed by a ketene-ketene coupling reaction, ring opening of the resulting lactone, and decarboxylation by refluxing the solution in aqueous sodium hydroxide (Figure 6.3). This yielded the desired 8-spacer ketone in high yields, which could be converted to the 8-spacer amine branched diene using the reductive amination method described above.

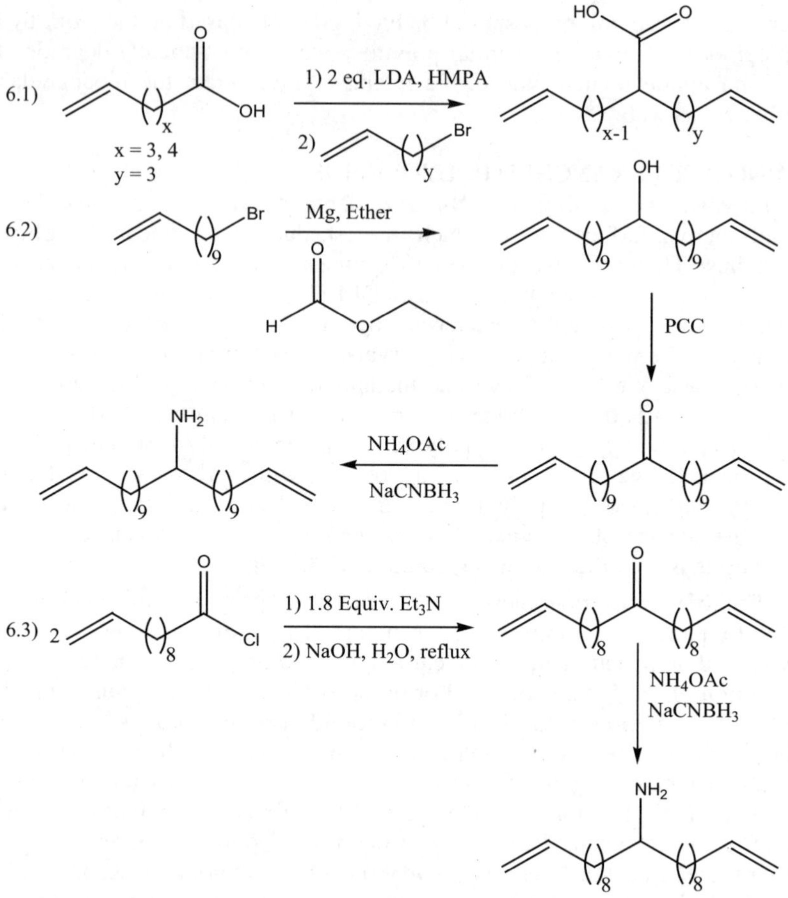

Figure 6. The preparation of the carboxylic acid brancheddiene (6.1), the 3 and 9-spacer amine branched diene (6.2),and the preparation of the 8-spacer amine branched diene (6.3).

The amino acid can be attached to the diene through the N or C terminus, which will result in a polymer with a negative or positive surface, respectively, once deprotected (Figure 7 and 8). Polymers **12** and **13** have the amino acid attached through the N terminus, whereas polymers **21-27** have the amino acid attached through the C-terminus. The polymers exhibit interesting thermal and physical properties; many of the polymers are strong, semicrystalline, film-forming materials (initial results have demonstrated moduli up to 220 mPa with elongation of up to 260 %) (Table 1).

Two polymers were prepared with the amino acid attached through the N terminus (Figure 7). These were prepared by coupling L-leucine methyl ester to the carboxyilic acid branched dienes shown in Figure 6.1 using 1-(3-dimethylaminopropyl)-3-ethylcarbodiimide hydrochloride (EDC·HCl), thus preparing symmetrical and asymmetrical monomers. The monomers were easily polymerized in $CHCl_3$ with an Ar purge to aid in the removal of ethylene. Both of the polymers were strong, film-forming, semicrystalline materials with melt transitions of 114 and 135 °C for polymers **13** and **14**, respectively.

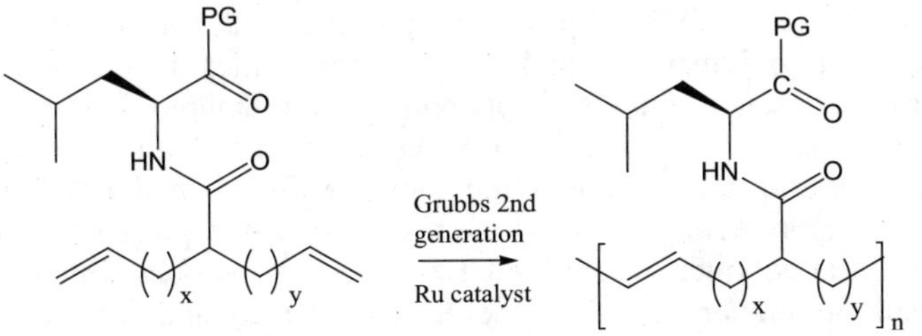

Figure 7. The synthesis of polymers with amino acids attached to a polyolefin backbone through the N-terminus of the amino acid via ADMET.

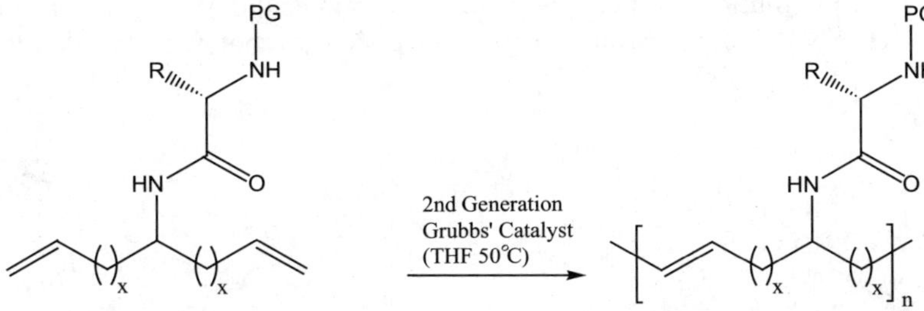

14) x = 9, R = $CH_2CH(CH_3)_2$, PG = BOC
15) x = 8, R = CH_3, PG = BOC
16) x = 8, R = CH_3, PG = CBz
17) x = 9, R = CH_3, PG = BOC
18) x = 9, R = $(CH_2)_4NHCBz$, PG = CBz
19) x = 9, R = $(CH_2)_4NHBOC$, PG = BOC
20) x = 9, R = CH_2SCBz, PG = BOC

21) x = 9, R = $CH_2CH(CH_3)_2$, PG = BOC
22) x = 8, R = CH_3, PG = BOC
23) x = 8, R = CH_3, PG = CBz
24) x = 9, R = CH_3, PG = BOC
25) x = 9, R = $(CH_2)_4NHCBz$, PG = CBz
26) x = 9, R = $(CH_2)_4NHBOC$, PG = BOC
27) x = 9, R = CH_2SCBz, PG = CBz

Figure 8. The synthesis of polymers with amino acids attached to a polyolefin backbone through the C-terminus of the amino acid via ADMET.

The amino acid branched dienes connected through the C-terminus were prepared by coupling the N-protected amino acids to the amine branched dienes (Figure 6.2 and 6.3) using 1,3-diisopropylcarbodiimide (DIC) and 1-hydroxybenzotriazole (HOBt) as the coupling reagents. The monomers were synthesized in high yields (up to 92 % after purification) and purified by simple recrystallization in methanol/water or ethanol/water. Initially, a variety of 8-spacer amino acid branched dienes of the type shown in Figure 8 were prepared; however, polymerization using the method described above for polymers 12 and 13 only yielded high polymer in the case of monomer 15. Yet, a simple change in solvent from $CHCl_3$ to THF allowed successful polymerization of monomers 14-20 to form polymers 21-27. We believe this phenomenon is due to one of two possible reasons: the THF breaks up the complexation between the catalyst and monomer functionality, or it is simply due to a conformational change in monomer structure when going from $CHCl_3$ to THF.

As shown in Table 1, many of the polymers are semicrystalline. Analysis of the results show that the polymers with CBz protecting groups are more prone to crystallizing than those with BOC protecting groups, as demonstrated for polymers 22 and 23. However, polymers 25 and 26 are both semicrystalline regardless of the protecting group chosen. Thus, it is apparent that the nature of the amino acid itself plays the largest role as to whether the polymers are semicrystalline or amorphous with the more polar amino acids crystallizing more readily than the hydrocarbon based amino acids. The protecting group chosen as well as the polarity of the side chain affect the T_m of the polymer, where the CBz-protected and polar amino acid-branched polymers result in higher T_ms than the BOC-protected and nonpolar amino acid-branched polymers, respectively (Table 1).

The successful polymerization of α, ω dienes functionalized with one amino acid as the branch, led us to the next challenge; the polymerization of longer peptides, specifically dipeptides. The first dipeptide used was CBz-Alanine-Valine branch (Figure 9), which upon polymerization using THF as the solvent resulted in what appeared to be an insoluble gel after three days. Polymer 28 was only soluble in hot DMF and DMSO solutions which prevented easy molecular weight determination; however, ^{1}H NMR demonstrated high conversion to polymer by the lack of observable

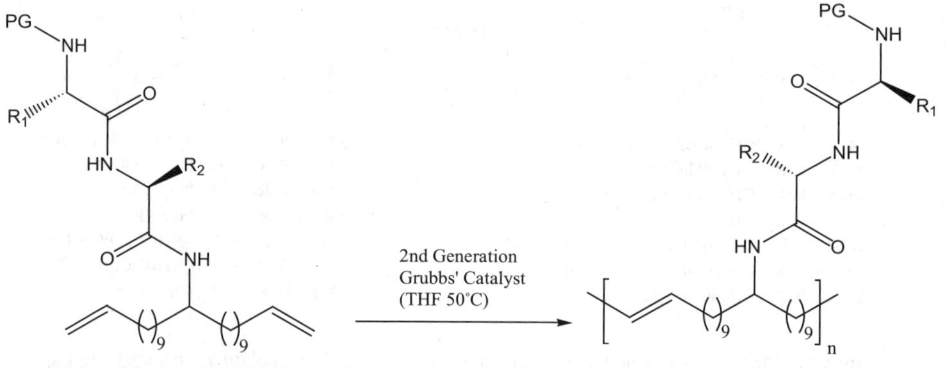

28) $R_1 = CH_3$, $R_2 = CH(CH_3)_2$, PG = CBz
29) $R_1, R_2 = CH_2CH(CH_3)_2$, PG = BOC

30) $R_1 = CH_3$, $R_2 = CH(CH_3)_2$, PG = CBz
31) $R_1, R_2 = CH_2CH(CH_3)_2$, PG = BOC

Figure 9. The synthesis of dipeptide branched polyolefins by ADMET

external olefin resonances, which is a characteristic of high molecular weight ADMET polymers. The polymer was semicrystalline with the highest T_m observed yet for these polymers, 150 °C (Table 1). The methodology was then used to polymerize a BOC-Leucine-Leucine branched dipeptide, which yielded an amorphous polymer. We believe the large hydrocarbon side chains as well as the bulky BOC protecting group were responsible for the lack of crystallinity. Most importantly, the ability to polymerize dipeptide branched dienes leads us to believe that the polymerization of longer peptides, *e.g.* the biologically active RGD sequence, should be achievable using this methodology.

Polymer	$\overline{M}_n{}^{a}$ (g/mol)	PDI^{a}	T^{b}(°C)	T^{b}(°C)
Linear				
6	4700	1.73	C	d
7	33,000	1.64	38	d
8	27,000	1.77	39	d
9	53,000	1.71	39	d
Branched				
12	31,000	2.02	114	d
13	27,000	2.10	135	d
21	20,000	1.96	C	34
22	37,000	1.85	C	7
23	21,000	1.89	46^{e}	d
24	13,000	2.11	C	18
25	25,000	1.92	96	d
26	11,000	1.92	60	d
27	14,000	1.76	110	d
30	f	f	150	d
31	8,000	1.66	C	40

Table 1. ^{a} values were calculated by GPC versus polystyrene standards. ^{b}Data obtained using a Perkin Elmer DSC 7. ^{c}No T_m observed over the scanned range of -80 °C to 190 °C. ^{d}No T_g observed over the scanned range of -80 °C to 190 °C. ^{e}The T_m reported is that of the solvent crystallized sample; no T_m was observed from the melt crystallized sample. ^{f}The insolubility of the polymer in THF and CHCl$_3$ inhibited the molecular weight determination by GPC.

3. Conclusions

Linear amino alcohol based polymers, amino acid branched polyolefins, and dipeptide branched polyolefins have been prepared using ADMET chemistry. The monomers were polymerized in the bulk for the linear amino alcohol-containing polymers, in a small amount of THF for the amide terminated amino acid branched monomers, or in a small quantity of CHCl$_3$ for the ester terminated amino acid branched monomers. The molecular weights obtained resemble that of typical polycondensation polymers, *i.e.* ~ 12,000 g/mol for Nylon and PET. In addition, the resulting polymers have the tendency to be strong, film-forming polymers (moduli up to 220 mPa with up to 260 % elongation), properties that could lend them useful for a variety of biomedical applications. In addition the polymers have shown the tendency to crystallize, which is uncommon for Poly(acrylamides) with T_ms up to 150 °C. The general trend in determining the capability for the polymers to crystallize is that the larger more polar amino acid branched polymers, *i.e.* lysine branched, tend to be semicrystalline where the nonpolar amino acid branched polymers, *i.e.* leucine branched, appear to be amorphous.

188

4. Acknowledgments

We would like to thank the National Science Foundation, Division of Materials Research and the Army Research Office for financial support of this work. We would like to thank Dr. Ravi Orugunty, Mr. Claud Robotham, Prof. Jean M. J. Fréchet, Prof. Dennis Wright, and Prof. Eric Enholm for helpful comments and discussions during the course of this work. We acknowledge Mr. Brian Cuevas, Mr. Clay Bohn, Prof. Anthony Brennan, and Prof. Eugene Goldberg for help in obtaining initial material property data. We would like to acknowledge Dr. Jason Smith and Mr. John Sworen for their help in obtaining thermoanalysis data for the polymers discussed.

5. References

1. Pino, P. Adv. Polym. Sci. 1965, 4, 393.
2. Sanda, F, Endo, T. Macromol. Chem. Phys. 1999, 200, 2651.

3. (a) J. Morcellet-Sauvage, M. Morcellet, C. Loucheux, Makromol. Chem. 1981, 182, 949. (b) J. Morcellet-Sauvage, M. Morcellet, C. Loucheux, Makromol. Chem. 1982, 183, 821.(c) J. Morcellet-Sauvage, M. Morcellet, C. Loucheux, Makromol. Chem. 1982, 183, 831.(d) C. Methenitis, J. Morcellet, M. Morcellet, Polym. Bull. 1984, 12, 133. (e) C. Methenitis, J. Morcellet, M. Morcellet, Polym. Bull. 1984, 12, 141. (f) C. Methenitis, J. Morcellet, G. Pneumatikakis, M. Morcellet, Macromolecules, 1994, 27, 1455.

4. (a) Murata, H.; Sanda, F.; Endo, T. Macromolecules 1996, 29, 5535. (b) Katakai, R.; Nakayama, Y.; J. Chem. Soc., Chem. Commun. 1977, 22, 805. (c) Murata, H.; Sanda, F.; Endo, T. Macromolecules 1997, 30, 2902. (d) Sanda, F.; Ogawa, F.; Endo, T. Polymer 1998, 39, 5543. (e) Sanda, F.; Kamatani, J.; Handa, H.; Endo, T. Macromolecules 1999, 32, 2490. (f) Kudo, H.; Sanda, F.; Endo, T. Macromol. Chem. Phys. 1999, 200, 1232.

5. (a) S. M. Bush, M. North, Polymer 1996, 37, 4649. (b) S. M. Bush, M. North, Polymer 1998, 39, 933. (c) S. M. Bush, M. North, S. Sellarajah, Polymer 1998, 39, 2991.

6. H. Murata, F. Sanda, T. Endo, J. Polymer. Sci:Part A: Polym. Chem. 1998, 36, 1679.

7. A. C. Birchall, S. M. Bush, M. North, Polymer 2001, 42, 375.

8. (a) Maynard, H. D.; Okada, S. Y.; Grubbs, R. H. Macromolecules 2000, 33, 6239-6248. (b) Maynard, H. D.; Grubbs, R. H. Macromolecules 1999, 32, 6917-6924.

9. Maynard, H. D.; Okada, S. Y.; Grubbs, R. H. J. Am. Chem. Soc. 2001, 123, 1275-1279.

10. Koyama, E.; Sanda, F.; Endo, T. Macromolecules 1998, 31, 1495-1500.

11. Koyama, E.; Sanda, F.; Endo, T. Macromolecules 1998, 31, 1495-1500.

12. Brzezinska, K. R.; Deming, T. J. Macromolecules 2001, 34, 4348.

13. Brzezinska, K. R.; Curtin, S. A. Macromolecules 2002, 35, 2970.

14. (a) Lindmark-Hamburg, M.; Wagener, K. B. Macromolecules 1987, 20, 2949. (b) Wagener, K. B.; Nel, J. G.; Konzelman, J.; Boncella, J. M. Macromolecules 1990, 23, 5155. (c) Smith J. A.; Brzezinska K. R.; Valenti D. J.; Wagener K. B. Macromolecules 2000, 33, 3781-3794.

15. (a) A. C. Church, J. H. Pawlow, and K. B. Wagener, "ADMET Polymerization as a Route to Functionalized Polycarbosilanes", Accepted in Macromolecular Chemistry and Physics, February, 2002. (b) A. Cameron Church, Jason A. Smith, James H. Pawlow, and Kenneth B. Wagener, "Non-Traditional Step-Growth Polymerization-ADMET", a chapter in Synthetic Methods and Step-Growth Polymers, Eds., M. E. Rogers and T. Long, Wiley & Sons, New York, NY, In Press, March 2002.

16. Hopkins, T. E.; Pawlow, J. H.; Deters, K. S.; Solivan, S. M.; Davis, J. A.; Koren, D. L.; Gómez, F. J.; Wagener, K. B. Macromolecules, 2001, 34, 7920.

17. (a) Hopkins, T. E.; Pawlow, J. H.; Tep, F.; Wagener, K. B. Polym. Prep. 2002, 43, 281. (b) Hopkins, T. E.; Wagener, K. B. Macromolecules, Submitted October, 2002. b) Hopkins, T. E.; Pawlow, J. H.; Tep, F.; Wagener, K. B. Polym. Prep. 2002, 43, 281. c)T. E. Hopkins and K. B. Wagener. Poly. Prep. 2002, 43, 729. d) T. E. Hopkins, J. H. Pawlow, K. S. Deters, S. M. Solivan, J. A. Davis, D. L. Koren, F. J. Gómez, and K. B. Wagener. Poly. Prep. 2000, 41(2), 1533. f) Hopkins, T. E.; Wagener, K. B. Adv. Mat. 2002, 14, 1703.

18. 1 (a) Ghosh, A. K.; Cappiello, J.; Shin, D. Tetrahedron Lett. 1998, 39, 4651-4654. (b) Choi, T. L.; Chetterjee, A. K.; Grubbs, R. H. Angew. Chem. Int. Ed. 2001, 40, 1277-1279. (c) Wright, Dennis Current Organic Chemistry 1999, 3, 211-240. (d) Furstner, Alois; Langemann, K. J. Am. Chem. Soc. 1997, 119, 9130-9136.

GRAFT COPOLYMERS ATTAINED BY ATRP AND ADMET

Patrick M. O'Donnell and Kenneth B. Wagener
The George and Josephine Butler Polymer Research Laboratory, Department of Chemistry, University of Florida, Gainesville, FL 32611-7200, USA.

1 Introduction

1.1 OLEFIN METATHESIS

The history of olefin metathesis dates back to over 50 years ago with the advent of Ziegler-Natta transition metal catalysts for olefin polymerization.[1] In turn, it has developed into a useful modern synthetic tool of specialty polymers, fine chemicals, natural products which continues to be developed for several other applications. As shown in Figure 1, olefin metathesis is an equilibrium process in which there is the exchange of carbon moieties between a pair of double bonds in two alkene molecules (Figure 1).

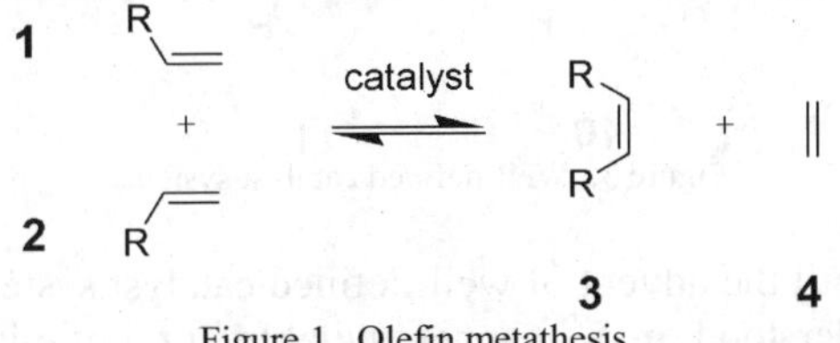

Figure 1. Olefin metathesis.

From the mid-1950s to early 80s, olefin metathesis was performed with ill-defined, multi-component, homogenous and heterogeneous catalyst systems that were not always so easily understood. Yet, even though the exact nature of the mechanism was unknown during this time, olefin metathesis developed into a powerful organic tool for the transformation of monomers to high molecular weight polymers.

Eventually, Chauvin proposed a mechanism involving the interconversion of a metal alkylidene (**5**) and an olefin (**6**) in which olefin containing compounds were divided at the carbon-carbon double bond into carbene fragments and then redistributed into new olefins.[2] The key step in this mechanism was the formation of the metallobutacycle (**7**), which was formed by the interaction of a metal carbene with an olefin (Figure 2). The formation of this key intermediate resembles a [2+2] cycloaddition from which the olefin coordinates with the alkylidene. Cleavage of the ring can occur in either a vertical or horizontal manner. However, the productive cleavage of the metallocycle generates a new alkylidene (**8**) that is reactive upon exposure to a second olefin. Stereoselectivity of these systems are mainly determined by steric interactions between the ligand set of the alkylidene and the olefin substituents.

Y. Imamoglu and L. Bencze (eds.), Novel Metathesis Chemistry: Well-Defined Initiator Systems for Specialty Chemical Synthesis, Tailored Polymers and Advanced Material Applications, 191–202.
© 2003 *Kluwer Academic Publishers. Printed in the Netherlands.*

5
6
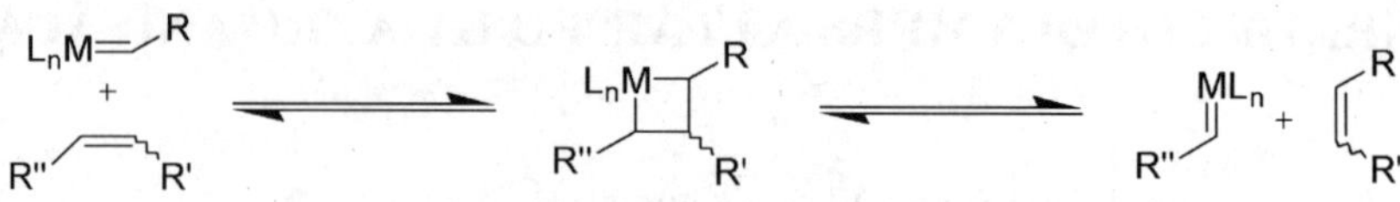
7 8 9

Figure 2. Formation of the metallobutacycle.

1.2 METATHESIS CATALYSTS

As previously discussed, the presence of a metal carbene is crucial to the metathesis reaction by forming the required metallobutacycle intermediate. The development of modern, well-defined catalysts was initiated in 1964 with Fischer carbenes (**10**), commonly known as carbenes, and expanded to Schrock carbenes (**11**), or alkylidenes, during the 1980's as seen in Figure 3. The fundamental difference between these two types of carbenes is Fischer carbenes show electrophilic behavior while Schrock carbenes exemplify nucleophilic behavior.[1a, 3]

Fischer Schrock

M is low oxidation state
R = amines, alkoxides

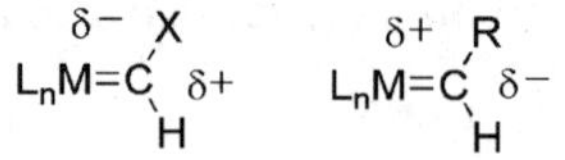

M is high in oxidation state
R = alkyl, aryl, silyl

10 11

Figure 3. Well-defined catalyst systems.

It was not until the advent of well-defined catalyst systems that the metathesis reaction became understood in a manner suitable for optimization conditions. It became apparent that proper catalyst choice is essential in order to avoid side reactions such as cation formation, vinyl addition, and the formation of multiple catalytic species.[4] The most common single site-catalyst/initiator systems for metathesis conditions typically require Schrock's carbene or Grubbs' alkylidene catalyst in order to proceed. These catalysts offer the advantage of higher activities under milder conditions. By varying the ligand environment of the metal center, a tunable catalyst system can be developed. Also, by differing the steric and electronic nature of the alkoxide ligands, the reactivity of the metal alkylidene can be engineered to different degrees. However, these types of catalysts are also limited by their high oxophilicity of the metal centers which causes them to be highly sensitive to oxygen and moisture.

Schrock's catalyst systems were developed through variations of the ligand groups and metals with high oxidation states. Bulky imido and alkoxy ligands were attached to four coordinate W(VI) or Mo(VI) metal centers (Figure 4). Within these systems, various electronic and steric factors were accounted for while allowing the approach of the olefinic moiety. It was eventually realized that the highest degree of reactivity was accomplished when the alkoxy groups were hexafluoro-t-butoxy.[5] Even though metathesis conditions were optimized in this manner, the W analogues (**12**) would not tolerate any functionality. It was not until the development of the Mo analogues (**13**) that the catalyst became more tolerant of functionality. It was these Mo catalysts that allowed for ROMP and RCM substrates to include a broad range of functionalities.[6] Eventually, these catalysts were developed to be suitable for acyclic diene metathesis (ADMET), and the first report of Schrock's catalyst under ADMET conditions was reported in 1991.[7] Even though Schrock's catalyst systems were a significant achievement, they still were not tolerant of functionality within the polymer. In addition, they are highly unstable during storage and are sensitive to air and moisture and easily deactivated. It was not until the advancement of Grubbs' ruthenium based systems that some of these problems were overcome.

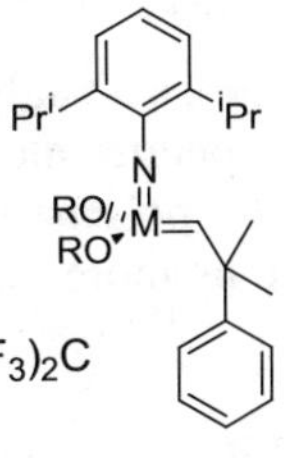

R = CH$_3$(CF$_3$)$_2$C

12, M = W
13, M = Mo

Figure 4. Schrock's carbene for olefin metathesis.

In the 1990's, ruthenium based catalysts were explored in order to compensate for functionality of the substrate or solvent binding to the metal center and deactivating the catalyst in Schrock-type systems.[8] Ruthenium was chosen because of its higher reactivity to olefins than any other functional group. Therefore, Grubbs developed robust benzylidene catalyst systems. The first reports of discrete ruthenium carbene complexes (**14**) were published in 1993 and again later with significant improvements (**15**) in 1996 and most recently (**16**) in 1999.[9] The development of these catalyst structures were a result of changing molecular features to acquire higher degrees of activity. There have also been several other publications over the years that have contributed to the development of ruthenium complexes such as Herrmann's[10] and Nolan's catalyst.[11]

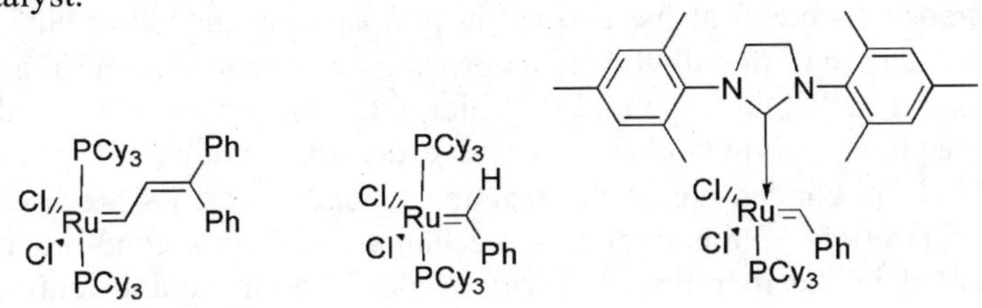

14   **15**   **16**

Figure 5. Grubbs' ruthenium benzylidenes.

These ruthenium catalyst systems are well-defined structures that are efficient under less stringent conditions. Although some are not as reactive as Schrock's alkylidenes, these catalysts can be handled under bulk conditions in the presence of air while remaining active. The electronic properties of the ruthenium systems are different from Schrock's catalyst. The ruthenium is not in a high oxidation state, and the halogens have been included as electron withdrawing ligands. In addition, the phosphine ligands are excellent σ-donors. The sterics of these systems were constantly examined and fine-tuned to identify the affects of ligand structure. The initial ruthenium complex (**14**) was not as reactive as Schrock's Mo catalyst, but it was the first well-defined complex tolerant of functionality due to high stability with respect to the vinylidene benzylidene. It is a very robust catalyst that is air-stable as a solid and retains activity when exposed to water, alcohols, or acids. Upon combining bulky and strong electron donating PCy_3 ligands, the activity of the benzylidene ruthenium catalyst (**15**) was greatly increased. The benzylidene catalyst became the preferred form of the alkylidene because of the ease of synthetic preparation.

Both Schrock's and Grubbs' catalyts have proven to be useful in ADMET polymerizations for unique and specific monomer properties. The Schrock's systems have greater activities due to faster polymerization kinetics. However, systems such as Grubbs' carbenes have proven to be more versatile and enduring for systems with polar functionalities.

1.3 ACYCLIC DIENE METATHESIS (ADMET)

Acyclic diene metathesis polymerization is a step polycondensation to produce a linear, unsaturated hydrocarbon polymer. Terminal α,ω-dienes are connected via a stepwise manner to produce a polyolefin (**24**) and a molecule of ethylene (**27**) in each propagation step (Figure 6). Since the reaction occurs under equilibrium, this process is driven to completion by removal of the condensate. Therefore, by definition,[12] ADMET can be considered a step-propagation condensation-type polymerization.

The initiation step of the mechanism starts with the coordination of the metal alkylidene with a olefin site within the diene monomer (**17**). As a result, a π-complex (**18**) forms between the metal alkylidene and the olefin site, and eventually forms the metallobutacycle (**19**) previously discussed. The metallobutacycle then disproportionates to form a new alkene (**20**) while regenerating the alkylidene (**21**), placing the transition metal at the end of the propagating monomer chain. Another diene then interacts with the alkylidene as previously described to form a π-complex, rearranging into a metallobutacycle (**23**), which then disproportionates and condenses into the propagating polymer chain (**24**) in order to continue the cycle. Finally, ethylene (**27**) is the condensate of the reaction when α,ω-dienes are used in a step-growth polymerization. Support of this mechanism and its second-order kinetics is well documented in the literature.[13] Since ADMET is an equilibrium reaction, the reverse reaction is a statistical possibility, in which the by-product may react with the main chain to form either smaller species, closed systems, and/or lower molecular weight products. However, it is also this equilibrium process that allows the reaction of all chains in the system to randomize the product distribution as described by the polydispersity (PDI = M_w/M_n) of 2.0.

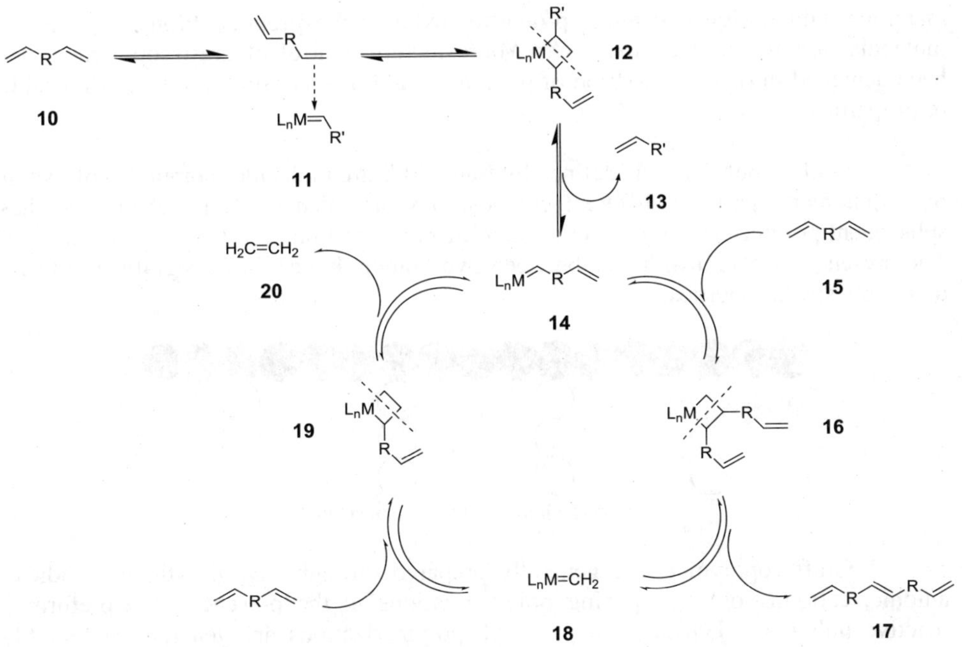

Figure 6. Acyclic diene metathesis (ADMET) mechanism.

Recent reviews on ADMET have described the theory and experimental conditions.[14] ADMET has developed into a powerful tool for the synthesis of organic polymers. This mechanism allows for the synthesis of hydrocarbons and placement functionality within or along the polymer backbone simply through design of the monomer unit.

There are other factors that play a significant role in the mechanism. Sterics of the α,ω-diene monomer alkylidene are important factors in the kinetic rate of the mechanism. Depending upon the steric incumberance around the π-complexation, the rate of the reaction can be hindered or halted altogether. Also, the efficiency of a step polymerization is highly dependent upon monomer purity. Any type of contamination present during the metathesis reaction is detrimental to accomplishing a high molecular weight polymer. Finally, the presence of functionality can modify the reaction rates since electronics associated with the transition metal are attached to the propagating alkylidene chain. These factors will be left for a later discussion.

2 Graft Copolymers

The design and synthesis of polymeric materials has developed into a useful and essential area of modern chemistry. Polymer science examines the demand for everyday challenges requiring the use of efficient polymerization techniques. However, such techniques require that viable routes toward specific structures are constantly developed and explored. The fact that this area of science can yield materials with designed genuine properties exhibits the progress brought fourth by molecular architectural engineering. Microstucture control of graft copolymers has been achieved through a variation of monomers and mechanisms resulting in tuneable of properties.

Graft copolymers, as defined by Battaerd,[15] are molecules comprised of two or more polymeric parts of different compositions in which one type of polymer has subsequent polymerization of another kind of polymer along the backbone (Figure 7). The presence of chemical bonds between two homopolymers has a significant impact on the physical properties.

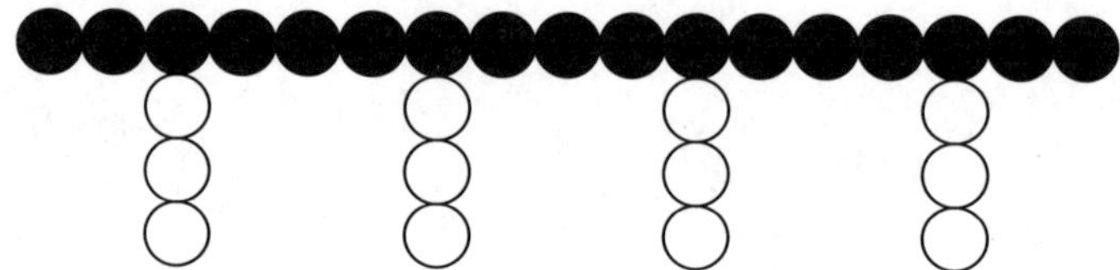

Figure 7. Graft copolymer architecture.

Graft copolymers are generally prepared through step-growth, free radical, anionic, cationic, or ring-opening polymerizations in the presence of a preformed reactive polymer. Typically, step-growth polymerizations are generally not highly considered due to their di- or multi-functional nature which allows for cross-linked systems to occur. Therefore, commercial graft copolymers are typically prepared through free radical polymerizations involving the polymerization of an olefinic monomer in the presence of a preformed polymer bearing a labile hydrogen. Then, initiation occurs by abstraction of the labile hydrogen producing radicals on the polymer backbone. The link between the graft and the backbone can be formed via monomer initiation by the backbone radical or by recombination reactions. Fischer was able to accomplish this by grafting styrene onto polybutadienes.[16] Another category of preparing graft copolymers through free radical initiators is hydroperoxidation.[15] The monomer is initiated by hydroperoxide or functional groups already on the preformed polymer backbone. Even though these are powerful techniques for commercial applications, they are still characterized with broad compositional heterogeneity and contaminated with homopolymers. A better control of the structure with less homopolymer contamination can be gained through anionic methods which lower the propensity for spontaneous termination.[17]

All of the previous synthetic methods involve grafting from an initiator along a multifunctional polymeric precursor. In addition to the "grafting from" technique, other strategies are available for the synthesis of graft copolymers. Jerome has discussed "grafting onto" and "macromonomer" techniques.[18] "Grafting onto" involves the coupling of functional polymers with mutually reactive groups on each homopolymer. The "macromonomer" technique places the graft in the monomer prior to the polymerization which allows for the preparation of a wide variety of macro- and co-monomers. However, the extent of the reaction is often limited by the branch character and ill-defined architectures as a result of radical initiation. The macromonomer approach of graft copolymers will be later discussed.

Since a graft copolymer is the combination of different homopolymers, it is commonly considered a hybrid molecule. Even though it is a single chemical species, the behavior of this molecule can display the properties of both components rather than an average. Graft copolymers are often insoluble in the solvents of either corresponding homopolymeric material. When examining thermal behaviors, a single-phase morphology is possible, but a two-phase morphology is much more commonly observed.[19] Typical transitional thermal behaviors display two distinct glass transition temperatures and/or melt transition temperatures. However, due to intersegmented linkages, they display a finer morphology depending upon the chemical matrix. Thus, at times an amorphous system would show good optical clarity in the absence of homopolymer contamination. Yet, it is also possible to blend a two-phase graft copolymer with their homopolymer. This feature has a utility whether the components are rigid or elastomeric.

3 Atom Transfer Radical Polymerization (ATRP)

3.1 CONTROLLED/"LIVING" MECHANISMS

The use of transition metal catalysts has been known for a long time in radical polymerizations. However, these catalyst systems often lead to ill-defined structures. It was not until the development of a copper complex that "living" radical polymerizations, produced defined structures by radical methods.[20] In 1995, two separate groups independently reported the use of atom transfer radical polymerization (ATRP) as a reversible metal catalyzed atom transfer mechanism to generate the propagation of radicals.[21] This is in contrast to conventional controlled/"living" radical methods that utilize photochemical or thermal homolytic cleavage.

The ATRP cycle proceeds in a manner that regenerates the propagating radical. During polymerization (Figure 8), the metal complex (**28**) undergoes a one-electron oxidation by abstraction of the halogen atom from the starting compound (**30**). This reaction generates an organic radical (**31**) and copper (II) complex (**29**). The substituents on the organic halide facilitate the reaction by stabilizing the radical. The radical then adds to an unsaturated group (**32**) in an inter- or intra-molecular fashion resulting in a terminal propagating terminal radical (**33**). Re-abstraction of the halogen atom from the copper (II) complex then occurs to reform the original copper (I) complex terminating the radical resulting in the product (**34**). The initial cycle is known as atom transfer radical addition (ATRA). ATRP occurs upon continuation of the atom transfer equilibrium that regenerates (**34**) and deactivates the propagating radicals from the initiator or a polymer chain end. The role of the initiator (**30**) during the polymerization cycle is to determine the number of dormant chains and to provide the structure of the initiating end of the polymer chain. After activation of a dormant chain, more than one monomer can add to the chain and the number of additions that occur before termination determines the degree of polymerization. Even though the persistent radical effect will minimize the total number of chains terminated by coupling or disproportionation, irreversible chain terminations must still be accounted for.[22]

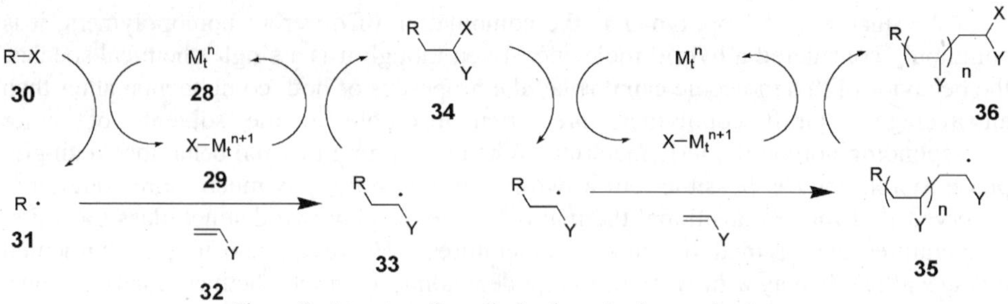

Figure 8. Atom transfer radical polymerization mechanism.

It was reported that the homogenous ATRP of styrene, acrylate, and methacrylate allows for excellent degrees of polymerization up to DP ~ 100 and polydispersities, M_w/M_n, as low as 1.04 to 1.05.[23] ATRP minimizes side reactions such as termination, elimination of HX from the chain end, and oxidation/reduction of the radicals to the corresponding cation/anion.[24] Unlike chain polymerizations of conventional radical conditions, controlled/"living" polymerizations are first order kinetics with respect to monomer concentration and are typically in close agreement with the calculated values based upon $([M]_o - [M]_t)/[ln]_o$. The same conversions were also reported for polymerizations conducted in solvents such as diphenyl ether, p-dimethoxybenzene, or benzophenone. As expected, conversion rates were slower than bulk polymerizations as expected due to the lower concentration of the initiating system.

3.2 ATRP CATALYSTS

There are several requirements that must be meet in order for ATRP catalysts to perform efficiently. A proper catalyst should be selective, efficient, and be able to change between two oxidation states, robust and inexpensive.[25] Ideally, ATRP catalysts should be involved in the reversible atom transfer but not in any other reactions. This goal can be accomplished through the appropriate reaction conditions and adjustment of the steric and electronic properties of the complex by the appropriate ligands.

The equilibrium constant is one of the most important parameters of the catalyst since it defines the proportion of radicals generated by the transition metal complex. Essentially a higher equilibrium constant is more favorable since the same amount of radicals can be formed with a lower concentration of the transition metal complex. However, if the equilibrium constant is too high, an excessive amount of radicals may lead to an exceedingly high contribution of radical termination and loss of control. Most systems operate with equilibrium constants in the range of $10^{-9} < K < 10^{-6}$.

An ATRP catalyst exists in two oxidation states where the activator is M_t^n/L in the lower oxidation state and $X\text{-}M_t^{n+1}/L$ is the deactivator in the higher oxidation state. These catalysts are easily oxidized to the higher oxidation state. The equilibrium constant between the two states defines the overall amount of radicals and polymerization rate. The number of monomer units added during each activation step is determined by the rate ratios of propagation to termination. Only a few monomer units should add during each step for a controlled ATRP process. Thus, the deactivation process may be as fast as diffusion control.

Finally, robustness and stability of ATRP catalysts is critical for the catalyst to retain its original activity after each cycle. If these requirements are not achieved during the polymerization process, the catalyst may become complexed by the monomer or polymer which may lead to the displacement of the original ligand from the coordination sphere and reduce catalytic activity. The released ligand may then become involved in other unwanted side reactions.

4 Controlled Architectures by ATRP and ADMET

The 'living' free radical conditions of ATRP can be employed to generate monodisperse macromonomers prior to metathesis condensation. This can be quite significant since free radical polymerization, in general, is of enormous commercial interest.[26] Furthermore, free radical polymerizations allow for a large diversity of macromonomers to be prepared and polymerized under simple experimental conditions.[27] Examples of ATRP/ADMET chemistry are given below.

4.1 MACROMONOMER SYNTHESIS

One of the most demanding challenges associated with creating precisely grafted copolymers has to do with locating the graft on the polymer backbone itself. Most often the graft is randomly placed along the backbone through typical polymerization methods. Random placement can be avoided using the macromonomer approach if the "graft" can be symmetrically disposed along the main chain of the macromonomer.[28] This is the approach we have taken, in essence by placing the graft site (the ATRP initiator) in the center of an α,ω-diene (Figure 10, Route A)). The first step consists of building an ATRP initiator site located symmetrically within an α,ω-diene. Primary alcohols consisting of the diene funicitionality (**37**) were synthesized and further used for the development of ATRP initiators (**38**) where the alcohol precursor was converted to a radical stabilizing group simply through a DCC (1,3-dicyclohexylcarbodiimide) coupling reaction with 1-bromo-2,2'-dimethyl propionic acid. Following the quantitative polymerization of styrene to the desired graft lengths (**39**), the ATRP catalysts were removed by simply passing over a bed of neutral alumina and precipitation into methanol yielding a pure macromonomer.

Since ATRP of styrene resulted in solid macromonomers (**39**), ADMET polymerizations could not be performed by standard bulk methods under high vacuum; thus a different approach was required to successfully achieve metathesis polycondensation. ADMET chemistry was conducted using solution polymerization conditions as the amorphous polymers are solids at reaction temperature. After several attempts of various solvents, toluene was found to yield the best reaction solvent due to the corresponding nonpolar nature of both the macromonomers (**39**) and toluene. In addition to proper solvent choice, the reaction required a slight argon purge to remove ethylene from the system for 10 days at raised temperatures of 75°C to drive the equilibrium of the polycondensation. These high temperatures and extended reaction times are strenuous on the lifetime of the metathesis catalyst and require that the solvent and catalyst be recharged throughout the duration of the reaction. Analysis by ^{1}H NMR of **40** indicates the conversion of the terminal dienes to internal olefin at 5.39 ppm. Also, this lower molar concentration contributes to the extended reaction times such that the bulky metal alkylidene could encounter the sterically hindered terminal olefin. Gel permeation chromatography (GPC) measurements indicated uncharacteristically greater polydispersities for ADMET conditions. The broad molecular weight distributions can be attributed to variations in solvent concentration throughout the reaction. If optimum concentrations were not achieved, a true equilibrium can not take place allowing other reactions to occur.

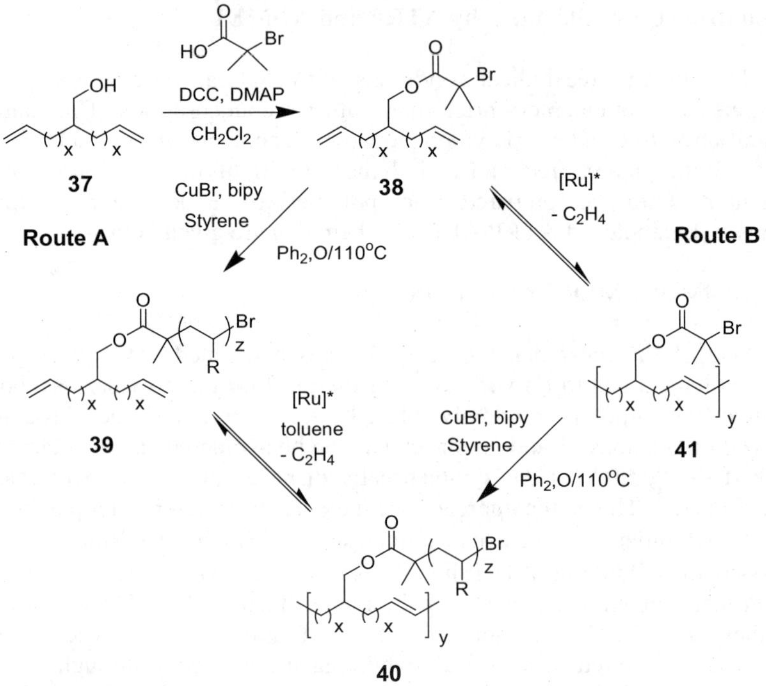

Figure 10. Polystyrene graft copolymer synthesis.

4.2 MACROINITIATOR SYNTHESIS

The macroinitiator route also can be used to synthesize of graft copolymers. This method allows for branch growth from a linear multifunctional precursor. The synthesis of well-defined graft copolymer structures is possible by this technique when living conditions are utilized with vinyl or heterocyclic comonomers. It is important to note that each functional site needs to be initiated for these copolymers to be considered "well-defined". The macroinitiator route (Figure 10, Route B) involved the polycondensation of the ADMET precursor (monomer **38**), resulting in an ATRP macroinitiator (polymer **41**) prior to growing the grafts by 'living' methods. This technique provides the advantange of performing metathesis in typical bulk conditions under high vacuum yielding a high molecular weight polymer. Observation of the terminal α,ω-dienes converting to internal olefin peaks was trivial by ^{1}H NMR. This measurement indicated that high polymer was achieved as the peaks at 4.98 and 5.81 ppm subsided in intensity while the peak at 5.37 ppm became more pronounced.

After complete characterization of the macroinitiator (**41**) and removal of the benzylidene catalyst, proper initiator:catalyst:monomer ratios were calculated from GPC measurements for the 'living' polymerization of styrene. With copper bromide and 2,2'-bipyridine as the catalyst system, both the number average molecular weight (M_n) and polydispersity index (PDI) slightly increased, correlating to the initiation of each site polymerizing only 2 repeat units of polystyrene per site. It is obviously better to do ATRP chemistry prior to ADMET chemistry in order to attain extended grafting.

5 Conclusions

Atom transfer radical polymerizations has shown to be capable of preparing different, well-defined chain lengths of polystyrene graft copolymers suitable for metathesis polymerization. Specifically, ATRP has been found to be useful in synthesizing macromonomers that can be fully characterized prior to polycondensation resulting in absolute certainty of the graft nature, length, and placement in the final copolymer structure when combined with the ADMET reaction. With the ability to synthesize these defined and controlled structures, the macromonomer technique offers helpful insight to molecular engineering of desired polymer properties.

6 References

1. (a) Crabtree, R.H. (1994) *The Organometallic Chemistry of the Transition Metals*, 2nd edition; Wiley and Sons: New York. (b) Ivin, K.J.; Mol, J.C. (1997) *Olefin Metathesis and Metathesis Polymerization*; Academic Press: San Diego.

2. Herrison, J.L.; Chauvin, Y. (1970) *Makromol. Chem.*, 141.

3. Elsenbroich, C.; Salzer, A. (1992) *Organometallics: A Concise Introduction*; VCH: New York.

4. Wagener, K.B.; Boncella, J.M.; Nel, J.G.; Duttweiler, R.P.; Hilmeyer, M.A. (1990) *Makromol. Chem.* **191**, 365.

5. Schaverien C.J.; Dewan, J.C.; Schrock, R.R. (1986) *J. Am. Chem. Soc.* **108**, 2771.

6. Schrock, R.R. (1990) *Acc. Chem. Res.* **23**, 158.

7. Wagener, K.B.; Boncella, J.M.; Nel, J.G. (1991) *Macromolecules* **24**, 2694.

8. Trnka, T.M.; Grubbs, R.H. (2001) *Acc Chem. Res.* **34**, 18.

9. (a) Nguyen, S.T., Grubbs R.H.; Ziller, J.W. (1993) *J. Am. Chem. Soc.* **115**, 9858. (b) Schwab, P.E.; Grubbs, R.H..; Ziller, J.W. (1996) *J. Am. Chem. Soc.* **118**, 100. (c) Scholl, M.; Ding, S.; Lee, C.W.; Grubbs, R.H. (1999) *Org. Lett.* **1**, 953.

10. (a) Herrmann, W.A.; Kratzer, R.M.; Fischer, R.W. (1997) *Angew. Chem. Int. Engl. Ed.* **36**, 2652. (b) Weskamp, T.; Kohl, F.J.; Hieringer, W.; Gleich, D.; Herrmann, W.A. (1999) *Angew. Chem. Int. Engl. Ed.* **38**, 2416.

11. Huang, J.; Stevens, E.D.; Nolan, S.P.; Petersen, J.L. (1999) *J. Am. Chem. Soc.* **121**, 2674.

12.	Odian, G. (1991) *Principals of Polymerization*, 3rd edition; Wiley Interscience: New York.
13.	Wagener, K.B.; Brzezinska, K.; Anderson, J.D.; Younkin, T.R.; Steppe, K.; DeBoer, W.(1997) *Macromolecules* **30**, 7363.
14.	(a) Tindall, D; Wagener, K.B.; Brzezinska, K.R. (1999) *Polym. Prepr. (Am. Chem. Soc. Div. Polym. Chem.)* **218** (2), 413. (b) Davidson, T.A.; Wagener, K.B. (1999) Acyclic Diene Metathesis (ADMET) Polymerization. In *Synthesis of Polymers*; Schluter, A.D., Ed.; Materials Science and Technology Series; Wiley-VCH: Wienheim, p.105. (c) Church, A.C.; Smith, J.A.; Pawlow, J.H.; Wagener, K.B. Non-Traditional Step-Growth Polymerization-ADMET. In *Synthesis Methods in Step-Growth Polymers*; Long, T.E.; Rodgers, M., Ed.; Wiley & Sons: New York, Ch. 9, in press.
15.	Battaerd, H.A.J.; Tregear, G.W. (1967) *Graft Copolymers*; Wiley & Sons: New York.
16.	Fischer, J.P. (1973) *Angew. Chem., Int. Ed. Engl.* 12, 428.
17.	Kennedy, J.P.; Charles, J.J.; Davidson, D.L. (1974) *Recent Advances in Polymer Blends, Grafts and Blocks*; Sperling, L.H., Ed.; Plenum: New York.
18.	Jerome, R. (1999) *Macromol Chem. Phys.* **200**, 156.
19.	Stannett, V.J. (1977) *Macromol. Sci., Chem.* **4**, 1177.
20.	Bellus, D. (1985) *Pure Appl. Chem.* **57**, 1827.
21.	(a) Wang, J.S.; Matyjaszewski, K. (1995) *J. Am. Chem. Soc.* **117**, 5614. (b) Kato, M.; Kamigaito, M.; Sawamoto, M.; Higashimura, T. (1995) *Macromolecules* **28**, 1721.
22.	Patten, T.E.; Matyjaszewski, K. (1999) *Acc. Chem. Res.* **32**, 895.
23.	Patten, T.E.; Xia, J.; Abernathy, T.; Matyjaszewski, K. (1996) *Science* **272**, 866.
24.	Matyjaszweski, K. (1997) *Tetrahedron* **53**, 15321.
25.	Matyjaszewski, K. (2000) *Polym. Prepr. (Am. Chem. Soc., Div. Polym. Chem.)* **41**(1), 411.
26.	Patten, T.E.; Matyjaszewski, K. (1998) *Adv. Mater.* **10**, 901.
27.	Moad, G.; Solomon, D. (1995) *The Chemistry of Free-Radical Polymerization*; Pergamon: Oxford.
28.	O'Donnell, P.M.; Brzezinska, K.; Powell, D.; Wagener, K.B. (2001) *Macromolecules* **34**, 6845.

SYNTHESIS, CONFORMATION, AND FUNCTIONS OF HELICAL POLY(*N*-PROPARGYLAMIDES):

R. NOMURA, J. TABEI, K. YAMADA, AND T. MASUDA
*Department of Polymer Chemistry, Graduate School of Engineering,
Kyoto University, Kyoto 606-8501, Japan*

1. Introduction

The studies on synthetic helical polymers were initiated by the discovery of helical structures in naturally occurring macromolecules. Natural macromolecules form three-dimensionally ordered supramolecules by self-organization process. The design strategy of nature to form helical polymers results from the appropriate arrangement of intermolecular and/or intramolecular noncovalent interactions. α-Helical polypeptides are representative examples, in which intramolecular hydrogen bonds induce and stabilize their helical conformation.

Unfortunately, apart from the well-defined oligomers [1–3], the arrangement of noncovalent interactions, especially that of the hydrogen bond, along the whole of the polymer backbone is very difficult for synthetic macromolecules. The first attempt to arrange hydrogen bonding in synthetic polymer was made by Nakahira et al. using stereoregular isotactic poly(methacrylamides).[4,5] The polymers have proven to possess a helical conformation that is stabilized by the hydrogen bonds between the amide groups. The IR spectra of the polymers, however, indicated the existence of both hydrogen-bonded and non-hydrogen bonded amide groups, which means that the hydrogen bonds are not regularly ordered. Another example was recently reported by Tang et al. using a Rh-based stereoregular poly(phenylacetylene) bearing amino acids on the phenyl rings.[6] The polymers were evidenced by CD spectra to adopt a helical conformation, and pH dependence of the CD effects suggests the significant contribution of hydrogen bonds to the secondary conformation. However, no detailed information on the hydrogen bond is available, and thus, regularity of the hydrogen bonds is unknown for this polymer. Successful results were recently obtained by Nolte et al. [7] and by us independently.[8] Nolte and his coworkers synthesized poly(isocyanides) having oligopeptide pendants. The preference of the side chain to form β-sheet-like structure works as a driving force to twist the main chain into a helical conformation. The IR data suggest that the hydrogen bonds are well-arranged and located intramolecularly. We introduced amide groups to polyacetylene, and succeeded in constructing an ordered hydrogen bonding. Specifically, the pendant amide groups in stereoregular (cis) poly(*N*-propargylamides), poly(**1**), intramolecularly hydrogen-bond, which simultaneously rigidifies the polymer backbone. Polymers from monosubstituted acetylenes, unless they have very bulky substituents, are generally flexible so that they cannot adopt a helical conformation. However, the backbone rigidity of poly(**1**), enhanced by hydrogen bonding, enables the main chain to take a helical structure (Scheme 1).

Y. Imamoglu and L. Bencze (eds.), Novel Metathesis Chemistry: Well-Defined Initiator Systems for Specialty Chemical Synthesis, Tailored Polymers and Advanced Material Applications, 203–213.
© 2003 *Kluwer Academic Publishers. Printed in the Netherlands.*

Scheme 1

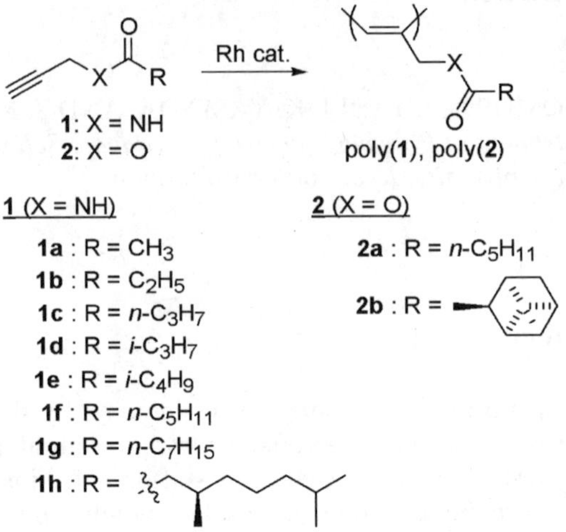

1: X = NH
2: X = O

poly(**1**), poly(**2**)

1 (X = NH)

1a : R = CH_3
1b : R = C_2H_5
1c : R = n-C_3H_7
1d : R = i-C_3H_7
1e : R = i-C_4H_9
1f : R = n-C_5H_{11}
1g : R = n-C_7H_{15}
1h : R =

2 (X = O)

2a : R = n-C_5H_{11}
2b : R =

(Rh cat. = $(nbd)Rh^+[(\eta^6$-$C_6H_5)B^-(C_6H_5)_3]$, $[(nbd)RhCl]_2$-Et_3N)

In this review article, we summarize our recent efforts on the conformational study of poly(N-propargylamides). First, we discuss the primary and secondary structures of poly(N-propargylamides). Next, we show the relationship between the stability of the helical conformation and the side chain structure. We also show that the thermodynamic parameters governing the stability of the helical conformation can be readily determined by UV-vis. spectroscopy. Finally we demonstrate some applications of poly(N-propargylamides) as stimuli-responsive functional materials.

2. Primary and Secondary Structure of Poly(N-propargylamides)

2.1. SYNTHESIS AND PRIMARY STRUCTURE

Rh complexes such as $(nbd)Rh^+[(\eta^6$-$C_6H_5)B^-(C_6H_5)_3]$ (nbd = 2,5-norbornadiene) and $[(nbd)RhCl]_2$–Et_3N are highly active to give good yields of poly(N-propargylamides).[8] The polymerization is completed within a few minutes at ambient temperature. The poly(N-propargylamides) prepared have perfect stereoregularity (cis), as with most of Rh-based substituted polyacetylenes.[9]

Poly(N-propargylamides) are characterized by their enhanced backbone rigidity.[8] Rh-based polymers from monosubstituted acetylenes, except for poly(propiolic esters),[10] generally possess flexible backbone unless they have bulky pendant groups. Thus, many substituted polyacetylenes stereoregularly prepared with Rh catalysts show well resolved NMR signals for the main chain olefinic protons around 6−7 ppm. [9] In contrast, as exemplified by Figure 1a, the ^{1}H NMR spectrum of poly(**1f**) in $CDCl_3$ exhibits very broad signals at a low temperature (19 °C). Well resolved signals are obtained only at high temperature (55 °C). The viscosity index (α) in the Mark−Houwink−Sakurada plot is also a good indication of the semiflexibility of the main chain. The α value for poly(**1f**) is 0.98 (in THF at 30 °C),

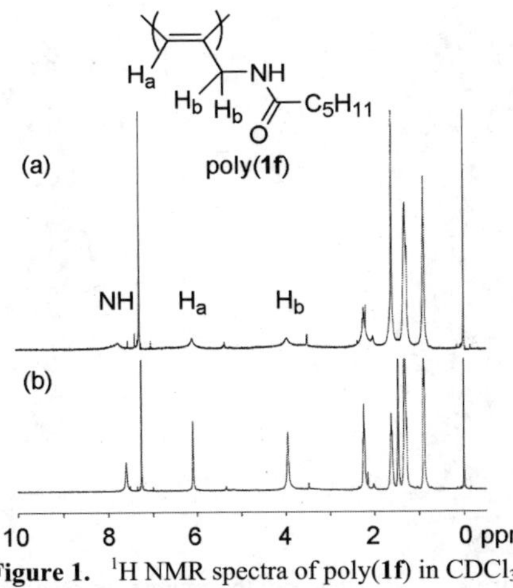

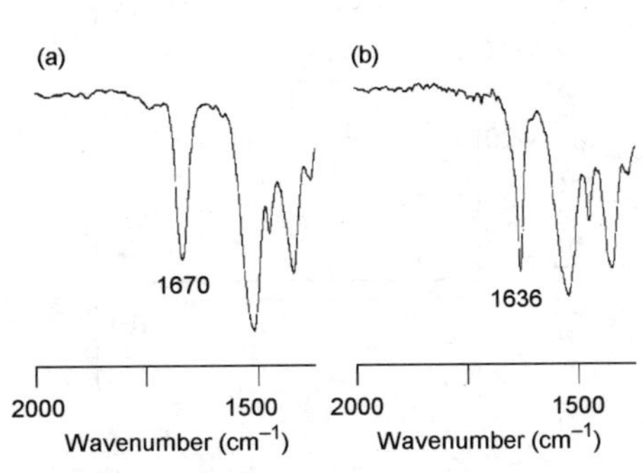

Figure 1. ^{1}H NMR spectra of poly(**1f**) in CDCl$_3$ at (a) 19 °C and (b) 55°C.

Figure 2. IR spectra of (a) monomer **1h** and (b) poly(**1h**) in CHCl$_3$.

while the ester version of poly(**1f**), the Rh-based poly(**2a**) (88% cis content) having no hydrogen-bonding donor group, displays clear ^{1}H NMR signals even at 22 °C and low α value (0.68). Thus, it is concluded that the pendant amide groups rigidify the polymer backbone. In other words, the hydrogen bonding between the amide groups improves the main chain rigidity.

The IR spectra have clearly proven the construction of the well-ordered intramolecular hydrogen bonding between the pendant groups.[8] As shown in Figure 2, the amide I absorption of monomer **1h** is observed at 1670 cm^{-1} in CHCl$_3$, and the wavenumber of this IR band is independent of concentration between 48−0.05 mM. This means that the monomer does not hydrogen-bond in solution. In contrast, under similar conditions, the wavenumber of the amide I absorption of poly(**1h**) in CHCl$_3$ (1636 cm^{-1}) is almost identical to that in the solid state (1638 cm^{-1}) and is shifted by 34 cm^{-1} to that of the monomer. These results mean that the pendant amide groups of poly(**1h**) are hydrogen-bonded. The concentration independence of this band in concentration from 0.05 to 48 mM leads to a conclusion that the intramolecular hydrogen bonds are constructed between pendant amide groups in poly(**1h**) even in very diluted solution, in which the corresponding monomer cannot form hydrogen bonding.

2.2. SECONDARY STRUCTURE

The perfect stereoregularity and the rigidity of poly(*N*-propargylamides), enhanced by the intramolecular hydrogen bonding, enable the polymer to self-organize into a helical conformation.[8] For example, the CD spectra of a poly(*N*-propargylamide), poly(**1h**), having chiral side chains displays very intense, concentration independent CD effect in the main chain absorption region (Figure 3). Thus, the main chain twists from the coplanar conformation to adopt the helical conformation. On the other hand, the chiral ester version polymer, poly(**2b**), prepared with a Rh catalyst, displays much lower optical rotation ([α]$_D$ = +50°) and quite weak CD effects although it has bulky pendant groups. These results indicate that, like α-helical polypeptides, the helical conformation of poly(*N*-propargylamides) is stabilized by the intramolecular hydrogen

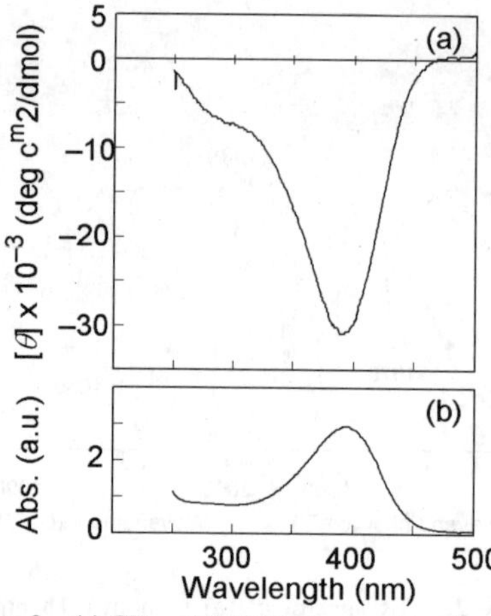

Figure 3. (a) CD and (b) UV-vis. spectra of poly(**1h**) in CHCl$_3$.

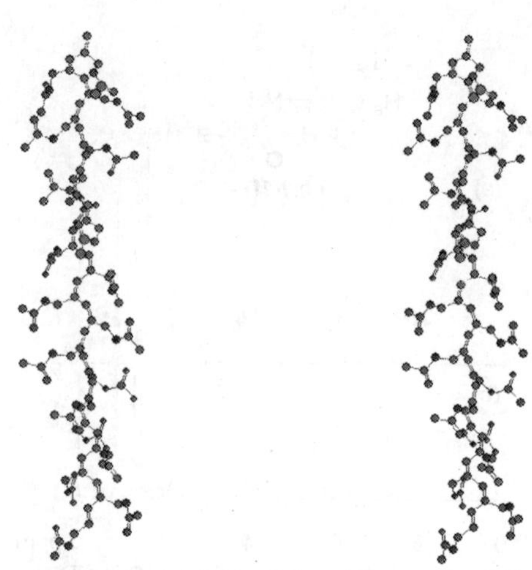

Figure 4. A stereoview of the optimized structures of 20-mer of **1a** terminated by hydrogen.

bonding.

The secondary conformation of poly(*N*-propargylamides) has been estimated on the basis of the CD, IR, and UV spectroscopies, the worm-like touched-bead model theory, and the semiempirical molecular orbital calculations (MM2 and AM1 methods).[11] The computationally optimized structure of the 20-mer of **1a** is illustrated in Figure 4. The main chain twists from coplanarity and takes a helical conformation with the averaged dihedral angle around the single bond of 139°. This value is larger than that of the tightly twisted helical poly(propiolic esters) (107−129°) [12] but smaller than that of helical poly(phenylacetylene) (147−155°).[13] The hydrogen bonds are constructed between the nth and n+2th repeating segments. Interestingly, the side chain amide groups are aligned and accumulated along the helix axis. The side chain amide groups form two helical hydrogen-bond strands that surround the helical conjugated backbone. Each hydrogen-bond strand has approximately 10 amide groups per turn.

3. Stability of the Helix

3.1. EFFECTS OF THE SIDE CHAIN STRUCTURE

The structure of the side chain has proven to markedly influence the rigidity and thus, stability of the helix.[14] For example, as tabulated in Table 1, the increase in the length of the alkyl chain results in increase in the half-height linewidths ($v_{1/2}$) of the vinyl and N−H protons, that is, the rigidity of the backbone. Introduction of a branch at the position β to the carbonyl group more significantly rigidifies the polymer backbone. For example, the N−H and vinyl protons of poly(**1e**) cannot be clearly detected at 19 °C, and well-resolved resonances are not attainable even at 55 °C. Similarly, the half-height linewidth of the vinyl proton of poly(**1h**), which possesses long alkyl chains with a β-branched structure, is almost one order larger than that of poly(**1b**) (Table 1), and the ^{1}H NMR measured at 60 °C shows broadened signals.

Table 1. Homopolymerization of *N*-propargylalkylamides (**1**) having various substituents.

Monomer	Yield (%)	$M_n{}^a$	$M_w/M_n{}^a$	$\nu_{1/2}{}^b$ (Hz) Vinyl	N–H	α^c
1a	99^d	7100^e	9.96^e	–	–	–
1b	89^d	10000	1.44	11	20	–
1c	90^d	8700	2.14	12	24	–
1d	51^d	8700	2.23	–	–	0.74
1e	86^f	9500	1.66	–	–	0.83
1f	77^g	18000	2.24	30	69	0.76 (0.98)
1g	63^g	14000	1.26	33	58	–
1h	62^g	8100	1.68	98	–	0.91

^{a}Estimated by GPC (THF, PSt standards). ^{b}Half-height linewidths of ^{1}H NMR resonance in CDCl$_3$ at 19 °C. ^{c}Viscosity index in THF at 40 °C. The value in the parenthesis was measured in THF at 30 °C. ^{d}Ether-insoluble part. ^{e}Water-soluble part. Estimated by GPC (H$_2$O, PEO standards). ^{f}Hexane-insoluble part. ^{g}Methanol-insoluble part.

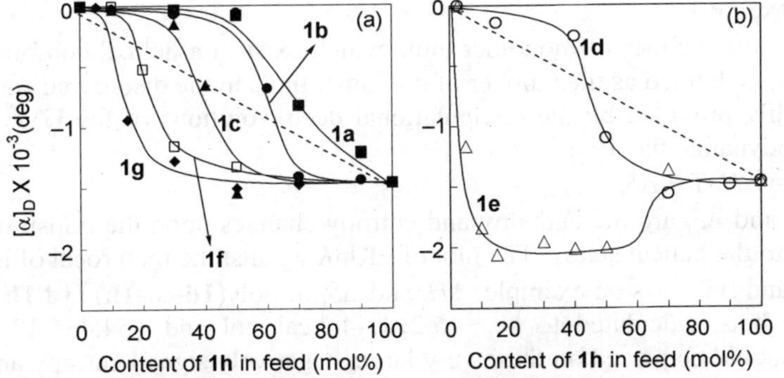

Figure 5. Plot of the optical rotation of the copolymers of achiral monomers (**1a–1g**) with chiral monomer (**1h**) against the feed content of **1h** (in CHCl$_3$ at room temperature).

These results mean that the main chain of poly(**1e**) and poly(**1h**) is stiffer than that of the other polymers carrying linear or *i*-propyl side chains.

The effects of the side chain structure on the rigidity of the polymer backbone directly relates to the stability of the helix, which has been evidenced by the chiral/achiral copolymerizations [15] with a chiral monomer **1h** (Figure 5).[14] In the case of less hindered comonomers **1a** and **1b**, no significant chiral amplification is recognized. On the other hand, the copolymerization using linear alkyl pendants longer than the ethyl group display the chiral amplification phenomenon to some extent. Specifically, when comonomers **1f** and **1g** having longer alkyl groups are employed, the copolymers maintain their rich chiroptical properties even at the low feed content of **1h**. However, even these copolymers also lack chirality when the feed content of **1h** is lower than 10%. Comonomer **1d** behaves similarly (Figure 5b). In contrast, the copolymers derived from **1e** bearing a β-branched alkyl chain clearly exhibit a positive, nonlinear relationship between the optical rotation and the feed ratio. A very large optical rotation (ca. −2000°) that is much larger than that of poly(**1h**) is attainable when the feed content of **1h** was 15−50%. Even a 5% feed of comonomer **1h** gives a copolymer having a quite large optical rotation (−1160°).

3.2. THERMODYNAMIC OF HELIX

It is quite interesting that the secondary conformation, *i.e.*, helix or random coil, can be distinguished by UV-vis. spectroscopy because the helical backbone displays an absorption centered at 400 nm, while the absorption of the polymer in the disordered state is located at 400 nm.[14] The molar absroptivities of the disordered and helical conformations are 4040 and 6600 $M^{-1}cm^{-1}$, respectively. From the variable temperature UV-vis. spectra, the temperature dependence of the ratio of the helical to the disordered state is estimated, which allows the determination of the thermodynamic parameters that govern the helix stability. The free energy difference between the helical and disordered states, ΔG_r, can be defined as

$$\Delta G_r = -RTlnK$$

where K is the equilibrium constant for the transition process from the disordered to the helical conformation. The proportion of the helical to the disordered state determines the constant K. Thus, the constant K is given by

$$K = N_h/N_r$$

where N_h is the number of monomer units which exist in a helical conformation and, similarly, N_r is defined as the number of monomer units in the disordered state. N_h and N_r are readily provided by the computational deconvolutions of the UV-vis. spectra. The thermodynamic theory predicts

$$-RlnK = \Delta H_r/T - \Delta S_r$$

where ΔH_r and ΔS_r are the enthalpy and entropy changes upon the transition from the disordered to the helical state. The plot of $-RlnK$ against the reciprocal of temperature gives ΔH_r and ΔS_r. As an example, ΔH_r and ΔS_r of poly(**1d**-*co*-**1h**) (**1d/1h** = 50/50 in feed) have been calculated to be -15.2 ± 1.44 kcal/mol and -54.4 ± 5.17 cal/mol·K. Emphasis should be placed on these very large, negatively signed entropy and enthalpy changes. This is the first example of the determination of ΔH_r and ΔS_r for the transition process from the randomly coiled to a helical conformation of an artificial polymer.

Here, the secondary structure of stereoregular substituted polyacetylenes can be roughly but simply pictured by the function of the dihedral angle around the single bonds. The entropy difference, ΔS_r, relates to the probability of encountering the helical conformation, *i.e.*, the relative range of the dihedral angle allowed for giving the helical conformation. Semiflexible or rigid substituted polyacetylenes should originally possess a limited region of the dihedral angle available which consequentially yields a relatively high probability of the conformation representing the helical backbone. On the other hand, for substituted polyacetylenes with a flexible main chain such as the polymers from monosubstituted aliphatic acetylenes, a wide region of the dihedral angle about the single bond is allowed. The large entropy loss of poly(*N*-propargylamides) is, thus, explained by the low probability of encountering the region of the dihedral angle which offers the helical conformation. To give the helical backbone, this large, negatively signed entropy change is compensated by the negatively large enthalpy change. In other words, bonding energy should be provided to overcome the large entropy loss. In the present system, the binding energy is given not only by the van der Waals interaction but also by the hydrogen bond. This situation is identical to that of the folding process of proteins where the large entropy and binding energy contributions are well-balanced.

4. Funcitons of Poly(*N*-propargylamides)

4.1. CHROMISM

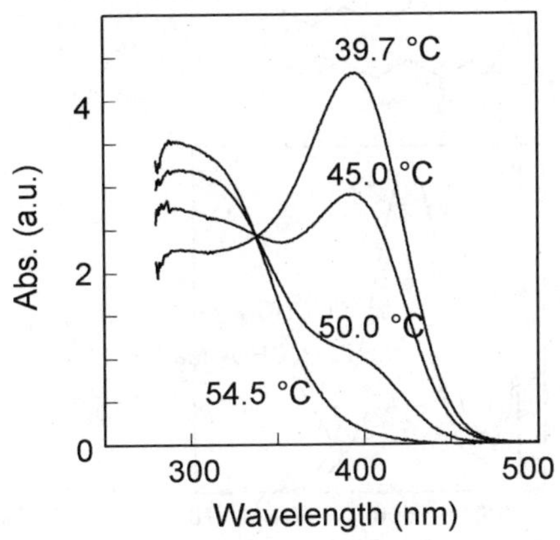

Figure 6. Variable temperature UV-vis. spectra of poly(**1e**) in CHCl₃.

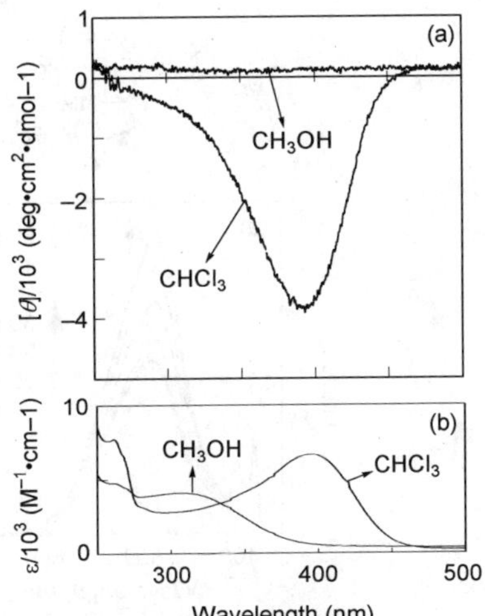

Figure 7. (a) CD and (b) UV-vis. spectra of poly(**1d**-*co*-**1h**) (**1d/1h** = 50/50 in feed) in CHCl₃ and methanol.

As shown above, the helical conformation of poly(*N*-propargylamides) is stabilized by the intramolecular hydrogen bonds, thus, external stimuli such as heating or adding polar solvents cleave the hydrogen bond, deforming the helical backbone. This process results in the change of the UV-vis. spectra, and also in the color of the solution.[14] For example, the absorption of poly(**1e**) centered at 400 nm decreases in intensity upon heating, and a new absorption appears at 320 nm that corresponds to the disordered conformation (Figure 6). An isosbestic point is observed upon this temperature change. Simultaneously, the color of the solution changes from yellow to colorless. When the temperature is decreased, the original spectrum and color are restored. Thus, this deformation process is reversible. Emphasis should be placed on the fact that, resulting from the negatively very large entropy and enthalpy changes in the random coil–to–helix transition process, the thermally induced conformational change occurs in a quite narrow temperature range (15 °C). Other polymers such as poly(**1f**) and poly(**1g**), which exist in a disordered state at room temperature, change color from achromic to yellow when the solution is cooled. Hence, poly(*N*-propargylamides) show thermochromism, which is driven by the conformational change between a helix and a random coil.

Poly(*N*-propargylamides) also achieve the solvatochromism.[14] An example is clearly demonstrated in Figure 7, where the solvent-driven conformational change occurs. In CHCl₃, poly(**1d**-*co*-**1h**) exists in the helical conformation and display intense CD and UV bands around 400 nm. Thus, the solution is yellow. In contrast, this copolymer looses the CD effects in methanol or after adding few drops of methanol. Simultaneously, the absorption at 400 nm disappears and new band appears

210

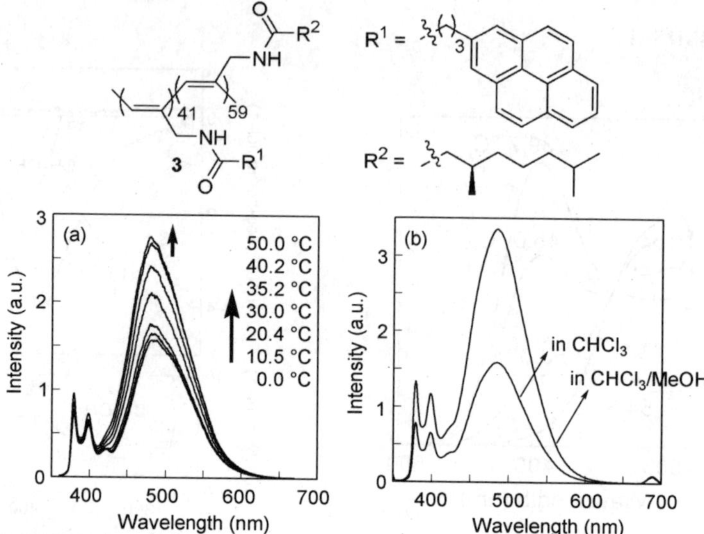

Figure 8. (a) Variable temperature fluorescence spectra of **3** in CHCl₃ (excited at 346 nm). (b) Fluorescence spectra of **3** in CHCl₃ in the presence or absence of methanol (excited at 346 nm).

at 320 nm, which corresponded to the change of the solution color to achromatic. Only 5 vol% of methanol is enough to completely deform the helical conformation.

The change of the secondary conformation of poly(N-propargylamides), induced by the external stimuli, leads to not only the change in the main-chain absorption but also the fluorescence property of the chromophores introduced to the side chain.[16]   For example, the copolymer **3** (Scheme 2), which possesses pyrene units in the side chain, exists in the helical conformation in CHCl₃ at low temperature.   This copolymer exhibits three major emission bands around 380, 395, and 480 nm upon photoexcitation at 346 nm as shown in Figure 8.   The former two bands originate from the isolated pyrene units, and the last one is the excimer fluorescence.   It is interesting that the increase in temperature promotes the deformation into a randomly coiled state and simultaneously increases the intensity of the excimer fluorescence, while that of the isolated pyrene unit stays almost unchanged.   This means that the transition from the helix to randomly coiled conformations results in the increase in probability of excimer formation.   In a similar way, the change in fluorescence intensity is observed when the conformational change is driven by adding a polar additive: the magnitude of the excimer fluorescence is approximately twice in the presence of methanol compared with that in the absence of methanol.   This means that population of the excimer forming site changes upon the conformational change that is driven by the change in external environment.

4.2. RESPONSIVE ORGANOGEL

The reversible volume change of gels upon alternation in the external environment was experimentally shown, for the first time, by Tanaka et al. in ionic gels.[17]   After this discovery, much effort has been made on the synthesis and application of the stimuli-

Scheme 3

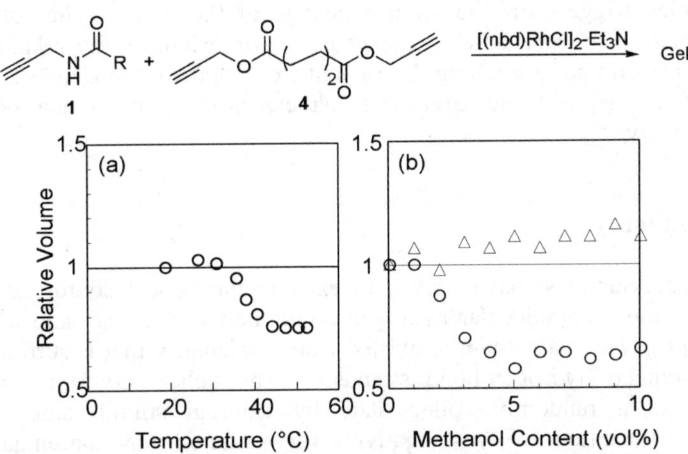

Figure 9. (a) Temperature dependence of the relative volume of poly(**1e**)-based gel in CHCl₃. (b) Solvent-composition dependence of the relative volume of (○) poly(**1e**)-based and (□) poly(**1b**)-based gels in mixed solvents of CHCl₃ and methanol.

responsive gels, especially those of hydrogels, in view of their scientific and technological importance.[18]   The frameworks of responsive gels are, however, unfortunately limited to water-soluble (meth)acrylic acids and acrylamides, and volume change of gels occurs only in aqueous solvents.   Furthermore, the response of these amorphous synthetic gels is quite slow.   However, the use of poly(*N*-propargylamides) as the framework of the gel has allowed the production of a quickly responsive organogel.[19]   That is, the poly(*N*-propargylamide)-based gel displays very quick response to change in temperature or solvent composition, which is accompanied with volume change in organic isotropic solvents.

The gel is readily prepared by the copolymerization of monomer **1e** with dipropargyl adipate (**4**) in the presence of a combined catalyst, [(nbd)RhCl]₂−Et₃N (Scheme 3).   The volume of a cylindrical poly(**1e**)-based gel (**1e/4** = 97.5/2.5 in feed) with the length and diameter of 7.1 and 2.2 mm, respectively, starts to shrink in CHCl₃ at approximately 35 °C, and the gel volume levels off at 50 °C as shown in Figure 9a. The temperature dependence of the gel volume agrees with that of the conformation of poly(**1e**) [19].   Emphasis should be placed on the very rapid volume change process of the gel; the temperature jump experiment from 35.2 to 37.5 °C for this relatively large cylindrical gel has indicated that the relaxation time is only 8.9 min.   This is in contrast to the very slow volume transition process of the conventional responsive hydrogels based on poly(acrylic acids) and poly(acrylamides).

The volume transition also occurs upon the change in solvent composition.   As illustrated in Figure 9b, the gel steeply shrinks with increasing methanol composition. The shrinkage stops when the methanol content reaches 5 vol%, which is identical behavior to that observed in the conformational change of the corresponding poly(**1e**). In contrast, the volume of the gel based on poly(**1b**) (**1b/4** = 97.5/2.5 in feed) remains almost unchanged under the same conditions.   Because poly(**1b**) exists in a randomly coiled state even in CHCl₃ at room temperature,[14] it is concluded that the helix-coil transition results in the volume change.   Specifically, the conformational change between the helix and random coil is accompanied with the change in hydrodynamic volume, which triggers off the volume change of the gels.   The solvent-induced volume change also very quickly proceeds.   For example, the relaxation time of volume expansion upon switching the solvent from methanol to $CHCl_3$ is only 43 min for a cylindrical gel with the length and diameter at the swollen state of 7.1 and 2.2 mm, respectively.

5. Summary

Poly(*N*-propargylamides) have proven to exist in the helical conformation which is stabilized by the intramolecular hydrogen bonds between the nth and n+2th pendant amide groups.   The main chain is twisted from coplanarity that is surrounded by two helically arranged hydrogen-bond strands.   The helical conformation is readily deformed into a randomly coiled state by external stimuli since the random coil−to−helix transition involves negatively very large entropy and enthalpy changes. In other words, the helical conformation of poly(*N*-propargylamides) is based on the well-balanced entropy and enthalpy contributions.   This characteristic can be applied to various stimuli-responsive functional materials such as chromic materials and responsive organogels with volume change.   The instability of polymers from monosubstituted acetylenes may render readers to conclude that poly(*N*-propargylamides) cannot be used in practical applications.   However, our preliminary investigation has suggested that poly(*N*-propargylamides) are unexpectedly stable even in $CHCl_3$, in which polyacetylenes rapidly decompose.   Especially, poly(*N*-propargylamides) having bulky pendant groups are quite stable.   Therefore, we believe that poly(*N*-propargylamides) will serve as a new class of functional polymers for practical applications.

6. References

1.   Rowan, A.E. and Nolte, R.J.M. (1998) Helical molecular programming, *Angew. Chem. Int. Ed. Engl.* **37**, 63–68.
2.   Lawrence, D.S., Jiang, T., and Levett M. (1995) Self-assembling supramolecular complexes, *Chem. Rev.* **95**, 2229–2260.
3.   Kirshenbaum, K., Zuckermann, R.N., and Dill, K.A. (1999) Designing polymers that mimic biomolecules, *Curr. Opin. Struct. Biol.* **9**, 530–535.
4.   Nakahira, T., Lin, F., Boon, C.T., Karato, T., Annaka, M., Yoshikuni, M., and Iwabuchi, S. (1997) Intramolecular hydrogen bonding in isotactic poly(methacrylamide)s and its implications for control of side-chain orientation, *Polym. J.* **29**, 701–704.
5.   Nakahira, T., Fan, L., Boon, C.T., Fukada, Y., Karato, T., Annaka, M., and Yoshikuni, M. (1998) Effects of side-chain structure and solvent on intramolecular hydrogen bonding in isotactic poly(methacrylamide)s, *Polym. J.* **30**, 910–914.
6.   Li, B.S., Cheuk, K.K.L., Salhi, F., Lam, J.W.Y., Cha, J.A.K., Xiao, X., Bai, C., and Tang, B.Z. Tuning the chain helicity and organization morphology of an L-valine-containing polyacetylene by pH Change, (2001) *Nano Lett.* **1**, 323–328.
7.   Cornelissen, J.J.L.M., Donners, J.J.J.M., de Gelder, R., Graswinckel, W.S., Metselaar, G.A., Rowan, A.E., Sommerdijk, N.A.J.M., and Nolte, R.J.M. (2001) β-Helical polymers from isocyanopeptides, *Science* **293**, 676–680.
8.   Nomura, R., Tabei, J., and Masuda, T. (2001) Biomimetic stabilization of helical structure in a synthetic polymer by means of intramolecular hydrogen bonds, *J. Am. Chem. Soc.* **123**, 8430–8431.
9.   Tabata, M., Sone, T., and Sadahiro, Y. (1999) Precise synthesis of monosubstituted polyacetylenes using

Rh complex catalysts. Control of solid structure and π-conjugation length, *Macromol. Chem. Phys.* **200**, 265–282.

10. Nomura, R., Fukushima, Y., Nakako, H., and Masuda, T. (2000) Conformational study of helical poly(propiolic esters) in solution, *J. Am. Chem. Soc.* **122**, 8830–8836.

11. Nomura, R., Tabei, J., Nishiura, S., and Masuda, T (2002) A helix in helices: a helical conjugated polymer that has helically arranged hydrogen bond-strands, Submitted to Macromolecules.

12. Nakako, H., Nomura, R., and Masuda, T. (2001) Helix inversion of poly(propiolic esters), *Macromolecules* **34**, 1496–1502.

13. Yashima, E., Huang, S., Matsushima, T., and Okamoto, Y. (1995) Synthesis and conformational study of optically active poly(phenylacetylene) derivatives bearing a bulky substituents, *Macromolecules* **28**, 4184–4193.

14. Nomura, R., Tabei, J., and Masuda, T. (2002) Effect of side chain structure on the conformation of poly(*N*-propargylalkylamide), *Macromolecules* **35**, 2955–2961.

15. Green, M.M., Park, J.-W., Sato, T., Teramoto, A., Lifson, S., Selinger, R.L.B., and Selinger, J.V. (1999) The macromolecular route to chiral amplification, *Angew. Chem. Int. Ed.* **38**, 3138–3154.

16. Nomura, R., Yamada, K., and Masuda, T. (2002) A chromophores-labeled poly(*N*-propargylamide): a new strategy for a stimuli-responsive conjugated polymer, *Chem. Commun.* 478–479.

17. Tanaka, T. (1978) Collapse of gels and the critical endpoint, *Phys. Rev. Lett.* **40** 820–823.

18. Dušek, K. (1993) *Responsive Gels: Volume Transition I and II*, Springer: Berlin.

19. Nomura, R., Yamada, K., Masuda, T., Takakura, Y., and Takigawa, T. (2002) Responsive Organogel with Volume Change, *J. Am. Chem. Soc.* submitted.

SYNTHESIS OF BLOCK COPOLYMERS AND STAR POLYMERS THROUGH LIVING POLYMERIZATION OF SUBSTITUTED ACETYLENES

K. KANKI, F. SANDA AND T. MASUDA
Department of Polymer Chemistry, Graduate School of Engineering, Kyoto University, Kyoto 606-8501, Japan

1. Introduction

Living polymerization consists of the initiation and propagation reactions and includes neither chain transfer nor termination reaction. Therefore, it is easy to elucidate polymerization mechanism and possible to strictly control molecular weight and molecular weight distribution (MWD) of polymer by using living polymerization. The molecular weight of the formed polymer is proportional to the monomer/initiator ratio. The MWD becomes narrow when the initiation reaction is faster than propagation and the propagating species is stable. Therefore, the initiating species that have structures similar to the propagating species are preferred. Living polymerization enables the synthesis of well-defined polymers such as block copolymers, end-functionalized polymers, and star polymers.

Group 5 and 6 transition metal catalysts and Rh catalysts are known to be effective in the polymerization of substituted acetylenes.[1–4] In general, group 6 catalysts such as $MoCl_5$ and WCl_6 are effective for monosubsutituted acetylenes, while group 5 catalysts such as $TaCl_5$ and $NbCl_5$ are effective for disubsutituted acetylenes. These polymerizations proceed by the metathesis mechanism, in which the propagating species are metal carbene complexes. On the other hand, the polymerization of substituted acetylenes with Rh catalysts proceeds by the insertion mechanism. Rh catalysts effectively induce the polymerization of monosubsutituted acetylenes, especially phenylacetylene (PA), propiolic esters, and propargylamides, to give stereoregular polymers with cis-transoidal structure selectively. Substituted polyacetylenes are usually soluble in various organic solvents and stable in air. Further, substituted polyacetylenes have the alternating double bond structure along the main chain, and hence they display properties and functions different from those of vinyl polymers; e.g., photoconductivity, nonlinear optical property, and electroluminescence.

Living polymerization of substituted acetylenes was first achieved by Masuda and coworkers in 1987.[5] Thus, 1-chloro-1-octyne (ClOc) polymerized in a living fashion with the $MoOCl_4/n\text{-}Bu_4Sn/EtOH$ (mole ratio = 1:1:0.5) ternary catalyst in toluene to give polymer with narrow MWD, where the initiation efficiency was about 10%. This catalyst system induced the living polymerization of other 1-chloro-1-alkynes, *tert*-butylacetylene and ortho-substituted PAs as well.[6-8] Recently, Hayano and Masuda have reported that the initiation efficiency reached 40% in the polymerization of [(*o*-trifluoromethyl)phenyl]acetylene (*o*-CF3PA) with $MoOCl_4/n\text{-}Bu_4Sn/EtOH$ in anisole, which completely solves $MoOCl_4$ and possesses a moderate

Y. Imamoglu and L. Bencze (eds.), Novel Metathesis Chemistry: Well-Defined Initiator Systems for Specialty Chemical Synthesis, Tailored Polymers and Advanced Material Applications, 215–227.

basicity.[9] By use of the MoOCl$_4$/n-Bu$_4$Sn/EtOH catalyst in anisole, the living polymerization of a variety of substituted acetylenes proceeded; e.g., 1-chloro-2-phenylacetylene, *tert*-butylacetylene and internal alkynes.[9,10] In anisole, n-BuLi, Et$_2$Zn and Et$_3$Al are also effective as cocatalyst.[11-14] The features of these catalyst systems include easy preparation and applicability to a large number of monomers.

Living polymerization of substituted acetylenes with metal alkylidene complexes, so-called Schrock carbenes, has also been achieved. In 1989, Schrock and coworkers reported the living polymerization of 2-butyne with a tantalum carbene.[15] However, this catalyst was ineffective for other substituted acetylenes. The living polymerization of ethynylferrocene, α,ω-diynes, and ortho-substituted PAs was accomplished by using molybdenum carbenes in 1994.[16-18] Schrock carbenes are well-defined catalysts, show quantitative initiation efficiency, and are able to strictly control the molecular weight and MWD.

Noyori and coworkers have reported that PA polymerizes in a living fashion with Rh catalyst systems, (Ph$_3$P)$_2$Rh(C≡CC$_6$H$_5$)(nbd)/4-Me$_2$N-pyridine (DMAP) and [(nbd)Rh(OCH$_3$)]$_2$/Ph$_3$P/DMAP and the initiation efficiencies were 37 and 72%, respectively.[19-21] The living polymerization of PA using a [(nbd)RhCl]$_2$/Ph$_2$C=C(Ph)Li/Ph$_3$P ternary catalyst was reported by Misumi and Masuda.[22,23] This system quantitatively initiates living polymerization of PA, and the resulting poly(PA) quantitatively possesses a triphenylvinyl group at the initiating chain end, which can be detected by ^{1}H NMR. Further, it has been revealed that an Rh complex, (nbd)Rh(CPh=C(Ph)$_2$)P(4-Cl-C$_6$H$_4$)$_3$, which is isolated from the Rh ternary catalyst, quantitatively induces the living polymerization of PA.[24]

In this article, we summarize our recent efforts on the application of the living polymerization of substituted acetylenes with MoOCl$_4$/n-Bu$_4$Sn/EtOH and [(nbd)RhCl]$_2$/Ph$_2$C=C(Ph)Li/Ph$_3$P catalyst systems to the precise synthesis of a variety of block and star polymers. First, we discuss the synthesis of di- and triblock copolymers using the MoOCl$_4$/n-Bu$_4$Sn/EtOH catalyst. Then, the synthesis of a novel star polymer with the same catalyst by means of linking method is described. In the second part, we describe the block copolymerizations of PA and (*p*-trityloxycarbonylphenyl)acetylene (*p*-TrOCOPA) by use of an Rh catalyst and then the hydrolysis of the obtained block copolymers to convert them into amphiphilic copolymers. Further, we introduce the synthesis of poly(PA) having a hydroxyl group at the initiating chain end by the living polymerization using a Rh catalyst and examined anionic polymerization of β-propiolactone (βPL) initiated by the end-functionalized poly(PA) as macroinitiator. Finally we demonstrate the synthesis of a star polymer, whose arms are composed of poly(PA); this was achieved by the living polymerization of PA with [(nbd)RhCl]$_2$/Ph$_2$C=C(Ph)Li/Ph$_3$P ternary catalyst followed by the linking reaction of the living polymer.

2. Synthesis of Various Block Copolymers with MoOCl$_4$-Based Catalysts [25,26]
2.1. SYNTHSIS OF DIBLOCK COPOLYMERS

The block copolymerizations between two monomers among ClOc, [*o*-(trimethylsilyl)phenyl]acetylene (*o*-Me$_3$SiPA), and *o*-CF$_3$PA (6 combinations in total) were examined with use of $MoOCl_4$/n-Bu$_4$Sn/EtOH (mole ratio 1:1:2) in anisole, the results of which are summarized in Table 1. Quite interestingly, block copolymers selectively formed irrespective of the order of monomer addition in every combination. That is, all of the produced polymers exhibited unimodal GPC profiles (M_n ca. 9 000–15 000), and no peak attributable to the homopolymer from the first monomer was detected. The MWDs of the produced block copolymers were quite narrow (M_w/M_n 1.06–1.12). It is especially worth noting that the polydispersities are low (M_w/M_n 1.06–1.08) in the o-Me$_3$SiPA/o-CF$_3$PA block copolymers (runs 3 and 6).

Table 1. Block copolymerization of substituted acetylenes by $MoOCl_4$/n-Bu$_4$Sn/EtOH(1:1:2)[a]

run	1st monomer[b]	2nd monomer[c]	M_n[d]	M_w/M_n[d]	[P*]/[Cat], %[e]
1	ClOc	o-Me$_3$SiPA	15 200	1.10	23.6
2	ClOc	o-CF$_3$PA	11 900	1.09	23.6
3	o-Me$_3$SiPA	o-CF$_3$PA	9 280	1.06	37.3 (28.9)
4	o-Me$_3$SiPA	ClOc	11 000	1.12	37.3 (28.9)
5	o-CF$_3$PA	ClOc	9 210	1.12	41.5 (28.2)
6	o-CF$_3$PA	o-Me$_3$SiPA	9 410	1.08	41.5 (28.2)

[a]Polymerized in anisole at 30 °C; [$MoOCl_4$] = 10 mM; all the monomer conversions were quantitative. [b][M]$_0$ = 0.10 M. [c][M]$_{added}$ = 0.10 M. [d]Determined by GPC using a polystyrene calibration. [e]Estimated from the first-stage polymerization. The values without parentheses were calculated on the basis of the relative M_n values by GPC, while those in parentheses were based on the absolute M_n values by VPO [13].

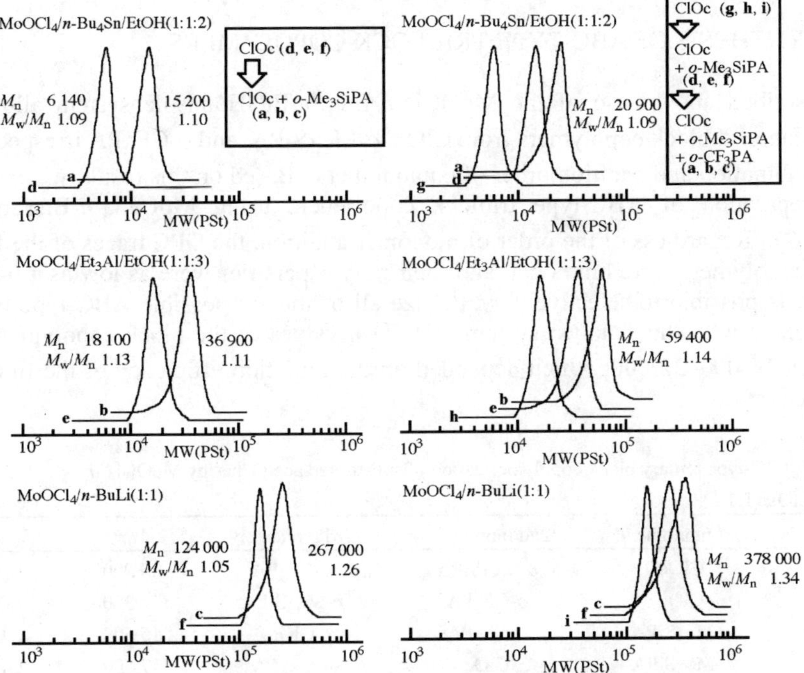

Figure 1. GPC curves of poly(ClOc)-*block*-poly(o-Me$_3$SiPA)s (curves **a**, **b** and **c**). Curves **d**, **e** and **f** are the corresponding poly(ClOc)s formed in the first stage.

Figure 2. GPC curves of poly(ClOc)-*block*-poly(o-Me$_3$SiPA)-*block*-poly(o-CF$_3$PA)s (curves **a**, **b**, and **c**). Curves **d**, **e**, and **f** and curves **g**, **h**, and **i** are for the first- and second-stage polymers, respectively.

The initiation efficiencies of the first-stage polymerizations were in a range of ca. 24–42% according to GPC (probably, 20–30% according to VPO), while those of the second-stage polymerization were quantitative.

As described in previously published papers [11-13], the kind of cocatalyst greatly influences the initiation efficiency of the $MoOCl_4$-catalyzed living polymerization of substituted acetylenes. Therefore, the change of cocatalyst may give an alternative method to control the M_n of block copolymers. As an example, the GPC profiles for the block copolymerization of ClOc and o-Me$_3$SiPA with the three $MoOCl_4$-based catalysts are illustrated in Figure 1, where o-Me$_3$SiPA was polymerized sequentially after ClOc. When the second monomer had been completely consumed in the $MoOCl_4$/n-Bu$_4$Sn/EtOH system, the GPC elution peak shifted towards a high molecular weight region maintaining its narrow MWD, which indicates the selective formation of a block copolymer (curve **a**). Similarly, $MoOCl_4$/Et$_3$Al/EtOH achieved the selective block copolymerization (curve **b**), and the M_n of the resulting block copolymer was 36 900 which is more than two times as large as that obtained with $MoOCl_4$/n-Bu$_4$Sn/EtOH. Emphasis should be placed on the fact that the selective production of a block copolymer with extremely high molecular weight (M_n 267 000) is possible by use of $MoOCl_4$–n-BuLi (curve **c**) without serious broadening of the MWD.

2.2. SYNTHESIS OF ABC-TYPE TRIBLOCK COPOLYMERS

As described above, use of the $MoOCl_4$/n-Bu$_4$Sn/EtOH/anisole system allows the formation of diblock copolymers from ClOc, o-Me$_3$SiPA, and o-CF$_3$PA irrespective of the combination and addition order of comonomers. Based on this result, we examined the preparation of ABC-type triblock copolymers using $MoOCl_4$/n-Bu$_4$Sn/EtOH (Table 2). Regardless of the order of monomer addition, the GPC traces of the formed block copolymers were unimodal, and their polydispersities were as low as 1.04–1.17. Thus it is possible to selectively synthesize all of the six possible ABC-type triblock copolymers with this catalyst system. The M_n values of the block copolymers were between 16 000–22 000, which depended on the initiation efficiency of the first-stage polymerization.

Table 2 ABC-type ternary block copolymerization of substituted acetylenes by $MoOCl_4$/n-Bu$_4$Sn/EtOH(1:1:2)^a

run	1st monomer^b	2nd monomer^c	3rd monomer^c	M_n^{d}	M_w/M_n^{d}
1	ClOc	o-Me$_3$SiPA	o-CF$_3$PA	20 900	1.09
2	ClOc	o-CF$_3$PA	o-Me$_3$SiPA	22 300	1.08
3	o-Me$_3$SiPA	o-CF$_3$PA	ClOc	15 500	1.12
4	o-Me$_3$SiPA	ClOc	o-CF$_3$PA	17 800	1.18
5	o-CF$_3$PA	ClOc	o-Me$_3$SiPA	16 800	1.16
6	o-CF$_3$PA	o-Me$_3$SiPA	ClOc	16 200	1.17

^{a}Polymerized in anisole at 30 °C; [$MoOCl_4$] = 10 mM; all the monomers were converted quantitatively. ^{b}[M]$_0$ = 0.10 M. ^{c}[M]$_{added}$ = 0.10 M. ^{d}Determined by GPC using a polystyrene calibration.

In the case of MoOCl$_4$/Et$_3$Al/EtOH and MoOCl$_4$/n-BuLi systems, diblock copolymers were exclusively formed when the monomer addition followed the order of ClOc, o-Me$_3$SiPA, and then o-CF$_3$PA. On the basis of this finding, an attempt was made to synthesize triblock copolymer by employing the three MoOCl$_4$-based catalysts and keeping this order of addition. It is concluded from Figure 2 that ABC-type triblock copolymers were selectively obtained with any of the three MoOCl$_4$–based catalysts. Similarly to the diblock copolymerization, the M_n values of the copolymers can be roughly controlled by varying the cocatalyst.

3. Synthesis of a Star Poly(o-CF$_3$PA) with MoOCl$_4$-Based Catalyst [27]

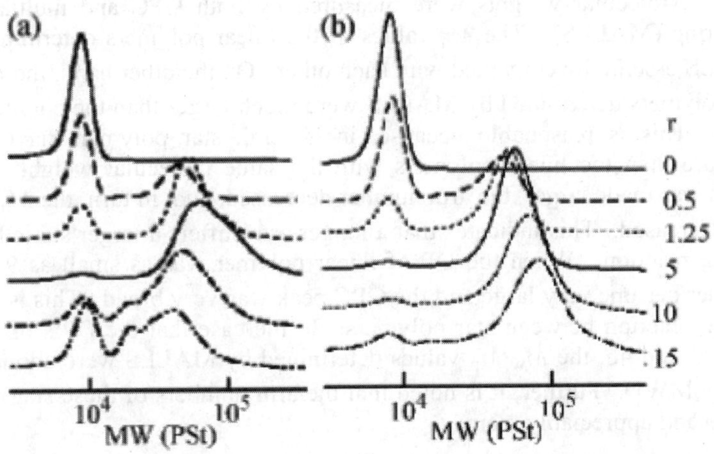

Figure 3. Effect of (a) [DEB] and (b) [DEDMB] on the GPC profile of star poly(o-CF$_3$PA) (polymerized in anisole, 30 °C, 15 min; [MoOCl$_4$] = 10 mM, [o-CF$_3$PA] = 200 mM, [P*]/[MoOCl$_4$] = 40%. Linking reaction 30 °C, 24 h; r = [DEB]/[P*] or [DEDMB]/[P*].).

A linear living polymer (M_n = 8 100 by GPC) was at first prepared by the polymerization of o-CF$_3$PA with MoOCl$_4$/n-Bu$_4$Sn/EtOH in anisole at 30 °C for 15 min. Then, 1,4-diethynylbenzene (DEB) was added as a linking agent, and the effect of DEB concentration was examined (Figure 3a). When DEB was allowed to react with the living polymer (r = [DEB]/[P*] = 5), the conversion of DEB was 100% after 24 h according to GC, and the linear living polymer was almost completely linked, as is seen from a unimodal peak of star polymer in a higher molecular weight region in the GPC curve. As r was appreciably increased (10, 15), the molecular weight of star polymer decreased and the GPC showed a multimodal profile. This is probably because the propagating end having DEB was gradually inactivated, as is presumed from the fact that PA does not polymerize in a living fashion with this MoOCl$_4$-based catalyst.

In order that the linking agent is highly effective, it should react in a living fashion. It is known that (o-methylphenyl)acetylene polymerizes in a living manner with $MoOCl_4$-based catalysts, whereas PA does not.[8] Thus, 1,4-diethynyl-2,5-dimethylbenzene (DEDMB) was next examined as a linking agent (Figure 3b). DEDMB was completely consumed at all the r ([DEDMB]/[P*]) values examined. When the r value was 0.5 and 1.25, 58% and 22% of the linear polymer remained, respectively, whereas no linear polymer was observed in GPC curves when r was larger than 5. A unimodal peak of the formed star polymer shifted to higher molecular weight regions with increasing r. This indicates that DEDMB polymerizes in a living fashion and is very effective as a linking agent.

Table 3 shows the effect of the degree of polymerization (DP) of living polymer (arm). Molecular weights were measured by both GPC and multiangle laser light scattering (MALLS). The M_n values of the linear polymers determined by GPC and MALLS essentially coincided with each other. On the other hand, the M_n values of the star polymers determined by MALLS were much larger than the counterparts based on GPC. This is reasonable because, in general, star polymers have more compact structure than the linear polymers with the same molecular weight. When the arm length was made larger, the arm number decreased, and, in turn, the M_n of star polymer also decreased. This indicates that a longer arm suffers a larger steric hindrance in the linking reaction. When the DP of linear polymer was as small as 9, the M_n of star polymer became very large and the GPC peak was very broad. This is attributed to the linking reaction between star polymers. In the case that the DP's of linear polymers were 23 and 46, the M_w/M_n values determined by MALLS were about 1.1, indicating narrow MWD. Further, it is noted that the arm numbers of these star polymers are 72 and 29 and appreciably many.

Table 3. Effect of the DP of linear living polymer on the synthesis of star polymer from o-CF$_3$PA and DEDMB[a, b]

linear living polymer (arm)				star polymer				
[o-CF$_3$PA]$_0$, mM	$M_n/10^3$		DP [c]	$M_n/10^3$		M_w/M_n	f [d]	
	GPC	MALLS		GPC	MALLS		GPC	MALLS
50	1.7	1.5	9	38	8800	very broad	16	4300
100	4.2	3.9	23	31	330	1.12	6	72
200	8.1	7.8	46	53	250	1.04	6	29

[a]Polymerized in anisole at 30 °C for 1 day; [$MoOCl_4$] = 10 mM, [P*]/[$MoOCl_4$] = 40%. [b]Linking reaction: 30 °C, 24 h; r = [DEDMB]/[P*] = 5. [c]Determined by MALLS. [d]The number of arms: f = (weight fraction of o-CF$_3$PA in star polymer)xM_n(star)/M_n(linear).

4. Synthesis of an Amphiphilic Conjugated Block Polymer with Rh Catalyst [28]

4.1. BLOCK COPOLYMERIZATION OF p-TrOCOPA WITH PA, AND HYDROLYSIS

Block copolymerization of (p-trityloxycarbonylphenyl)acetylene (p-TrOCOPA) and PA was examined by using [(nbd)RhCl]$_2$/Ph$_2$C=C(Ph)Li/Ph$_3$P ternary catalyst (Scheme 1). PA was at first polymerized and then p-TrOCOPA was added. The molecular weight of the product of block copolymerization clearly increased relative to that of poly(PA), while its polydispersity ratio remained nearly equal to that of the first stage. These results manifest the formation of poly(PA)-*block*-poly(*p*-TrOCOPA).

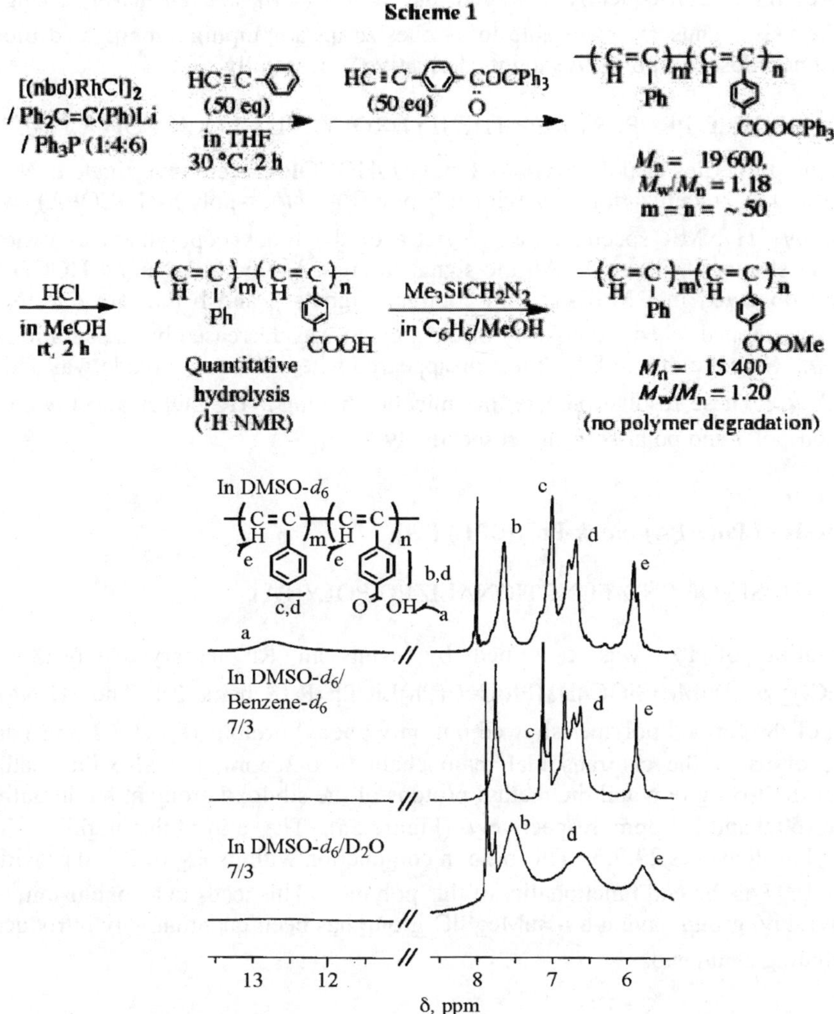

Figure 4. ¹H NMR spectra of poly(PA)-*block*-poly(HOCOPA) [block composition =26:32] in various solvents.

Next, we examined the conversion of trityloxycarbonyl groups of the block copolymer to carboxyl groups by hydrolysis. At first, the hydrolysis of the homopolymer (poly(*p*-TrOCOPA)) was examined. We used here the homopolymer whose DP was 75. When hydrochloric acid was added as a catalyst, hydrolysis proceeded within 1 h and a homogeneous orange solution was obtained. Here, 98% of the ester group was hydrolyzed; the DP of the resulting poly(*p*-HOCOPA) hardly decreased, while the polydispersity was kept low. This indicates that the hydrolysis of trityloxycarbonyl groups easily proceed with hydrochloric acid.

We examined the hydrolysis of poly(PA)-*block*-poly(*p*-TrOCOPA) having two blocks composed of 50 units each (see Scheme 1). Then methyl esterification was carried out to measure GPC. Consequently it proved that both M_W/M_n and DP hardly changed after hydrolysis. Thus we were able to synthesize an amphiphilic conjugated block copolymer composed of poly(PA) and its derivative successfully.

4.2. AMPHIPHILIC PROPERTY OF THE HYDROLYZED BLOCK COPOLYMER

Amphiphilic properties of poly(PA)-*block*-poly(*p*-HOCOPA) were investigated. More specifically, the micellization behavior of poly(PA)-*block*-poly(*p*-HOCOPA) was examined by ^{1}H NMR spectroscopy. Spectra of the block copolymers in various solvents are shown in Figure 4. All the signals assigned to both PA and *p*-HOCOPA units were observed in DMSO-d_6. In contrast, signals a and b that are due to *p*-HOCOPA weakened when the polarity of the solvent was decreased by the addition of benzene-d_6. Signal c due to PA almost disappeared when deuterium oxide was added to DMSO-d_6. These results indicate that micelles having *p*-HOCOPA and PA cores form in non-polar and polar solvents, respectively.

5. Synthesis of Poly(PA)-*block*-Poly(βPL) [29]

5.1. SYNTHESIS OF END-FUNCTIONALIZED POLY(PA)

Polymerization of PA was examined by using an Rh ternary catalyst, i.e., [(nbd)RhCl]$_2$/*p*-*t*-BuMe$_2$SiOC$_6$H$_4$(Ph)C=C(Ph)Li /Ph$_3$P (Scheme 2). The ^{1}H NMR spectrum of the formed polymer showed not only phenyl protons (δ 6.7-7.2 ppm) and olefinic protons of the cis-transoidal main chain (δ 6.2 ppm) but also the methyl protons of the *t*-Bu group and the methyl protons of the silyloxy group at the initiating chain end (δ0.9 and 0.1 ppm, respectively) (Figure 5a). The ratio of the olefinic, *t*-Bu, and methyl protons was 22:9:6. This ratio in conjunction with its M_n of 2 520 provides a value of 0.97 as the end functionality of this polymer. This leads to a conclusion that a triphenylvinyl group having a *t*-BuMe$_2$SiO group has been quantitatively introduced at the initiating chain end.

Scheme 2

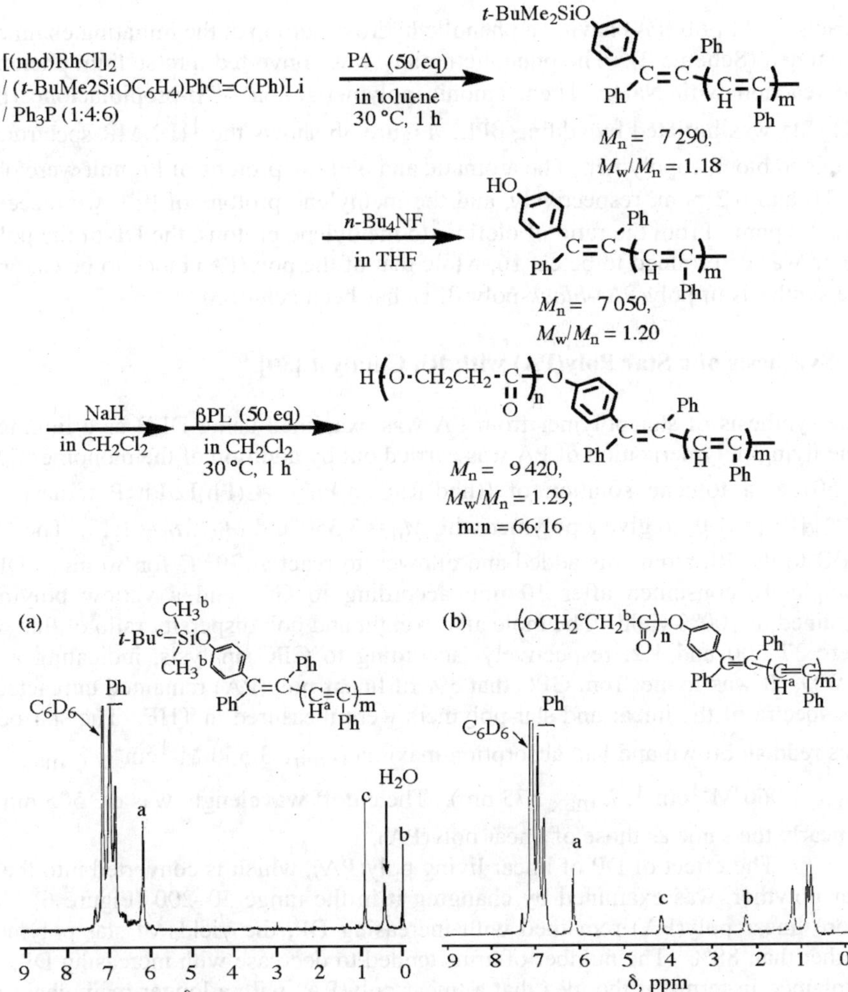

Figure 5. ^{1}H NMR spectra (in C_6D_6) of (a) poly(PA) having a t-BuMe$_2$SiO group at the initiating chain end and poly(PA)-*block*-poly(βPL).

The elimination of t-BuMe$_2$Si group catalyzed by Bu$_4$NF was examined. A yellow poly(PA) was obtained in a virtually 100% yield. The molecular weights and polydispersity ratios of the polymer before and after desilylation were almost the same, which indicates that virtually no degradation occurred during desilylation. The signals at δ 0.9 and 0.1 ppm, attributable to t-BuMe$_2$Si group disappeared in ^{1}H NMR spectrum. These results confirm that a polymer having a phenolic hydroxyl group at the initiating chain end has been obtained by this deprotection reaction.

5.2. SYNTHESIS OF BLOCK COPOLYMER

Reactivity of poly(PA) having a phenolic hydroxyl group at the initiating chain end was examined (Scheme 2). The phenolic moiety was converted into sodium phenoxide by the reaction with NaH. Then, anionic polymerization of β-propiolactone (βPL) in CH_2Cl_2 was initiated by adding βPL. Figure 5b shows the ^1H NMR spectrum of the obtained block copolymer. The aromatic and olefinic protons of PA unit were observed at 7.0 and 6.2 ppm, respectively, and the methylene protons of βPL were seen at 4.2 and 2.2 ppm. From the ratio of olefinic to methylene protons, the DP of the poly(βPL) block was determined to be ca. 16, while that of the poly(PA) block to be ca. 66. Thus the synthesis of poly(PA)-*block*-poly(βPL) has been achieved

6. Synthesis of a Star Poly(PA) with Rh Catalyst [30]

The synthesis of star polymer from PA was examined using DEB as a linking agent. The living polymerization of PA was carried out by addition of the monomer ($[M]_0$/Rh = 50) to a toluene solution of $[(nbd)RhCl]_2$/$Ph_2C=C(Ph)Li$/Ph_3P ternary catalyst ($[P^*]$/$[Rh]$ = 1.0) to give a polymer with M_n = 5 350 and M_w/M_n = 1.13. Then 5 eq. of DEB to the Rh atom was added and allowed to react at 30 °C for 30 min. DEB was completely consumed after 10 min according to GC, and a yellow polymer was obtained in 100% yield. The molecular weight and polydispersity ratio of this polymer were 37 500 and 1.2, respectively, according to GPC analysis, indicating a narrow MWD. It was found from GPC that 3% of linear poly(PA) remained unreacted. UV-vis spectra of the linear and star polymers were measured in THF. The star poly(PA) was reddish brown and had absorption maxima (ε_{max} 3 550 $M^{-1}cm^{-1}$, λ_{max} 325 nm; ε_{max} 3 060 $M^{-1}cm^{-1}$, λ_{max} 375 nm). The cutoff wavelength was ca. 505 nm, which is nearly the same as those of linear poly(PA).

The effect of DP of linear living poly(PA), which is converted into the arm of star polymer, was examined by changing it in the range 50–200 (Figure 6). Though more linear poly(PA) remained with increasing DP, the yields of star polymer were higher than 88%. The number of arms tended to decrease with increasing DP. This is explained in terms of the idea that a linear poly(PA) with a longer main chain is more difficult to approach the core of star polymer.

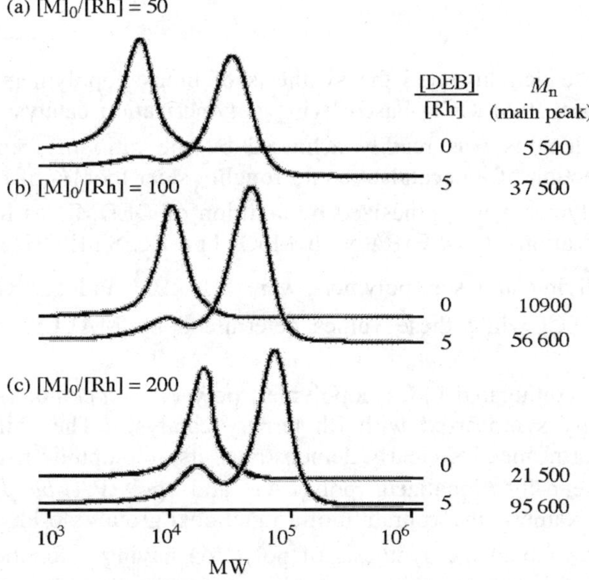

Figure 6. Effect of the DP of living poly(PA) on the GPC profile of star poly(PA) (polymerized in toluene at 30 °C for 30 min; [Rh] = 4.0 mM; linking reaction performed at 30 °C for 30 min; [DEB] = 20 mM).

It is clear from GPC profiles that star poly(PA) has been formed practically selectively by using DEB as the linking agent. It is well known that the molecular weight of star polymers determined by GPC is much smaller than the absolute values because of their rather small hydrodynamic volume. This means that the number of arms evaluated by GPC is not strict. Thus, the molecular weights of a few star poly(PA)s were measured by means of MALLS which gives absolute values (Table 4). The two star polymers that were prepared under standard conditions (different batches) and had about 6 arms according to GPC turned out to possess 23-25 arms on the basis of MALLS. It is noted that the M_w/M_n ratios of the star polymers (ca. 1.2) determined by MALLS were close to those based on GPC. In the case of a star polymer with longer arms (Table 4, Run 3) as well, the arm number based on MALLS was larger than that based on GPC, showing the same tendency.

Table 4. M_n and M_w/M_n of star poly(PA) measured by GPC and MALLS

Run	GPC			MALLS		
	M_n	M_w/M_n	f^a	M_n	M_w/M_n	f^a
1[b]	37 500	1.17	6.5	145 100	1.21	25
2[c]	34 800	1.21	6.3	136 000	1.27	23
3[d]	95 600	1.10	4.5	246 000	1.10	11

[a] f is the number of arms per molecule. f = (wt fraction of PA) x M_n(star)/M_n(arm). [b] Fig. 6(a), M_n(arm) = 5 250. [c] Star polymer were prepared under the same conditions as in Run 1 (different batches), M_n(arm) = 5 270. [d] Fig. 6(c), M_n(arm) = 21 600.

7. Summary

At first, we have demonstrated the synthesis of block copolymers from substituted acetylenes by use of MoOCl$_4$-based living polymerization catalysts. Control of the ratio of block lengths was readily achieved by the ratio of feed monomers, and appropriate selection of a cocatalyst could roughly steer the M_n of block copolymers. Next, a star polymer was synthesized by addition of DEDMB as linking agent after living polymerization of o-CF$_3$PA with MoOCl$_4$/n-Bu$_4$Sn/EtOH catalyst. The M_n values of the living and star polymers were 8.1 x10^3 and 5.3 x10^4, respectively, according to GPC, while these values determined by MALLS were 7.8x10^3 and 2.5x10^5.

A new conjugated block copolymer, poly(PA)-*block*-poly(p-HOCOPA), has been successfully synthesized with Rh ternary catalyst. The ability of this block copolymer to form micelles clearly demonstrates its amphiphilic nature. Further, the synthesis of end-functionalized poly(PA) and poly(PA)-*block*-poly(βPL) was accomplished. Namely, incorporation of a functional group into Ph$_2$C=C(Ph)Li in Rh ternary catalyst enabled the synthesis of poly(PA) having a functional group at the initiating chain end, and consequently a block copolymer was prepared by sequential polymerization of PA and a different type of monomer, βPL by use of end-functionalized poly(PA) as macroinitiator. Finally, a novel star polymer based on poly(PA) was selectively synthesized by the linking reaction of living poly(PA) with DEB.

8. References

1. Choi, S.-K., Gal, Y.-S., Jin, S.-H., Kim, K. (2000) Poly(1,6-heptadiyne)-Based Materials by Metathesis Polymerization, *Chem. Rev.* **100**, 1645-1682.
2. Masuda, T. (1997) *Catalysis in Precision Polymerization: Chap. 2.4*, Wiley, Chichester, U.K..
3. Ivin, K. J. (1997) *Olefin Metathesis and Metathesis Polymerization: Chap. 10*, Academic Press, London, U.K..
4. (1998) *Metathesis Polymerization of Olefins and Polymerization of Alkynes*, Kluwer Academic, Dordrecht.
5. Masuda, T., Yoshimura, T., Fujimori, J., Higashimura, T. (1987) Living Polymerization of Subsutituted Acetylenes by MoCl$_5$- and MoOCl$_4$-Based Catalysts, *J. Chem. Soc., Chem. Commun.* 1805.
6. Yoshimura, T., Masuda, T., Higashimura. (1989) Living Polymerization of 1-Chloro-1-alkynes by MoOCl$_4$-n-Bu$_4$Sn-EtOH Catalysts, *Macromolecules* **22**, 3804-3806.
7. Nakano, M., Masuda, T., Higashimura, T. (1994) Stereospecific Living Polymerization of *tert*-Butylacetylene by Molybdenum-Based Ternary Catalyst Systems, *Macromolecules* **27**, 1344-1348.
8. Mizumoto, T., Masuda, T., Higashimura, T. (1995) Effects of ortho-Substituents on the Living Polymerization of Phenylacetylenes by MoOCl$_4$-Based Catalysts, *Macromol. Chem. Phys.* **196**, 1769-1778.
9. Hayano, S., Itoh, T., Masuda, T. (1999) Living Polymerization of Substituted Acetylenes by MoOCl$_4$-n-Bu$_4$Sn-EtOH in Anisole Which Features High Initiator Efficiency, *Polymer* **40**, 4071-4075.
10. Kubo, H., Hayano, S., Misumi, Y., Masuda, T. (2002) Living Metathesis Polymerization of Diethyl Di-2-butynyl Malonate by Molybdenum-Based Ternary Catalysts, *Macromol. Chem. Phys.* **203**, 279-283.
11. Masuda, T., Hayano, S., Iwawaki, E., Nomura, R. (1998) Living Metathesis Polymerization of substituted Acetylenes by MoOCl$_4$-Et$_2$Zn-EtOH and MoOCl$_4$-n-BuLi Systems, *J. Mol. Catal. A. Chemical* **133**, 213-220.

12. Hayano, S., Masuda, T. (1997) Living Polymerization of Substituted Acetylenes by a Novel Ternary Catalyst, MoOCl$_4$-Et$_2$Zn-EtOH, *Macromol. Chem. Phys.* **198,** 3041-3079.

13. Kaneshiro, H., Hayano, S., Masuda, T. (1999) Living Polymerization of [*o*-(Trifluoromethyl)phenyl]acetylene by a New Catalyst System, MoOCl$_4$-Et$_3$Al-EtOH (1 : 1 : 4), *Macromol. Chem. Phys.* **200,** 113-117.

14. Hayano, S., Masuda, T. (1998) Living Polymerization of Substituted Acetylenes by a Novel Binary Catalyst, MoOCl$_4$-*n*-BuLi, *Macromolecules* **31.** 3170-3174.

15. Wallace, L. C., Liu, A. H., Davis, W. M., Schrock, R. R. (1989) Living Polymerization of 2-Butyne Using a Well-Characterized Tantalum Catalyst, *Oranometallics* **8,** 644-654.

16. Buchmeiser, M., Schrock, R. R. (1995) Synthesis of Polyenes That Contain Metallocenes via the Living Polymerization of Ethynylferrocene and Ethynylruthenocene, *Macromolecules* **28,** 6642-6649.

17. Fox, H. H., Wolf, M. O., O'Dell, R., Lin, B. L., Schrock, R. R., Wrighton, M. S. (1994) Living Cyclopolymerization of 1,6-Heptadiyne Derivatives Using Well-Defined Alkylidene Complexes: Polymerization Mechanism, Polymer Structure, and Polymer Properties, *J. Am. Chem. Soc.* **116,** 2827-2843.

18. Schrock, R. R., Luo, S., Lee, J. C., Zanetti, N., Davis, W. M. (1996) Living Polymerization of (o-(Trimethylsilyl)phenyl)acetylene by Molybdenum Imido Alkylidene Complexes, *J. Am. Chem. Soc.* **118,** 3883-3895.

19. Kishimoto, Y., Eckerle, P., Miyatake, T., Ikariya, T., Noyori, R. (1994) Living Polymerization of Phenylacetylenes Initiated by Rh(C≡CC$_6$H$_5$)(2,5- norbornadiene)[P(C$_6$H$_5$)$_3$]$_2$, *J. Am. Chem. Soc.* **116,** 12131-12132.

20. Kishimoto, Y., Itou, M., Miyatake, T., Ikariya, T., Noyori, R. (1996) An Efficient Rhodium(I) Initiator for Stereospecific Living Polymerization of Phenylacetylenes, *Macromolecules* **29,** 5054-5055.

21. Kishimoto, Y., Eckerle, P., Miyatake, T., Kainosho, M., Ono, A., Ikariya, T., Noyori, R. (1999) Well-Controlled Polymerization of Phenylacetylenes with Organorhodium(I) Complexes: Mechanism and Structure of the Polyenes, *J. Am. Chem. Soc.* **121,** 12035-12044.

22. Misumi, Y., Masuda, T. (1998) Living Polymerization of Phenylacetylene by Novel Rhodium Catalysts. Quantitative Initiation and Introduction of Functional Groups at the Initiating Chain End, *Macromolecules* **31,** 7572-7573.

23. Misumi, Y., Kanki, K., Miyake, Y., Masuda, T. (2000) Living Polymerization of Phenylacetylene by Rhodium-Based Ternary Catalysts, (Diene)Rh(I) Complex/Vinyllithium/Phosphorus Ligand. Effects of Catalyst Components, *Macromol. Chem. Phys.* **201,** 2239-2244.

24. Miyake, Y., Misumi, Y., Masuda, T. (2000) Living Polymerization of Phenylacetylene by Isolated Rhodium Complexes, Rh[C(C$_6$H$_5$)=C(C$_6$H$_5$)$_2$](nbd)(4-XC$_6$H$_4$)$_3$P (X = F, Cl), *Macromolecules* **33,** 6636-6639.

25. Iwawaki, E., Hayano, S., Nomura, R., Masuda, T. (2000) Selective Synthesis of Various Di- and Tri-Block Copolymers from Substituted Acetylenes through Sequential Living Polymerization by MoOCl$_4$-Based Catalysts in Anisole, *Polymer*, **41,** 4429-4436.

26. Iwawaki, E., Hayano, S., Masuda, T. (2001) Synthesis of Block Copolymers Containing 1-Chloro-2-phenylacetylene, 2-Nonyne, and (*p-n*-Butyl-*o,o,m,m*-tetrafluorophenyl)acetylene through Sequential Living Polymerization by MoOCl$_4$-Based Catalysts, *Polymer*, **42,** 4055-4061.

27. Minaki, N., Kanki, K., Masuda, T. Synthesis of Star Polymer by Means of the Living Polymerization of [(*o*-Trifluoromethyl)phenyl]acetylene Using a MoOCl$_4$-Based Catalyst and the Linking Method, *Polymer*, in press.

28. Isomura, M., Misumi, Y., Masuda, T. (2001) Synthesis of an Amphiphilic Conjugated Polymer through Block Copolymerization of Phenylacetylene and (*p*-Trityloxycarbonylpheny)acetylene and the Subsequent Hydrolysis, *Polym. Bull.*, **46,** 291-297.

29. Kanki, K., Misumi, Y., Masuda, T. (2002) Synthesis of Poly(phenylacetylene)-*block*-poly(β-propiolactone) by Use of Rh-Catalyzed Living Polymerization of Phenylacetylene, *Inorg. Chim. Acta.*, **336,**.101-104

30. Kanki, K., Masuda, T. Synthesis of a Conjugated Star Polymer and Star Block Copolymers Based on the living Polymerization of Phenylacetylene with an Rh Catalyst, *Macromolecules*, in press.

WELL-DEFINED LINEAR AND CROSSLINKED MATERIALS VIA ROMP AND RTM PROCESSING

E. KHOSRAVI
Interdisciplinary Research Centre in Polymer Science and Technology,
Durham University,
Durham, DH1 3LE, UK.

1. Summary

The work reported here is a summary of some of our recent results on the synthesis and characterisation of polymeric materials via ROMP-RTM process. It describes the synthesis and characterisation of well-defined linear and crosslinked polymeric materials via ROMP-RTM. The process involves in-mould polymerisation of monofunctional imidonorbornene monomers, with different alkyl side chain lengths, to give a range of linear polymers. The process also involves in-mould copolymerisation of monofunctional imidonorbornene monomers, with different alkyl side chain lengths and difunctional monomers with different alkylene spacer lengths, to produce well-defined crosslinked polymers. The glass transition temperature (T_g) of the linear polymers was found to depend on the length of the alkyl side chain. For the crosslinked materials the results show that as the percentage of the difunctional, crosslinking unit, is increased (1, 5 and 10 molar percentage of the difunctional monomer) the glass transition shifts to a higher temperature, the height of the tanδ peak decreases and the plateau shear modulus above T_g increases. These results are as expected for an increase in the crosslink density of a polymer.

2. Introduction

The olefin metathesis has made remarkable strong developments with an incredible speed in various directions. New catalyst systems have been developed which have resulted in the synthesis of novel materials. Other fascinating developments have been the new catalysts for stereoselective metathesis and catalysts with considerable functional group tolerance. These new catalysts in addition to Ring Opening Metathesis Polymerisation (ROMP) and Acyclic Diene Metathesis (ADMET) are now powerful tools for Ring Closing Metathesis (RCM) and have found many applications in the synthesis of natural products. A lot of information has been established about all aspects of the olefin metathesis and there is a vast literature concerning the process, covering the initiators, mechanistic features and applications of this reaction in organic and polymer synthesis [1, 2].

We demonstrated that well-defined initiators allow the control of many aspects of the process, including *cis/trans* vinylene content and tacticity as well as molecular weight and its distribution [3, 4]. These living ROMP reactions have been

Y. Imamoglu and L. Bencze (eds.), Novel Metathesis Chemistry: Well-Defined Initiator Systems for Specialty Chemical Synthesis, Tailored Polymers and Advanced Material Applications, 229–235.

shown to result in the synthesis of a variety of interesting polymers such as stereoregular fluoropolymers, block copolymers and stereo-block copolymers [5, 6]. We also combined living ROMP and living anionic polymerisation to synthesise well defined comb-graft copolymers. This technique allows the control of the lengths of the grafts and the backbone chains through the use of the two living polymerisation methods [7, 8]. The advances achieved, during the last few years, in the design of ROMP initiators which are tolerant towards polar functional groups (aldehydes, acids, alcohols) and aqueous environments have provided the polymer community with access to a wide range of functional polymers not previously available. This major achievement has opened up a number of technological possibilities in the area of novel functional materials, such as water-soluble polymers, polar polymers and biocompatible materials [1, 2]. Here we report a summary of some of our recent results on the synthesis and characterisation of well-defined linear and crosslinked polymeric materials by ROMP in resin transfer moulding (RTM) processing.

3. Reaction injection moulding and ROMP processing

Dicyclopentadiene, DCPD, is cheap and can be polymerised by ROMP, which yields a crosslinked polymer. This polymerisation process can be tailored to have the characteristics which make it readily adaptable to either reaction injection moulding (RIM) or RTM.

The production of large moulded objects from DCPD-based feeds using RIM technology was developed mainly in the USA by BF Goodrich under the trade name TELENE and by Hercules under the trade name METTON. The catalysts used in METTON process is based on a combination of $WCl_6/WOCl_4$ and nonylphenol with Et_2AlCl. In the TELENE process the preferred catalyst is based on trialkylammonium molybdates and Et_2AlCl. The advantages of DCPD include a rapid reaction and good product mechanical properties including low density, low water absorption and excellent toughness. The disadvantages include a nauseating monomer smell, a cure exotherm, which can be difficult to control, uncertainty about the nature of the crosslinking process and difficulty in its regulation. This reduces the processability and the range of mechanical properties available. Nevertheless, poly(DCPD) is an interesting material and is finding market applications, as evident by the number of published papers and patents [9-15].

4. Resin transfer moulding and ROMP Processing

A major motivation for this work was a desire to synthesise alternative monomer systems to DCPD for RIM and RTM applications. The process that we have developed is very simple and involves in-mould polymerisation of monofunctional monomers to give a range of linear polymers. It also involves in-mould copolymerisation of monofunctional monomers with difunctional monomers to produce well-defined crosslinked polymers, see Scheme 1. A crucial factor in the development of the synthesis route was the availability of the ruthenium initiator, $Cl_2Ru(=CHPh)((PC_6H_{11})_3)_2$, developed by Grubbs and coworkers [16-19].

N-alkyldicarboxyimidonorbornenes, termed CnM, carrying pendant alkyl chains of different lengths (n=3,4,5) and bis(N-alkylenedicarboxyimidonorbornenes) , termed CmD, with an alkylene spacer (m=3,5,6,9,12) shown in Figure 1 were used as monofunctional and difunctional monomers respectievely [4, 20-22]. Monofunctional N-alkyldicarboxyimidonorbornenes were polymerised using RTM-ROMP processing to give linear polymers. The difunctional bis(N-alkylenedicarboxyimidonorbornenes) were co-polymerised with the monofunctional monomers to produce crosslinked polymers using the same processing method.

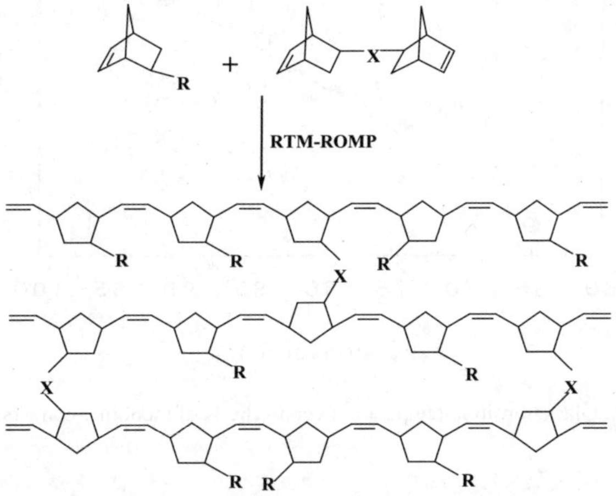

Scheme 1. Reaction Scheme for RTM-ROMP processing

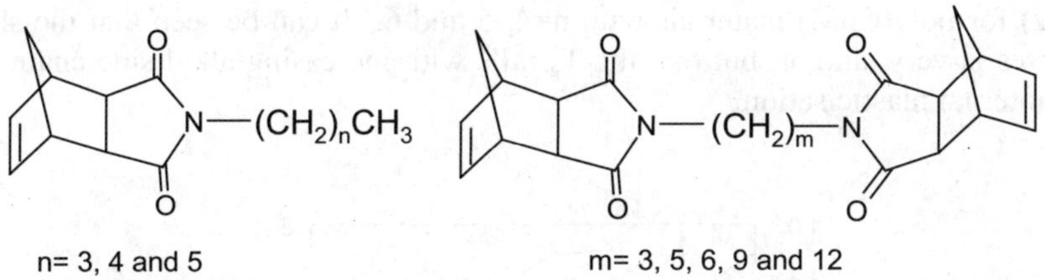

n= 3, 4 and 5   m= 3, 5, 6, 9 and 12

Figure 1. The chemical structures of mono- and di-functional N-alkyl dicarboxyimido norbornene monomers.

4.1.   MECHANICAL PROPERTIES OF LINEAR POLYMERS [POLY(CnM]

For the homopolymerisation processing of monofunctional we looked for the optimum monomer to initiator ratios to obtain a high monomer conversion since the unreacted monomer is expected to be a very efficient plasticiser for the polymer and can have a deleterious effect on thermal and mechanical properties. ^{1}H NMR was used to establish the extent of the monomer conversion by the comparison of the integration of signals due to the vinylic protons of the monomer and the polymer.

The importance of obtaining a high monomer conversion is indicated by the results shown in Figure 2, where the percentage degree of monomer conversion is plotted against the glass transition temperature, Tg, of the resulting material for various samples. The figure confirms the monomer as a very efficient plasticiser of the polymer, and that high degrees of conversion are required in order to obtain a high Tg.

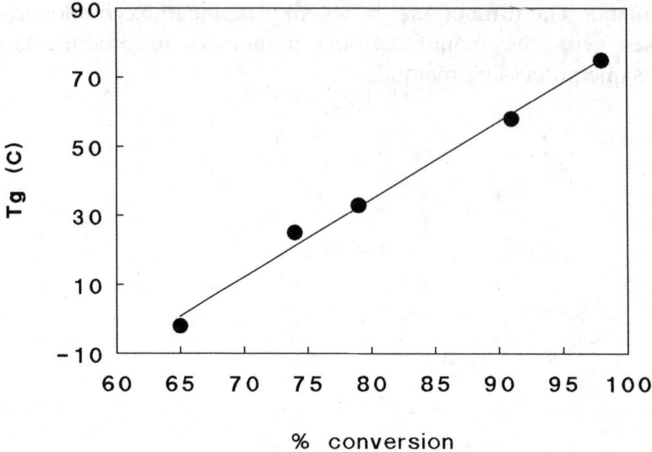

Figure 2. Glass transition temperature versus the % of monomer conversion C6M

The results showed that the optimum Monomer : Initiator ratio for the processing of C6M system was 4000:1.

Figure 3 shows a comparison of the dynamic torsion results (measured at a frequency of 1 Hz) for poly(CnM) materials with n=4, 5 and 6. It can be seen that the shape of the curves is very similar, but that the T_g falls with increasing alkyl side chain length, due to internal plasticisation.

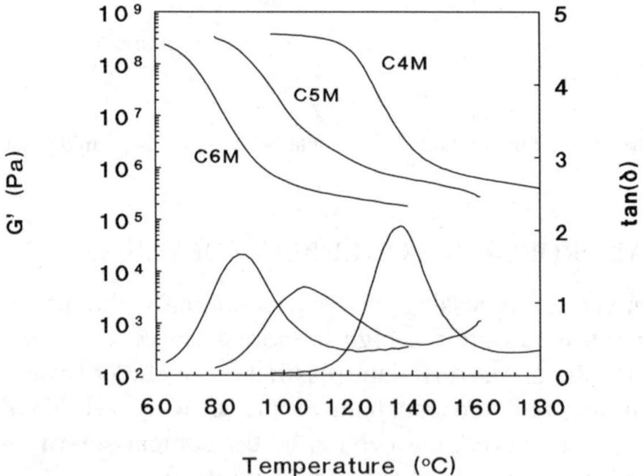

Figure 3. Torsion modulus G' and tanδ versus temperature for polyC4M, polyC5M and polyC6M.

Some preliminary mechanical properties were measured for the polyC4M, poly(C5M) and poly(C6M) and the results are shown in Table 1. The poly(C4M) had the highest density at 1150 kg/m^3, with the other two polymers showing a similar density of 1100 kg/m^3. The room temperature modulus was found to fall with increasing side chain length, a consequence of the different temperature differences between room temperature (20°C) and T_g for the three polymers, as mentioned earlier. Three point bend measurements showed that the yield stress fell with increasing side chain length, with an accompanying change from brittle to ductile behaviour. A much more detailed study of the effect of the side chain length on deformation behaviour will form the basis of future studies.

Table1. The mechanical properties of the linear polymers

Polymer	T_g (°C)	Density kg/m^3	Modulus (20°C) (Gpa)	Yield Strength (20°C) (Mpa)
C4M	136	1150	2.36	67.4 ± 5.7
C5M	106	1100	1.53	60.4 ± 2.1
C6M	87	1100	1.46	42 ± 2.0

4.2. MECHANICAL PROPERTIES OF CROSSLINKED POLYMERS [(POLY(CnM+ x% CmD)]

Copolymerising the monofunctional and difunctional monomers by this process will form crosslinked polymers. For the copolymerisation processing the optimum monomers to initiator ratios is necessary to obtain a high gel fraction. The gel fraction was determined by sol-gel technique.

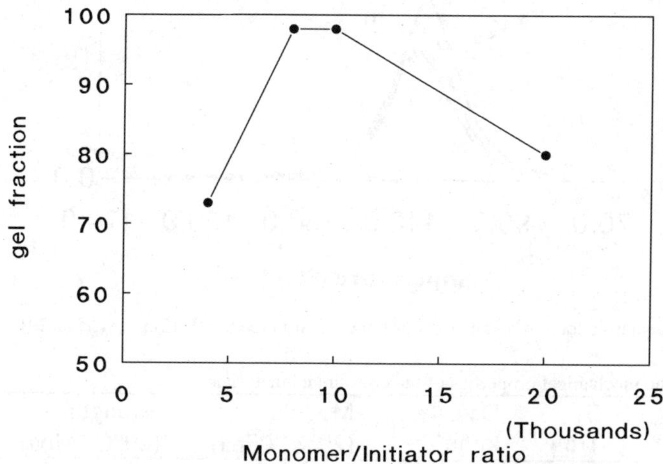

Figure 4. Gel fraction versus M:I ratio for C5M + 2mol% C12D

Figure 4 shows gel fractions for the series of samples obtained by the copolymerisation processing. It can be seen that for these monomer systems, the highest gel fraction ~97% are obtained for an M:I ratio of 8000:1-10,000:1. The results showed that the optimum Monomer : Initiator ratio for the processing of C5M/CmD system was 8000:1. Having established this fact, a range of samples was prepared using this ratio, to investigate the effect of varying the percentage of the difunctional component, and the length of the difunctional linkage. Figure 5 shows the dynamic shear modulus results for the C5M/C12D combination, with 1, 5 and 10 molar percentage of the difunctional component. The results show that as the percentage of the difunctional, crosslinking unit, is increased the glass transition shifts to a higher temperature, the height of the tanδ peak decreases and the plateau shear modulus above T_g increases. These results are as expected for an increase in the crosslink density of a polymer. Table 2 shows measured mechanical properties for the crosslinked materials and compares them with bis-phenol-A-polycarbonate and Poly(DCPD), TELENE. The room temperature moduli and ultimate bending strength of the crosslinked materials are, in general, higher than the equivalent thermosetting materials. Of particular interest are the high values for yield strength and toughness, which are comparable to known 'high toughness' materials such as polycarbonates.

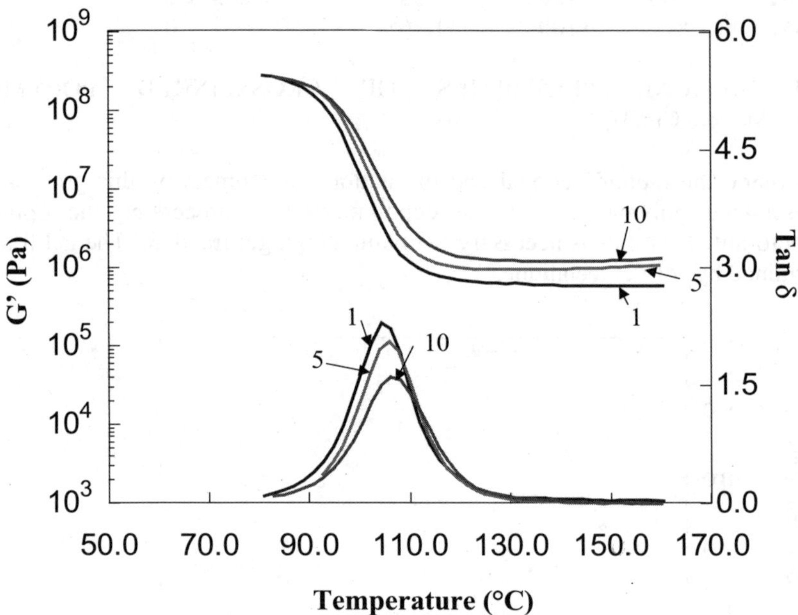

Figure 5. Dynamic torsion results for C5M + x molar percent of C12D: x=1, 5 and 10.

Table 2. The mechanical properties of the crosslinked materials

Polymer	T_g (°C)	Density kg/m³	Modulus (20°C) (Gpa)	Strength (20°C)(Mpa)
C4M/2%C12D	~120	1150	2.36±0.2	61 ± 15(F)
C5M/2%C12D	106	1100	1.53±0.2	72 ±2 (Y)
C6M/2%C12D	87	1100	1.46±0.1	59 ±3 (Y)
Polycarbonate	150	1200	2.3	62 (Y)
Poly(DCPD) TELENE 1100	145	1030	1.9	45

5. Conclusion

The results reported here established the synthesis and successful RTM processing polymerisation and copolymerisation of a range of mono- and difunctional N-alkylimidonorbornenes. We can conclude that this class of new polymers has very interesting mechanical properties and appears to open definite prospects for RTM as larger quantities of monomers could be made relatively easily and cheaply. More research is underway to extend this preliminary work to a much wider range of monomers.

6. References

1.   R. H. Grubbs, E. Khosravi, In *Synthesis of Polymers - a Volume of Materials Science and Technology Series*; Schluter, A. D., Ed.; Wiley-VCH: **1998**, p 63-104.
2.   E. Khosravi and T. Szymansk-Buzar, "ROMP and related chemistry: state of the art and visions for the new century", NATO ASI Series, Kluwer Academic Publishers, in print.
3.   G. C. Bazan, E. Khosravi, R. R. Schrock, W. J. Feast, V. C. Gibson, M. B. Oregan, J. K. Thomas, W. M. Davis, *J. Am. Chem. Soc.,* **1990**, *112*, 8378.
4.   E. Khosravi, A. A. Al-Hajaji, *Polymer,* **1998**, *39*, 5619.
5.   E. Khosravi, In *Modern Fluoropolymers, High Perfomance Polymers for Diverse Applications*; Scheirs, J., Ed.; Wiley & Sons: **1997**, p Chapter 8.
6.   W. J. Feast, E. Khosravi, *J. Fluor. Chem.,* **1999**, *100*, 117.
7.   W. J. Feast, V. C. Gibson, A. F. Johnson, E. Khosravi, M. A. Mohsin, *Polymer,* **1994**, *35*, 3542.
8.   A. C. M. Rizmi, E. Khosravi, W. J. Feast, M. A. Mohsin, A. F. Johnson, *Polymer,* **1998**, *39*, 6605.
9.   R. J. Minchak, US Patent 4,426,502, issued 17/01/**1984**.
10.   R. H. Grubbs, L. K. Johnson, S. T. Nguyen, US Patent No. 5,312,940, issued 5/17/**1994**.
11.   R. H. Grubbs, L. K. Johnson, S. T. Nguyen, US Patent No. 5,342,909, issued 8/30/**1994**.
12.   R. H. Grubbs, S. T. Nguyen, L. K. Johnson, US Patent No. 5,710,298, issued 1/20/**1998**.
13.   R. H. Grubbs, Jr. C. S. Woodson, US Patent No. 5,728,785, issued 3/17/**1998**.
14.   Jr. C. S. Woodson, R. H. Grubbs, US Patent No. 5,939,504, issued 8/17/**1999**.
15.   A. Bell, *Polym. Preprints, 1994, 35, 694.*
16.   S. T. Nguyen, R. H. Grubbs and J. W. Ziller, J. Am. Chem. Soc., **1993**, 115, 9858.
17.   P. E. Schwab, M. B. France, J. W. Ziller and R. H. Grubbs, *Angew. Chem., Int. Ed. Engl., 1995, 34, 2039.*
18.   P. E. Schwab, R. H. Grubbs and J. W. Ziller, *J. Am. Chem. Soc., 1996, 118, 100.*
19.   M. Hillmeyer, W. R. Laredo, R. H. Grubbs, *Macromolecules, 1995, 28, 6311.*
20.   E. Khosravi and A. A. Al-Hajaji, *Eur. Polym. J., 1998, 34, 153.*
21.   E. Khosravi, W. J. Feast, A. A. Al-Hajaji and T. J. Leejarkpai, *Mol. Catal. A: Chemical, 2000, 160, 1.*
22.   P. J. Hine, T. Leejarkpai, E. Khosravi, R. A. Duckett and W. J. Feast, Polymer, **2001**, 42, 9413.

RING OPENING METATHESIS POLYMERISATION (ROMP) OF CYCLOOCTENE AND SUBSTITUTED NORBORNENE DERIVATIVES USING A RUTHENIUM CATALYST CONTAINING A TRIAZOL-5-YLIDENE LIGAND

KAREN MELIS[a], DIRK DE VOS[b], PIERRE JACOBS[b] AND FRANCIS VERPOORT[a]*

[a]*Division of Organometallic Chemistry and Catalysis, Department of Inorganic and Physical Chemistry, Ghent University, Krijgslaan 281 (S-3),9000 Gent, Belgium,*
[b]*Center for Surface Chemistry and Catalysis, Katholieke Universiteit Leuven, Kasteelpark Arenberg 23, 3001 Heverlee, Belgium*

Abstract

The ring opening metathesis polymerisation (ROMP) of cyclooctene and substituted norbornene derivatives is performed in the presence of the Ru-alkylidene (**4**) bearing a triaol-5-ylidene ligand (**2**). Analysis of the formed polymers clearly indicates the low initiation efficiency of **4**. Only a small amount of the added Ru-alkylidene reacts with the monomer to generate the propagating species.

Keywords: Ruthenium/ ROMP/ triazol-5-ylidene

1. Introduction

Well-defined, single-component, single-site homogeneous catalysts for olefin metathesis have provided excellent tools for carbon-carbon bond formation in organic chemistry and polymer chemistry.[1-2] The development of the Ru-alkylidene, $Cl_2(PR_3)_2Ru=CHPh$, by Grubbs and co-workers initiated a revolution in the olefin metathesis.[3-4] The Ru-based complexes exhibit a remarkable tolerance towards oxygen and moisture and thus the transformation of a broad variety of functionalised substrates becomes possible.[5-8]

Since 1999, a new type of stable Ru-alkylidene has emerged as a powerful tool for the olefin metathesis carbon-carbon transformation reactions.[9-11] Due to the disadvantages of the phosphines, i.e. oxidation sensitivity, irreversible cleavage of phosphorus-carbon bond and inability to cover the entire range of electronic control of the catalytically active metal centres, their industrial application is limited. Exchange of one phosphine ligand by a strongly nucleophilic, i.e. electron-rich, ligand that is resistant to oxidising

237

Y. Imamoglu and L. Bencze (eds.), Novel Metathesis Chemistry: Well-Defined Initiator Systems for Specialty Chemical Synthesis, Tailored Polymers and Advanced Material Applications, 237–242.

238

agents and that possesses a stable bond with the metal produce a very stable and extremely active catalyst.

In 1995, a new type of heterocyclic carbene, 1,3,4-triphenyl-4,5-dihydro-1H-1,2,4-triazol-5-ylidene (2), which exhibited a marked nucleophilic reactivity, was synthesised.[12] In 2001, Fürstner *et al.* reported a new Ru-complex (4) which is synthesised from the reaction between the Grubbs' catalyst (3) and the triazol-5-ylidene ligand (2) (Scheme 1).[13]

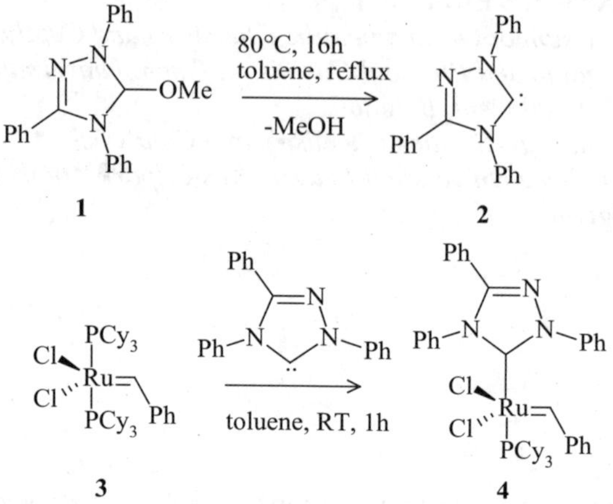

80°C, 16h
toluene, reflux
$\longrightarrow$
-MeOH

1   **2**

toluene, RT, 1h
$\longrightarrow$

3   **4**

Scheme 1

Complex 4 shows a trigonal bipyramidal geometry.[13] Investigation of the catalytic properties reveals a very active catalyst towards the ring closing metathesis of dienes. Surprisingly, 4 does not effect the cyclisation of enyne derivatives. Since no data are available in literature describing the ring opening methathesis (ROMP) activities of 4, we were encouraged to investigate this unexplored field of the olefin metathesis properties of 4.

2. Results and discussion

The activity of the Ru-complex 4 is tested for the ring opening metathesis polymerisation (ROMP) of cyclooctene and substituted norbornene derivatives (Table 1). The polymerisations are performed with a catalyst/monomer ratio of 1/10, since at increasing catalyst/monomer ratio only the formation of insoluble polymers is observed. All polymerisations proceed smoothly and reach quantitative conversion. Only the presence of a nitrile derivative results in a decrease in catalytic performance (run 6).

The broad molecular weight distributions clearly indicate that complex 4 does not promote a controlled metathesis polymerisation, even at a catalyst/monomer ratio of 1/10. The polydispersity indices (PDI's) of the synthesised polymers are high. The calculated initiation efficiencies indicate that less than 5% of the added catalyst is initiated before the ROMP is complete, except the polymerisation of 5-phenyl-bicyclo[2.2.1]hept-2-ene (run 2), where nearly 70% of the added Ru-complex is transformed into the propagating species. The slow initiation rate explains the high experimental molecular weights compared to the theoretical values based on the initial monomer/catalyts ratios. This is in agreement with the results obtained with Ru-alkylidenes which possess a similar N-heterocyclic carbene, i.e. imidazol-2-ylidene.[14]

Table 1: Ring Opening Metathesis Polymerisation (ROMP) catalysed by catalyst **4**

Run[a]	Monomer	Yield (%)[b]	M_w[c]	M_n[c]	$M_{n\ theor}$	M_w/M_n[c]	f (%)[d]
1		100	276602	60315	110000	4.59	1.8
2		98.1	33495	2508	166770	13.36	66.6
3		96.4	239933	104622	170520	2.29	1.6
4		100	362584	126315	186000	2.87	1.5
5		98.2	65203	11444	117720	5.70	10.3
6[e]		75.5	e	e	90600	e	e

[a]Reactions were carried out by using 0.079 mmol of monomer and 0.0079 mmol of catalyst in $CDCl_3$ (0.8 ml) under nitrogen. The reaction mixture was stirred at room temperature for 60 min. [b]Yield is determined gravimetrically and by ^{1}H-NMR. [c]M_n and M_w are determined by GPC analysis. [d]f = $M_{n\ theor}/M_{n\ exp}$ with $M_{n\ theor}$ = ([monomer]/[catalyst]) * yield (%) * $MW_{monomer}$ (g/mol). [e]Insoluble polymer.

To get an accurate comparison between the catalytic activity of the parent benzylidene **3** and **4**, the ring opening metathesis polymerisation of cyclooctene is monitored by ^{1}H-NMR spectroscopy.[15] Both reactions are performed with a catalyst/monomer ratio of 1/10 and under identical reaction conditions. The rate of the formation of the polyoctenamer for both catalysts is compared in figure 1. The ROMP initiated by **4** is clearly the most active process.

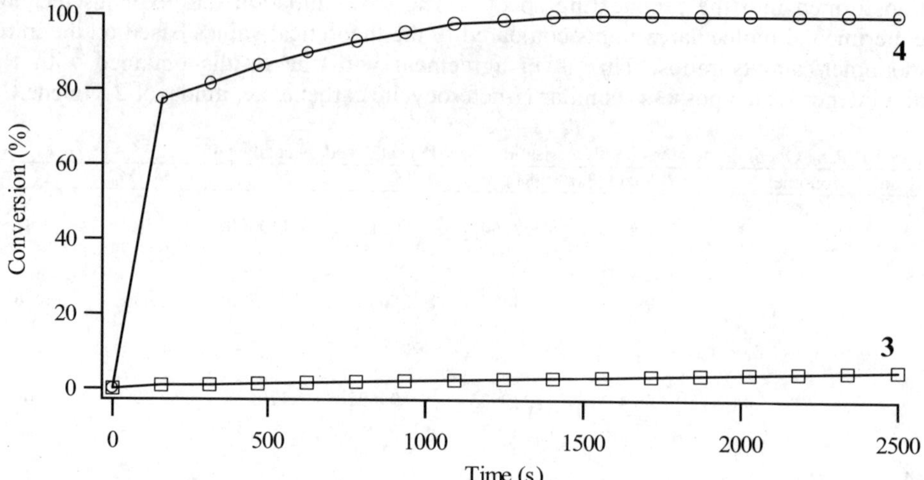

Figure 1: Kinetics of the ROMP reaction of cyclooctene. Comparison of polyoctenamer formation catalysed by **4** versus **3** as monitored by ^{1}H-NMR (25°C, 300 MHz, CDCl$_3$) with [Ru]/monomer: 1/10.

The polymerisation of cyclooctene with catalyst **4** shows good pseudo-first order kinetics (r^2 = 0,968) with respect to the monomer conversion (Figure 2).[16] To compare the relative activity of **4** to the parent benzylidene **3**, the kinetics of **3** is approximated by a first-order curve fit such that a pseudo first-order constant could be extracted. This means that an overestimate of the ROMP activity is withdrawn. The relative rate constant k$_{rel}$ obtained for **4** equals 147 when k$_{rel}$ of the parent alkylidene **3** equals 1. This means that complex **4** exhibits a polymerisation rate that is approximately 147 times greater than the parent complex **3** does.

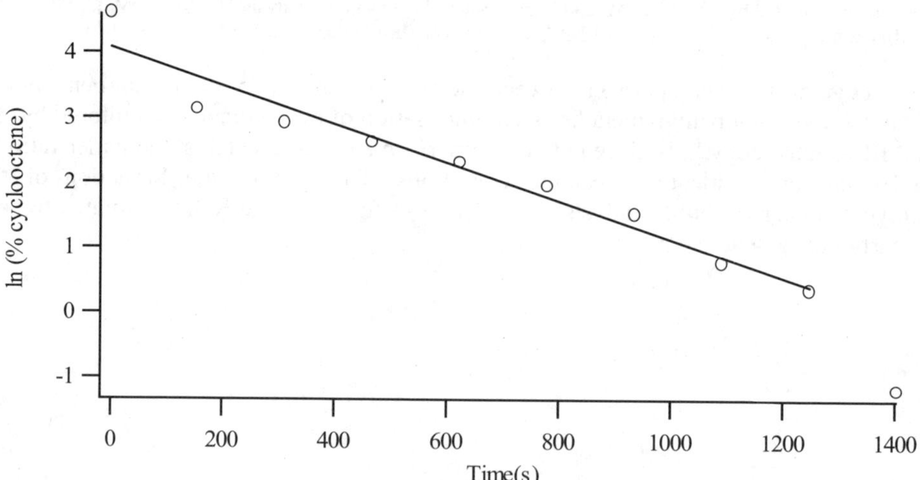

Figure 2: Plot of ln (percentage cyclooctene) versus time for catalyst **4** in the polymerisation of cyclooctene in CDCl$_3$ at roomtemperature, where [**4**] = 3,06 mM. The percentage cyclooctene is expressed as the amount of the monomer still present in the reaction mixture. Linear fit: y = 0,0029342x + 4,0785 with r^2 = 0,974.

3. Conclusion

The ring opening metathesis polymerisation (ROMP) reactions of cyclooctene and several norbornene derivatives performed in the presence of a **4** proceed smoothly and reach full conversion. However, the highy active ruthenium system **4** does not form well-defined polymeric structures, since less than 5% of the initial catalyst acts as the propgating species.

4. Acknowledgements

K. M. is indebted to the BOF (Bijzonder Onderzoeksfonds) of Ghent University for a research grant. We are indebted to the FWO (Fonds voor Wetenschappelijk Onderzoek-Vlaanderen) for a research grant (D. D. V., P. J., and F. V.). Financial support by The Research funds of Ghent University is gratefully acknowledged.

5. Experimental Section

5.1. GENERAL REMARKS:

All reactions were performed under inert atmosphere using Schlenck techniques. NMR spectra were recorded on a Varian Unity 300 MHz spectrometer. GPC (chloroform) conditions: a sequence of three columns: PSS SDV 1000Å, 10000Å and 100000Å (8 x 300 mm, 5µm) (obtained from Shimadzu); detector: differential refractive index detector RID-10A (obtained from Shimadzu) and flow = 1 ml/min (concentration = 20 mg/ml). Toluene is dried over Na. $CDCl_3$ (obtained from Acros) was dried over molecular sieves. $Cl_2(PR_3)_2Ru=CHPh$ (**3**) (obtained from Strem Chemicals), 1,3,4-triphenyl-4,5-dihydro-1H-1,2,4-triazol-5-ylidene (**2**) (obtained from Acros), substituted norbornene derivatives (obtained from Ineos) were used without further purification. Complex **4** is synthesised as described in literature.[14]

5.2. RING OPENING METATHESIS OF CYCLOOCTENE AND SUBSTITUTED NORBORNENE DERIVATIVES:

Reactions were carried out by using 0.079 mmol of monomer and 0.0079 mmol of catalyst **4** in $CDCl_3$ (0.8 ml) under nitrogen. The reaction mixture was stirred at room temperature. The polymerisation is followed in situ by ^{1}H-NMR. Yield is determined gravimetrically and by ^{1}H-NMR. M_n and M_w are determined by GPC analysis.

6. References

1. A. Fürstner, *Angew. Chem. Int. Ed.*, **2000**, *39*, 3012.
2. U. Frenzel, O. Nuyken, *J. Polym. Sci., A: Polym. Chem.*, **2002**, *40*, 2895.
3. P. Schwab, R. Grubbs, J. Ziller, *J. Am. Chem. Soc.*, **1996**, *118*, 100.
4. E. Dias, S. Nguyen, R. Grubbs, *J. Am. Chem. Soc.*, **1997**, *119*, 3887.
5. K. Ivin, *J. Mol. Catal. A: Chem.*, **1998**, *133*, 1.
6. M. Schuster, S. Blechert, *Angew. Chem. Int. Ed. Engl.*, **1997**, *36*, 2036.
7. K. Melis, D. De Vos, P. Jacobs, F. Verpoort, *J. Mol. Catal. A : Chem.*, **2001**, *169*, 47.

8.   B. De Clerq, F. Verpoort, *J. Mol. Catal. A : Chem.*, **2002**, *170*, 67.
9.   T. Weskamp, F. Kohl, W. Herrmann, *J. of Organomet. Chem.*, **1999**, *582*, 362.
10.   L. Ackermann, A. Fürstner, T. Weskamp, F. Kohl, W. Herrmann, *Tetrahedron Lett.*, **1999**, *40*, 4787.
11.   M. Scholl, T. Trnka, J. Morgan, R. Grubbs, *Tetrahedron Lett.*, **1999**, *40*, 2247
12.   D. Enders, K. Breuer, G. Raabe, J. Runsink, H. Teles, J.-P. Melder, K. Ebel, S. Brode, *Angew Chem. Int. Ed. Engl.*, **1995**, *34*, 1021.
13.   A. Fürstner, L. Ackermann, B. Gabor, R. Goddard, C. Lehmann, R. Mynott, F. Stelzer, O. Thiel, *Chem. Eur. J.*, **2001**, *7*, 3236.
14.   C. Bielawski, R. Grubbs, *Angew. Chem. Int. Ed. Engl.*, **2000**, *39*, 2903.
15.   S. Hansen, M. Volland, F. Romiger, F. Eisenträger, P. Hofmann, *Angew. Chem. Int. Ed., Engl.*, **1999**, *38*, 1273.
16.   E. Dias, R. Grubbs, *Organometallics*, **1998**, *17*, 2758.

ACYCLIC DIENE METATHESIS (ADMET) POLYMERIZATION BY ELECTROCHEMICALLY GENERATED TUNGSTEN-BASED ACTIVE CATALYST SYSTEM: OPTIMIZATION OF REACTION CONDITIONS

O. DERELİ, B. DÜZ AND Y. İMAMOĞLU*
Chemistry Department, Hacettepe University,
06532 Beytepe, Ankara, Turkey

1. Introduction

Dienes can undergo olefin metathesis reactions of two types: (i) intermolecular, and (ii) intramolecular. Intramolecular metathesis (RCM) reactions occur with great readiness whenever the product is a 6-membered ring. They are also often favoured for the production of 5-, 7-, and 8-membered rings, depending on the nature, number and location of any substituents. Intermolecular metathesis reactions lead eventually to high polymers and proceed very cleanly when initiated by metal carbene complexes; these are known as ADMET (acyclic diene metathesis) polymerizations (Figure 1). This class of polymerization reactions has been well established and comprehensively studied by Wagener and his group [1-6]. ADMET polymerization has been also a convenient route to linear polymers containing inorganic elements and functional groups for the preparation of new materials [7].

$$WCl_6 \xrightarrow{\text{Al} / \bar{e} / CH_2Cl_2} [W]{=}CH_2 \xrightarrow{} \ \ \underset{R}{\diagdown}\big(\diagup R\big)_n\diagup + \ n\,C_2H_4$$

Figure 1. ADMET Polymerization by electrochemically produced tungsten catalyst.

It was first reported that the electrochemical reduction of transition metal salts, such as WCl_6 and $MoCl_5$, under controlled potential at a platinum cathode with an aluminum anode, results in the formation of stable and active olefin metathesis catalysts [8-11]. The application of the WCl_6 / $\bar{e}$ / Al / CH_2Cl_2 system to ADMET polymerization of 1,9-decadiene was reported and the polyoctenamer formed is of a weight-average molecular weight of 9000 with a polydispersity of 1.92 [10]. IR and NMR spectral analyses indicate the retainment of the double bonds in the polymer structure with high trans content as expected from a step condensation reaction.

This presented study describes the optimization of the reaction conditions using the WCl_6 / $\bar{e}$ / Al / CH_2Cl_2 system in the ADMET polymerization of 1,9-decadiene. Optimum conditions for metathetic polymerization such as the catalyst/olefin ratio, reaction time and electrolysis time were investigated.

Y. Imamoglu and L. Bencze (eds.), Novel Metathesis Chemistry: Well-Defined Initiator Systems for Specialty Chemical Synthesis, Tailored Polymers and Advanced Material Applications, 243–247.
© 2003 *Kluwer Academic Publishers. Printed in the Netherlands.*

244

2. Experimental

1,9-decadiene was obtained from Aldrich and purified by refluxing over KOH followed by distillation over CaH_2 under nitrogen atmosphere. WCl_6 (Aldrich) was purified by sublimation at 220 °C under nitrogen to remove the more volatile WO_2Cl_2 and $WClO_4$ impurities [12]. Dichloromethane (Merck, $\varepsilon = 9.1$) was first washed with concentrated H_2SO_4 until the acid is colorless, then respectively with water, aqueous solution of NaOH (5% w/w) and water again. After drying over anhydrous $CaCl_2$ it was then distilled over P_2O_5 under nitrogen [13]. THF and MeOH were supplied from Merck and used as received.

The electrochemical equipment consisted of a POS Model 88 potentiostat and EVI 80 Model voltage integrator (coulometer). The measurements were carried out under nitrogen atmosphere in a three-electrode cell having a jacket through which water from a constant temperature bath was circulated. Exhaustive controlled potential experiments were carried out in an undivided cell with a macro working Pt foil electrode (2 cm^2) and a Al foil (2 cm^2) counter electrode. The reference electrode consisted of AgCl coated on a Ag wire in CH_2Cl_2 / 0.1 N $TBABF_4$ (tetra-n-butyl ammonium tetrafluoroborate) which was separated from the electrolysis solution by a sintered glass disc. Electrolysis was carried out without the supporting electrolyte due to its deleterious effect on the catalyst system. For this reason the distance between Pt working and Al counter electrode was kept constant and as small as possible (i.e. 2.0 mm) in order to keep the solution resistance at minimum.

All operations were performed under pure and dry nitrogen. WCl_6 (0.2 g, 0.50 mmol) was introduced into the electrochemical cell containing CH_2Cl_2 (20 mL) and a red solution was observed. Reductive electrolysis at 0.9 V was applied to the red solution. The color of the solution was darkened progressively. Aliquots from this catalytic solution were used in different runs of polymerization reactions and optimum conditions were determined where highest percentage conversion to the polymer was obtained.

All reactions were initiated in the bulk, at room temperature and under dry nitrogen atmosphere. Molecular weights were experimentally controlled by varying the monomer/catalyst ratio and reaction time. Reaction combinations ranged within 30:1 to 200:1 and 4 h to 32 h. A typical reaction is as follows: 1mL of the catalytic solution was taken with an automatical pipette from the cell and added onto the monomer (0.20 g) in a Schlenk tube containing a magnetic stir bar. A rapid gelation was observed and stirring was continued until prevented by the viscosity increase. The reaction was quenched by methanol addition. The polymers formed were washed with methanol, dissolved in THF and reprecipitated with methanol to remove the catalytic residues, dried and weighed. Percentage conversion of the monomer to the polymer was defined on weight basin.

3. Results and discussion

To optimize the reaction conditions a series of experiments were performed by varying the catalyst/olefin ratio, reaction time and electrolysis time.

Catalyst/olefin ratio:

In this experiment, 0.037 mmol of catalyst was added onto different amounts of olefin. Polymerizations were carried out at ambient temperature, for 24 hours. The polymers obtained were weighed, and the percentage conversions were calculated.

Figure 2 shows the effect of catalyst/olefin ratio on the conversion of 1,9-decadiene. Conversion to the polymer increases with the amount of catalyst used, and reaches a maximum at a certain catalyst/olefin ratio which is around 0.01726.

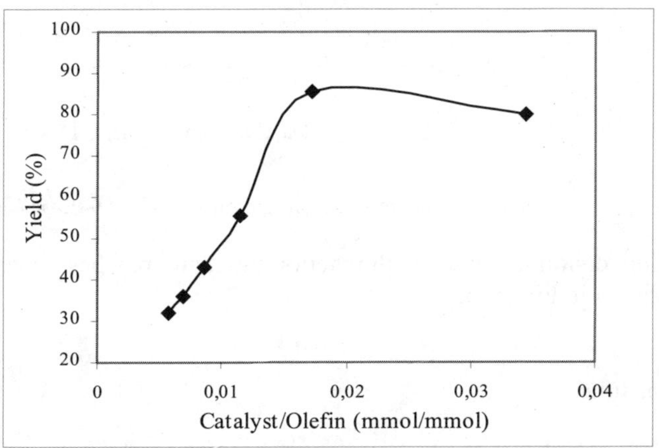

Figure 2. Effect of catalyst/olefin ratio on conversion of 1,9-decadiene.

At high catalyst concentrations deactivation of the active catalyst or degradation of the product may occur which both result in a decrease in conversion.

Reaction time:

At this stage, the effect of the reaction time on the conversion of 1,9-decadiene was studied. The catalyst/olefin ratio was kept at 0.01726, and the reaction was quenched by the addition of methanol after 4, 8, 12, 16, 24, and 32 hours from the start of reaction. The percentage conversions to the polymers obtained were calculated.

246

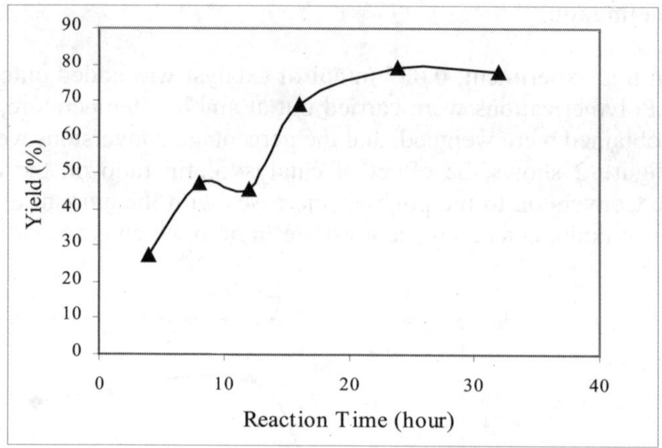

Figure 3. Effect of reaction time on conversion of 1,9-decadiene.

Conversion increases with reaction time, and reaches a maximum, in about 24 hours as shown in Figure 3.

Electrolysis time:

Effect of the electrolysis time on conversion of 1,9-decadiene was examined. The catalyst/olefin ratio was kept at 0.01726, and the active catalyst obtained in different electrolysis periods (30, 60, 90, 120, 150, and 180 minutes) was added onto the olefin. Polymerizations were quenched by the addition of methanol after about 24 hours from start of reaction. The percentage conversions to the polymers obtained were calculated. Figure 4 shows the effect of electrolysis time on conversion to poly(octenamer).

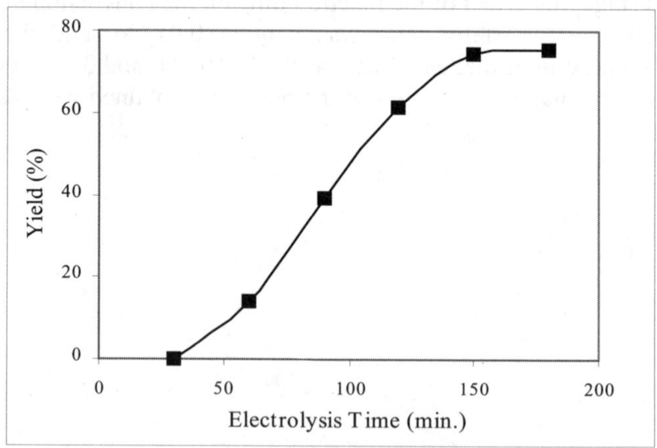

Figure 4. Effect of electrolysis time on conversion of 1,9-decadiene.

With prolonged electrolysis time, concentration of the active catalyst formed during the electrolysis, and conversion to the polymer increases, and maximum conversion was obtained with the active species produced in 3 hours of electrolysis time. This is confirmed also by accounting the charges during electrolysis.

4. Conclusions

The optimum reaction conditions for the ADMET polymarization of 1,9-decadiene using the WCl_6 / $\bar{e}$ / Al / CH_2Cl_2 system were found as 0.01726 mmol catalyst / mmol olefin, 24h reaction time and 3h electrolysis time. This ill-defined catalyst system is less sensitive to oxygen than most of the well-defined systems and retains its activity for about ten hours.

5. Acknowledgements

The financial support from Hacettepe University Research Fund (project no. 98 K 121 720) is greatly acknowledged.

6. References

[1]   Lindmark-Hamberg, M., Wagener, K.B. (1987) Macromolecules **24**, 2649.
[2]   Wagener, K.B., Boncella, J.M., Nel J.G., Duttweiler R.P., Hillmyer, M.A. (1990) Macromol. Chem. **191**, 365.
[3]   Wagener, K.B., Nel, J.G., Konzelman, J., Boncella, J.M. (1990) Macromolecules **23**, 5155.
[4]   Wagener, K.B., Boncella, J.M., Nel, J.G. (1991) Macromolecules **24**, 2649.
[5]   Konzelman, J., Wagener, K.B. (1996) Macromolecules **29**, 7657.
[6]   Wagener, K.B., Brzezinska, K., Anderson, J.D., Dilocker, S.J. (1997) Polym. Sci. **A35**, 3441.
[7]   Wagener, K.B. Formation of Hydrocarbon and Functionalized Polymers by ADMET Polymerization. In Metathesis Polymerization of Olefins and Polymerization of Alkynes, Imamoğlu Y (ed.). (1998) NATO ASI Series C506, Kluwer Academic Publishers: Dordrecht, 277-296.
[8]   Gilet, M., Mortreux, A., Nicole, J., Petit, F. (1979) J. Chem. Soc. Chem. Commun. 521.
[9]   Gilet, M., Mortreux, A., Folest, J.C., Petit, F. (1983) J. Am. Chem. Soc. **105**, 3876.
[10]   Dereli, O., Düz, B., Zümreoğlu-Karan, B. and Imamoğlu, Y. (2003) Appl. Organometal. Chem. **17**, 23-27.
[11]   Çetinkaya, S., Düz, B. and Imamoğlu, Y. (2003) Appl. Organometal. Chem. **17**, 232-235.
[12]   Uchida, A., Hamano, Y., Mukai, Y. and Masuda, S. (1971) Ind. Eng. Chem. Prod. Res. Develop. **10**, 372.
[13]   Calderon, N., Ofstead, E.A., Ward, J.P., Judy, W.A. and Scott, K.W. (1968) J. Am. Chem. Soc. **90**, 4133.

PROBING THE TACTICITY OF RING-OPENED METATHESIS POLYMERS OF NORBORNENE AND NORBORNADIENE DIESTERS BY NMR SPECTROSCOPY

L. DELAUDE, A. DEMONCEAU, AND A. F. NOELS*
Center for Education and Research on Macromolecules (CERM), Institut de Chimie (B6a), University of Liège, Sart-Tilman par B-4000 Liège, Belgium

1. Introduction

The obtainment of highly stereoregular macromolecular chains is crucial to ultimately control the bulk properties -and hence the practical and commercial values- of many polymeric materials [1]. In the case of polymers prepared by ring-opening metathesis polymerization (ROMP) of cyclic olefins, double bonds are retained in the products that can result in complex microstructure variations. The metathesis polymers prepared from norbornene and its derivatives, in particular, give rise to two independent types of isomerism. The first one comes from the configuration of the exocyclic double bonds and can be defined in terms of cis/trans ratio and distribution. The second stereochemical differentiation arises from the relative orientation that two neighboring cyclopentylene units can adopt in the polymer chain. Thus, meso or racemic dyads may be formed that correspond, respectively, to isotactic or syndiotactic segments. Since there is no direct relationship between double bond stereochemistry and dyad tacticity, four regular structures are therefore possible for symmetrically substituted polynorbornenes and polynorbornadienes (Figure 1).

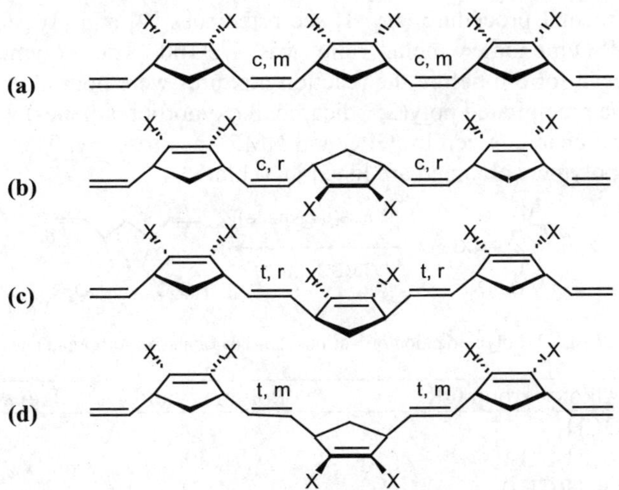

Figure 1. The four possible regular structures of poly(2,3-disubstituted norbornadienes): (a) cis, isotactic; (b) cis, syndiotactic; (c) trans, syndiotactic; (d) trans, isotactic.

Y. Imamoglu and L. Bencze (eds.), Novel Metathesis Chemistry: Well-Defined Initiator Systems for Specialty Chemical Synthesis, Tailored Polymers and Advanced Material Applications, 249–261.
© 2003 *Kluwer Academic Publishers. Printed in the Netherlands.*

Thanks to the developments and the availability of high magnetic field instruments, NMR spectroscopy has been used extensively to probe the molecular architecture of ROM polymers [2,3]. Yet, most literature reports focus only on cis/trans ratio determination, the microstructure feature which is usually responsible for the main spectral line splittings, while the more complex fine structure patterns due to tacticity are frequently overlooked. Systematic investigations on tacticity issues in ROMP reactions of norbornene and norbornadiene derivatives have nevertheless been carried out during the past decades. They are based principally on ^{13}C NMR spectroscopy and were first initiated by Ivin and Rooney in Belfast [4-7].

Work from our laboratory has evidenced that the ROMP of 2,3-dicarboalkoxy-norbornadienes and their 7-oxa analogues catalyzed by [RuCl$_2$(p-cymene)]$_2$ (p-cymene is 1-isopropyl-4-methylbenzene) in the presence of trimethylsilyldiazomethane (TMSD) yielded high-trans, highly tactic polymers [8,9]. Subsequent reactions with enantiomerically pure norbornadiene diesters and the reduction of unsaturated polynorbornenes and polynorbornadienes into the corresponding polynorbornanes afforded polymer samples suitable for an absolute tacticity determination by NMR spectroscopy [10]. This chapter summarizes the main stereochemical features of the polymers formed using the ruthenium-arene catalytic system and the NMR experiments that were used to establish them.

2. Results and Discussion

2.1. POLYMERIZATION OF 2,3-DICARBOALKOXYNORBORNADIENES

Various esters of norbornadiene-2,3-dicarboxylic acid were prepared by Diels-Alder reaction of cyclopentadiene and an appropriate dialkyl acetylenedicarboxylate. They were reacted with [RuCl$_2$(p-cymene)]$_2$ activated by TMSD in dry THF at 60 °C using a standard experimental procedure (Eq. 1, see references [9] and [10] for details). The ruthenium/ TMSD/monomer molar ratio was 1/2/100. The polymerizations were allowed to proceed for 6 h before the reaction mixtures were poured in a large volume of methanol. The precipitated poly(2,3-dicarboalkoxynorbornadienes) were dried under high vacuum and characterized by GPC and NMR spectroscopy. The yields and trans contents of the polymers obtained are listed in Table 1.

$$n \quad \overset{CO_2R}{\underset{CO_2R}{\diagup}} \quad \xrightarrow[\text{TMSD, 60 °C}]{[\text{RuCl}_2(p\text{-cymene})]_2} \quad \overset{}{\underset{RO_2C \quad CO_2R}{\diagdown}}_n \tag{1}$$

TABLE I. Polymerization of various 2,3-dicarboalkoxynorbornadienes

Monomer	Alkoxy group OR	Polymer yield (%)	Trans content (%)
1	OCH$_3$	86	100
2	OCH$_2$CH$_3$	78	100
3	OCH(CH$_3$)$_2$	71	85
4	(R)-OCH(CH$_3$)CH$_2$CH$_3$	17	84
5	(S)-OCH$_2$CH(CH$_3$)CH$_2$CH$_3$	38	82
6	OC(CH$_3$)$_3$	44	61

2.1.1. Analysis of 1D NMR Spectra

2,3-Dicarbomethoxynorbornadiene (**1**) served as a probe substrate to investigate the influence of the various experimental parameters on the polymer yield and microstructure (reaction time and temperature, nature of the solvent and catalyst, influence of the TMSD cocatalyst) [9]. An all-trans, highly tactic polymer was obtained under a wide variety of conditions. This was evidenced by $^{13}C\{^{1}H\}$ NMR analysis. In all cases, only single sets of signals were observed for each type of carbon atoms (Figure 2). Comparison of the chemical shifts with those reported for cis- [11] and trans-poly(**1**) [12] indicated that only the trans polymer was formed using the ruthenium dimer catalyst precursor [RuCl$_2$(p-cymene)]$_2$. ^{1}H NMR spectroscopy confirmed this analysis. Only trans olefinic protons and methine protons adjacent to trans double bonds (at 5.40 and 3.51 ppm in CDCl$_3$, respectively) were visible. GPC traces were unimodal and molecular weight distributions ranged between 1.7 and 1.9. Such a rather poor control over polydispersity is ineluctable in a system where the actual catalytic species are generated in situ from stable, readily available catalyst precursors. It constitutes the price to pay for the ease of implementation of our ROMP procedure.

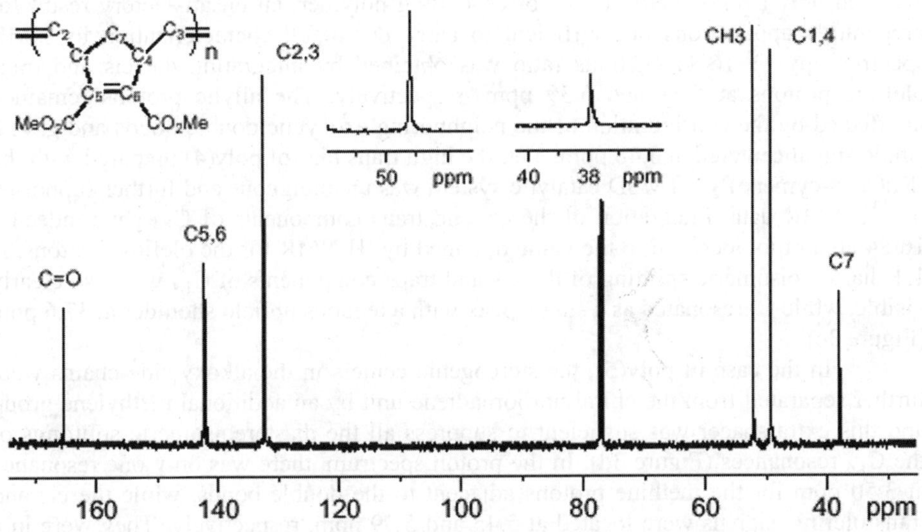

Figure 2. 100 MHz ^{13}C NMR spectrum in CDCl$_3$ of *all-trans*-poly(**1**) prepared using as catalyst [RuCl$_2$(p-cymene)]$_2$ in the presence of TMSD.

No major change of reactivity occurred when the diethyl ester (**2**) was substituted for the its lower homologue **1**. The polymerization rate was only slightly slower and, accordingly, the isolated yield after 6 h dropped down a few percent. ^{1}H NMR spectroscopy revealed the presence of only one sharp doublet at 5.40 ppm and one broad singlet at 3.50 ppm for olefinic and allylic resonances in CDCl$_3$, respectively, in agreement with an all-trans microstructure. The ^{13}C NMR spectrum of poly(**2**) comprised only one neat singlet at 38.0 ppm for the methylene carbon resonance (Figure 3a), thereby confirming the all-trans, highly tactic nature of the macromolecular chain. Examination of the C$_{1,4}$ signals resonating between 48 and 50 ppm proved also to be very useful, because the various lines present in this region are most sensitive to stereochemical variations and present the highest chemical shift differences. Furthermore, a survey of the literature pertaining to metathesis polymers revealed that allylic carbons adjacent to trans double bonds are always deshielded relative to their cis counterparts, thereby simplifying the assignments [1]. In the case of *all-trans*-poly(**2**), the presence of a predominant line at 49.3 ppm strongly suggests that the polymer is highly tactic (Figure 3a).

When the size of the alkyl substituent was further increased, the polymers formed lacked a stereoregular structure and the reactions became even slower. Indeed, poly(2,3-dicarboalkoxynorbornadienes) derived from monomers **3-6** contained a significant fraction of cis exocyclic double bonds and were isolated in less than quantitative yields under the standard reaction conditions at 60 °C (Table I). With the diisopropoxy derivative **3**, a decent 71% yield was reached after 6 h. A 15/85 cis/trans ratio was derived from the integrals calculated for the cis and trans olefinic resonances at 5.48 and 5.38 ppm in $CDCl_3$, respectively. Signals for methine protons adjacent to cis and trans double bonds (at 3.64 and 3.47 ppm, respectively) gave an identical value, while the separation proposed in Figure 3b between the cis and trans components of the $C_{1,4}$ signal gave a 14/86 ratio.

The optically active monomer **4** containing two (*R*)-1-methylpropoxy groups afforded only a 17% yield of the corresponding polymer, an unsatisfactory result for preparative applications but sufficient to carry out a full characterization by NMR spectroscopy. A 16/84 cis/trans ratio was obtained by integrating the cis and trans olefinic protons at 5.51 and 5.39 ppm, respectively. The allylic protons remained unaffected by the configuration of the neighboring exocyclic double bonds and gave a single signal centered at 3.48 ppm. Yet, the high trans bias of poly(**4**) prepared with the $[RuCl_2(p\text{-cymene})]_2$ + TMSD catalytic system was unambiguous and further supported by ^{13}C NMR data. Integration of the cis and trans components of $C_{1,4}$ gave indeed a 16/84 area ratio, identical to the value obtained by ^1H NMR for the olefinic protons. A 1:1 diastereoisomeric splitting of the cis and trans components of $C_{1,4}$ was also clearly visible, while C_7 resonated as a single peak with a tenuous upfield shoulder at 37.6 ppm (Figure 3c).

In the case of poly(**5**), the stereogenic centers in the alkoxy side chains were further separated from the chiral norbornadiene unit by an additional methylene group and this extra spacer was sufficient to suppress all the diastereoisomeric splittings of the $C_{1,4}$ resonances (Figure 3d). In the proton spectrum, there was only one resonance at 3.50 ppm for the methine protons adjacent to the double bonds, while the cis and trans olefinic signals were located at 5.45 and 5.39 ppm, respectively. They were in a 18/82 ratio. With the tentative distinction between the cis and trans components of the $C_{1,4}$ signal depicted in Figure 3d an identical value could be obtained from the deconvoluted integrals. Surprisingly, this ratio is slightly superior to the one measured in the polydiisopropoxydiester derived from monomer **2**, although the intercalation of a methylene spacer between the carboxy group and the chain ramification would have been expected to reduce the steric hindrance around the metal catalytic center, hence favoring a better stereochemical control. The close similarity between the trans contents in poly(**2**) and poly(**3**), on the other hand, fits nicely with the fact that the 1-methylpropoxy group is the immediate superior homologue of the isopropoxy radical. In both cases, the methyl branch lies next to the ester link and the chain length modification occurs only on a more remote position from the cycloolefin unit.

With the bulky *tert*-butoxy groups of monomer **6**, the fraction of cis double bonds within the polymer backbone reached 39%. In ^{1}H NMR spectroscopy, this proportion was deduced from the integrals of cis and trans olefinic protons resonating at 5.49 and 5.34 ppm, respectively, in CDCl$_3$. As in poly(**4**) and poly(**5**) the allylic hydrogen atoms remained

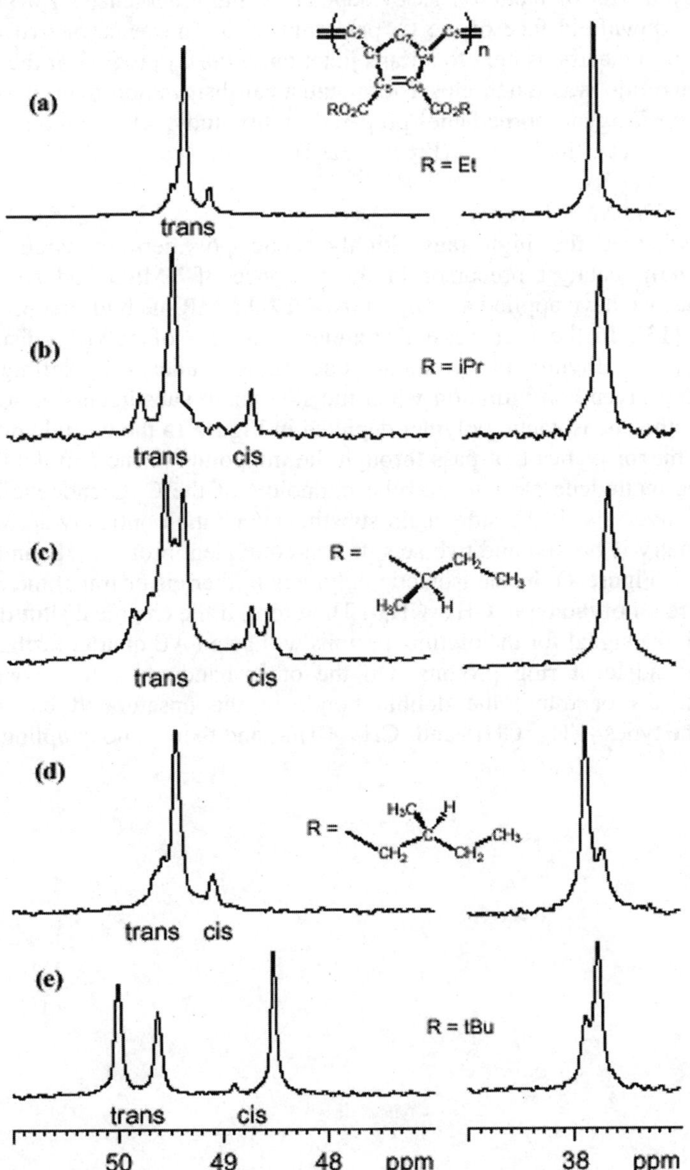

Figure 3. C$_{1,4}$ (left) and C$_7$ (right) regions of the 100 MHz ^{13}C NMR spectra in CDCl$_3$ of (a) *all-trans* poly(**2**), (b) 85% *trans*-poly(**3**), (c) 84% *trans*-poly(**4**), (d) 82% *trans*-poly(**5**), and (e) 61% *trans*-poly(**6**) prepared using as catalyst [RuCl$_2$(*p*-cymene)]$_2$ in the presence of TMSD.

unaffected by the configuration of the adjacent double bonds and gave a unique singlet at 3.41 ppm. In ^{13}C NMR spectroscopy the cis content of poly(6) calculated from the four components of the $C_{1,4}$ resonance (Figure 3e) reached 38/62 and was therefore highly consistent with the 39/61 value obtained by proton integration. The increase of the cis content in poly(6) compared to the polymers derived from monomers 1-5 was accompanied by a loss of tacticity, as evidenced by the emergence of a significant shoulder on the downfield face of the C_7 peak and by the presence of two lines in a 57/43 ratio for allylic carbons next to a trans junction in the $C_{1,4}$ region of the spectrum (Figure 3e). This ratio was much closer to a statistical distribution than in any of the other polydicarboalkoxynorbornadienes prepared in this study, which all displayed one predominant trans tactic line for $C_{1,4}$ (Figures 3a-d).

2.1.2. Analysis of 2D NMR Spectra

To establish whether the high-trans, highly tactic polymers prepared with the $[RuCl_2(p\text{-cymene})]_2$ catalyst precursor in the presence of TMSD had an iso- or a syndiotactic bias, we have applied a straightforward 2D NMR method first proposed by Schrock et al. [13]. In the four possible regular structures of poly(2,3-disubstituted norbornadienes) (cf. Figure 1) symmetry operations render the olefinic protons equivalent in any given configuration when the side chain substituents X are achiral. For instance, in the cis, isotactic polymer depicted in Figure 1a the olefinic protons are related by two mirror planes that pass through the midpoints of the c,m double bonds and through the methylene carbons and the midpoints of the C=C endocyclic bonds, respectively. Conversely, if the side chain substituents contain optically active groups (X*), the symmetry is broken and two sets of non-equivalent protons (H_A and H_B) can be distinguished (Figure 4). In the isotactic polymers (either cis or trans) the exocyclic double bonds are all of the type -CH_A=CH_B-. Therefore, if the chemical shift difference is not too small, the signal for the olefinic protons will be an AB quartet further split by coupling to the adjacent ring protons. On the other hand, in a fully syndiotactic polymer (either cis or trans) the double bonds in the unsaturated backbone are alternately of the types -CH_A=CH_A- and -CH_B=CH_B-, and there is no coupling between H_A and H_B.

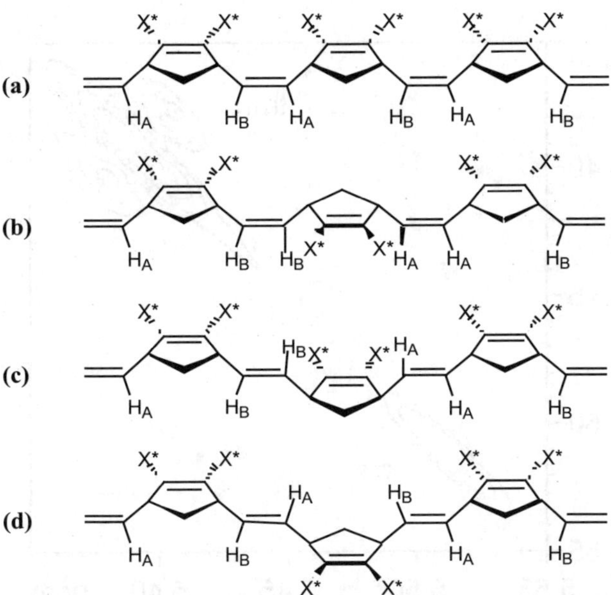

Figure 4. The four possible regular structures of poly(2,3-disubstituted norbornadienes) bearing optically active substituents: (a) cis, isotactic; (b) cis, syndiotactic; (c) trans, syndiotactic; (d) trans, isotactic.

The olefinic region of the 400 MHz COSY spectrum of 84% *trans*-poly(**4**) clearly showed the presence of intense cross-peaks between the main upfield trans protons at 5.39 ppm, while the minor downfield cis resonances at 5.51 ppm did not give any off-diagonal peak (Figure 5). A similar spectrum was obtained with 82% *trans*-poly(**5**) (not represented). Based on the coupling patterns summarized above (cf. Figure 4), one can therefore infer that the minor *cis* fraction of both polymers is *syndiotactic,* while the major *trans* fraction is *isotactic*. It should also be reasonably safe to carry over these conclusions to the various other poly(2,3-dicarboalkoxynorbornadienes) bearing methyl, ethyl, isopropyl, and *tert*-butyl groups that lacked chirality in their side chains and could not be subjected to an absolute tacticity determination.

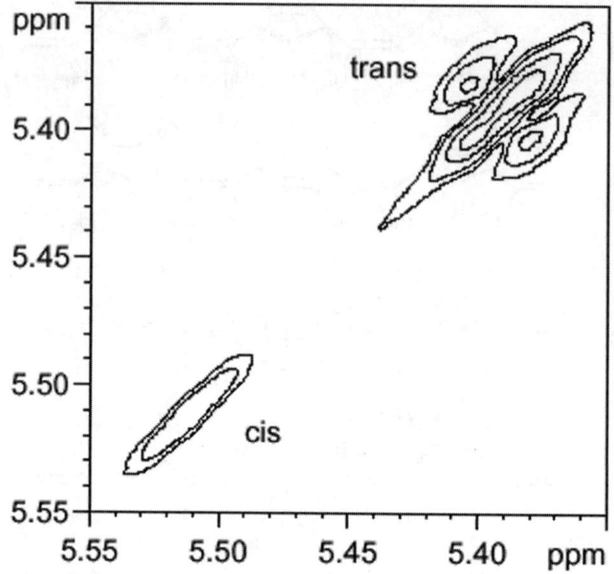

Figure 5. 400 MHz ^{1}H COSY spectrum of 84% *trans*-poly(4) in CDCl$_3$ showing the olefinic proton region.

2.2. POLYMERIZATION OF 2,3-DICARBOALKOXYNORBORNENES

To further probe the steroselectivity of the [RuCl$_2$(p-cymene)]$_2$ catalytic system activated by TMSD, the ROMP of norbornene-2,3-dimethyl esters was investigated. Whereas the corresponding norbornadiene derivative (**1**) existed as a single meso isomer, 2,3-dicarbomethoxy-5-norbornene gives rise to three diastereoisomers, depending on the relative orientations of the two carboxylate groups on the norbornene ring. The exo,exo and the endo,endo isomers (**7** and **8**, respectively) are achiral due to the presence of a symmetry plane in these molecules. Their ROM polymers are prone to cis/trans and meso/racemic isomerism like that of **1**. The *endo,exo*-diester (**9**) lacks symmetry and exists as a pair of enantiomers. If only one of them is subjected to ROMP instead of the racemic mixture, the tacticity of the resulting polymer can be deduced immediately from ^{1}H COSY experiments, as described in the previous section [13]. Unfortunately, only minute amounts of macromolecular products (ca. 1% yield) were obtained when monomers **7** and **9** (racemate) were reacted with the ruthenium dimer catalyst precursor activated by TMSD in THF at 60 °C for 6 h [10]. Under these standard experimental conditions the exo,exo derivative **7** was more reactive and afforded a modest 33% yield of polymer (Eqs 2-4). No further attempts were made to optimize the reaction conditions and to improve the polymer yield.

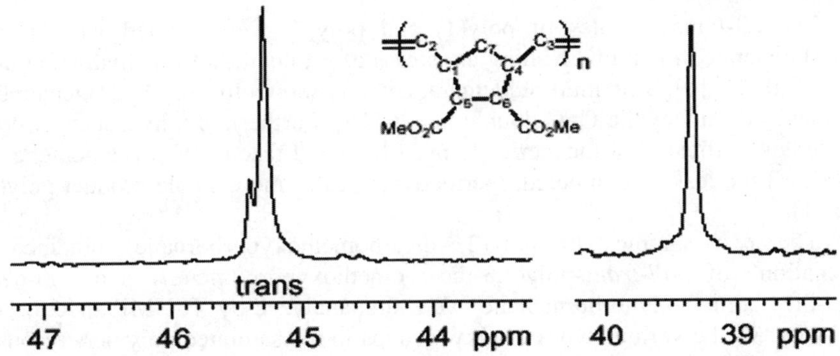

$$(3)$$

$$(4)$$

The ^{1}H and ^{13}C NMR spectra of poly(**7**) were recorded in CDCl$_3$ and assigned by comparison with literature data [5,14-16]. They clearly showed that an all-trans polymer was formed. Only one type of olefinic hydrogens was visible in the proton spectrum at 5.41 ppm, corresponding to a trans environment. The carbon spectrum also gave unequivocal evidence of a high stereoregularity. Six out of the seven different types of carbon atoms in the repeating unit resonated as sharp singlets. For example, the signal of C$_7$ at 39.3 ppm was assigned to a methylene group surrounded by two trans double bonds (Figure 6). Only for C$_{1,4}$ was a tacticity splitting visible by NMR. The high sensitivity of allylic carbon atoms to structural variations was already pointed out for poly(2,3-dicarboalkoxynorbornadienes) in Section 2.1.1. In the case of *high-trans*-poly(**7**) up to four lines could be distinguished for C$_{1,4}$ due to tacticity splitting at the triad level and their chemical shifts were found to follow the sequence $\delta_{rr} > \delta_{rm} > \delta_{mr} > \delta_{mm}$ [16]. Only two overlapping peaks were observed at 45.3 and 45.4 ppm for *all-trans*-poly(**7**) prepared using the [RuCl$_2$(*p*-cymene)]$_2$ catalytic system (Figure 6). The much higher intensity and the upfield location of the major line with respect to the minor one strongly suggests that the polymer has a strong isotactic (*m* or *mm*) bias.

Figure 6. C$_{1,4}$ (left) and C$_7$ (right) regions of the 100 MHz ^{13}C NMR spectra in CDCl$_3$ of *all-trans*-poly(7) prepared using as catalyst [RuCl$_2$(*p*-cymene)]$_2$ in the presence of TMSD.

2.3. REDUCTION OF POLY(DICARBOALKOXYNORBORN(ADI)ENES)

Thanks to the efforts of the Belfast school the reduction of polynorbornenes and polynorbornadienes into the corresponding polynorbornanes has emerged as a powerful tool for the determination and the interpretation of tacticity in ROM polymers [1,17]. Indeed, the hydrogenation of the C=C double bonds not only eliminates the cis/trans

258

isomerism but also often leads to enhanced splittings in the NMR signals, due to the sole effect of meso or racemic environments. The tacticity of the reduced products is directly related to that of the parent materials, because the absolute configurations of the asymmetric carbon atoms in the cyclopentyl rings remain unchanged upon hydrogenation (Scheme 1).

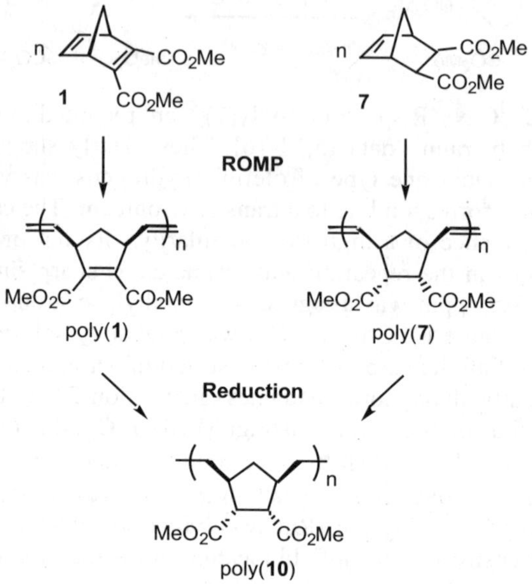

n 1 CO_2Me CO_2Me

n 7 CO_2Me CO_2Me

ROMP

MeO_2C CO_2Me poly(1)

MeO_2C CO_2Me poly(7)

Reduction

MeO_2C CO_2Me poly(10)

Scheme 1.

The all-trans samples of poly(**1**) and poly(**7**) were reacted with diimide (NH=NH) generated in situ by heating an excess of *p*-toluenesulfonyl hydrazine in *m*-xylene at 120 °C [10]. This mild reducing agent was shown to fully hydrogenate both the exo- and the endocyclic C=C double bonds [16]. Furthermore, hydrogen insertion occurs stereospecifically on the sterically most hindered face of the cyclopentene ring unit to afford the least encumbered, thermodynamically more stable product poly(**10**) (Scheme 1).

The two samples of poly(2,3-dicarbomethoxynorbornane) obtained by hydrogenation of *all-trans*-poly(2,3-dicarbomethoxynorbornene) and *all-trans*-poly(2,3-dicarbomethoxynorbornadiene) were first analyzed by ^{1}H NMR spectroscopy at 400 MHz. All the various types of alkyl groups in the saturated polymers resonated as rather broad singlets. No fine structure and no tacticity splitting were visible. The main feature of the proton spectra was the absence of any signal in the olefinic region that confirmed the occurrence of a complete reduction of both the exo- and endocyclic C=C double bonds. ^{13}C NMR spectra brought in more valuable information about the polymer microstructures. Carbon atoms linked to the ester groups (C=O, OCH_3, and $C_{5,6}$) appeared as sharp singlets in poly(**10**), as in poly(**1**) and poly(**7**). Thus, the stereoselective *cis*-hydrogenation of the endocyclic double bonds of the poly(norbornadiene-2,3-diester) by diimide did not cause new tacticity splittings in the corresponding polynorbornene or polynorbornane. On the other hand, the reduction of the exocyclic trans junctions led to significant changes in the chemical shifts of the neighboring carbon atoms. Indeed, the peaks of $C_{1,4}$, $C_{2,3}$, and C_7 in poly(**10**) displayed enhanced, well-resolved fine structures compared to their equivalents in poly(**1**) or poly(**7**) (Figure 7). The various components of each signal could be assigned to a specific tacticity by following the line orders determined by Rooney and coworkers at the diad ($\delta_m > \delta_r$) or triad ($\delta_{mm} > \delta_{mr} > \delta_{rr}$) level [16]. In both samples, the lines corresponding to a meso arrangement were the most intense, indicating that the fully hydrogenated derivatives had an isotactic bias and that the parent unsaturated polymers had an *all-trans highly isotactic* structure.

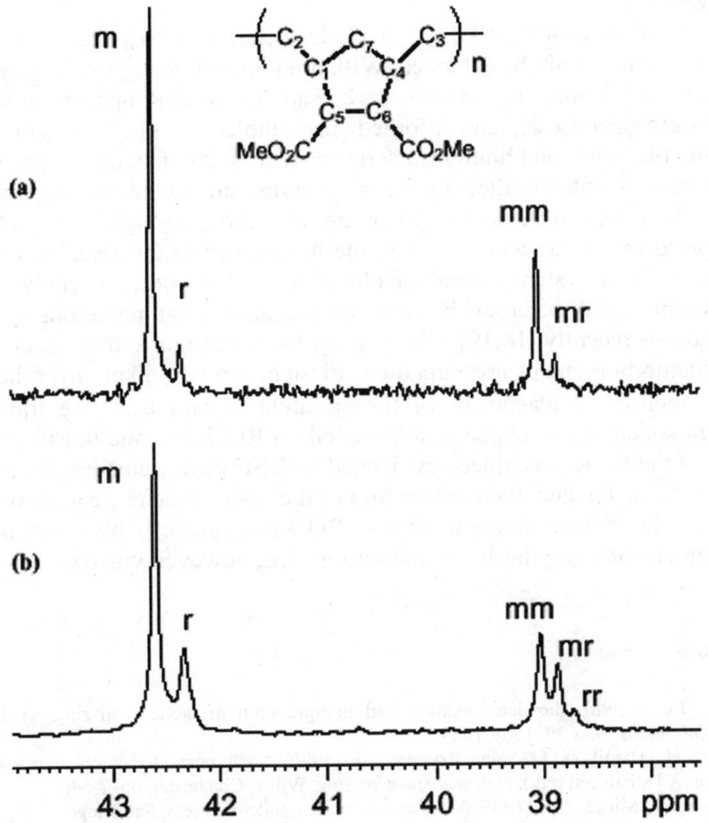

Figure 7. $C_{1,4}$ (left) and C_7 (right) regions of the 100 MHz ^{13}C NMR spectra in CDCl$_3$ of *all-trans*-poly(**10**) derived from (a) *all-trans*-poly(**1**) and (b) *all-trans*-poly(**7**) prepared using as catalyst [RuCl$_2$(p-cymene)]$_2$ in the presence of TMSD.

Although 2,3-dicarbomethoxynorbornadiene **1** and the *exo,exo*-norbornene diester **7** behaved in like manner toward our ruthenium catalytic system, a closer look at the spectra depicted in Figure 7 revealed that a higher degree of tacticity was achieved after polymerization and hydrogenation of the former monomer. Deconvolution and integration of the $C_{1,4}$ NMR signals indicated that for the polynorbornane derived from poly(**1**), the meso/racemic ratio reached 91/9, whereas the sample prepared from poly(**7**) gave a 69/31 value.

3. Conclusion

Various NMR techniques were applied to probe the stereoregularity of polymers prepared by ROMP of 2,3-dicarboalkoxynorbornenes and -norbornadienes catalyzed by the [RuCl$_2$(p-cymene)]$_2$ dimer in the presence of TMSD. Direct methods based on the examination of cross-coupling peaks in polymers derived from optically active monomers, and indirect methods based on the spectroscopic analysis of hydrogenated derivatives from the parent polyolefins led to the same conclusion. They both showed that high trans, highly isotactic polymers were formed using the ruthenium-arene catalyst precursor.

The bulkiness of the alkoxy side chains determined to a large extent the degree of stereocontrol that could be achieved. With monomers bearing small ester groups like the methoxy and ethoxy derivatives **1**, **2**, and **7**, the ring-opening polymerization proceeded stereospecifically and afforded macromolecular products with 100% trans exocyclic double bonds and highly isotactic sequences. An increase in the steric bulk of the alkoxy substituents resulted in the progressive emergence of cis double bonds, together with a loss of tacticity. Such an evolution strongly suggests that ester coordination to the metal center is of prime importance to the catalytic cycle. This is not surprising for a system devoid of phosphine or any other strongly coordinating ligands, like the new N-heterocyclic carbenes that have found numerous applications in olefin metathesis recently [18,19]. Chelation of the ruthenium active sites by the alkoxy groups of the norbornene or norbornadiene diesters is a very likely hypothesis to make up for the thermal displacement of the η^6 arene ligand from the transition-metal coordination sphere. In a related system based on RuCl$_2$(p-cyme-ne)(PR$_3$) complexes, the release of the arene was indeed evidenced by DSC measurements [20] and by NMR observations [21]. To date, the exact nature of the active species generated in situ upon activation of the ruthenium-arene dimer [RuCl$_2$(p-cymene)]$_2$ by a diazo compound cocatalyst and a chelating diester monomer remains, however, elusive.

4. References

1. Hamilton, J.G. (1998) The determination and interpretation of tacticity in ring-opening metathesis polymerization, *Polymer* **39**, 1669-1689.
2. Fawcett, A.H., Hamilton, J.G., and Rooney, J.J. (1996) NMR characterisation of macromolecules in solution, in A.H. Fawcett (ed.), *Polymer Spectroscopy*, Wiley, Chichester, pp. 7-54.
3. Bovey, F.A. and Mirau, P.A. (1996) *NMR of Polymers*, Academic Press, San Diego.
4. Ivin, K.J., Laverty, D.T., and Rooney, J.J. (1977) The ^{13}C NMR spectra of poly(1-pentenylene) and poly(1,3-cyclopentylenevinylene), *Makromol. Chem.* **178**, 1545-1560.
5. Ho, H.T., Ivin, K.J., Reddy, B.S.R., and Rooney, J.J. (1989) Metathesis polymerization of substituted norbornenes: microstructure of the polymers of some ester derivatives, *Eur. Polym. J.* **25**, 805-811.
6. Hamilton, J.G., Rooney, J.J., and Snowden, D.G. (1993) Ring-opening metathesis polymerization of 7-methylnorbornadiene, *Makromol. Chem.* **194**, 2907-2922.
7. Amir-Ebrahimi, V., Corry, D.A., Hamilton, J.G., Thompson, J.M., and Rooney, J.J. (2000) Characteristics of RuCl$_2$(CHPh)(PCy$_3$)$_2$ as a catalyst for the ring-opening metathesis polymerization, *Macromolecules* **33**, 717-724.
8. Noels, A.F. and Demonceau, A. (1998) Metathesis of low-strain olefins and functionalized olefins with new ruthenium-based catalyst systems, in Y. Imamoglu (ed.), *Metathesis Polymerization of Olefins and Polymerization of Alkynes*, Kluwer, Dordrecht, pp. 29-46.
9. Delaude, L., Demonceau, A., and Noels, A.F. (1999) Highly stereoselective ruthenium-catalyzed ring-opening metathesis polymerization of 2,3-difunctionalized norbornadienes and their 7-oxa analogues, *Macromolecules* **32**, 2091-2103.

10. Delaude, L., Demonceau, A., and Noels, A.F. (2003) Probing the stereoselectivity of the ruthenium-catalyzed ring-opening metathesis polymerization of norbornene and norbornadiene diesters, *Macromolecules*, **36**, 1446-1456.
11. McConville, D.H., Wolf, J.R., and Schrock, R.R. (1993) Synthesis of chiral molybdenum ROMP initiators and all-cis highly tactic poly(2,3-(R)$_2$norbornadiene) (R = CF$_3$ or CO$_2$Me), *J. Am. Chem. Soc.* **115**, 4413-4414.
12. Bazan, G.C., Khosravi, E., Schrock, R.R., Feast, W.J., Gibson, V.C., O'Regan, M.B., Thomas, J.K., and Davis, W.M. (1990) Living ring-opening metathesis polymerization of 2,3-difunctionalized norbornadienes by Mo(CH-*t*-Bu)(N-2,6-C$_6$H$_3$-*i*-Pr$_2$)(O-*t*-Bu)$_2$, *J. Am. Chem. Soc.* **112**, 8378-8387.
13. O'Dell, R., McConville, D.H., Hofmeister, G.E., and Schrock, R.R. (1994) Polymerization of enantiomerically pure 2,3-dicarboalkoxynorbornadienes and 5,6-disubstituted norbornenes by well-characterized molybdenum ring-opening metathesis polymerization initiators. Direct determination of tacticity in cis, highly tactic and trans, highly tactic polymers, *J. Am. Chem. Soc.* **116**, 3414-3423.
14. Ivin, K.J., Kress, J., and Osborn, J.A. (1988) Kinetics of initiation and propagation of the metathesis polymerization of the exo Diels-Alder adduct of cyclopentadiene and maleic anhydride initiated by the tungsten-carbene complex W[C(CH$_2$)$_3$CH$_2$](OCH$_2$CMe$_3$)$_2$Br$_2$, *J. Mol. Catal.* **46**, 351-358.
15. Bazan, G.C., Schrock, R.R., Cho, H.-N., and Gibson, V.C. (1991) Polymerization of functionalized norbornenes employing Mo(CH-*t*-Bu)(NAr)(O-*t*-Bu)$_2$ as the initiator, *Macromolecules* **24**, 4495-4502.
16. Amir-Ebrahimi, V., Corry, D.A.K., Hamilton, J.G., and Rooney, J.J. (1998) Determination of the tacticities of ring-opened polymers of symmetrical 5,6-disubstituted derivatives of norbornene and norbornadiene from the ^{13}C NMR spectra of their hydrogenated derivatives, *J. Mol. Catal. A: Chem.* **133**, 115-122.
17. Al-Samak, B., Amir-Ebrahimi, V., Carvill, A.G., Hamilton, J.G., and Rooney, J.J. (1996) Determination of the tacticity of ring-opened metathesis polymers of norbornene and norbornadiene by ^{13}C NMR spectroscopy of their hydrogenated derivatives, *Polym. Int.* **41**, 85-92.
18. Bourissou, D., Guerret, O., Gabbaï, F.P., and Bertrand, G (2000) Stable carbenes, *Chem. Rev.* **100**, 39-91.
19. Herrmann, W.A. (2002) N-heterocyclic carbenes: a new concept in organometallic catalysis, *Angew. Chem. Int. Ed.* **41**, 1290-1309.
20. Hafner, A., Mühlebach, A., and van der Schaaf, P.A. (1997) One-component catalysts for the thermal and photoinduced ring opening metathesis polymerization, *Angew. Chem., Int. Ed.* **36**, 2121-2124.
21. Demonceau, A., Stumpf, A.W., Saive, E., and Noels, A.F. (1997) Novel ruthenium-based catalyst systems for the ring-opening metathesis polymerization of low-strain cyclic olefins, *Macromolecules* **30**, 3127-3136.

METATHESIS AND POLYOLEFIN GROWTH ON CADMIUM SELENIDE SURFACES USING RUTHENIUM-BASED CATALYSTS

M. FIRAT ILKER, HABIB SKAFF, TODD EMRICK,* E. BRYAN COUGHLIN*
University of Massachusetts, Department of Polymer Science and Engineering 120 Governors Drive, Amherst, Massachusetts, 01003, U.S.A.

1. Introduction

The development of highly active and well-defined catalyst systems has driven metathesis chemistry to its current status as a versatile technique for the synthesis of both small organic molecules and polymeric materials.[1] As a part of this progress, metathesis polymerization has become a major technique for the preparation of tailored polymers with desired properties. Through advances in ruthenium-based catalyst systems introduced by Grubbs and coworkers, a high level of synthetic control over polymer architecture can be achieved.[2] This includes the preparation of block copolymers, alternating copolymers, control over functional pendant groups and chain-ends, and tolerance toward functional groups and various reaction media. The robust nature of this class of catalysts has allowed numerous types of monomers to be polymerized, thus providing access to a large range of polymeric materials. Metathesis reactions in heterogeneous media such as emulsions[3] and on various organic or inorganic surfaces have been achieved.[4] The ability to polymerize designer monomers from various material surfaces is an attractive technique for surface modification and preparation of new composite materials.

Studies on composite materials at the nanoscale constitute an important area of research and require novel interdisciplinary techniques. Materials with well-defined nanometer-scale features are expected to achieve enhanced performance in numerous applications. Inorganic colloidal nanoparticles (e.g., Au, Pd, ZnS, CdSe, Co, and Fe_2O_3) afford access to a size scale of 1 to 10 nm and consequently exhibit unique properties intermediate between those of molecular and bulk material.[5] Properties such as photoluminescence (PL), magnetism and electrical transport make them potentially useful for nanoparticle-based devices (biological probes, LED displays, tunable lasers, photovoltaic cells, etc.). However their poor mechanical properties and tendency toward oxidation and aggregation currently limit their usefulness. Preparation of nanoparticle-polymer composites holds the promise of combining the useful properties of both materials into robust devices for a myriad of applications.

For many applications, nanoparticles must be dispersed in a matrix. One approach to achieve this task is to prepare polymers bearing ligands, or side groups, compatible with the surface of the inorganic particle. Bawendi has reported that the use of monomers containing long aliphatic chains that are compatible with the tri-*n*-octyl phosphine oxide (TOPO) ligands covering CdSe nanoparticles resulted in enhanced dispersion in the polymer matrix.[6] In another report Schrock and Thomas described the

Y. Imamoglu and L. Bencze (eds.), Novel Metathesis Chemistry: Well-Defined Initiator Systems for Specialty Chemical Synthesis, Tailored Polymers and Advanced Material Applications, 263–270.

preparation and metathesis polymerization of phosphine and phosphine oxide functionalized norbornene derivatives to prevent aggregation of CdSe nanoparticles within a polymer matrix.[7]

Our approach detailed here centers on the preparation of functional ligands for CdSe nanoparticles, followed by growth of polyolefins through metathesis chemistry that provide a general route for composite preparation, and greatly expands the choice of polymer matrices (Figure 1).[8] The use of ROMP on Au nanoparticle surfaces to prepare Au-polymer hybrids has been reported by Mirkin, Nguyen and coworkers, where they used a norbornene functionalized thiol based ligand on the Au surfaces for the attachment of ruthenium based catalyst and subsequent growth of ferrocene functionalized block copolymers from these surfaces.[9] In our work the use of CdSe nanoparticles required development of a new ligand system, and an exploration of the compatibility of ruthenium based metathesis catalyst with CdSe nanoparticles.

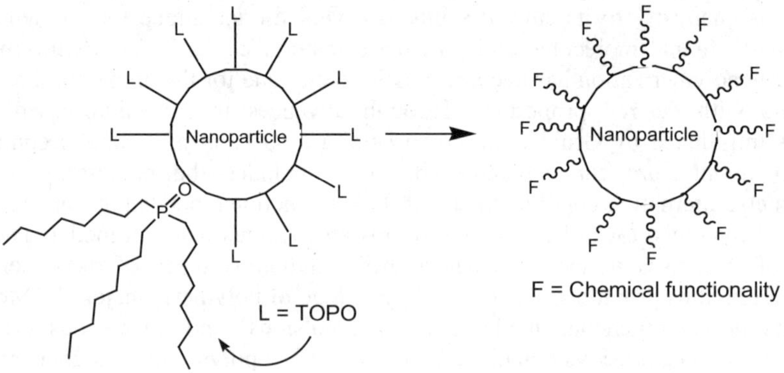

Figure 1. Conventional TOPO (tri-*n*-octyl phosphine oxide) covered nanoparticle (on the left), and nanoparticle with a chemically functional periphery (on the right) that can provide control over ligand-environment interactions such as solubility, miscibility within polymer matrix, etc.

2. Results and Discussion

A general route to grafting polymers to CdSe nanoparticle surfaces is described. CdSe nanoparticles can be prepared by a variety of published reports using high-temperature synthetic methods and phosphorous-based ligands.[10] The resulting nanoparticles are nearly uniform in size, and are typically encapsulated by aliphatic phosphine oxides (e.g., TOPO) that provide surface passivation and solubility in organic solvents.

$$\underset{\text{H}}{\overset{\text{O}}{\text{P}}}(n\text{-octyl})_2 \; + \; [\text{4-vinylbenzyl chloride}] \quad \xrightarrow[\text{THF, 65 C}]{\text{NaH}} \quad \mathbf{1}$$

Equation 1.

Functionalization of the nanoparticle surface with polymerizable moieties was achieved by preparing an olefin functionalized phosphine oxide ligand (**1**), (Equation 1). Prior to attaching the catalyst onto surfaces, the benzylidene exchange equilibrium reaction between first generation Grubbs' catalyst (**2**) and compound **1** was investigated (Equation 2). ^{1}H NMR spectra of mixtures of **1** and **2** showed that the equilibrium shifts toward the ligand functionalized ruthenium benzylidene **3**. This reaction is suggesting that attachment of **2** to CdSe nanoparticles covered by **1** will be efficient. Additionally, ^{31}P NMR studies do not indicate coordination of the phosphine oxide ligand (**1** or TOPO) to the ruthenium metal center.

Ligand **1** was introduced on nanoparticle surfaces by ligand exchange chemistry. Through this procedure TOPO-functionalized nanoparticles were stripped of their TOPO periphery by precipitating into methanol, dissolved in excess pyridine to replace the TOPO ligand, followed by another precipitation into hexanes, resulting in insoluble pyridine-passivated crystals. When the crystals were slurried in a refluxing dry THF solution of ligand **1** for several hours they were carried back to a homogeneous solution state after being covered by the new ligand. The presence of **1** and the absence of TOPO were confirmed using ^{1}H and ^{31}P NMR spectra. It was possible to precipitate these nanoparticles into methanol and subsequently dissolve in organic solvents while preserving the functional ligand periphery.

The attachment of Grubbs' catalyst onto styrenic units of ligand **1** on the nanoparticle surfaces was achieved by the addition of **2** to nanoparticles dissolved in a suitable solvent (e.g., chloroform, methylene chloride, or toluene). Approximately 20 minutes after this addition, ^{1}H NMR spectroscopy clearly showed

1 : 2	3 : 2	
(feed ratio)	10 min	30 min
0.5	0.53	0.54
1	1.41	1.52
2	2.73	3.10
50	>100	>100

Equation 2. Equilibrium reaction between compound **1** and catalyst **2** resulting in compound **3** and a styrene unit. The table provides the ratio of compound **3** to catalyst **2** at different reaction times at different reaction feed ratios. According to this data the equilibrium favors the formation of compound **3**.

the ratio of free catalyst, to surface-bound catalyst and the ratio of ligand **1** with or without the attached catalyst (Figure 2). At high catalyst loadings it was possible to cover the entire periphery with ruthenium catalyst. However, the high activity of Grubbs' catalyst system and the presence of cross-metathesis allowed for the use of small amounts of **2** relative to nanoparticle and ligand concentration. When a desired amount of cyclic olefin monomer, in this case cyclooctene was added to this solution, the ligand functionalized surface of nanoparticle is then covered by polycyclooctene. ^{1}H NMR spectra showed the absence of end-group methylene units on the ligand **1** revealing the full coverage of the nanoparticles by polymer chains due to successive chain transfer to styrenic units on the surface (Figure 3). When second generation Grubbs' catalyst[11] is employed in the same procedure, much smaller amounts of catalyst loading was sufficient to obtain similar results, due to the higher activity of this catalyst.[12]

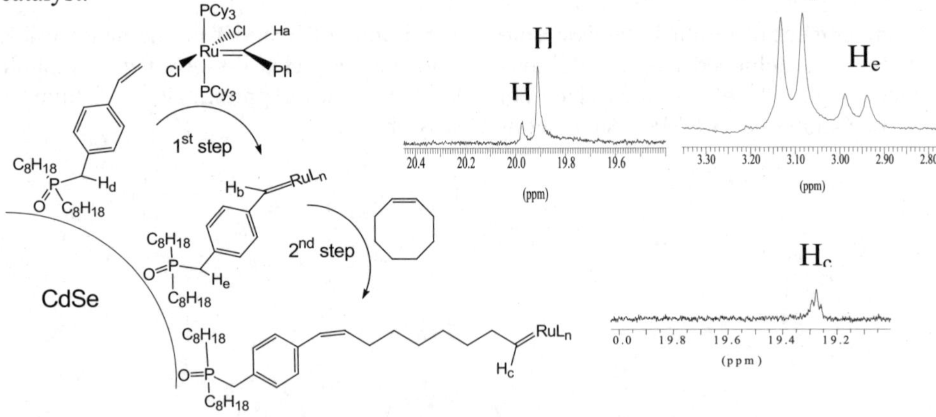

Figure 2. Selected regions of ^{1}H NMR spectra showing the shift of the ruthenium benzylidene proton from free catalyst **2** (H_a), to catalyst attached to **1** (H_b) on the first step, and catalyst at the growing chain end (H_c) on the second step. The methylene unit on the compound **1** shows the free of catalyst and catalyst bound states (H_d and H_e respectively).

Ethyl vinyl ether was used to quench the polymerization to avoid extensive chain-transfer and crosslinking. The resulting product after quenching was precipitated by slow addition of methanol to remove residual catalyst that interferes with the emission spectra of the nanoparticles. By changing the catalyst, monomer and nanoparticle concentrations, it was possible to prepare polymers of number average molecular weights ranging from 10,000-50,000 g/mol attached to nanoparticle surfaces with polydispersity index (PDI) values near 2.

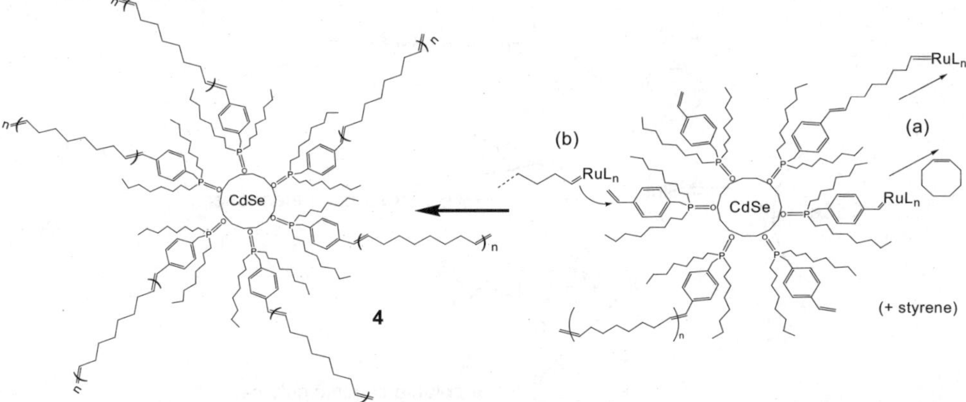

Figure 3. The complete attachment of polycyclooctene on the ligands of nanoparticle through polymer growth radially outward (a), and chain transfer to ligands on the nanoparticle surface (b). The presence of styrene comes from the benzylidene exchange-catalyst attachment step of the catalyst **1** prior to monomer addition.

The resulting nanoparticle-poly(cyclooctene) composite could be spin or solution cast into homogeneous thin films. Transmission Electron Micrography (TEM) analysis of these films showed that there was no aggregation of nanoparticles inside the polymer matrix (Figure 4). However a control experiment by polymerizing cyclooctene in the presence of TOPO covered non-functionalized nanoparticles resulted in strong aggregation of particles in the solid polymer matrix. The photoluminescence spectrum of the nanoparticle-polycyclooctene composite (**4**) showed a narrow luminescence profile. The above-

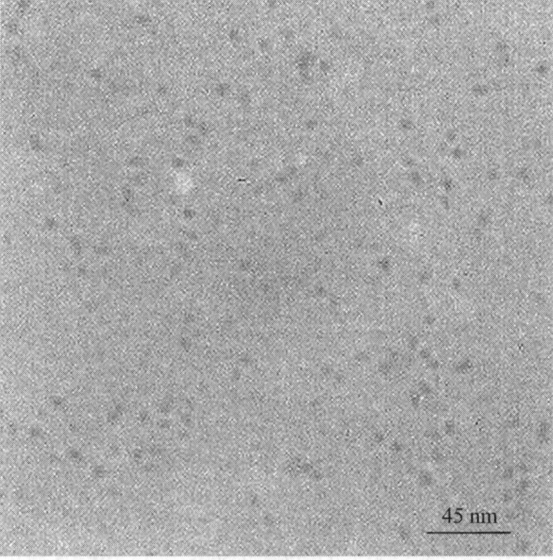

Figure 4. TEM micrograph of composite **4**.

described procedure has shown good versatility, and has been employed for the polymerization of various cyclic olefins to prove the generality of this method. In the literature, mono-substituted or 2,3-disubstituted norbornene derivatives have been used extensively to prepare polymers with various functional groups. To display the generality of introducing chemical functionalities into CdSe-polyolefin composites, several 2,3-disubstituted norbornene derivatives bearing anhydride or imide groups

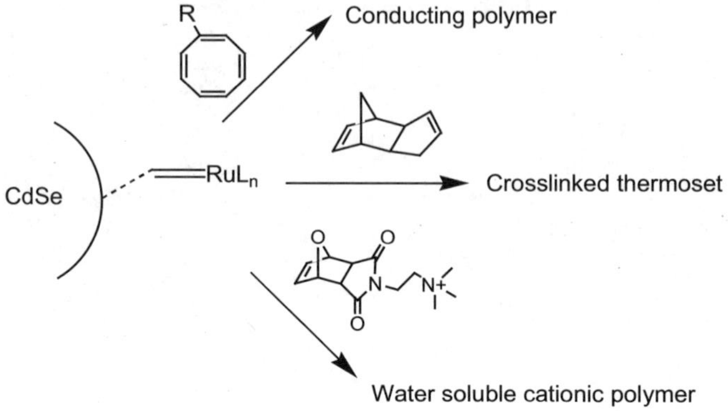

Figure 5. Examples of functional polymers that can be attached on nanocrystal surfaces using ROMP.

have been polymerized from nanoparticle surfaces (Figure 5). In another example dicyclopentadiene has been used to dissolve olefin functionalized nanoparticles and polymerized via addition of metathesis catalyst to produce robust crosslinked matrices with luminescent CdSe nanoparticles dispersed throughout. There have been many reports describing preparations of polyacetylene or polyphenylenevinylene derivatives using metathesis polymerization[13] that can be adapted to the procedure used above. The ongoing research in our laboratories includes the preparation of composite materials with the polymer component exhibiting semiconductor or luminescent properties.

3. Conclusion

In summary, we have demonstrated the attachment of ruthenium-based metathesis catalysts on olefin functionalized CdSe nanoparticle surfaces and polymer growth radially outward from these surfaces. The well-defined chemical nature of the ligand and the catalyst system allowed each step of this preparation technique to be monitored by analytical techniques such as NMR spectroscopy, TEM, UV adsorption and emission spectra. Hence it was possible to probe the nature of the growth process and the resulting composite material. This method provides control over the periphery of nanoparticle that dictates the interactions of nanoparticle with its surrounding such as solubility, resistance against oxidation, miscibility in a matrix, etc. It is important to note that many other nanoparticles are prepared using similar phosphine oxide based ligand environments. The simplicity of the described procedure and the robust nature of ruthenium-based metathesis catalyst system may allow for the application of this methodology to various nanoparticles. On the other hand, ROMP provides access to a wide range of polymeric materials that exhibit interesting mechanical, optical or electronic properties. When these factors are considered, the method reported here may be considered as a general route to prepare nanoparticle-polymer composites and explore the outcome of combining the complimentary properties of these components.

Acknowledgement

The authors thank the University of Massachusetts, Amherst and the NSF-sponsored Materials Research Science and Engineering Center (DMR 9809365) at UMass for financial support.

References

1   Furstner, A. (2000) Olefin metathesis and beyond, *Angewandte Chemie International Edition* **39**, 3012-3043.

2   (a) Buchmeiser, M.R. (2000) Homogeneous metathesis polymerization by well-defined group VI and group VIII transition-metal alkylidenes: Fundamentals and applications in the preparation of advanced materials, *Chemical Reviews* **100**, 1565-1604. (b) Trnka, T.M. and Grubbs, R.H. (2001) The development of $L_2X_2Ru=CHR$ olefin metathesis catalysts: An organometallic success story, *Accounts of Chemical Research* **34**, 18-29. (c) Frenzel, U. and Nuyken, O. (2002) Ruthenium based metathesis initiators: Development and use in ring-opening metathesis polymerization, *J. Polymer Science Part A: Polymer Chemistry* **40**, 2896-2916.

3   Claverie, J.P., Viala, S., Maurel, V. and Novat, C. (2001) Ring-opening metathesis polymerization in emulsion, *Macromolecules* **34**, 382-388.

4   (a) Sinner, F. and Buchmeiser, M.R. (2000) A new class of continuous polymer supports prepared by ring-opening metathesis polymerization: A straightforward route to functionalized monoliths, *Macromolecules* **33**, 5777-5786. (b) Kim, N.Y., Jeon, N.L., Choi, I.S., Takami, S., Harada, Y., Finnie, K.R., Girolami, G.S., Nuzzo, R.G., Whitesides, G.M. and Laibinis, P.E. (2000) Surface initiated ring-opening metathesis polymerization on Si/SiO_2, *Macromolecules* **33**, 2793-2795.

5   Trinade, T., O'Brien, P. and Pickett, N.L. (2001) Nanocrystalline semiconductors: Synthesis, properties, and perspectives, *Chemistry of Materials* **13**, 3843-3858.

6   Lee, J., Sundar, V., Bawendi, M.G. and Jemsen, K.F. (2000) Full color emission from II-VI semiconductor quantum dot-polymer composites, *Advanced Materials* **12**, 1102-1105.

7   Fogg, D.E., Radzilowski, L.H., Blanski, R., Schrock, R.R. and Thomas, E.L. (1997) Fabrication of quantum dot/polymer composites: Phosphine-functionalized block copolymers as passivating hosts for Cadmium Selenide nanoclusters, *Macromolecules* **30**, 417-426.

8   Skaff, H., Ilker, M.F., Coughlin, E.B. and Emrick, T. (2002) Preparation of Cadmium Selenide-polyolefin composites from functional phosphine oxides and ruthenium-based metathesis, *J. American Chemical Society* **124**, 5729-5733.

9   Watson, K.J., Zhu, J., Nguyen, S.T. and Mirkin, C.A. (1999) Hybrid nanoparticles with block copolymer shell structures, *J. American Chemical Society* **121**, 462-463.

10   (a) Murray, C.B., Norris, D.J. and Bawendi, M.G. (1993) Synthesis of characterization of nearly dispersed CdE (E = sulfur, selenium, tellurium) semiconductor nanocrystallites, *J. American Chemical Society* **115**, 8706-8715. (b) Bowen Katari, J.E., Colvin, V.L. and Alivisatos, A.P. (1994) X-ray photoelectron spectroscopy of CdSe nanocrystals with applications to studies of the nanocrystal surface, *J. Physical Chemistry* **98**, 4109-4117. (c) Manna, L., Schere, E. and Alivisatos, A.P. (2000) Synthesis of soluble and processable rod-, arrow-, teardrop-, and tetrapod-shaped CdSe nanocrystals, *J.American Chemical Society* **122**, 12700-12706. (d) Peng, Z.A. and Peng, X.G. (2001) Formation of high-quality CdTe, CdSe and CdS nanocrystals using CdO as precursor, *J. American Chemical Society* **123**, 183-184.

11 (a)-(tricyclohexylphosphine)(1,3-dimesitylimidazolydene-2-ylidene)benzylideneruthenium dichloride.

12 Bielawski, C.W. and Grubbs, R.H. (2000) Highly efficient ring-opening metathesis polymerization (ROMP) using new ruthenium catalysts containing N-heterocyclic carbene ligands, *Angewandte Chemie International Edition* **39**, 2903-2906.

13 (a) Scherman, O.A. and Grubbs, R.H. (2001) ROMP of 1,3,5,7-cyclooctatetraene (COT) with a ruthenium olefin metathesis catalyst coordinated with a N-heterocyclic carbene ligand, *Abstracts of Papers of the American Chemical Society, PMSE Part-2* **221**, 331. (b) Wagaman, M.W. and Grubbs, R.H. (1997) Synthesis of organic and water soluble poly(1,4-phenylenevinylenes) containing carboxyl groups: Living ring-opening metathesis polymerization (ROMP) of 2,3-dicarboxybarrelenes, *Macromolecules* **30**, 3978-3985. (c) Miao, Y. and Bazan, G.C. (1994) Paracyclophene route to poly(*p*-phenylenevinylene), *J. American Chemical Society* **116**, 9379-9380.

ACYCLIC DIENE METATHESIS CONDENSATION (ADMET) OF 1,2-DIVINYLFERROCENE (DVFC)

Christine Wirth-Pfeifer, Armin Michel, K. Weiss
University of Bayreuth, Department of Inorganic Chemistry,
D-95447 Bayreuth, Germany
e-mail: karin.weiss@uni-bayreuth.de

Summary

We found a route to synthesize poly(1,2-divinylferrocene) via **A**cyclic **D**iene **Met**hathesis (**ADMET**) condensation. The lewis acid free Schrock type Mo(VI) carbene complex 2,6-diisopropylphenylimido-neophylidene-molybdenum-bis(hexafluor-t-butoxid) [**Mo**] proved to be the most active catalyst for the metathesis reaction of the olefin substituted ferrocene monomer 1,2-divinylferrocene (**DVFC**). The ADMET reaction yields at 0.1hPa pressure a dark brown metallic product with $P_n = 6$.

We also tested the copolymerisation of **DVFC** with 1,9-decadiene. The condensation with 1,9-decadiene gives a high molecular weighted polymer ($M_n = 26000$) which is completely soluble in polar and nonpolar solvents.

I. Introduction

In the last 20 years metal atom containing polymers have become important classes of polymers [1]. Properties like high thermic stability, electric and photo conductometry make them very interesting for producing films, fibres and coatings [2].
Many of these compounds can be synthesized by conventional methods [3]. For synthesizing poly(vinyl)ferrocene radicalic, cationic, anionic and Ziegler-Natta initiated polymerisation of vinylferrocene and 1,1'-divinylferrocene were used [4-7].
Ken Wagener was the first to use Acyclic Diene Metathesis Condensation (ADMET) for condensation of dienes [8].
 In 1993 Boncella et al. succeeded in synthesizing poly(ferrocenylene)vinylenes (**PFV**) via ADMET condensation of 1.1'-divinylferrocene and 1,1'-di-tert-butyl-3,3-divinylferrocene with the lewis acid free methesis catalyst [W=CHC(CH$_3$)$_2$R(N-2,6-C$_6$H$_3$-^{I}Pr)$\{$OCCH$_3$(CF$_3$)$_2\}_2$] [**W**] (R = CH$_3$, Ph) [9]. The polymerisation yielded an oligomer product with $P_n = 4$.

II. Results

In 1997 we found a synthetic route for poly(1,2-divinylferrocenes) using ADMET condensation reactions of the monomer 1,2-divinylferrocenen (**DVFC**).

Y. Imamoglu and L. Bencze (eds.), Novel Metathesis Chemistry: Well-Defined Initiator Systems for Specialty Chemical Synthesis, Tailored Polymers and Advanced Material Applications, 271–276.

Monomer synthesis

The monomer was synthesized by U.H.F. Bunz et al., Max-Planck-Institut (MPI) für Polymerforschung in Mainz [10].

1,1'-Ferrocenedicarbaldehyde was added to a suspension of *instant ylide* (a mixture of (methyltriphenyl)phosphonium bromide and sodium amide) in anhydrous thf. Aqueous workup and distillation yields **DVFC** as a dark orange coloured oil after 18h reaction time.

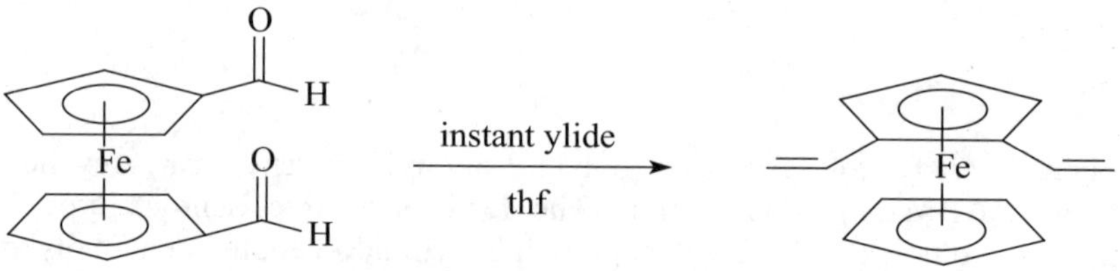

Fig. 1. monomer synthesis of 1,2-divinylferrocene (**DVFC**)

The Schrock type catalyst [M=CHC(CH$_3$)$_2$Ph(N-2,6-C$_6$H$_3$-^{i}Pr$_2$){OCCH$_3$(CF$_3$)$_2$}$_2$] (M = W oder Mo) proved to be the most capable catalysts for the metathesis condensation of olefin substituted ferrocenyl compounds. For the condensation of 1,2-divinylferrocene the lewis acid free Mo(VI) carbene complex Mo=CHC(CH$_3$)$_2$Ph(N-2,6-C$_6$H$_3$-^{i}Pr$_2$){OCCH$_3$(CF$_3$)$_2$}$_2$ [**Mo**] was used as catalyst.

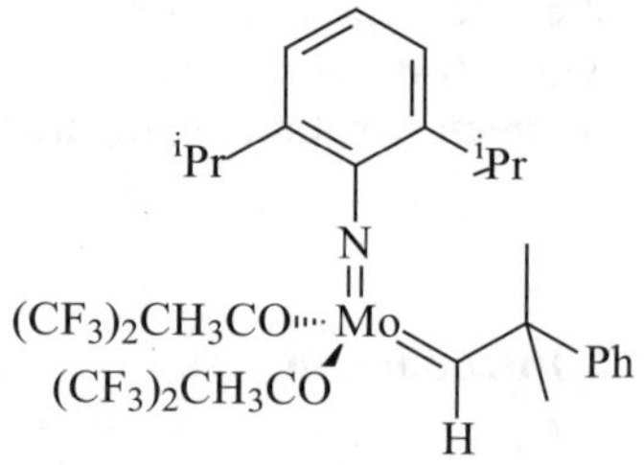

Fig. 2. Mo(VI) complex [**Mo**]

The stepwise ADMET condensation of 1,2-divinylferrocene **DVFC** with the Schrock catalyst [**Mo**] at 40°C formed 52% of dark brown metallic glittering product within 72h. The polymer is partially soluble in polar solvents. The GPC measurements (PS standard) provides the molecular weight M_n = 1200 g/mol (P_n = 6) with the polydispersity D = 2.7 [11] (see table).

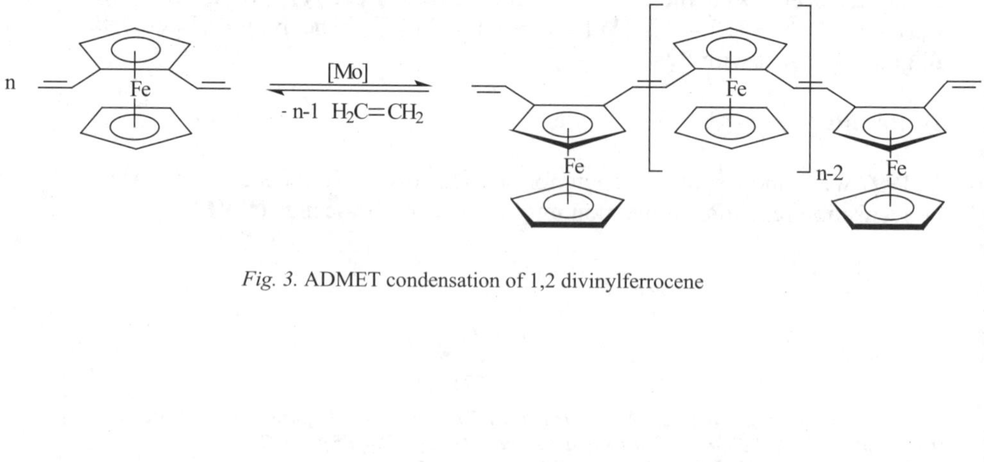

Fig. 3. ADMET condensation of 1,2 divinylferrocene

In the olefin area of the ^{13}C-NMR of the polymer product the signals of the endstanding vinyl groups (δ = 132.9ppm) as well as the signals of the internal double bond between the Cp rings (δ = 123.2ppm, *trans*-Cp-CH=CH-Cp; δ = 122.8ppm, *cis*-Cp-CH=CH-Cp) can be detected.

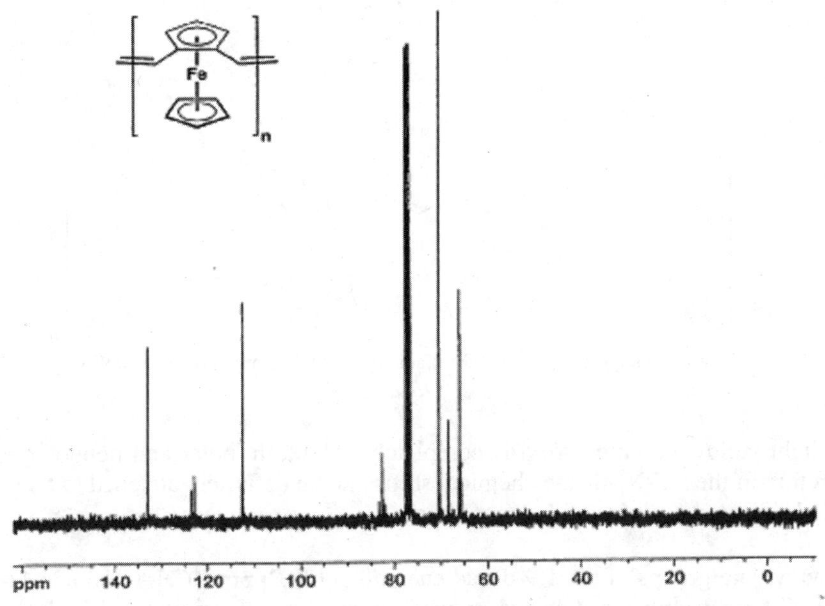

Fig. 4. ^{13}C-NMR spectra of poly(1,2-divinylferrocene) (*63MHz, CCCl₃, 25°C*)

The low *trans* rate of 48% is untypical for methathesis condensation with the catalyst [**Mo**]. The ADMET polymerisation with 1,9-decadiene with this catalyst shows a *trans* rate of above 80%. This results in the catalytic isomerisation of the internal olefins during the reaction.

The oligomers built during the ADMET condensation of 1,2-divinylferrocene are unsoluble in the monomer and precipitate irreversibly before the isomerisation from *cis* to *trans* can take place. Because of the 1,2-position of the divinyl ligands of the monomer the structure is more rigid which causes a steric hindrance for the isomerisation.

The MALDI-TOF spectra shows mass peaks up to 3200 mass units which derive from polymer chains with up to 16 repeating units.

We also tested an ADMET condensation of 1,2-divinyl-ferrocene together with 1,9-decadien. The condensation of **DVFC** with 1,9-decadien (molar ratio: 1 : 20) produces after 16h reaction time a high molecular weighted polymer (M_n = 26000g/mol, D = 2.2) [11].

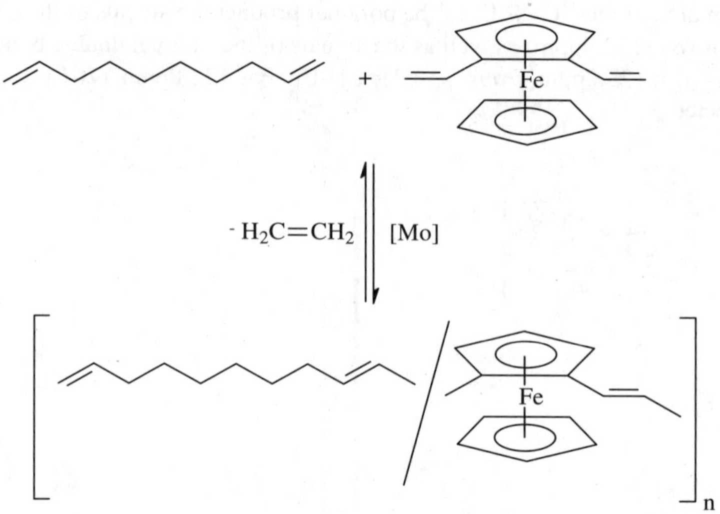

$$- H_2C=CH_2 \quad [Mo]$$

Fig. 5. copolymerisation of 1,9-decadiene with 1,2-divinylferrocene (**DVFC**)

The light yellow coloured solid is completely soluble in polar and nonpolar solvents. Therefore in the ^{13}C-NMR the chemical shifts can be definitely attached to the different sp^2-hybridizated C atoms of the copolymer.

The vinyl end groups of the 1,9-decadiene at $\delta = 139.2$ppm ($-CH=CH_2$) and 114.1ppm ($-CH=CH_2$) can only be detected as weak signals which points to a high molecular weighted product. The internal double bonds of the diene can be attached to the signals at $\delta = 130.3$ppm (*trans*-CH=CH-) and 129.9ppm (*cis*-CH=CH-). The resulting trans rate is 86%.

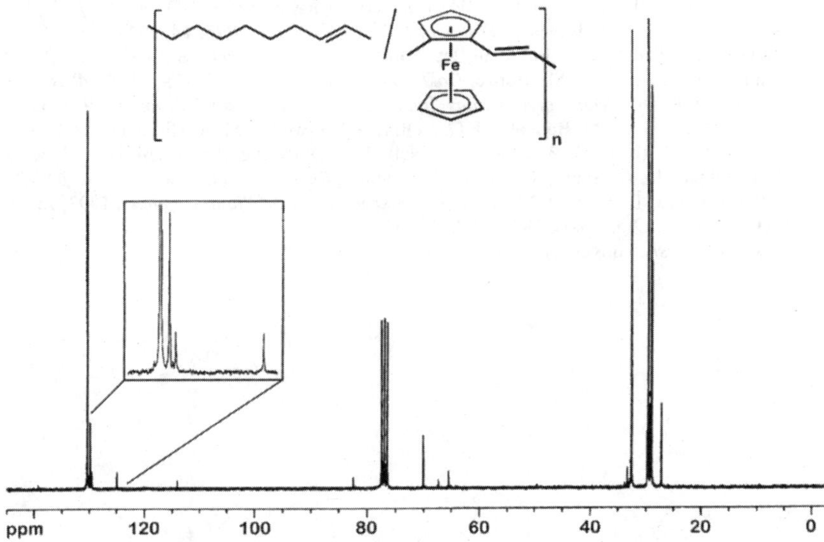

Fig. 6. ^{13}C-NMR spectra of the 1,9-decadiene/1,2-divinylferrocene copolymer (*63MHz, CCCl$_3$, 25°C*)

In the ^{1}H-NMR spectra the integrals of the connecting bond and the internal CH=CH hydrogen atoms exhibit a ratio 1,2-divinylferrocene : 1,9-decadiene of 1 : 29 in the copolymer. The 1,2-divinylferrocene has a rate of 2.2% in the copolymer.

Table. ADMET condensation of 1,2-divinylferrocene and Co-ADMET-condensation with 1.9-decadiene with the Mo(VI) carbene complex [**Mo**].

DVFC/[Mo]/ 1,9-decadiene	solvent	pressure [hPa]	temp. [°C]	time [h]	yield [%] polymer	M$_w$(PS) polymer	M$_n$(PS) polymer	D
30 / 1 / -	-	0.1	25-45	24	35	340	280	1.2
50 / 1 / -	-	0.1	40	72	51	3200	1200	2.7
50 / 1 / 1000	pentane	0.1	25-60	16	88	56500	26000	2.2

References:

[1]   J. E. Sheats, C. E. Carraher, C. U. Pittman, *Metal Containing Polymer Systems*, Plenum Press, New York **1985**, p. 1.

[2]   (a) J. Bill, F. Aldringer, *Adv. Mater.* **1995**, *7*, 775. (b) J. P. Pillot, *Chem. Rev.* **1995**, *95*, 1443. (c) M. Peuckert, T.Vaahs, M. Brück, *Adv. Mater.* **1990**, *9*, 399.

[3]   E. W. Neuse, *Adv. Macromol. Chem.* **1968**, *1*, 1.

[4]   F.S. Arimoto, A.C. Haven, Jr., *J. Am. Chem. Soc.* **1955**, *77*, 6295.

[5]   C. Aso, T. Kunitake, T. Nakashima, *Macromol. Chem.* **1969**, *124*, 232.

[6]   O. Nyken, V. Burkhardt, C. Hübsch, *Macromolecular Reports* **1996**, A33 (*SUPPL. 1*), 13.

[7]   (a) M. H. George, G.F. Hayes, *J. Pol. Sci., Polym. Chem. Ed.* **1975**, *13*, 1049. (b) Y. Sasaki, L. L. Walker, E. L. Hurst, C. U. Puittman, Jr., *J. Pol. Sci., Polym. Chem. Ed.* **1973**, *11*, 1213. (c) S. L. Sosin, V. V. Korshak, T. M. Frunze, *Dokl. Akad. Nauk SSSR* **1968**, *179*, 1124. (d) S. L. Sosin, L. V. Dzhashi, *Kinet. Mech. Polyreactions, Int. Symp. Macromol.Chem. Prepr.* **1969**, *1*, 327.

[8]   (a) K. B. Wagener, J. M. Boncella, J. G. Nell, R. P.Duttweiler, M. A. Hillmeier, *Makromol. Chem.* **1990**, *191*, 365. (b) K. B. Wagener, J. G. Nell, R. P. Duttweiler, M. A. Hillmeier, J. M. Boncella, J. Konzelman, D. W. Smith, R. Puts, L. Willoughby, *Rubber Chem. Technol.* **1991**, *64*, 83.

[9]   A. S. Gamble, J. T. Patton, J. M. Boncella, *Makromol. Chem. Rapid Commun.* **1993**, *13*, 109.

[10]   U.H.F. Bunz, *J. Org. Chem.* **1995**, *494*, C8 – C11.

[11]   A. Michel, *thesis*, University Bayreuth, **1997**.

ACYCLIC DIYNE METATHESIS CONDENSATION OF 3-ALKYL-2,5-DI (1-PROPYNYL)-THIOPHENES

KARIN WEISS, ARMIN MICHEL AND KATRIN SATTLER

Department of Chemistry I, University of Bayreuth

D-95440 Bayreuth, Germany

e-mail: karin.weiss@uni-bayreuth.de

Summary

The Schrock Type tungsten(VI) carbyne complex $(^tBuO)_3W\equiv C^tBu$ catalyzes the Acyclic Diyne Methasesis Condensation (ADIMET) of conjugated and nonconjugated diynes. Substituted conjugated monomers formed polymers with conjugated $C\equiv C$-bonds. These products possess very interesting optical and electronic qualities. Especially the synthesis of poly(phenylene) ethynylenes (PPE) und poly(thiophenylene) ethynylenes (PTE) were studied.

Introduction

Schrock et al. synthesized many carbene und carbyne complexes which contain transission metals in a high oxidation state. The very strong metall carbene carbon π backbonding is responsible for the nucloephilic character of the carbene and carbyne carbon atoms. As there is no $-I$ substituent the nucleophilic character is strengthened. Because of this fact these complexes have high metathesis activity. Schrock et al. were engaged intensivly in stöchiometric and catalytic metathesis reactions of disubstituted alkynes with the tungsten(VI) carbyne complex $(^tBuO)_3W\equiv C^tBu$ as catalyst.

$$2\ R_1\!-\!\!\equiv\!\!-R_2 \quad \underset{\longleftarrow}{\overset{\text{Kat}}{\longrightarrow}} \quad R_1\!-\!\!\equiv\!\!-R_1 \ + \ R_2\!-\!\!\equiv\!\!-R_2$$

Figure 1: *Metathesis of Alkynes.*

They discovered, that monosubstituted alkynes can produce polyalkynes by metathesis condensation.

Weiss and Auth found, that 1-alkynes predominantly polymerize with tungsten(VI) carbyne complexe $(^tBuO)_3W\equiv C^tBu$ as catalyst [i,ii]. In 1995 Auth started to study the polymerization of nonconjugated diynes with this catalyst. She devolped a similar reaction to the ADMET condensation, the acyclic diyne metathesis condensation (ADIMET). Diyens with methyl and ethyl endgroups were tested for acyclis diyne metathesis. The reaction conditions for ADIMET were the same as for the ADMET reaction [iii].

Y. Imamoglu and L. Bencze (eds.), Novel Metathesis Chemistry: Well-Defined Initiator Systems for Specialty Chemical Synthesis, Tailored Polymers and Advanced Material Applications, 277–284.
© 2003 *Kluwer Academic Publishers. Printed in the Netherlands.*

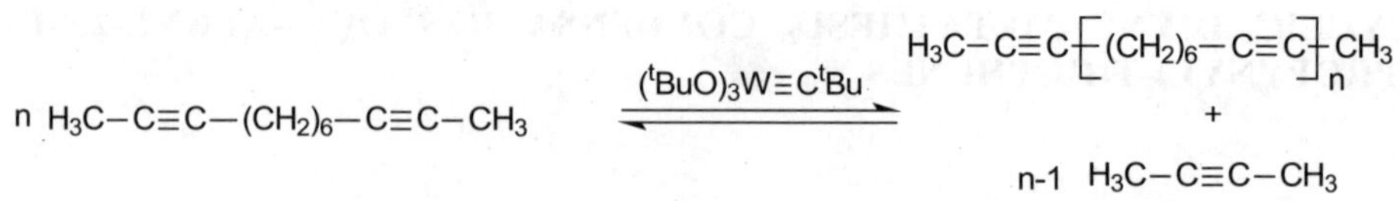

$$n \; H_3C-C\equiv C-(CH_2)_6-C\equiv C-CH_3 \quad \xrightarrow{(^tBuO)_3W\equiv C^tBu} \quad H_3C-C\equiv C\!-\!\!\left[\!(CH_2)_6-C\equiv C\right]_n\!\!-CH_3$$

$$+ \quad n\text{-}1 \; H_3C-C\equiv C-CH_3$$

Figure 2: ADIMET Condensation of 2,10-dodecadiyne.

A. Michel started in cooperation with K. Müllen and U. Bunz et al. the syntheses of poly(phenylene) ethynylene. Conjugated bispropynylbenzene derivatives were polymerizied in an ADIMET condensation with $(^tBuO)_3W\equiv C^tBu$ as catalyst. Poly(phenylene) ethynylenes with alkylgroups in position 1 and 4 were produced by this reaction [iv].

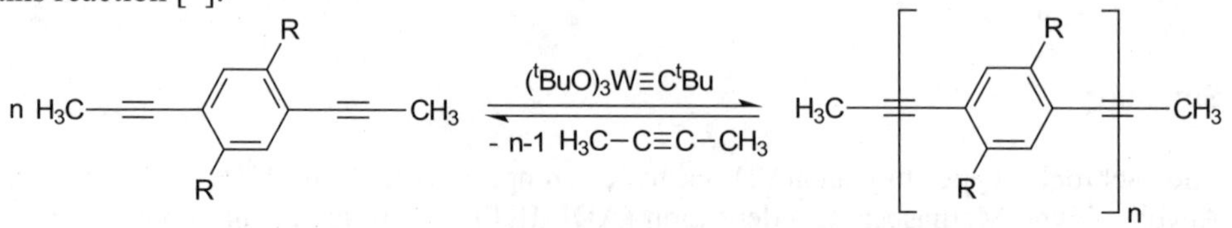

Figure 3: Synthesis of poly(phenylene) ethynylenes with ADIMET condensation.

This synthetic route has some advantages to the conventionall methods of producing poly(phenylene) ethynylenes. Because the Heck Cassar Sonogashira reaction can produce sidereactions and palladium and phosphorus impurities may be found in the products [v, vi, vii].

Other methods for the syntheses of poly(phenylene) derivatives are electrochemical polymerization which results thin polymer films [viii, ix], the autopolymerization of 1-copper-1-ethynyl-4-iodobenzene [x, xi] and the dehydrobromination of PPV films [xii]. All these methods produced polymer products with insoluble impurities. In contrast to them the acyclic diyne metathesis condensation yields purer poly(phenylene) derivatives.

Michel also studied the ADIMET condensation poly(thiophenylene) ethynylenes. The reaction conditions were similar to the acyclic diyne methatesis condensation of the bispropynylbenzene derivatives [8, xiii, xiv, xv]. But with this monomer only an insoluble and low moleculare polymer was produced.

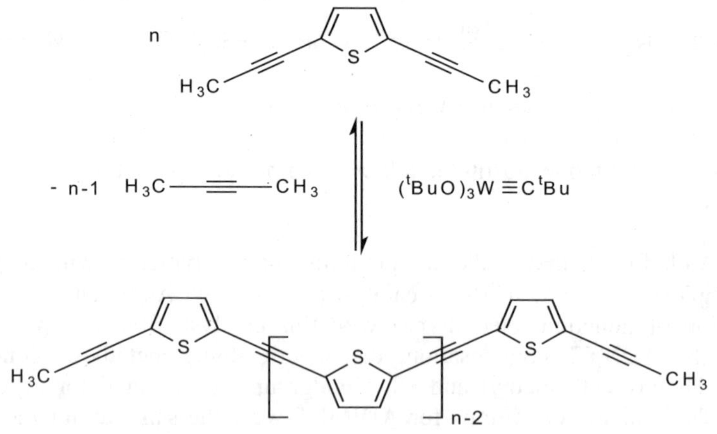

Figure 4: ADIMET Condensation of 2,5-di(1-propynyl) thiophene with $(^tBuO)_3W\equiv C^tBu$.

Results

Poly(thiophenylene) ethynylenes are used for antielectrostatic coatings, light emitting diodes und molecular wires, because of their optical and electronic qualities [xvi,xvii,xviii].

We synthesized new monomers with propyl and octyl groups in position 3 of the thiophene to get soluble poly(thiophenylene) derivatives.

Monomer Synthesis

3-Alkylthiophenes **1**, **2** were synthesized by coupling the Grignard reagent of 1-bromoalkyl with 3-bromothiophene using [1,3-bis-(diphenylphosphino)propane] nickel(II) chlorid as catalyst [xix].

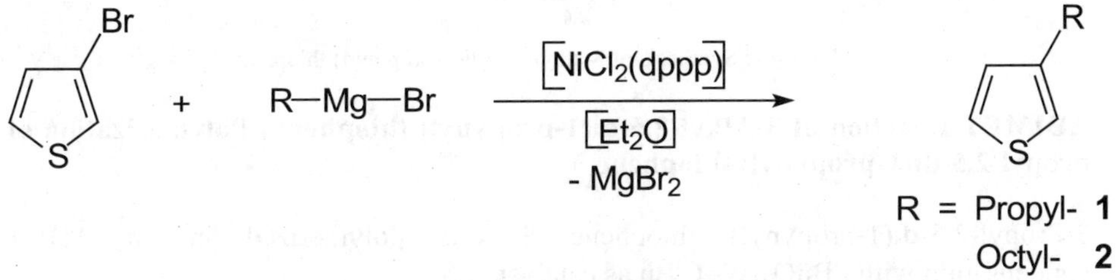

Figure 5: Synthesis of 3-alkylthiophene.

3-Propylthiophene **1** and I_2 were added to a mixture of HIO_3, acetic acid, CCl_4, H_2SO_4 and H_2O. The mixture was refluxed for 4h [xx]. 2,5-Diiod-3-propylthiophene **3** was yielded.

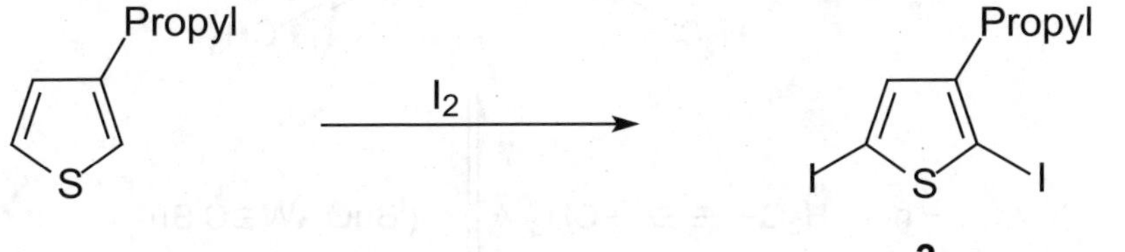

Figure 5: Synthesis of 2,5-diiod-3-propylthiophene.

2,5-Diiod-3-octylthiophene **4** was synthesized by the reaction of 3-octyl thiophene **2** and I_2. They were added to a mixture of chloroforme, HNO_3 and H_2O and refluxed for 16h [xxi,xxii,xxiii].

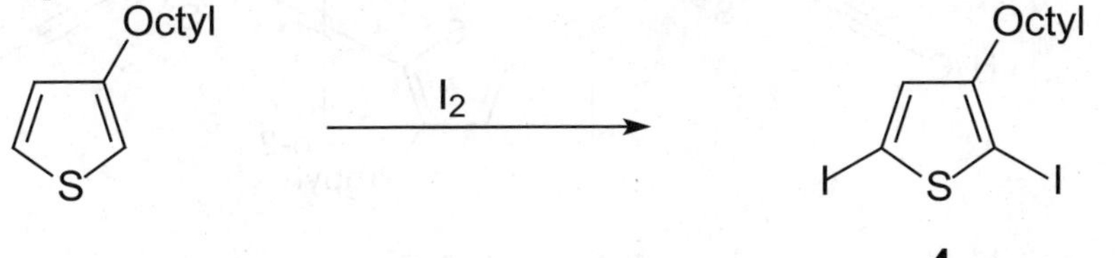

Figure 6: Synthesis of 2,5-diiod-3-octylthiophene.

280

2,5-Diiod-3-alkylthiophene **3**, **4**, CuI and HI were dissolved in NEt$_3$. Propyne was passed in for 4h, while the mixture was heated at 40°C by reflux (-40°C) [6,7,8]. 3-Propyl-2,5-i(1-propynyl) thiophene **5** and 3-octyl-2,5-di(1-propynyl) thiophene **6** were yielded.

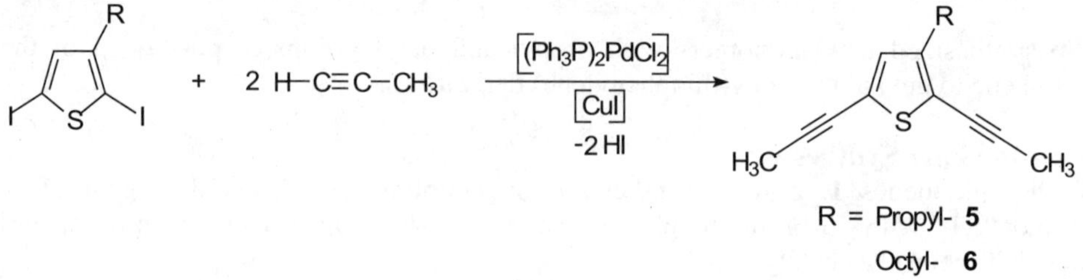

Figure 7: Synthesis of 3-alkyl-2,5-di(1-propynyl) thiophene.

ADIMET Reaction of 3-Alkyl-2,5-di(1-propynyl) thiophenes Polymerization of 3-propyl-2,5-di(1-propynyl) thiophene 5

3-Propyl-2,5-di(1-propynyl) thiophene **5** was polymerized in an ADIMET condensation with (^{t}BuO)$_3$W≡C^{t}Bu as catalyst.

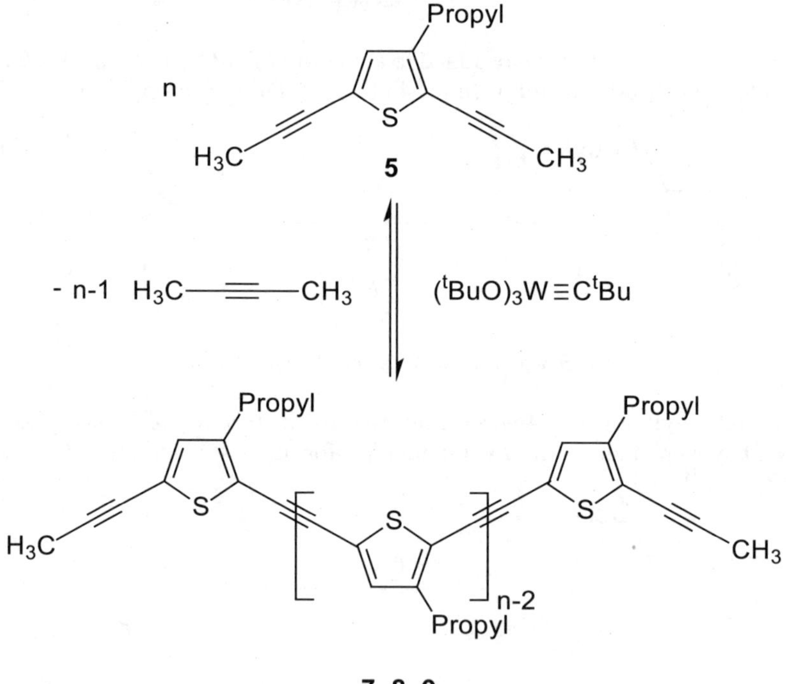

Figure 8: ADIMET condensation of 3-propyl-2,5-di(1-propynyl) thiophene with (^{t}BuO)$_3$W≡C^{t}Bu.

In a bulk reaction at 35°C and a pressure of 12hPa for 48h the reaction proceed. The molar ratio catalyst/monomer was 1/30. A small amount of catalyst was added after 24h, than the reaction was stopped by the addition of methanol and the product washed with methanol. Monomeric and dimeric products were solved in methanol. Product **7** (yield: 12.5%) was insoluble in methanol but completly soluble in hexane.

At higher reaction temperatures, 55°C and 80°C, higher moleculare weights were formed. Two polymers **8** (T = 55°C, yield: 17.1%) and **9** (T = 80°C, yield: 27.3%) were yielded, which are both insoluble in hexane but completly soluble in toluene. All products were investigated by GPC-measurement.

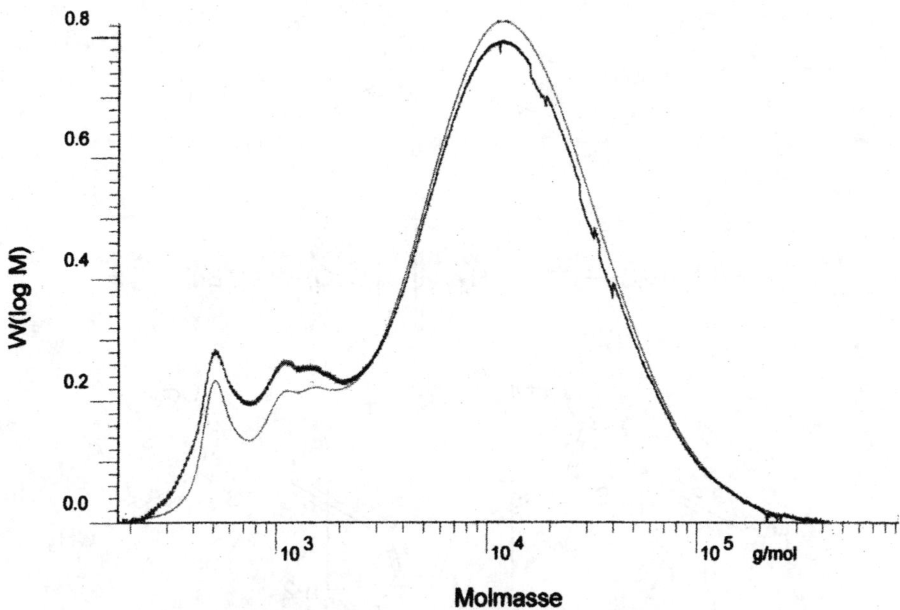

Figure 9: GPC of poly[2,5-(3-propyl-thiophenylene) ethynylene] **9**.

TABLE 1: ADIMET of 3-propyl-2,5-di(1-propynyl) thiophene; in bulk; molar ratio diyne/catalyst 30/1; catalyst: $(^tBuO)_3W\equiv C^tBu$; pressure: 12hPa; M_n, M_w = GPC-measurement, PS-Standard.

polymer	T [°C]	yield [%]	M_w [g·mol⁻¹]	M_n [g·mol⁻¹]	D
7	35	12.5	3 990	1 563	2.5
8	55	17.1	7 544	3 022	2.5
9	80	27.7	18 977	3 733	5.1

The results line out show, that the increasing of the temperature leads to higher polymerisation dergrees and enhances the yields.

Polymerization of 3-octyl-2,5-(1-propynyl) thiophene

I contrast to 3-propyl-2,5-(1-propynyl) thiophene **5**, 3-octyl-2,5-(1-propynyl) thiophene **6** yielded only low moleculare polymers. Therefore the reaction temperature was increased for further polymerisations.

3-Octyl-2,5-(1-propynyl) thiophene **6** was polymerized in an ADIMET condensation at temperatures of 115°C, 125°C and 130°C with (^{t}BuO$_3$)W≡C^{t}Bu as catalyst. The other reaction conditions were the similar to those of the polymerization of 3-Propyl-2,5-di(1-propynyl) thiophene **5**.

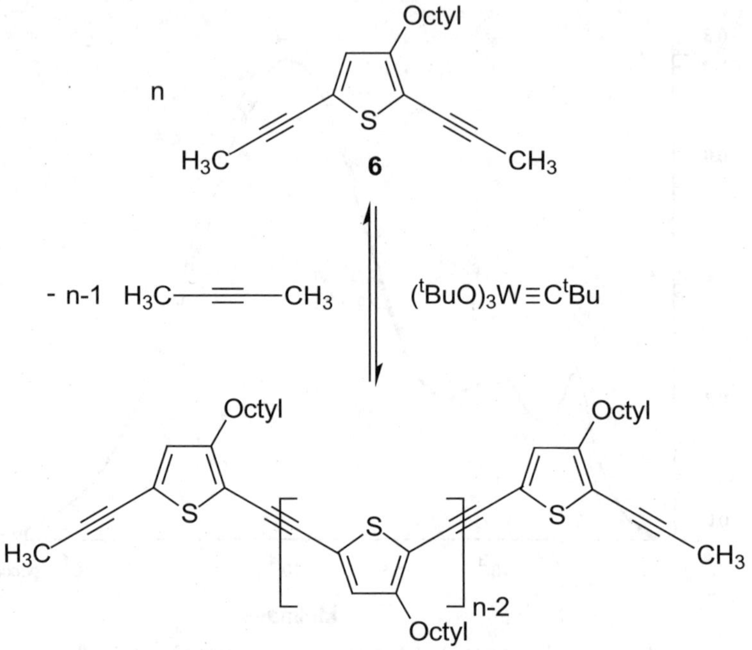

10, 11, 12

Figure10: ADIMET condensation of 3-octyl-2,5-di(1-propynyl) thiophene with (^{t}BuO)$_3$W≡C^{t}Bu.

This reactions yielded three polymer **10, 11, 12**. Polymere **10** (yield: 14.9%) and **11** (yield: 42,8%) are totally soluble in toluene. A completly insoluble polymer **12** (yield: 48.5%) was produced by ADIMET condensation of 3-octyl-2,5-di(1-propynyl) thiophene **6** at 130°C. The soluble products **10** and **11** were analyzed by GPC-measurement,

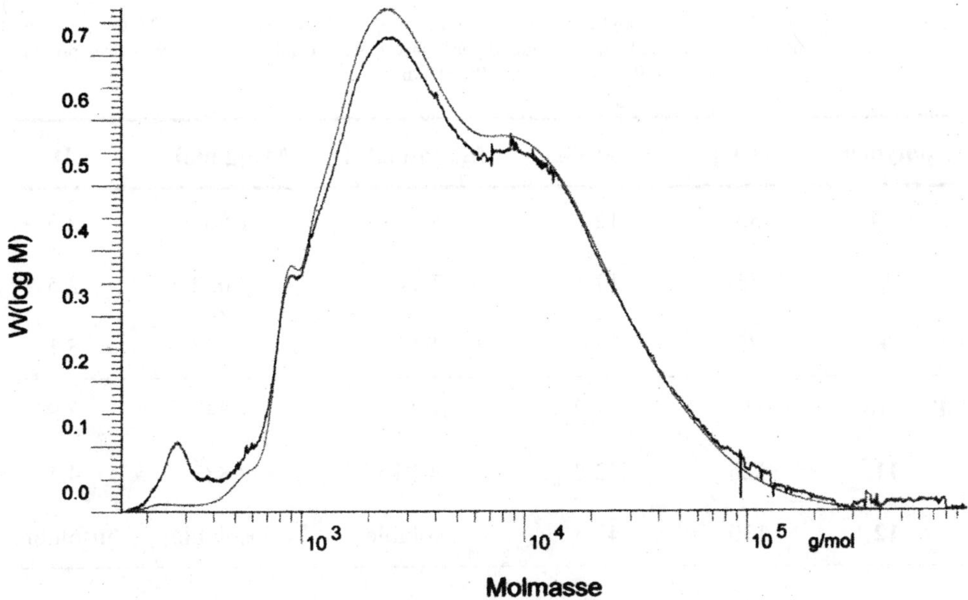

Figure 11: GPC of poly[2,5-(3-octyl-thiophenylene) ethynylene] **11**.

TABLE 2: ADIMET of 3-octyl-2,5-di(1-propynyl) thiophene; in bulk; molar ratio diyne/catalyst 30/1; catalyst: (^{t}BuO)$_3$W≡C^{t}Bu; pressure: 12hPa; M_n, M_w = GPC-measurement, PS-Standard

polymer	T [°C]	yield [%]	M_w [g·mol^{-1}]	M_n [g·mol^{-1}]	D
10	115	14.9	16 513	1 847	8.9
11	125	42.8	11 916	2 883	4.1
12	130	48.5	insoluble	insoluble	insoluble

We get the same results as with monomer **5**. Increasing of the temperature leads to higher polymerisation dergrees and the yield is enhanced.

Conclusion

The Schrock type carbene complex (^{t}BuO$_3$)W≡C^{t}Bu is not only a very avtiv catalyst in the polymerisation of alkylsubstituted 1-alkynes and for bispropynyl benzene but also for the ADIMET condensation of alkylsubstituted thiophenes.
3-Propyl-2,5-di(1-propinyl)-thiophen **5** and 3-octyl-2,5-di(1-propinyl)-thiophen **6** produce poly(thiophnylen ethynylenes (PTE) with (^{t}BuO$_3$)W≡C^{t}Bu as catalyst. These polymers have very interesting qualities becaus of their conjugated C≡C bonds. The alkylsubstituents are reliable for the good solubility of the polymere products. We found also that increasing the temperature causes a higher molecular weight and a yields.

284

TABLE 3.   ADIMET of 3-propyl-2,5-di(1-propynyl) thiophene **A** and 3-octyl-2,5-di(1-propynyl) thiophene **B**; in bulk; molar ratio diyne/catalyst 30/1; catalyst: (^{t}BuO)$_3$W≡C^{t}Bu; pressure: 12hPa; M_n, M_w = GPC-measurement, PS-Standard

polymer		T [°C]	yield [%]	M_w [g·mol^{-1}]	M_n [g·mol^{-1}]	D
A	**7**	35	12.5	3 990	1 563	2.5
	8	55	17.1	7 544	3 022	2.5
	9	80	27.7	18 977	3 733	5.1
B	**10**	115	14.9	16 513	1 847	8.9
	11	125	42.8	11 916	2 883	4.1
	12	130	48.5	insoluble	insoluble	insoluble

References

1   K. Weiss, R. Goller, G. Lößel, *J. Mol. Catal.* **1988**, 46, 267.
2   K. Weiss in F. R. Kreißl, *Transition Metal Carbyne Complexes*, VHC Verlag Weinheim **1988**, 206.
3   K. Weiss, A. Michel, E.-M. Auth, U. H. F. Bunz, T. Mangel, K. Müllen, *Angew. Chem.* **1997**, 109, 522.
4   A. Michel, *Dissertation*, Universität Bayreuth, **1997**.
5   R. F. Heck, *Palladium Reagents in Organic Syntheses*, Academic Press, New York, **1990**, 298.
6   L. Cassar, *J. Organomet. Chem..* **1975**, 93, 253.
7   K. Sonogashira, Y. Thoda, N. Haghiara, *Tetrahedon Lett.* **1975**, 4467.
8   M. Tateishi, H. Hishihara, K. Aramaki, *Chem. Lett.* **1987**, 1727.
9   T. Kahata, T. Osawa, *Chem. Abstr* **1989**, 110, 26700.
10   M. V. Lakshmikantham, J. Vartikar, K. Y. Jen, M. P. Cava, W. J. Huang, A. G. McDiramid, *Am. Chem. Soc. , Polym. Prepr.* **1983**, 24, 75.
11   R. D. Stephens, W. C. Krey, Jr., *Chem. Abstr.* **1973**, 74, 146990.
12   B. R. Hsieh, *Polym. Bull.* **1991**, 25, 177.
13   R. D. Stephens, C. E. Castro, *J. Org. Chem.* **1963**, 28, 3313.
14   J. Suffert, R. Ziessel, *Tetrahedron Lett.* **1991**, 32, 757.
15   Z. Xu, J. S. Moore, *Angew. Chem., Int. Ed. Engl.* **1993**, 32, 1354.
16   F. Würthner, *Angew. Chem.* **2001**, 113, 1069.
17   A. Kraft, A. C. Grimsdale, A. B. Holmes, *Angew. Chem.* **1998**, 110, 417.
18   J. L. Bredas, R. R. Chance, Eds., *Conjugated Polymeric Material: Opportunieties in Electronics, Optoelectronics and Molecular Electronics,* Kluwer Academic Press, Dordrecht, Netherlands **1990**.
19   M. Kumada, K. Tamao, K. Sumitani, *Organic Syntheses* **1978**, 58, 127.
20   Z. Bao, W. Chan, L. Yu, *Chem. Mat.,* **1993**, 5, 2.
21   H. Mao, S. Holdcroft, *Macromolecules* **1992**, 25, 554.
22   H. Mao, B. Xu, S. Holdcroft, *Macromolecules* **1993**, 26, 1163.
23   J. M. Barker, P. R. Huddlestone, M. L. Wood, *Synth Commun.* **1975**, 5(

RING CLOSING VERSUS CYCLIC ISOMERIZATION OF 1,6-DIENES BY RUTHENIUM ALLENYLIDENE COMPLEXES

BEKIR ÇETINKAYA[a], ISMAIL ÖZDEMIR[b], ENGIN ÇETINKAYA[a], CHRISTIAN BRUNEAU[c], PIERRE DIXNEUF[c]

[a] Department of Chemistry, Ege University, 35100-Bornova-İzmir,Türkiye, [b]Department of Chemistry, Inönü University, Malatya, Türkiye,[c] Institut de Chimie (CNRS UMR 6509) University of Rennes, Campus de Beaulieu, 35042 Rennes, France,

E-mail: bekircetinkaya@ hotmail.com

ABSTRACT

Treatment of electron rich carbene precursor olefins, L_2^R (R= CH_2CH_2OMe, CH_2Mes), bearing a 2,4,6-trimethylbenzyl group on the N atom(s), with $[RuCl_2(\eta^6\text{-arene})]_2$ (arene = p-cymene, hexamethylbenzene) can afford one of two products. Thus, cleavage of the bridge occurs to yield the expected carbene ruthenium(II) complex **5** or further a displacement reaction can occur to give chelated η^6-mesitylene carbene complexes **6** and **7** (Scheme 1). In a separate but complementary research area we also have interests in the rapidly developing chemistry of (η^6-arene)(N-alkylbenzimidazole)ruthenium (II), **8**.

Compounds **6** and **7** were transformed into the corresponding cationic allenylidene complexes (**6 → 6'**, **7 → 7'**, **8→8'**) by AgOTf and propargylic alcohol $HC\equiv CCPh_2OH$. In the same cases the *in situ* generated intermediates were found to be active and selective catalysts for ring closing metathesis (RCM) or cycloisomerization reactions of 1,6-dienes.

1. INTRODUCTION

Well-defined, single component metal-carbene complexes of the type $[RuCl_2(=CHPh)(PCy_3)_2]$ (**1**) are highly efficient ring closing metathesis (RCM) catalyst precursors, moderately sensitive to air and moisture and tolerant of many different organic functional groups.[1] However, they decompose at high temperatures. It has been shown that the replacement of one of the phosphine ligands in **1** with a sterically demanding NHC (*e.g. N,N'*-bis(mesityl)imidazol-2-ylidene (IMes) to yield **2** leads to an increased thermal stability of the catalyst precursor and in an activity towards tri- and tetrasubstituted substrates.[2] More recently, the saturated imidazol-2-ylidenes (SIMes) (**3**) have been prepared. It was proposed that the higher basicity of the saturated imidazol ligand compared to that of its unsaturated analogue leads to an

Y. Imamoglu and L. Bencze (eds.), Novel Metathesis Chemistry: Well-Defined Initiator Systems for Specialty Chemical Synthesis, Tailored Polymers and Advanced Material Applications, 285–293.
© 2003 Kluwer Academic Publishers. Printed in the Netherlands.

increased catalyst activity profile. **3** displayed increased RCM and cross metathesis (CM) activite compared to those of the their phosphine analogues.[3]

1   2   3

Dixneuf *et.al.* has recently introduced alternative 18-electron cationic allenylidene complexes of the type $[(arene)(R_3P)Ru=C=C=CPh_2]PF_6$ (**4**) which constituted a new class of RCM catalyst in the same range of activity as the 16-electron complexes **1-3**.[4] Nolan *et.al.* has reported allenylidene precatalysts bearing imidazol-2-ylidene ligands which were found to be efficient catalyst for RCM.[5] In order to compare the influence of ancillary ligands on the olefin metathesis activity of **4** new complexes of type **5-7** and **8**, containing saturated NHCs and *N*-alkylbenzimidazole ligands, respectively, have been prepared. After converting them into cationic allenylidene derivatives. Their activity in RCM catalysis was studied in the transformation of *N,N*-bis(allyl)toluene-*p*-sulfonamide. A preliminary report, related with saturated NHCs, has already appeared.[6]

2. RESULTS AND DISCUSSION

2.1. NHC complexes and intramolacular chelate formation

For the preparation of the desired ruthenium NHC complexes we used a similar procedure originally developed by Lappert et al[7] (Scheme 1). 4,5-Dihydroimidazolium salts, bearing at least one mesitylbenzyl substituent on the N atom are appropriate starting materials for the synthesis of electron rich olefins (L_2^R) (R= CH_2CH_2OMe or CH_2Mes) by deprotonation with NaH or Bu^tOK.

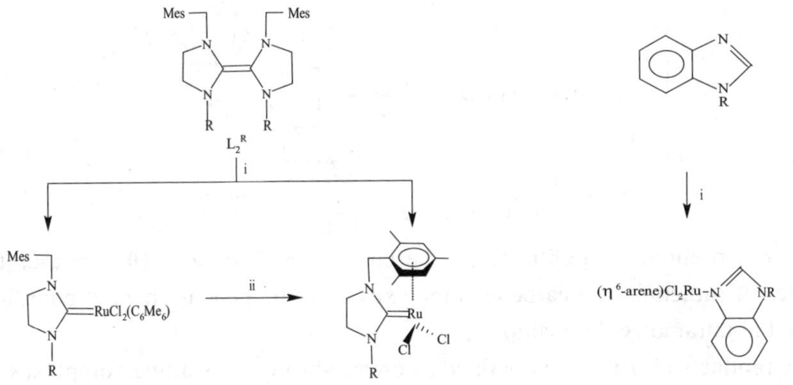

5 R = CH_2CH_2OMe

6 R = CH_2CH_2OMe

7 R = CH_2Mes

8 R = CH_2Mes

Scheme 1 (i) $[RuCl_2(\eta^6\text{-arene})]_2$ (arene = *p*-$MeC_6H_4CHNMe_2$, C_6Me_6), 100-110 °C, PhMe, (ii) *p*-Xylene, 140 °C.

The reaction of the dinuclear $[RuCl(\mu\text{-}Cl)(\eta^6\text{-arene})]_2$ (arene= *p*-$MeC_6H_4CHMe_2$ and C_6Me_6) with the electron rich olefins $\mathbf{L_2^R}$ in boiling toluene afforded monomeric complexes (**5-7**). Depending on the nature of the arene and olefin used, the derived carbene behaves as both a monodentate ligand, bonded to the metal directly through the carbene carbon atom (**5**) and as a bidentate ligand, bonded to the metal atom through both the carbene carbon and the arene carbons of the mesityl group, acting as a chelating ligand (**6** and **7**). It is noticeable that *p*-cymene is displaced more readily than C_6Me_6. Thus, the reaction of the bulky dimer $[RuCl_2(\eta^6\text{-}C_6Me_6)]_2$ with $\mathbf{L^R}_2$(R= CH_2CH_2OMe) affords complex **5** which contains a pendant trimethylbenzyl side-chain. The complex **5** could however be converted to the chelating product **6** by refluxing in *p*-xylene.

This observation clearly indicates the mode of the reaction: the first step of the procedure consists in the conversion of the dimeric starting material to the monomeric (carbene)(arene)ruthenium complexes, such as **5** which upon heating afforded the chelated carbene complexes **6** and **7**. Upon comparison with earlier work, a dramatic contrast is noticed between the olefins $\mathbf{L_2}^{CH2Ph}$ and $\mathbf{L_2}^R$ (R= CH_2Mes) in terms of their coordinative behavior toward $[RuCl_2\ (\eta^6\text{-arene})]_2$. Thus, the carbene derived from $\mathbf{L_2}^{CH2Ph}$ is not able to react in intramolecular manner[8] yet the carbenes derived from $\mathbf{L_2}^R$ (R =CH_2Mes) reacts so rapidly that the unchelated carbenes are unaccesible. On the other hand, preliminary studies on the carbene complexes of the type **9** has shown that these complexes can be converted to **10**.

288

X = NMe$_2$, OMe

Prior to our preliminary publication,[6] the complexes **5, 6** and **10** are thus the first examples of ruthenium(II) carbene complexes with the pendant benzyl coordinated to the metal in intramolecular fashion.

All new products (**5-10**) were obtained as orange-brown crystalline complexes in good yields. They are air-stable and soluble in dichloromethane. Product formulations were confirmed by NMR and mass spectroscopy, elemental analysis, and X-ray diffraction for **5** and **7**.[6]

The deprotonation reactions of the salts can be monitored by ^{13}C NMR. Thus, the resonance for the carbon at position 2 (*ca* 160 ppm) vanishes and the relatively high-field shifted resonance for the olefinic carbon atom appears at *ca* 135 ppm. This observation excludes the possibility of existence of corresponding free carbene since their typical values fall in the 210-240 ppm range. ^{13}C NMR spectra were particularly diagnostic as to the nature of the bonding in these new complexes, establishing them firmly to be either the mesityl-ruthenium bound or pendant benzyl. Thus, the chemical shifts of the metal-bound arene protons in complexes **6** and **7** and are found at higher fields ($\Delta\delta \sim$ 1.4 ppm) than in the pendant mesityl. The ^{1}H NMR spectra of **5** and **7** display two resonances for the arene ring protons and the ^{13}C NMR spectra six signals for the corresponding ring carbon atoms. ^{1}H-^{1}H COSY and HETCOR NMR studies were required to assign unexpectedly complicated methylene signals in the complexes **5** and **7**. The non-equivalence of each proton in both CH$_2$ groups indicates the absence of any symmetry element in the complex, in perfect agreement with the solid state structure. The remarkable high field ^{13}C NMR resonances of the carbene carbon atoms, in the 200-210 ppm range, are similar to those for other ruthenium carbene complexes.[6,9] The typical Ru=C singlet for **7** is at lower field (δ 210.26 ppm) than for the corresponding complex **6** (δ 200.14 ppm). Furthermore, CH$_2$Mes substituent on the second N atom of **7** does not show a noticeable effect on the chemical shift (δ 199.95 ppm).

2.2. N-alkylbenzimidazole complexes of ruthenium

Recently, we have reported the straightforward preparation of a series of (η^6-arene)ruthenium complexes of *N*-coordinated benzole derivatives, such as **8**. Catalytic activity of these complexes for intramolacular cyclization of (Z)-3-methylpent-2-en-4-

yn-1-ol into 2,3-dimethylfuran was studied. In the present study we prepared new Ru(II) complexes, bearing a variety of alky substituents on the benzimidazole N.[10]

The desired complexes, **8a – 8g,** were prepared in 83-90 % yield by simply heating the appropriate 1-substituted benzimidazole with $[RuCl_2(\eta^6\text{-arene})]_2$ in toluene (Scheme 1). The compounds **8a – 8g** were precipitated as red-brown crystalline solids. They are perfectly stable in the solid state and their spectroscopic properties indicate that all are N(3)-bonded. The imino carbon appeared as a typical singlet in the [1]H-decoupled mode in the 142.9- 146.2 ppm range. The ^{1}H NMR spectra of the complexes further supported the assigned structures, the resonances for C(2)-H were observed as sharp singlets in the 7.70 – 8.44 ppm range.

2.3. Ring closing metathesis and cyclicisomerization reactions catalyzed by ruthenium allenylidene carbene and benzimidazole complexes

Unsaturated cycloalkanes are an integral part of many natural products and their preparation is of permanent interest in organic synthesis. One important method under the conditions of homogeneous catalysis concerns the (atom economically) interesting cycloisomerisation of dienes.[11]

Transition metal prompted cyclization of 1,6-diens (**11,** Y= TsN, $(EtOCO)_2C$, EtOCOCCN) produces a mixture of several cyclic isomers (**12-14**), depending on the employed substrate and the catalyst.[12] Among the products *exo*-methylenecycopentenes (**12**) are important synthetic intermediates because they can be readily transformed to other functionalities. We recently have reported related procedures for the cyclization / ring closing metathesis (RCM) of 1,6-diens catalyzed by ruthenium-allenylidene complexes with chelating arene carbene ligand.[6]

11   **12**   **13**   **14**   **15**

Y; TsN (**a**), $(EtOCO)_2C$ (**b**), EtOCOCCN (**c**)

The high stabilities of **6** and **7**, owing to the chelate effect and the tunable nature of the R group on nitrogen of **8** provides a new perspective on the potential of such complexes as catalysis. Therefore we tested these complexes as catalysts in an RCM reaction with diallyltosylamide at 80 °C, using 2 mol% of **5-10**. However, they showed no activity for RCM reactions. To activate these neutral complexes for use in RCM, it was necessary to remove the strongly bound chloride ligand and form allenylidene. This was carried out by treatment with propargylic alcohol in the presence of AgOTf at ambient temperature. One such example was shown in Scheme 2.

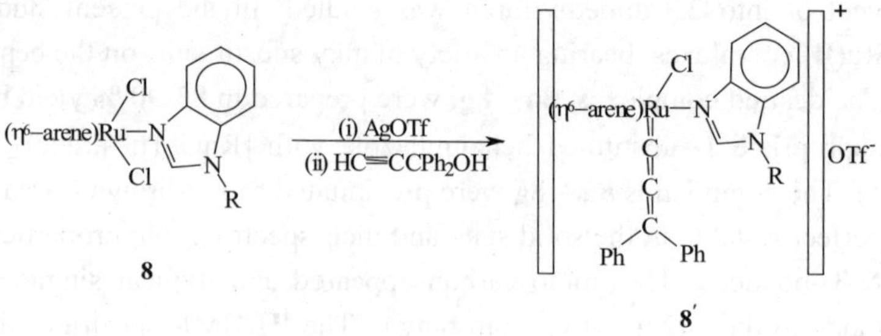

$$(i) \ AgOTf$$
$$(ii) \ HC{\equiv}CCPh_2OH$$

8

8'

a	R = CH$_2$CH$_2$OCH$_3$; arene = p-MeC$_6$H$_4$-Pr^i
b	R = CH$_2$CH$_2$OCH$_3$; arene = C$_6$(CH$_3$)$_6$
c	R = CH$_2$-2,4,6-(CH$_3$)$_3$-C$_6$H$_2$; arene = p-MeC$_6$H$_4$-Pr^i
d	R = CH$_2$-2,4,6-(CH$_3$)$_3$-C$_6$H$_2$; arene = C$_6$(CH$_3$)$_6$
e	R = (CH$_2$)$_3$CH$_3$; arene = p-MeC$_6$H$_4$-Pr^i
f	R = (CH$_2$)$_3$CH$_3$; arene = C$_6$(CH$_3$)$_6$
g	R = CH(CH$_3$)$_2$; arene = p-MeC$_6$H$_4$-Pr^i

Scheme 2 Synthetic route to benzimidazole complexes of ruthenium

The *in situ*-generated intermediates **5'** – **10'** have been evaluated for the catalytic cycloisomerisation of the diallyltosylamide in chlorobenzene or toluene at 80 °C. It was found that all catalyst precursors (2.5 mol%) were active. These results are summarized in Table 1. They show that the most efficient complexes are **8'c**, **8'e** and **8'g** (Table 1). The best yields were obtained at 80 °C for 4 – 18 h. Comparison of the catalysts containing p-cymene and hexamethylbenzene indicates that the presence of the p-cymene ligand gives a much better catalyst precusor. This may be due to the higher lability of the p-cymene ligand.

TABLE 1. Catalytic RCM / cycloisomerisation of 1,6-dienes, **11**.

Diene	Catalyst[a]	Solvent	Time/h	Conv.(%)[b]	12 (%)[b]	14 (%)[b]	15 (%)[b]
a	5'	Toluene	6	20	15	-	5
a	5'	C_6H_5Cl	5	30	10	-	10
a	6'	Toluene	4	100	100	-	-
a	6'	C_6H_5Cl	4	84	84	-	-
a	7'	Toluene	4	25	21	-	4
a	7'	C_6H_5Cl	4	100	94	-	6
a	8'a	C_6H_5Cl	10	82	82	-	-
a	8'a	Toluene	12	37	37	-	-
a	8'b	Toluene	10	58	58	-	-
a	8'b	C_6H_5Cl	20	100	100	-	-
a	8'c	Toluene	4	100	100	-	-
a	8'c	Toluene	10	100	-	100	-
a	8'c	C_6H_5Cl	6	98	60	38	-
a	8'c	C_6H_5Cl	10	100	-	100	-
a	8'd	C_6H_5Cl	20	79	40	39	-
a	8'e	Toluene	10	48	48	-	-
a	8'e	Toluene	13	76	76	-	-
a	8'e	C_6H_5Cl	10	100	100	-	-
a	8'f	C_6H_5Cl	18	-	-	-	-
a	8'g	C_6H_5Cl	18	100	100	-	-
a	10'	Toluene[c]	30	26	-	4	-
a	10'	C_6H_5Cl[c]	38	35	-	8	-
[d]							100

[a] The catalyst was prepared *in situ* from **8a** – **8g** (2.5 mol%), in 2.5 mL of solvent (chlorobenzene or toluene), AgOTf (2.5 mol%), $HC{\equiv}CCPh_2OH$ (2.6 mo%), 20 min. Stirring to give 8'**a** – 8'**g**. The substrate was then added and the solution heated at 80 °C for 4-18 h. [b] Determined by ^1H NMR. [c] Y = OMe, [d]

The efficiency and selectivity of transition metal-catalyzed processes can be strongly affected by the nature of the ancillary ligands on the metal complex. When **8'c** was reacted with diallyltosylamide for 4 h At 80 °C, **11**, was the highly selective product due to cycloisomerisation, whereas the same compound, **8'c**, for 10 h. at 80 °C give the main product **13**; the yield was 100% in all cases. It is clearly evident that the compound **11** is transformed to **13** by intramolecular isomerisation.

3.   CONCLUSION

In the present work we focused on the characterization and catalytic behavior of chelated carbene and N-alkylbenzimidazole ruthenium complexes.

The ambivalent reactivity of the electron rich olefins L_2^R was demostrated by virtue of their interaction with the ruthenium(II) dimers, $[RuCl_2\ (\eta^6\text{-arene})]_2$, as illustrated in Scheme 1. Sterically demanding CH_2Mes and electron donating substituent(s) at N atom(s) of the imidazolidine ring faciltates the displacement of the arene unit to yield chelated complexes **6**, **7** and **10** which have been fully characterized. The compounds **6,7** and **10** are the first examples of carbene complexes with tethered arene functionality. Concerning intramolecular coordination a hemilabile interaction is of special interest for catalytic purposes weak interactions are a prerequisite to use corresponding compounds as precursors in the related processes.

[RuCl$_2$(η^6-arene)(1-alkylbenzimidazole)], **8** was easily prepared and extremely stable in air so that it can be stored for a long period and handled without special precautions.

Although the new complexes, **5-10** are inactive, their *in situ* formed allenylidene derivatives (**5'-10'**) behave as catalyst for RCM or cycloisomerization of 1,6-diens. The transformation into the corresponding allenylidene complexes, **5'–10'** were carried out on reaction at room temperature with 1 equivalent silver triflate in CH$_2$Cl$_2$ and then 1.1 equiv. of propargyl alcohol, HC≡CCPh$_2$OH, for 20 min according to procedure described. The allenylidene complexes (**5'-10'**) are dark-violet, air-sensitive and decompose quite rapidly in solution. Although they were not stable enough for analysis and ^{13}C NMR studies.

Catalytic reactions proceed with very good efficiency and selectivity under relatively mild conditions (80 °C in tolune or chlorobenzene). In the case of **8'** synthetically useful *exo*-methylene products were obtained selectively.

Acknowledgment. We are grateful to TUBITAK MISAG COST D17 WG for financial support.

REFERENCES

1   Dias, E.L., Nguyen, S.T. and Grubbs, R.H. (1997) Well-defined Ruthenium olefin metathesis catalysts: Mechanism and activity, *J.Am.Chem.Soc.* 119, 3887-3897.

2   (a) Scholl, M., Trnka, T.M., Morgan, J.P. and Grubbs, R.H. (1999) Increased Ring Closing Metathesis activity of Ruthenium-based olefin metathesis catalysts coordinated with imidazolin-2-ylidene ligands, *Tedrahedron Lett.* 40, 2247-2250.

     (b) Huang, J., Stevens, E.D., Nolan, S.P. and Peterson, J.L. (1999) Olefin metathesis-activite ruthenium complexes bearing a nucleophilic carbene ligand, *J.Am.Chem.Soc.* 121, 2674-2678.

3   Scholl, M., Ding, S., Lee, C.W. and Grubbs, R.H. (1999) synthesis and activity of a new generation of ruthenium based olefin metathesis catalysts coordinated with 1,3-dimesityl-4,5-dihydroimidazol-2-ylideneligands, *Org. Lett.,* 1, 953-956.

4   (a) Picquet, M., Touchard, D., Bruneau, C. and Dixneuf, P.H. (1999) Room temparature operating allenylidene precatalyst [L$_n$Ru=C=C=CR$_2$]$^+$X$^-$ for olefin metathesis: dramatic influence of the counter anion X$^-$, *New J.Chem.,* 141-143.

     (b) Fürstner, A., Picquet, M., Bruneau, C. and Dixneuf, P.H. (1998) Cationic ruthenium allenylidene complexes as a new class of performing catalysts for ring closing metathesis, *Chem. Commun,* 1315-1316.

5   Jafarpour, L., Huang, J., Stevens, E.D. and Nolan, S.P. (1999) (*p*-cymene)RuLCl$_2$ (L = 1,3-bis(2,4,6-trimethylphenyl)imidazol-2-ylidene and 1,3-bis(2,6-diisopropylphenyl)imidazol-2-ylidene) and related complexes as ring closing metathesis catalyst, *Organometallics,* 18, 3760-3763.

6   Çetinkaya, B., Demir, S., Özdemir, I., Toupet, L., Semeril, D., Bruneau, C., Dixneuf, P.H. (2001) First Ruthenium   Complexes with a Chelating Arene Carbene Ligand as Catalytic Precursors for Alkene Metathesis and Cycloisomerisation, *New J. Chem.,* **25**, 519-521.

7   Lappert., M. F., (1988) The coordination chemistry of electron-rich alkenes (enetetramines), *J. Organomet. Chem.* **358**, 185-213.

8   Küçükbay, H., Çetinkaya, B., Guesmi, G.,and Dixneuf, P.H.(1996) New Carbene Ruthenium-Arene Complexes: Preparation and use in Catalytic Synthesis of Furans *Organometalics*, **15**, 2434-2439.

9   Çetinkaya, B., Ozdemir, I., Dixneuf,P.H.(1997) Synthesis and Catalytic Properties of *N*-Functionalized Carbene Complexes of Rhodium(I) and Ruthenium(II) *J. Organomet. Chem.* **534**, 153.

10   Çetinkaya, B., Özdemir, I., Bruneau, C., and Dixneuf, P.H. (2000) Benzimidazole, benzothiazole and benzoxale ruthenium(II) complexes; catalytic Synthesis of 2,3-dimethylfuran, *Eur.J.Inorg.Chem.*, 29-32.

11   B. Trost, B., and Krische, M. J. (1998) Transition metal catalyzed cycloisomerizations, *Synlett*, 1-16.

12   Kisanga, P., Goj,L. A. and Widenhoefer, R. A. (2001) Cycloisomerization of funtionalized 1,5-and 1,6-dienes catalyzed by cationic palladium phenantroline complexes, *J.Org.Chem.* **635**, 635-637; (b) Yamamoto, Y. Nakagai, Y Ohkoshi, N., and Itoh, K (2001) Ruthenium(II)-catalyzed isomer-selective cyclization of-dienes leading to exo-methylenecyclopentanes: unexpected cycloisomerization mechanism involving ruthenacyclopentane(hydrido) intermediate *J.Am.Chem.Soc.* **123**, 6372-6381; (c) Bray, K.L., Charmant, J.P.H., Fairlamb,I.J.S., Jones., G.C.L., (2001) Structural and mechanistic studies on the activation and propagation of a cationic allylpalladium procatalyst in 1,6-diene cycloisomerization, *Chem.Eur.J.* **7**, 4205-4215.

APPLICATION OF UNIFORM MACROPOROUS POLYSTYRENE PARTICLES AS SUPPORT IN W(CO)$_6$/CCl4/hv PHOTOCATALYTIC OLEFIN METATHESIS SYSTEM

BÜLENT DÜZ*, CEMİL AYDOĞDU AND YAVUZ IMAMOĞLU

Hacettepe University, Chemistry Dept., 06532, Ankara, Turkey

1.Introduction

Olefin metathesis has made a great impact in synthetic organic chemistry as a result of the preparation of many new polymers and valuable low molecular compounds. In this reactions, active species are metal carbenes that they can be generated in situ (classical systems) or added as well defined single components,which provide potential for enlarged application of metathesis in various fields[1]. Immobilization of the metathesis catalysts on solid supports will facilitate the work up and make the metahesis reactions more attractive for both pratical and industrial applications.

During the recent years, intensive efforts have been devoted to develope polymer-supported catalysts in order to introduce the advantages of heterogenous catalysts, such as easy separations from the products and facile recover for recycling into homogenous catalyst system [2-12]. The use of polymer as support in organometallic catalysis is well documented and has been extensively reviewed [13,14]. Polymer supported Mo and W catalysts of the first generation (classical system) have been found to possess olefin metathesis activities [15-17].

In the present study, we report the synthesis and characterization of the uniform macro porous poly(styrene-co-divinylbenzene) particles (PS-DVB) and immobilization of photocatalytically produced active species onto the well characterized uniform macro porous poly(styrene-co-divinylbenzene) particles and the catalytic activities of the both free and PS-DVB adduct photocatalytic system (W(CO)$_6$ /CCl$_4$/ PS /hv) in terminal olefins metathesis reactions comparatively.

2. Experimental

Preparation of macroporous-uniform polystyrene(PS) particles

The macro porous uniform PS latex particles were produced by a two step seeded polymerization method[18] starting from the seed lattices. The seed latex preparation and the characterization procedures were described elsewhere[19]. In the first stage of synthesis, 2.1 mL dibuthyl phtalate (DBP) was emulsified by ultrasonication within 60 mL of aqueous medium including 0.25 % (w/w) sodium dodecyl sulfate (SDS). 10 mL of aqueous dispersion including 1.4 g of polystyrene seed particles was added into the aqueous

Y. Imamoglu and L. Bencze (eds.), Novel Metathesis Chemistry: Well-Defined Initiator Systems for Specialty Chemical Synthesis, Tailored Polymers and Advanced Material Applications, 295–301.
© 2003 *Kluwer Academic Publishers. Printed in the Netherlands.*

DBP emulsion. The resulting dispersion was stirred with 250 rpm at +4°C for 24 h for the complete adsorption of DBP by the seed particles. In the second stage, DBP-swollen seed particles were reswollen with the monomer phase. For this purpose, 10 mL of monomer phase including 7.5 mL of divinylbenzene (DVB), 2.5 mL of styrene and 0.24 g benzoyl peroxide (BPO) was emulsified by ultrasonication within 60 mL of aqueous medium including 0.25% (w/w) SDS. The monomer emulsion was then mixed with the aqueous dispersion of DBP swollen seed particles. The adsorption of monomer phase by the DBP-swollen seed particles was performed at +4°C for 24h at 250 rpm stirring rate. 10 mL of 10 % (w/w) polyvinylalcohol (PVA) solution was added into the resulting dispersion transferred into a sealed pyrex polymerization reactor. The reactor was purged with bubbling nitrogen for 10 minutes at room temperature. The repolymerization of monomer phase within the swollen seed particles was conducted at 70°C for 24 h at 120 cpm shaking rate. By the repolymerization, large uniform and macroporous latex particles, together with some amount of small particles (i.e. about 1 μm) as a by product. The final particle dispersion was washed with ethanol and water to remove small particles from the dispersion medium, absorbed DBP, and physically bound emulsifier from the large macroporous particles by applying a serum replacement method, including suggestion centrifugations. The detailed washing procedure was described elsewhere[19].

Typical catalytic reaction procedure

All reactions were carried out in a dry nitrogen atmosphere in 5ml water-jacketed pyrex cells placed around a 150 W Hanau high-pressure mercury lamp. Schlenk technique was applied. After the required amount of $W(CO)_6$ (0.25 mmol) was introduced, CCl_4 (4mL) and the certain amount of α-olefins were then added while the system was still under nitrogen flow. Magnetic stirring was employed during the irradiations as a general reguirement for photochemical reactions. Irradiations were continued for 2 hours or more and then the catalytic reaction were monitored by GC-MS using n-heptadecane ($C_{17}H_{36}$) as internal standard. The same procedure was followed for PS supported photocatalytic system after adding of the calculated amount of PS support.

Analysis

Reaction mixtures were analyzed with a Shimadzu GCMS-QP5050A model gas chromatograph and mass spectrometer dedector. Seperations were performed in 25 mm x 50 m glass capillary filled with dimethylpolysiloxane (Permobond SE-30, 0.25 μm). The following instrumental conditions were used: inlet temperature of 275°C, injection volume of 1 μl (split ratio: 200), oven programmed from 50 to 250°C with the rate of 20 °C min^{-1} the carried gas with a flow rate of 1 mL. min^{-1} and interface temperature of 280°C. The internal standard method was used to calculated the yield (%) of metathesis products formed.

3. Results and Discussion

<u>Preparation and characterization of the uniform polystyrene particles</u>

Monodisperse-macroporous particles in the form of styrene-divinylbenzene copolymer was obtained by multistage-seeded polymerization method. The effects of polymerization conditions on the average size and porosity properties of monodisperse-macroporous particles were investigated in details elsewhere [20]. The SEM (scanning electron microscope) photographs showing the size distribution properties and the detailed surface morphology of the particles used in this study are given in Figure 1A and 1B, respectively. The size distribution properties of the particles are given in Table I. The size-polydispersity index calculated based on the SEM photograph given in Figure 1A indicated that the particles were in the highly monodisperse form. On the other hand, the SEM photograph in Figure 1B showed that the particles had a sponge like porosity. In other words, the average pore size on the particle surface was less than 100 nm.

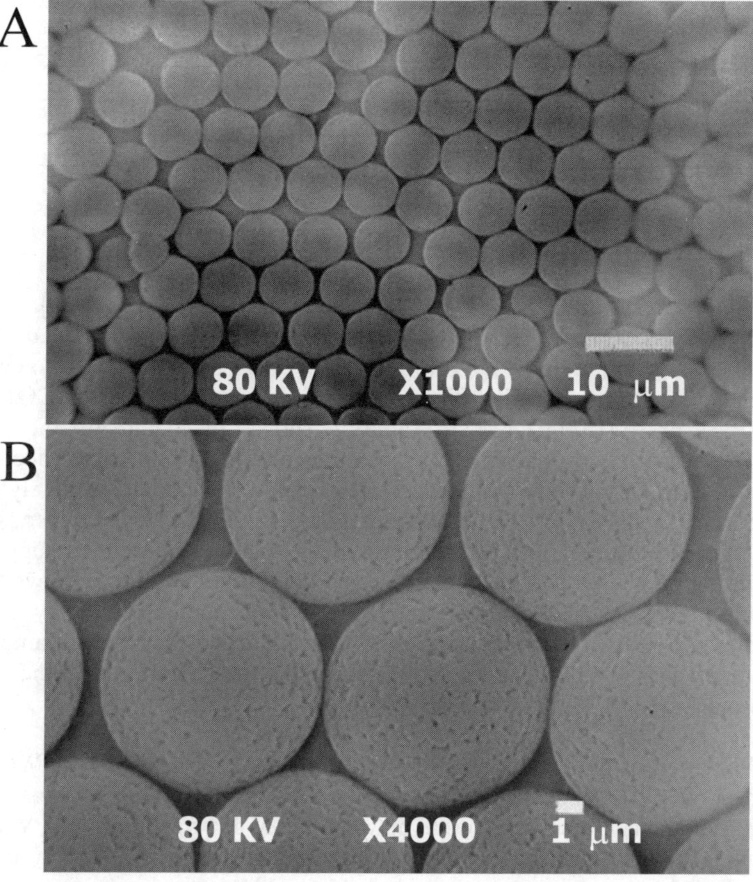

Figure 1. The SEM photographs of uniform macroporous particles, (A) particle-size   distribution, (B) surface structure.

The bulk and surface chemistries of extensively cleaned monodispese and macroporous poly(styrene-divinylbenzene) particles were investigated by FTIR and FTIR-DRS spectroscopy. For this purpose, monodisperse-macroporous poly(styrene-divinylbenzene) particles were produced by using and without using PVA as a stabilizer. The As expected, no hydroxyl band was detected both in the FTIR and FTIR-DRS spectra of the particles produced in the absence of PVA. However, the hydroxyl bands appeared both in the FTIR and FTIR-DRS spectra of poly(styrene-divinylbenzene) particles produced by using PVA indicated the presence of PVA both in the bulk structure and on the surface of the particles. The relative intensity of the hydroxyl band appeared on the FTIR-DRS spectrum taken for the particle surface was reasonably stronger to that of FTIR showing the bulk characteristics of the particles. This finding also indicated that the surface concentration of PVA was higher relative to that in the bulk of the particle. In our study, since the monodisperse poly(styrene-divinylbenzene) particles were obtained by using PVA as a stabilizer, covalently bound or strongly entrapped PVA chains should be also present on the particle surface.

Table I. The size distribution properties of macroporous particles

Number-average diameter (μm)	6.31
Weight-average diameter (μm)	6.36
Polydispersity Index	1.008

Comparison photocatalytic activities of free and PS adduct systems

Irradiation of $W(CO)_6$ in CCl_4 causes the metathesis of olefins [21-26]. One CO molecule dissociates from $W(CO)_6$ to form the coordinatively unsaturated $W(CO)_5$ intermediate upon irradition by UV light. In CCl_4, $W(CO)_5$ undergoes reaction with the solvent to generate the catalytically active species (precursors) such as $W(CO)_5Cl$[21], $W(CO)_4Cl_2$[22] or $W(CO)_4CCl_2$[23]. Table 2 shows the yield of the primary metathesis products of 6-dodecene(C_{12}), 7-tetradecene(C_{14}) and 8-hexadecene(C_{18}) obtained from terminal olefins, 1-heptene,1-octene and 1-nonene, respectively in both free and PS supported systems. In PS supported system, reaction yield depends on the amount of PS. To illustrate this effect, the amount of macroporous PS support was varied in the range of 0.0 to 0.10 g. at the same conditions. Figure 2 shows that the product yield increases when immobilized form of the photocatalytically produced active species are used. However, the yield was reduced when higher amounts of polymer support was used due to the slower diffusion of substrate on to the polymer matrix (colloidal effect).

The immobilization of photocatalytically produced active species, such as $W(CO)_5Cl$, $W(CO)_4Cl_2$ or $W(CO)_4CCl_2$ complexes, onto the monodisperse-macroporous particles should occur via the chemical interaction between the hydroxyl groups of PVA chains on the particle surface and the chlorine groups linked to the tungsten atom of the complexes. The exhange of these groups probably results in the binding of complexes onto the particle surface (Scheme 1).

Scheme 1;

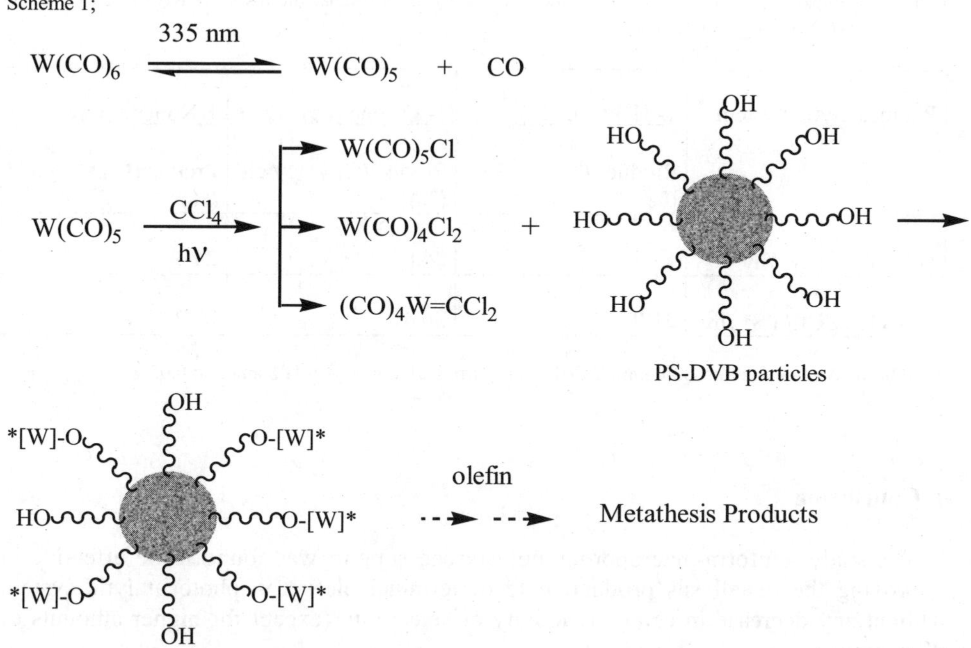

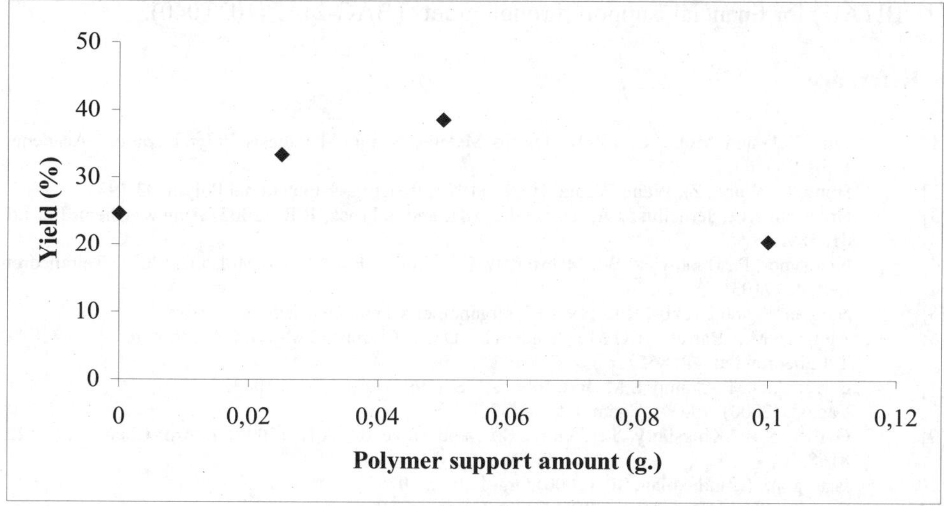

([W]*= metathetic active tungsten sites)

The uniform-macroporous polystyrene (PS) particles were found to be effective to improve the product yield (Table 2) of the metathesis of terminal-olefins in photocatalytic system (W(CO)$_6$ /CCl$_4$/ PS /hν). The results indicated the important function of the polystyrene support as a sumultanous immobilizing /catalyzing agent.

Figure 2. Influence of the polymer support quantity on the primary metathesis product of 1-octene in the photocatalytic system. W(CO)$_6$:0.16 mmol; olefin: 3.20 mmol; CCl$_4$ 3 mL; irradiation time: 4 h.

Table 2. Comparison of The Primary Product Yields(%) of Terminal olefins with $W(CO)_6$/hv/CCl_4 and $W(CO)_6$/hv/PS/CCl_4 Systems

Photocatalytic System	1- Heptene (C_7) Product(C_{12}) yield (%)	1- Octene (C_8) Product(C_{14}) yield (%)	1-Nonene (C_9) Product(C_{16}) yield (%)
$W(CO)_6$ /CCl_4/ /hv	13.7	24.6	24.3
$W(CO)_6$ /CCl_4/ PS* /hv	21.0	36.4	21.2

$W(CO)_6$: 0.16 mmol; olefins: 3.20 mmol; CCl_4 3 mL; irradiation time: 4 h.*PS amount: 0.05 g.

4. Conclusion

In this study, uniform-macroporous polystyrene support was found to be effective in improving the metathesis product yield of terminal olefins in photocatalytic system without any decrease in catalytic activity or selectivity (except for higher amounts of PS support).

Our experiments aimed at the re-useability of the PS supported photocatalytically produced active species are in progress.

5. Acknowledgement

We are grateful to The Scientific and Technical Research Council of Turkey (TUBITAK) for financial support through grant TBAG-2147 (102T040).

6. Reference

[1]   Ivin, K.J. and Mol, J.C. (1997) Olefin Metathesis and Metathesis Polymerization, Academic Press,London.

[2]   Hong, L., Wang, Z., Wang, Y. and He, B. (1997) Reactive & Functional Polym. **33**,193.

[3]   Hultzsch, KC., Jernelius, J.A., Hoveyda, A.H. and Schrock, R.R. (2002) Angew. Chem. Int. Ed. **41**, 589.

[4]   Nieczypor, P., Buhowicz, W., Meester, W.J.N., Rutjes, F.P.J.T. and Mol, J.C. (2001) Tetrahedron Lett. **42**, 7103.

[5]   Nguyen, S. and Grubbs, R.R. (1995) J. Organomet. Chem. **497**, 195.

[6]   Ahmed, M., Barret, A.G.M., Braddock, D.C., Cramp, S.M. and Procopiou, P.A. (1999) Tetrahedron lett. **40**, 8657.

[7]   Barret, A.G.M., Cramp, S.M. and Roberts, RS. (1999) Org. Lett.**1**, 1083.

[8]   Yao, Q. (2000) Angew. Chem. Int. Ed. **39**, 3896.

[9]   Garber, S.B., Kingsbury, J.S., Gray, B.L. and Hoveyda, A.H. (2000) J. Am. Chem. Soc. **122**, 8168.

[10]   Jarfarpour, L. and Nolan, SP. (2000) Org. Lett. **2**, 4075.

[11]   Dowden, J. and Savovic, J. (2001) Chem. Commun. 37.

[12]   Schürer, S.C., Gessler, S., Buschmann, N. and Blechert, S. (2000) Angew. Chem. Int. Ed. **39**, 3898.

[13]   Grubbs, R.H. (1977) Chemtech; 512.
[14]   Pittman, C.U. Polymer Supported Catalysts. (1982) In Comprehensive   Organometallic Chemistry, Wilkinson, G., Stone, F.G.A. and Abel, E.A. (eds.). Pergamon; Oxford, **8**, 553-605.
[15]   Basset, J.M., Mutin, R., Desotes, G. and Sinou, D. (1975) C.R. Acad. Sci. Ser. C , **280**, 1181.
[16]   Grubbs, R.H., Swetnick, S. and Su, S.C.H. (1978) J. Mol. Catal. **3**, 11.
[17]   Hong, L., Wang, Z. and He, B. (1999) J. Mol. Catal. A: Chmical **147**, 83.
[18]   Galia, M., Svec, F. and Frechet, J.M.J. (1994) J. Polym. Sci.Part A: Polym. Chem. **32**, 2169.
[19]   Tuncel, A., Tuncel, M. and Salih, B. (1999) J. Appl. Polym. Sci. **71**, 2271.
[20]   Tuncel, A., Tuncel, M. and Salih, B. (1999) J. Appl. Polym. Sci. **71**, 2271.
[21]   Agapiou, A. and McNeils, E. (1975) J. Organometall. Chem. **99**, C 47.
[22]   Krausz, P., Garnier, F. and Dubois, J.E. (1976) J. Organometall. Chem. **108**, 197.
[23]   Garnier, F., Krausz, P. and Rudler, H. (1980) J. Organometall. Chem. **186**, 77.
[24]   Krausz, P., Garnier, F. and Dubois, J.E. (1975) J. Am. Chem.Soc. **97**, 437.
[25]   Agapiou, A. and McNeils, E. (1975) J. Chem. Soc. Chem. Commun. 187.
[26]   İmamoğlu, Y., Zümreoğlu, B. and Amass, A.J . (1986) J. Mol. Catal. **36**, 107.

REACTIONS OF ATOMIC CARBON WITH 2-NORBORNENE

FATMA SEVİN* AND BÜLENT DÜZ
Department of Chemistry, Hacettepe University,
06532, Beytepe-Ankara, Turkey

Abstract

Reaction of atomic carbon with 2-norbornene, **1** generates octa-3,7-dien-1-yne, **5** in a reaction over bicyclo[3.2.1]octa-2,3-diene, **4** which fragments to **5**. **4** has also been trapped by THF. The potential energy surfaces of C_8H_{10} have been evaluated by means of the AM1 semiempirical method.

1. Introduction

The carbon arc is used as a source of carbon vapor which is produced as a result of the high temperature attained by the arc. When organic compounds react with carbon vapor, they give a variety of products containing an extra carbon atom. Both triplet (ground state) and singlet (excited state) carbon atoms insert into carbon-hydrogen bonds and/or add to double carbon-carbon bonds to make triplet and singlet carbenes [1]. These carbenes have high energy and they facilitate the formation of many new, high-energy molecules. Furthermore, their chemistry corresponds to pure carbene chemistry without precursor.

On the other hand, bicyclic allenes are of considerable interest in chemistry because of their high strain and reactivity [2]. Enormous effort has been devoted toward the synthesis of cyclohexa-1,2-diene and a number of its derivatives [3]. The addition-ring expansion is efficient with norbornenes or norbornadienes and has been used quite often as a simple inexpensive preparative method for construction of useful functionalized bicyclo[3.2.1]octanes. The highly strained bicyclo[3.2.1]octa-2,3-diene, **4** has been trapped either as an enol ether or as its styrene [2+2] cycloadduct by Mohanakrishnan et al [4].

In this study, we report the first experimental investigation of the reaction of arc-generated carbon atoms with 2-norbornene **1** to determine their likely pure carbene products, and we evaluate the energetics of this reaction by means of the AM1 semiempirical method.

Y. Imamoglu and L. Bencze (eds.), Novel Metathesis Chemistry: Well-Defined Initiator Systems for Specialty Chemical Synthesis, Tailored Polymers and Advanced Material Applications, 303–308.

2. Experimental Section

The experimental setup is described in reference [5]. The reactions were carried out by cocondensing **1** with arc-generated carbon at 77K in an evacuated reactor. The substrate was bled into the reactor and condensed on a 77K surface while an arc was struck intermittently between two high-purity graphite rods. At the conclusion of the reaction, the reactor was allowed to warm to room temperature. Chloroform was added in the reactor, then the filitered reaction mixtures were analyzed by GC/MS using a Shimadzu QP5050A mass spectrometer interfaced with a Shimadzu GC17A gas chromatograph equipped with a 50m x 0.32mm ID OPTIMA-5-1.0m capillary column. Also, the trapping experiment has been done with tetrahydrofuran (THF). In this experiment, THF has been bled into the reactor after each arc strike.

Computational Method

Semiempirical methods have progressed over the past few years to a surprising level of accuracy and reliability, considering the limitations of the underlying approximations [6]. Even simple calculations at a moderate level of theory such as semiempirical procedures give useful information on the molecular mechanism for different chemical reactions. Accordingly, the AM1 semiempirical method [7] in the RHF formalism was selected to carry out the calculations with the PC SPARTAN'02 Essential program [8]. The geometries were optimized without constraint in the final run of the calculation. Vibrational frequencies were then computed to characterize each stationary structure. Imaginary frequencies (transition vectors) were animated graphically for all transition structures to ensure that the motion was appropriate for converting reactants to products.

3. Results and Discussion

The possible carbenes which will be formed via double bond addition (*path a*) and C-H insertion (*paths b-g*) from the reaction of **1** with the carbon atom are given in Scheme1.

Scheme 1

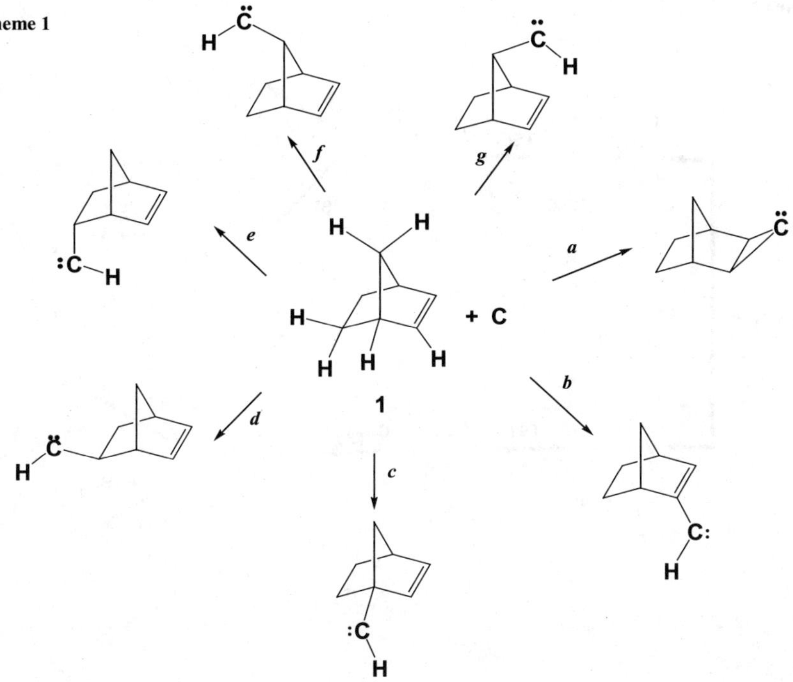

We have done theoretical calculations for the paths *a, d,* and *e* since in the experiments, we did not detect the products from the other paths. In addition, we made theoretical investigations only for the singlet carbene state. The calculated structures are given in Scheme 2 and their heat of formation values and relative energies with respect to the reactants are presented in Table 1.

Firstly, we investigated a possible complex between the reactants to create an activation barrier from the complex to the transition state. We found lower energy complexes (**2CMP** and **7CMP**) which are more stable than the reactants by −0.82 and −2.48 kcal/mol, respectively. In a recent work, Jursic [9] has suggested that methylene insertion into a polar bond, such as hydrogen fluorine, can occur owing to the fact that a relatively strong (around 7.7kcal/mol) complex is formed from the reactants.

Scheme 2

Table 1. AM1 calculated Heat of Formation values and relative enthalpies for reaction of atomic carbon and **1** shown in Scheme-2(kcal/mol).

structures	ΔH_f	relative energy	structures	ΔH_f	relative energy
1	25.96		TS1	239.69	-0.06
1 +C	239.74	0.0	TS2	137.97	-101.77
2CMP	238.92	-0.82	TS3	129.74	-109.56
3	130.79	-108.95	TS4	103.64	-136.10
4	85.19	-154.56	TS5	239.08	-0.66
5	71.23	-168.51	TS6	137.99	-101.75
6	-14.70	-254.44	TS7	155.47	-84.27
7CMP	237.26	-2.48	TS8	103.35	-136.17
8	126.07	-113.68	4 + THF	88.15	
9	51.12	-188.62	THF	2.96	
10	72.65	-167.09	C	213.78	
11	50.87	-188.88			

The complex **2CMP** resulting from double bond addition is less stable than **7CMP,** which results from C-H insertion. However, in the theoretical calculations the double bond addition is competitive with C-H insertion at the AM1 level. There is almost 0.6kcal/mol energy difference between the energies of **TS1** and **TS5**.

The bicyclo[2.2.1]hept-5-en-2-yl methyl carbene intermediate, **8** is 4.7 kcal/mol more stable than the **3** carbene intermediate. In the case of a transition from these carbenes to the products, the lowest activation energy corresponds to the

3→TS2→4 conversion that has taken place via double bond addition (7.21 kcal/mol). The other conversions, **8→TS6→9** and **8→TS7→10** which correspond to C-H insertion products, are at 11.93 and 28.98 kcal/mol, respectively. The conversion **3→TS2→4** gives the strained bicyclic allene, bicyclo[3.2.1]octa-2,3-diene and its reaction enthalpy is -45.58 at the AM1 level. We found a transition state, **TS3**, for a new reaction over **4** : fragmentation to (3Z)-octa-3,7-dien-1-yne, **5**. The activation energy of **TS3** is 45.0 kcal/mol and higher than **TS6** and **TS7**. However, **TS3** on the potential energy surfaces of C_8H_{10} has an energy value of 7.81 kcal/mol and is preferred relative to **TS6**. From GC/MS analysis, we see that **4** decays to two **5** isomers at the highest yield [10 a,b,c,d]. There is as yet no clear evidence for the other possible products, **9** and **10,** due to unmatched mass ions. Although we feel that the small peak at 13.57 minute looks like the product **11** [11]. The tricyclo[3.3.0.0^{2,7}]oct-3-ene, **10** gives by fragmentation to **11**. The activation energy of **11** is 30.92kcal/mol.

The strained allene **4** can be trapped by trapping agents[4]. We can report that in the reaction described here, **4** is efficiently trapped by THF as trapping agent. GC/MS analysis of the products reveals *endo* and *exo* isomers of **6** and dimers of **1**. The other products of **1** with a carbon atom have not been observed when the reaction with THF is done. The relative yields of the isomers of **6** which are given same mass ion fragments [12 a,b] with different intensities are 5.29 / 7.23. However, we could not determine which peak is corresponding to which isomer. Our calculations for this reaction on a concerted pathway have predicted a low energy barrier (15.46 kcal/mol). It is well known that in semiempirical methods, there is some inherent inaccuracy in the prediction of activation energies. Accurate energetic results, however, can be obtained by using post-Hartree-Fock or density functional-based methods [13-15]. We stress that the present investigation focuses on the trends rather than the absolute values. Our future work will include calculations with *ab initio* or density functional-based methods to elucidate the role of interactions in determining the values of the activation energies.

4. Acknowledgements

We thank the TUBITAK (The Scientific and Technical Research Council of Turkey, Project No. TBAG-AY/284(102T142)) for generous support.

5.   References

[1]   Skell, S.P., Havel, J.J., McGlinchey, M. (1973), *J. Acc. Chem. Res.*, **6**, 97-105.

[2]   (a) Balci, M., Taskesenligil, Y. Advances in Strained and Interesting Organic Molecules; Halton, B., Ed., (2000) **8**, pp 43-81. (b) Johson, R.P. (1989), *Chem.Rev.* **89**, 1111-1124.

[3]   For most recent papers see: (a) Sevin, F., Sökmen, İ., Düz, B. and Shevlin, P.B. (2003) *Tetrahedron Lett.* **44**, 3045-3407. (b) Drinkuth, S., Groetsch, S., Peters, E. M., Petrs, K., Christl, M. (2001), *Eur. J. Org. Chem.* 2665-2670. (c) Fernandez-Zertuche, M., Hernandez-Lamonrda, R., Ramirez-Solis, A. (2000), *J. Org. Chem*, 1871-1874. (d) Christl, M., Groetsch, S. (2000), *Eur. J. Org. Chem.* **56**, 4163-4171.

[4]   Mohanakrishnan, P., Tayal, S.R., Vaidyanathaswamy, R., Devaprabhakara, D. (1972), *Tetrahedron Lett.*, **28**, 2871-2872.

[5]   The carbon atom reactor is modeled after that describe by Skell, P.S.; Wescott, L.D., Golstein, J.P., Engel, R.R. (1965), *J. Am.Chem.Soc.* **87**, 2829.

[6]   Stewart, J.J.P., Lipkowitz, K.B., Boyd, D.B.V. (Eds.), (1990), *Reviews in Computational Chemistry*, **1**, Wiley-VCH, New York, 45. (b) Zerner, M.C. in Lipkowitz, K.B., Boyd, D.B.V. (Eds.), (1991), *Reviews in Computational Chemistry*, **2**, Wiley-VCH, New York, 313.

[7]   Dewar, M.J.S., Zoebisch, E.G., Healy, E.F., Stewart, J.J.P. (1985) *J. Am. Chem. Soc.* **107**, 3902.

[8]   Hehre, W.J., Deppmeier, B.J., Klunzinger, P.E., Wavefunction, Inc. Irvine, C.A. SPARTAN'02 Windows, Version 1.0 Essential Eddition 2002.

[9]   Jursi, B. S. (1999), *Theochem* , **467**, 103-113.

[10]   a) The new peaks from GC result (R.Time (%Area)) of **1** and C atom are 10.93(5.78%), 11.54(5.05%), 11.73(12.10%), 12.81(3.11%), 13.36(2.29%), 13.57(1.59%), 16.40(2.98%), 18.24(12.14%) b) At 11.73min; MS(70ev) m/z 103(M-3 17.82%), 67(100%), 66(80%), 54(40%), 81(28%), 68(27%), 79(21%), 53(21%), 55(14.93 %), 51(11.0 %), 52(6.02 %); c) At 10.93 min; MS(70ev) m/z 109(M+3 5%), 81(100%), 79(87%), 54(40%), 67(30%), 53(27%), 95(23%), 66(23%) ; d) The m/z 39 is diagnostic peak for terminal alkyne, the m/z 67 comes from 106(MW)-39, the m/z 79 (-CH_2CH from MW).

[11]   At   13.57 min; MS(70ev) m/z 107 (M+1 22.90%), 67(100%), 66(54%), 79(23%), 77(17%), 77(17%).

[12]   a) The new peaks from GC result (R.Time(%Area)) of **1** + C-atom + THF are 10.93(21.20%), 22.17(7.52%), 22.26(31.77%), 23.95(7.78%), 24.08(5.29%), 24.45(19.20%), 25.03(7.23%) b) At 25.04min; MS(70ev) m/z 175(M+1 6.58 %), 115(100 %), 91(99 %), 117(60 %), 81(24 %), 145(42 %), 128(35 %), 65(24.48 %), 133(27.82 %), 93(16 %), 156(16 %).

[13]   Barone, V., Arnaud, R. (1996), *Chem.Phys.Lett.* **251**, 393.

[14]   Parr, R.G., Yang, W. (1989, *Density Functional Theory of Atoms and Molecules*, Oxford University Press, Oxford.

[15]   Labanowski, J.K., Andzelm, J.W., (Eds.), (1991), *Density Functional Methods in Chemistry*, Springer, New York.

INTRAMOLECULAR TRAPPING OF STRAINED BICYCLIC ALLENE IN CARBON ATOM REACTIONS

FATMA SEVİN* AND BÜLENT DÜZ
Department of Chemistry, Hacettepe University,
06532, Beytepe-Ankara, Turkey

Abstract

We report here the intermolecular trapping of endo-bicyclo[3.2.1]octa-2,3-dien-6-ol generated by reacting C with endo-bicyclo[2.2.1]hept-5-en-2-ol. The fragmentation reaction through endo-bicyclo[3.2.1]octa-2,3-dien-6-ol favours product (5Z)-octa-1,5-dien-7-yn-3-ol rather than the intramolecular trapping product 2-oxatricyclo [4.2.1.0^{3,8}] non-4-ene. We also describe this reaction theoretically by means of the AM1 semiempirical method.

1. Introduction

One-carbon ring expansions attract ever-growing interest[1]. Nowadays, this method constitutes an important approach which has been incorporated in some elegant synthetic strategies[2]. After its discovery, the rearrangement of gem-dihalocyclopropanes observed with strained cyclic olefins was extrapolated to the norbornane series[3]. The highly strained bicyclo[3.2.1]octa-2,3-diene obtained by this way has been trapped with t-ButOK and styrene as [2+2] cycloadduct[4]. On the other hand, the one-carbon ring expansion can be made by reaction of atomic carbon[5]. However, thus far ring-expansion by reaction of atomic carbon with norbornene and its derivatives has not been reported in the literature.

During our investigations of reactions with atomic carbon, we became interested in the intermolecular trapping of endo-bicyclo[3.2.1]octa-2,3-dien-6-ol, **3** which is generated by reacting C with endo-bicyclo[2.2.1]hept-5-en-2-ol, **1**. We also used the AM1 semiempirical method to make a theoretical calculation for this reaction.

Y. Imamoglu and L. Bencze (eds.), Novel Metathesis Chemistry: Well-Defined Initiator Systems for Specialty Chemical Synthesis, Tailored Polymers and Advanced Material Applications, 309–312.

2. Experimental Section

The reactions were carried out by cocondensing **1** with arc-generated carbon at 77K in an evacuated reactor[6]. The substrate was bled into the reactor and condensed on a 77K surface while an arc was struck intermittently between two high-purity graphite rods. At the conclusion of the reaction, the reactor was allowed to warm to room temperature. Chloroform was added in the reactor, and then the filtered reaction mixtures were analysed by GC/MS using a Shimadzu QP5050A mass spectrometer interfaced with a Shimadzu GC17A gas chromatograph equipped with a 50m x 0.32mm ID OPTIMA-5-1.0m capillary column.

3. Computational Method

The AM1 semiempirical method [7] in RHF formalism was selected to carry out the calculations with PC SPARTAN'02 Essential program [8]. Geometries were optimized without constraint in the final run of the calculation. Vibrational frequencies were then computed to characterize each stationary structure. Imaginary frequencies were animated graphically for all transition structures to ensure that the motions were appropriate for converting reactants to products.

4.   Results and Discussion

The highly strain bicyclic allene **3** can be generated via double bond addition from the reaction of **1** with the carbon atom. Therefore, we focus on this reaction in our calculations. The calculated structures and their relative energies with respect to the reactants are given in Scheme 1. Their heat of formation values are also presented in parentheses.

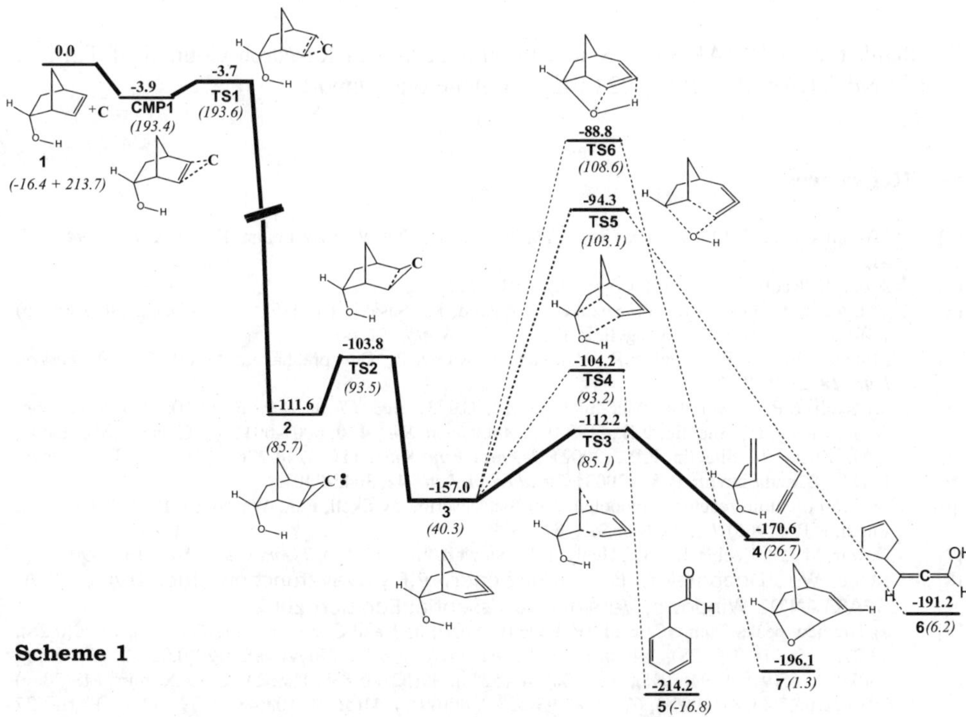

Scheme 1

As seen in Scheme 1, we found a low-energy complex **CMP1** which is more stable than the reactants. **CMP1** gives **2,** a carbene intermediate over a low activation energy. With an activation energy of 7.8 kcal/mol, **2** turns into bicyclic allene **3** with an energy of approximately –45.4kcal/mol.

The conversion of **3** to **4**, to **5**, to **6** or directly to the intramolecular trapping product **7** requires the calculated activation energies of 44.8, 52.9, 62.8, 68.3 kcal/mol, respectively. At the AM1 level, although the activation energies of these conversions are so high, the fragmentation reaction through allene **3** appears to favour product **4** rather than the intramolecular trapping product **7**.

In an IR spectrum of a sample taken from the reaction mixture of **1** with C atoms, no aldehyde and etheric bands were found. Even though **5** and **7** have not been seen in the IR spectrum, the mass fragments of **5** and **7** are not expected to be very different from those of product **4**. In a GC/MS analysis, five new peaks have been obtained. One of them has a yield of 12.85 %; it originates in fragmentation of product **4**. The other peaks correspond to trace amounts[9]. The mass fragment ions of these peaks indicate that they are caused by isomers of product **4** except for the peak at 14.2 minute. It fits the mass ions of the dimer of bicyclo[2.2.1]heptan-2-ol which may be formed by a radical reaction encountered often in C atom reactions [5].

Our results clearly indicate that it is desirable to reexamine of the products resulting from the **1** + C atom reaction both experimentally and theoretically; in more sophisticated experiments and at a higher level of theory. We plan to undertake this reexamination. As is often the case, the application of computational methods to interpretation of the experimental results points the way toward additional experiments.

5. Acknowledgements

We thank the TUBITAK (The Scientific and Technical Research Council of Turkey, Project No. TBAG-AY/284 (102T142) for generous support.

6. References

[1] a)Wolinsky, J. (1961), *J. Org. Chem.*, **26**, 704 ; b) Hückel, W. ; Hartmann, K. *Chem. Ber.* 1947, 80, 41.

[2] Alder, K. Reubke, R. (1958), *Chem. Ber.*, **91**, 1525

[3] a) Eguchi, S., Furukuwa, Y., Suzuki, T., Kondo, K., Sasaki, T.(1985), *J.Org.Chem.*, **50**, 1895; b) Jefford, C. W., Brun, P., Waegell, B. (1971), *Org. Synth.*, **51**, 60.

[4] Mohanakrishnan, P., Tayal, S.R., Vaidyanathaswamy, R., Devaprabhakara, D. (1972), *Tetrahedron Lett.* **28**, 2871-2872.

[5] (a) Skell, S.P., Havel, J.J., McGlinchey, M.J. (1973), *Acc. Chem. Res.*, **6**, 97-105. (b) Armstrong, B.M. , Zheng, F., Shevlin, P.B., (1998) *J. Am. Chem. Soc.*, **120**, 6007-6011. (c) Geise, C.M., Hadad, C.M., Zheng, F., Shevlin, P.B., (2002), *J. Am. Chem. Soc.* , **116**, 3248-3262. (d) Sevin, F., Sökmen, İ., Düz, B. and Shevlin, P.B., (2003) *Tetrahedron Lett.* **44**, 3045-3407.

[6] The carbon atom reactor is modeled after that describe by Skell, P.S., Wescott, L.D., Golstein, J.P., Engel, R.R. (1965) *J. Am.Chem.Soc.* **87**, 2829.

[7] Dewar, M.J.S., Zoebisch, E.G., Healy, E.F., Stewart, J.J.P. (1985), *J. Am. Chem. Soc.*, **107**, 3902.

[8] Hehre, W.J., Deppmeier, B.J., Klunzinger, P.E., Wavefunction, Inc. Irvine, C.A. SPARTAN'02 Windows, Version 1.0 Essential Eddition 2002.

[9] a) The new peaks from GC result (R.Time (%Area)) of **1** and C atom are 11.76(12.85), 13.20(0.28), 13.27(0.35), 14.07(0.28), 14.20(0.42); b) At 11.76min; MS(70ev) *m/z* 122(0.02%), 67(100%), 79(92%), 65(91%), 94(59%), 68(54%), 41(52%), 39(52%), 83(14%); c) At 13.20 min; MS(70ev) *m/z* 122(0.47%), 81(100%), 67(81%), 93(55%), 80(60%), 41(56%), 108(48%), 39(41%); d) At 13.27 min; MS(70ev) *m/z* 124(M+2.09%), 67(100%), 93(74%), 41(58%), 80-82(54%), 39(47%), 79(32%), 57(23%), 31(38%); e) At 14.07 min; MS(70ev) *m/z* 124(M+2, 3.58%), 80(100%), 79(60%), 78(22%), 77(19%), 39(19%), 41(10%) ; f) At 14.2min; MS(70ev) *m/z* 126(M+2, 4.25%), 66(100%), 43(55%), 79(43%), 91(31%), 39(22%), 106(20%).

INDUSTRIAL APPLICATIONS OF OLEFIN METATHESIS

J. C. MOL
Institute of Molecular Chemistry, Faculty of Science,
Universiteit van Amsterdam,
Nieuwe Achtergracht 166, 1018 WV Amsterdam, The Netherlands
E-mail: jcmol@science.uva.nl.

1. Introduction

Olefin metathesis opens up new industrial routes to important petrochemicals, oleochemicals, polymers, and specialty chemicals. Industrial applications of the olefin metathesis reaction known until 1997 have been described elsewhere, including processes that have not been commercialized and processes that are no longer commercial [1]. Here we will discuss the present situation.

2. Industrial Processes for the Production of Normal Olefins

2.1. PRODUCTION OF PROPENE

Propene is obtained mainly from naphtha steam crackers (globally about 65%) as a coproduct with ethene, and from fluid catalytic cracking (FCC) units. Relatively small amounts are produced by propane dehydrogenation and by coal gasification via Fischer-Tropsch chemistry. Strong global demand for propene, however, presently outpaces supply from these conventional sources.

2.1.1. *The Phillips triolefin process.*
An alternative route to propene is by applying the metathesis reaction for the conversion of a mixture of ethene and 2-butene into propene. This process, called the Phillips triolefin process, was originally developed by Phillips Petroleum Co., USA, and was commercialized in 1966 for the conversion of propene into ethene and butene, due to less propene demand at that time. The Phillips process in the reverse direction, equation (1), is now offered by ABB Lummus Global, Houston (USA), for license as olefins conversion technology (OCT) for the production of propene.

$$CH_2{=}CH_2 \ + \ CH_3CH_2{=}CH_2CH_3 \ \rightleftharpoons \ 2\ CH_3CH_2{=}CH_2 \tag{1}$$

Figure 1 shows a simple process flow diagram of the OCT process. Fresh C_4's plus C_4 recycle are mixed with ethene feed plus recycle ethene and sent through a guard bed to remove trace impurities from the mixed feed. The feed is heated before entering the

Y. Imamoglu and L. Bencze (eds.), Novel Metathesis Chemistry: Well-Defined Initiator Systems for Specialty Chemical Synthesis, Tailored Polymers and Advanced Material Applications, 313–322.
© 2003 *Kluwer Academic Publishers. Printed in the Netherlands.*

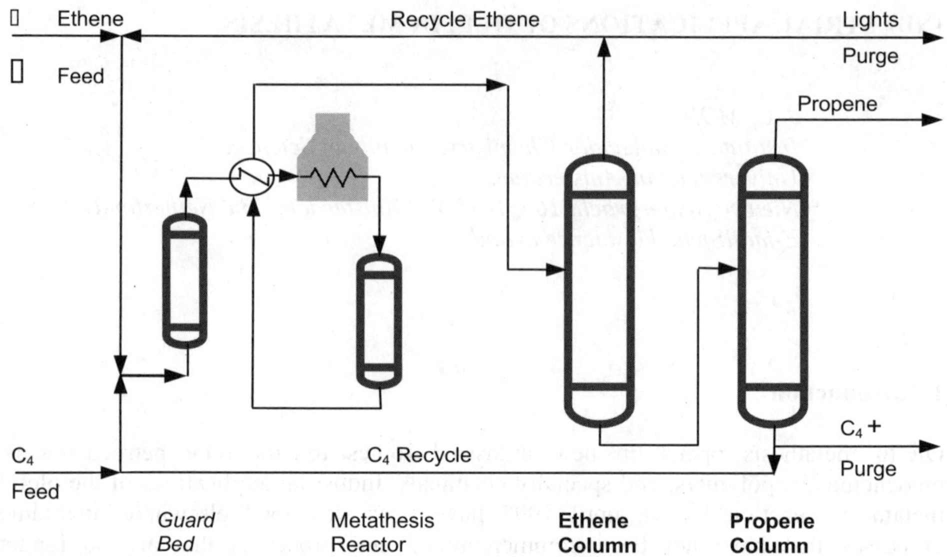

Figure 1. The OCT Process (Reversed Phillips Triolefin Process) [3].

metathesis reactor. The reaction takes place in a fixed-bed reactor over a mixture of WO_3/SiO_2 (the metathesis catalyst) and MgO (an isomerization catalyst) at >260°C and 30-35 bar [2]. 1-Butene in the feedstock is isomerized to 2-butene as the original 2-butene is consumed in the metathesis reaction. The conversion of butene is above 60% per pass and the selectivity for propene is >90%. The reactor is regenerated on a regular basis [3].

At the end of 1985 Lyondell Petrochemical Co. started to operate a 136,000 t/yr (tons per year) OCT plant in Channelview, Texas (USA) for the production of polymerization-grade propene via cross-metathesis between ethene and 2-butene. In their process, part of the ethene from cracking units is dimerized to 2-butene, using a homogeneous nickel catalyst developed by Phillips, which reacts with the rest of the ethene to produce propene [4]. The process has been retrofitted to the company's steam cracker in Channelview [5].

In particular, naphtha steam crackers with an integrated metathesis unit are an interesting alternative for producing more propene. In December 2001, BASF Fina Petrochemicals (a 60-40 joint venture between BASF and Atofina) brought on stream a world-scale steam cracker plant in Port Arthur, Texas (USA), which will integrate the OCT process to enhance the production of propene in relation to ethene. This plant produces 920,000 t/yr of ethene and 550,000 t/yr of propene, but when the metathesis unit is added (at the end of 2003), it will adjust the output to 830,000 t/yr of ethene and 860,000 t/yr of propene [6]. Raw C_2 and C_4 feedstocks are supplied directly by the steam cracker, but additional C_4's will be taken from the Sabena Chemicals' C_4 splitter being planned by a joint venture of Shell Chemicals, BASF and Atofina; by-products are recycled to the cracker.

Mitsui Chemicals will install the OCT technology to increase the propene capacity at it's olefins plant at its Osaka works in Japan by 140,000 t/yr to 420,000 t/yr.

Completion of the project is expected in August 2004 [7]. The OCT process will also be used at Shanghai Secco Petrochemical, a joint venture of BP Chemical, Sinopec and Shanghai Petrochemical Corporation, which is building a 900,000 t/yr naphtha cracker integrated with an OCT unit to produce a total of 590,000 t/yr of propene at Caojing, China. The complex is scheduled for start-up in the first half of 2005 [8]. PCS (Petrochemical Corp. of Singapore), a joint venture between Shell Chemicals and Sumitomo Chemical, is studying an increase in propene capacity at its olefin units by 200,000–300,000 t/yr using the OCT technology [9].

2.1.2. *The Meta-4 Process.*

The Institut Français du Pétrole (IFP) and the Chinese Petroleum Corporation (Kaoshiang, Taiwan) have jointly developed a process for the production of propene, called Meta-4. In their process, ethene and 2-butene react with each other in the liquid phase in the presence of a Re_2O_7/Al_2O_3 catalyst at 35°C and 60 bar. The (equilibrium) conversion is 63% per pass.

The process features semi-continuous countercurrent contact between catalyst and the liquid reactant medium. The feed stream enters the bottom of the metathesis reactor and leaves at the top. As the catalyst descends through the reactor, it becomes less active and must be regenerated. A small fraction of the catalyst is withdrawn periodically from the bottom of the reactor and transferred to the top of a regenerator. The fully regenerated catalyst leaves at the bottom of the regenerator and is returned to the top of the metathesis reactor. Unconverted ethene and butene are recycled to the reactor [10,11].

The process is not yet commercialized, mainly because of the cost of the catalyst and the requirement of a high purity of the feed stream. This metathesis technology is presently offered by France's Axen, a subsidiary of IFP, formed in 2001 through the merger of IFP's licensing division with Procatalyse Catalysis & Adsorbents.

2. 2. PRODUCTION OF 1-HEXENE

A semi-works unit using the OCT process for butene metathesis to produce 1-hexene is under construction at Sinopec's ethene plant in Tianjin, and will be started up in early 2003 [12].

2.3. PRODUCTION OF NEOHEXENE

Neohexene (3,3-dimethyl-1-butene) is an important intermediate in the synthesis of Tonalide®, a synthetic musk perfume. It is also used to make Terbinafine®, an anti-fungal agent. A neohexene unit located within Chevron Phillips Chemical Company LP's Houston Chemical Complex was built in 1980 with a capacity of 1400 t/yr [13].

The process is based on the dimer of isobutene, which consists of a mixture of 2,4,4-trimethyl-2-pentene and 2,4,4-trimethyl-1-pentene. Cross-metathesis of the former with ethene yields the desired product; equation (2). The latter is not wasted since a dual catalyst can be used to ensure that it is isomerized to 2,4,4-trimethyl-2-pentene as this gets used up by metathesis.

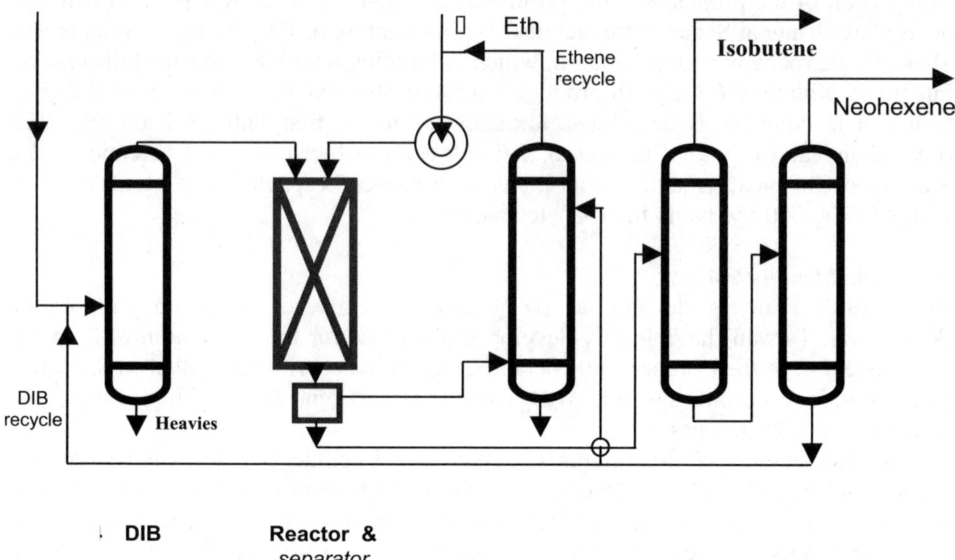

Figure 2. The Neohexene process. DIB = di-isobutene [11].

$$(2)$$

With a 1:3 catalyst mixture of WO_3/SiO_2 and MgO an average conversion of the di-isobutene of 65-70% and a selectivity to neohexene of approx. 85% is achieved at 370°C and 30 bar (molar ratio ethene/di-isobutene = 2). The coproduct isobutene is recycled to an isobutene dimerization reactor.

Figure 2 shows the process scheme of the neohexene process. Commercial di-isobutene is first fractionated to remove an oxidation inhibitor, which would otherwise poison the dual catalyst system. The fractionated di-isobutene along with the ethene stream enters the top of the reactor containing the catalysts. The ethene, consisting of make-up and recycled ethene, is compressed to the required pressure before it enters the reactor. The separation of reactants and products is achieved by stripping and fractionation. The catalyst is reactivated from time to time, using a mixture of air and inert gas to control the temperature of the coke burn-off.

2.4. THE SHELL HIGHER OLEFINS PROCESS (SHOP)

A large-scale industrial process incorporating olefin metathesis is the Shell Higher Olefins Process (SHOP) for producing linear higher olefins from ethene. The process takes place in three stages.

In the first step, ethene is oligomerized in the presence of a homogeneous Ni-phosphine catalyst (at 90-100°C and 100-110 bar) in a polar solvent (1,4-butanediol) to give a mixture of linear even-numbered alpha olefins ranging from C_4 to C_{40} with a Schulz-Flory type of distribution; equation (3). The catalyst is prepared *in situ* from a nickel salt, *e.g.*, nickel chloride, and a chelating phosphorous-oxygen ligand by reduction with sodium boron hydride.

$$(n + 2)\ CH_2{=}CH_2 \longrightarrow \quad\quad\quad (3)$$

$$n = 0 - 36$$

The olefins formed are immiscible with the solvent; product and catalyst phases are thereby readily separated so that the Ni catalyst can be recycled repeatedly. The C_6-C_{18} 1-alkenes are separated from the product mixture by distillation. This fraction can be further fractionated into individual compounds, which can be used as comonomer in polyethene production or converted into products such as synthetic lubricants, plasticizer alcohols, detergent alcohols, synthetic fatty acids etc. The remaining lighter ($<C_6$) and heavier ($>C_{18}$) alkenes go to purification beds, which remove catalyst and solvent residues that would otherwise deactivate the isomerization catalyst.

In the second step, these lighter and heavier alkenes undergo double-bond isomerization over a solid potassium metal catalyst to an equilibrium mixture of internal alkenes; equation (4).

$$R{-}CH{=}CH_2 \ \rightarrow \ R_1{-}CH{=}CH{-}R_2 \quad\quad\quad (4)$$

In the third step, this mixture is passed over an alumina-supported molybdate metathesis catalyst, resulting in a statistical distribution of linear internal alkenes with both odd and even numbers of carbon atoms via cross-metathesis reactions such as reaction (5). This yields about 10-15wt% of the desired C_{11}-C_{14} linear internal alkenes per pass, which are subsequently separated by normal distillation.

$$CH_3CH{=}CHCH_3 \ + \ CH_3(CH_2)_7CH{=}CH(CH_2)_9CH_3 \ \rightleftharpoons$$

$$CH_3CH{=}CH(CH_2)_7CH_3 \ + \ CH_3CH{=}CH(CH_2)_9CH_3 \quad\quad (5)$$

The isomerization and metathesis catalysts operate at 100-125°C and 10 bar. The remaining lower ($<C_{11}$) and higher ($>C_{14}$) alkenes are recycled. The product consists of $>96\%$ of linear internal C_{11}-C_{14} alkenes. These can then be converted into detergent alcohols, via a hydroformylation process, or into detergent alkylates [14.15].

Shell Chemicals operates a SHOP unit at Stanlow (UK) with a capacity of 270,000 t/yr and two large-scale SHOP units at Geismar, Louisiana (USA) with a total capacity of 600,000 t/yr of higher olefins. In 2002 Shell has brought on stream a third SHOP unit at their Geismar location for the production of another 320,000 t/yr of higher olefins. This expansion brings Shell Chemicals' total worldwide production capacity to

1,190,000 tons of linear alpha and internal olefins per year; these are sold under the trade name Neodene[16].

3. Industrial Processes for the Production of Polymers

In the polymer field ring-opening metathesis polymerization (ROMP) of cycloalkenes is an attractive process for making polyalkenamers when based on cheap monomers or possessing special properties compensating for a high price. Several industrial processes involving homogeneously catalyzed ROMP have been developed and brought into practice. See also the book of Dragutan and Streck [17].

3.1. POLYOCTENAMER

Since 1980 Degussa-Hüls has been producing Vestenamer 8012, the metathetical polymer of cyclooctene. This polymer also goes under the name TOR (*trans*-polyoctenamer). The polymerization is performed in hexane as a solvent in the presence of a WCl_6-based catalyst, giving almost 100% yield. The polymer consists of two distinct parts: a high-molecular-weight fraction (molecular weight $>10^5$) and a low-molecular-weight fraction consisting of a series of cyclic oligomers. This is readily explained in terms of a competition between a propagation reaction (6a) and an intramolecular backbiting metathesis reaction (6b).

$$
\begin{array}{c}
M=CHR \\
+ \\
HC=CH
\end{array}
\longrightarrow
\;\;\;
\longrightarrow
\;\;\;
M\text{--}CH\,\left[\;HC\text{--}CHR\;\right]_p
\qquad (6a)
$$

$$
\left[\begin{array}{c} M\text{--}CH \\ RHC\text{--}CH \end{array}\right]_n
\longrightarrow
M=CHR \;+\; HC\text{--}CH\,\left[\,HC\text{--}CH\,\right]_n
\qquad (6b)
$$

The product has a purity of 99.5%. The *cis* to *trans* ratio, which determines the degree of crystallinity, is controlled by the polymerization conditions. The *trans* double bond content of Vestenamer 8012 is 80%, the crystallinity 30%, and the molecular weight 75000. Used as a blending material, it offers possibilities for the improvement of properties of rubber compounds and for use in rubberized cement. An additional type, Vestenamer 6213, with a lower *trans* content (60%) and, therefore, lower crystallinity (10%), has been developed to provide for low-temperature applications where the admixture of the standard type would lead to excessive stiffening [18].

3.2. POLYNORBORNENE

The first commercial metathesis polymer was polynorbornene, which was put on the market in 1976 by CdF-Chimie in France, and in 1978 in the USA and Japan, under the trade name Norsorex[19].

The polymer is obtained by ROMP of 2-norbornene (bicyclo[2.2.1]-2-heptene), which is made from dicyclopentadiene and ethene, and gives a 90 % *trans* polymer with a very high molecular weight ($> 3 \cdot 10^6$ g/mol) and a glass transition temperature (T_g) of

37°C; eq. (7). The process uses a $RuCl_3$ catalyst in butanol, operates in air, and produces a useful elastomer, to be used for oil spill recovery, as a sound barrier, or for damping.

$$n \quad \longrightarrow[\text{RuCl}_3 \text{ / butanol}]{\text{ROMP}} \quad \tag{7}$$

Norsorex® is presently produced in France by Elf Atochem, and developed worldwide by the Japanese company Nippon Zeon.

3.3. POLYDICYCLOPENTADIENE

Much interest has been shown in the ROMP of *endo*-dicyclopentadiene (DCPD), obtained as a by-product from naphtha crackers. If only the highly strained norbornene ring opened, a linear polymer should be formed; eq. (8a). However, under certain conditions the double bond in the disubstituted cyclopentene ring may also undergo metathesis, thereby giving rise to cross-linking, eq. (8b).

Equation (8)

The product is a tough, rigid, thermoset polymer of excellent impact strength. Quite large objects can be produced via a reaction injection molding (RIM) process. The commercial production of molded objects from DCPD-based feed using RIM technology has been developed mainly by the BFGoodrich Co., under the trade name Telene[®], and by Hercules Inc. under the trade name Metton[®]. The latter is now produced by Metton America, Inc. at La Porte (USA), who have also licensed their process to the Teijin-Metton Co. in Japan.

In the RIM technique, two monomer streams are used. In the Metton[®] liquid molding resin (LMR) system developed by Hercules, one stream contains DCPD monomer, catalyst (WCl_6 + $WOCl_4$), nonylphenol (to solubilize the tungsten compounds in the monomer), additives (such as antioxidants), and fillers. The other stream contains DCPD monomer, cocatalyst ($EtAlCl_2$), retarder, additives and fillers. The two streams pass first into a mixing chamber and then into the mold, where an exothermic polymerization takes place at a high rate after a short induction period. The solutions of the individual catalyst components in the monomer are stable, and the length of the induction period can be controlled [20].

In the Telene[®] process the procatalyst is a tetrakis(tridodecylammonium) octamolybdate, activated with a mixture of Et_2AlCl, propanol and $SiCl_4$ Up to 10% trimer of cyclopentadiene is added to the monomer to increase cross-linking in the polymer, while the trimer also lowers the melting point of DCPD. The T_g of the product is typically 150°C. BFGoodrich Co. have licensed their process to Nippon Zeon, which produces it under the trade name Pentem[®]. In the USA Telene[®] RIM polymers are presently produced by Cymetech, LLC.

Poly(DCPD) has won several marine, recreational vehicle and off-road utility vehicle applications around the world, such as tonneau covers, snowmobile hoods, tractor fenders, and heavy truck panels.

3.4. HYDROGENATED POLYMERS

Since 1991 Nippon Zeon has been producing the polymer Zeonex[®], of the general type **1,** obtained by ROMP of norbornene and related (multi-ring) monomers, followed by partial or total hydrogenation.

1

Equation (9) gives an example starting with tetracyclododecene (TCD). Zeonex[®] is an amorphous, colorless and transparent polymer with a high T_g (140°C) and low moisture absorption. These properties make it very suitable for optical applications (disks, lenses, and camera prisms).

$$\text{(9)}$$

Zeonor® (**2**) is an amorphous hydrogenated co-polymer and has been commercialized since 1998. It has been accepted for use in optical, electronics, and automobile applications.

(2)

4. Conclusions

In the chemical process industry, olefin metathesis has now become a process with large-scale applications using heterogeneous as well as homogeneous catalysts systems. More commercial applications are to be expected, in particular considering the recent development of highly active ruthenium metathesis catalysts that are more tolerant to functional groups and resistant towards moisture and oxygen. In the area of fine chemicals, interesting products will be synthesized in the (near) future via metathesis reactions, such as biologically active compounds (*e.g.* pharmaceuticals, insect pheromones, prostaglandins, etc) and advanced polymeric materials via ROMP. Moreover, the metathesis reaction has also favourable perspectives for the oleochemical industry [21].

5. References

1.   Ivin, K.J. and Mol, J.C. (1997) *Olefin Metathesis and Metathesis Polymerization*, Academic Press, London, Chapter 17.
2.   Parkinson, G. (2001) *Chemical Engineering,* **108** (8) August, p. 27.
3.   Kantorowicz, S.I. (2002) *C₄ Processing Options to Upgrade Steam Cracker and FCC Streams*, Lecture
        presented at the 2nd Asian Petrochemicals Technology Conference, 7-8 May, Seoul, Korea.
4.   *Chemical Week* (1985), 20 November, p. 54.
5.   Scott, A. (1999) *Chemical Week*, 3 November, p. 41.
6.   *European Chemical News* (2002), 25 March, p. 20.
7.   *Chemical Week* (2002), 16 October, p. 16.
8.   Wood, A. (2002) *Chemical Week*, 27 February, p. 40; *News Release*, ABB Lummus Global, 15 April, 2002.
9.   N. Alperowicz (2002) *Chemical Week*, 6 March, p. 16.
10.   Amigues, P., Chauvin, Y., Commereuc, D., Lai, C.C., Liu, Y.H. and Pan, J.M. (1990) *Hydrocarbon Process.,* **69**, October, p. 79;
11   Cosyns, J., Chodorge, J., Commereuc, D. and Torck, B. (1998) *Hydrocarbon Process.,* **77**, March, p. 61.
12.   Wood, A. (2002) *Chemical Week*, 13 February, p. 32.
13.   Banks, R.L., Banasiak, D.S., Hudson, P.S. and Norell, J.R. (1982) *J. Mol. Catal.* **15**, 21.
14.   Freitas, E.R. and Gum, C.R. (1979) *Chem. Eng. Progr.* **75** (1), 73.
15.   Sherwood, M. (1982) *Chem. Ind. (London)*, 994.
16.   *Shell Chemicals Information Handbook* (2002). Shell Chemicals Limited.

17.   Dragutan, V. and Streck, R. (2000) *Catalytic Polymerization of Cycloolefins,* Elsevier Science B.V., Amsterdam, The Netherlands.
18.   Diedrich, K.M. (1993) in *Ullmann's Encyclopedia of Industrial Chemistry*, 5th ed. Vol. A23, VCH, Weinheim, p. 302.
19.   Marbach, A. and Hupp, R. (1989) *Rubber World,* June, p. 30.
20.   Breslow, D.S. (1993) *Prog. Polym. Sci.* **18**, 1141.
21.   Mol, J.C. (2002) *Green Chemistry* **4**, 5.

SYNTHESIS OF CYCLOBUTANE HYDROCARBONS BY COMBINATION 0F (2+2)$_\pi$-CYCLOADDITION AND OLEFIN METATHESIS. THEIR ABILITIES AS EFFECTIVE PROPELLANTS.

E.SH. FINKELSHTEIN[a], V.S.ANUFRIEV[b], B.S.STRELCHIK[c], S.P.CHERNYKH[b], M.L. GRINGOLTS[a], E.B.PORTNYKH[a], A.B.AMERICK[d], F.YU.CHELKIS[e]

[a]*Topchiev Institute of Petrochemical Synthesis, Russin Academy of Science.*
29, Leninsky prospect, Moscow, 119991, Russia, E-mail: fin@ips.ac.ru
[b]*"VNIOS" 12, Radio street, 107005, Moscow, Russia*
[c]*Novokuibyshevsk Petrochemical Company, 446214, Novokuibyshevsk, Russia.*
[d]*LUKOIL Company, 101000 Moscow, Sretenskii bul., 11*
[e]*N.P.O. "Energomash", Burdenko street, Moscow, Khimki, Russia.*

1. Introduction

Chemistry of cyclobutane hydrocarbons and their derivatives has been a subject of scientific attention science the beginning of the last century up to present time. The reasons of such a long attention is in special features of highly strained 4-membered ring structure and peculiar reactivity of cyclobutane compounds [1]. In application sense the compounds of this type demonstrate a range of interesting properties such as biological activity [2,3], capability to store energy [4,5], monomers activity to polymerization [1], etc. They are used as insecticides, in particular insect pheromone components [2,3], highly effective propellants [4.5] and energy accumulators [6].

Wide application of cyclobutane derivatives including hydrocarbons is limited by difficulties in their syntheses requiring the use of multi-step techniques and expensive reagents [1]. In full measure this situation falls into bi- and polycyclic hydrocarbons. We believe that catalytic approaches to syntheses of highly strained hydrocarbons can make them accessible raw materials and semi-products for laboratory and technological goals.

Our contribution to the chemistry 3- and 4-membered carbocycles has been always connecting with catalytic reactions such as dimerization, cycloaddition and olefin metathesis.

Unfortunately, the latter reaction can't be used as cyclo-forming one in the case of small cycles, including cyclobutanes. All our attempts to realize ring-closing metathesis of various 1,5- dienes failed even at very high delution. On the other hand, olefin metathesis is more preferable for doubling and multiplying a number of small rings in molecules providing high energy store, in comparison to di- or oligomerization.

2. Synthesis of cyclobutane hydrocarbons.

We started with methylenecyclobutane (MCB) as a substrate for metathesis because it has both 4-membered ring and a semi-cyclic double bond, in principle

Y. Imamoglu and L. Bencze (eds.), Novel Metathesis Chemistry: Well-Defined Initiator Systems for Specialty Chemical Synthesis, Tailored Polymers and Advanced Material Applications, 323–340.

available for catalytic conversions. On the other hand MCB was known as an accessible by-product formed in the course of a large scale industrial process of isoprene production by " dioxane method" from isobutylene and formaldehyde [7,8].

Together with MCB we have studied behavior of a series of 1,1-disubstituted ethylenes in the presence of typical heterogeneous Mo- and Re-containing catalysts in combination with alkyltins and alkylleads as promotors [9-14].
Non-promoted Mo- and Re- oxide catalysts showed activity in cleavage of the 4-membered rings performing probably by cationic mechanism.
The use of alkyltins and alkylleads as promotors in combination with renium-on-alumina catalyst allowed to realize first a smooth metathesis of MCB [9,15]. The most activity and selectivity has been achieved by the use of tetraethyllead. It seems the latter was first successfully used by us in olefin metathesis. The scheme 1 demonstrates reaction ways of various 1,1-disabstituted ethylene in the presence of Re_2O_7/A_2O_3-$SnBu_4$ (or $PbEt_4$) at mild conditions (0-60°C).

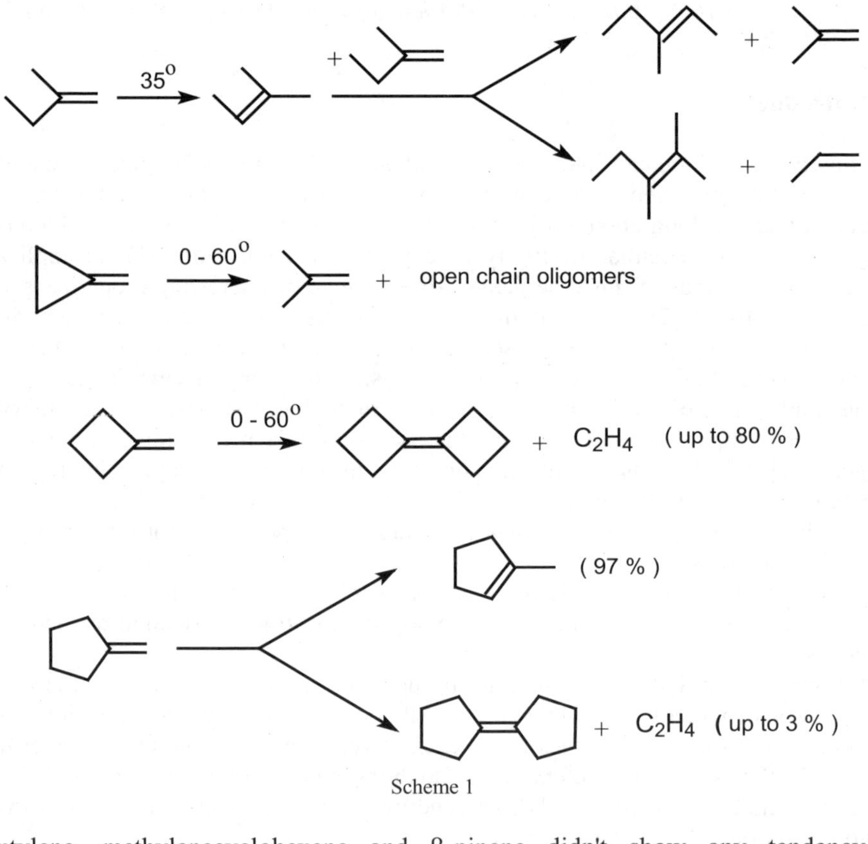

Scheme 1

i-Butylene, methylenecyclohexene and β-pinene didn't show any tendency to metathesis. 2-Methylbyt-1-ene, methylenecyclopentene, methylenecyclohexene and β-pinene converted actively into internal olefins by the double bond shift from terminal position. Methylenecyclopropane underwent a cleavage of highly strained 3-membered ring. MCB turned out to be the only 1,1-disubstituted ethylene capable to effective exchange by alkylidene groups. One of the reasons of such behavior is in a less tendency to double bond shift due to an increase of a strain in 4-membered ring at the cost of arised extra sp^2 C-atom [16]. In other cases (except i-butylene and methylenecyclopropane) isomerization of this type performs smoothly as the main reaction because of thermodynamic profit.

Possible structure reasons of this phenomena are on the Scheme 2 demonstrating results of electronographic study of some 1,1-disubstituted ethylenes and their possible metathesis products.

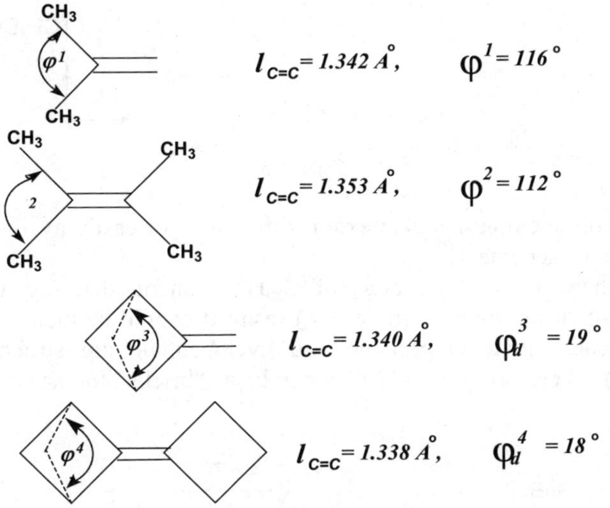

$$l_{C=C} = 1.342\ \mathring{A}, \qquad \varphi^1 = 116°$$

$$l_{C=C} = 1.353\ \mathring{A}, \qquad \varphi^2 = 112°$$

$$l_{C=C} = 1.340\ \mathring{A}, \qquad \varphi_d^3 = 19°$$

$$l_{C=C} = 1.338\ \mathring{A}, \qquad \varphi_d^4 = 18°$$

Scheme 2

In the first pair of molecules the marked difference in molecular parameters is observed. It seems that four methyl groups induce increased steric hindrances in tetramethylethylene. In contrast to it, in the second pair these parameters values are close. Therefore, transformation of MCB into dicyclobutylidene doesn't have thermodynamic limitations.

Because of hyperconjugation, $>=$ Re is too stable to transform for primery carbene $CH_2=$Re and tetramethylethylene. On the other hand it seems that $\diamondsuit=$Re doesn't have a conjugation effect. In addition, this metallocarbene it is not so stable because of a strain in the cycle and double bond [16].

There weren't any problem in cometathesis with participation of MCB. In the case of cometathesis with trimethylallylsilane we have realized a unique technique of elongation of α-olefin chains, based on a special peculiarity of Si-C-bond in allylsilanes (scheme 2, [17]). The presence of σ(Si-C)-π conjugation provides a selective rupture of Si-C_{All}-bond in the cometathesis product under influence of a proton (H$^+$). It leads to 100% selective formation of α-olefin (here vinylcyclobutane) elongated by one C-atom in comparison to the starting one (here MCB).

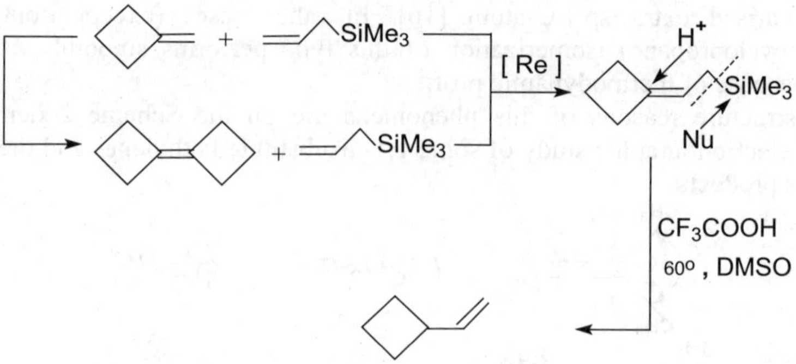
SiMe3
[Re]
H+
SiMe3
SiMe3
Nu
CF3COOH
60o , DMSO

Scheme 3

Cometathesis with cycloolefins also extends the range of easily available cyclobutane hydrocarbons (see scheme 4).
Among them there are hydrocarbons of $C_{10}H_{16}$ content, that say artificial terpens having typical smell of forestry (pine-tree) resin. It should be mentioned that MCB , methylcyclobutene, vinylcyclopropane and cyclopentene are structure isomers of isoprene (C_5H_8). Therefore, formally they can be as "bricks" for assembly of terpenoid structures.

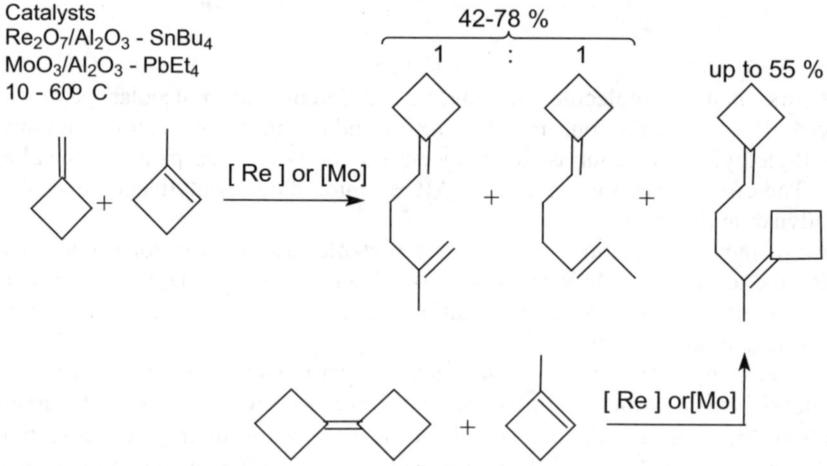
Catalysts
Re2O7/Al2O3 - SnBu4
MoO3/Al2O3 - PbEt4
10 - 60o C
42-78 %
1 : 1
up to 55 %
[Re] or [Mo]
[Re] or[Mo]

Scheme 4

Physico-chemical parameters of cometathesis products obtained from MCB have been published in [18].
Hydrocarbons containing two and more 4-membered rings in molecular structure (including terpenoids $C_{10}H_{16}$ and $C_{15}H_{24}$) can be achieved not only by cometathesis but also according to classical cationic oligomerization (Scheme 5). In the Table 1 experimental results unpublished earlier are presented.
We could easily prepare rather large-scale samples of hydrogenated dimers for tests on energy parameters together with dicyclobutyl prepared by metathesis.

Scheme 5

TABLE 1. Oligomerization of methylencyclobutane.

№	Catalyst	Mass ratio MCB/cat. г/г	T, °C	Reaction time, h	Oligomer yield, % mass.			Dimer Content		Dimers selectivity, %
					Altogether	$C_{10}H_{16}$ (dimers)	$C_{15}H_{24}$ (trimers)	I	II	
1	CaNiY	5	60	3	76	26	12	55	25	34
2	CaNiLaY	5	40	40	58	38	10	60	28	66
3	CaNiLaY	10	60	12	64	34	16	59	26	53
4	CaNiLaY	5	80	3	89	43	19	52	35	49

So, MCB is a promising raw material for preparation of a family of cyclobutane hydrocarbons practically impossible-to-prepare by any other methods.

However, synthetic abilities of MCB are limited. That is why a general strategy of effective synthesis of cyclobutane hydrocarbons containing semi-cyclic or exo-cyclic double bonds suitable for metathesis or dimerization should be elaborated.

From technological point of view it has to be a catalytic cyclo-forming way based on simple starting olefins of petrochemical origin and accessible catalysts.

Capability of double bonds in [2+2]-cycloaddition has long been known [19,20]. This is the most general way to 4-memberd cycles. In some cases cyclobutane compounds obtained as a result of cycloaddition, have semi- or exo-cyclic double bonds available for metathesis.

In 1972 L.Cannel realized the cross reaction between butadiene and ethylene in the presence of Ti-containing homogeneous catalysts (120-180°C, TiBr$_4$-2,2-bipyridyl or TiOBu$_4$) [21]. Later we have used our own homogeneous catalysts in this reaction (130-140°C, Ti(OSiMe$_3$)$_4$-Et$_3$AlkMgX) [22]. The yields of vinylcyclobutane have been improved up to 45%. But the reaction was expanded for a range of highly strained cyclobutane derivatives used as starting olefin comonomers.

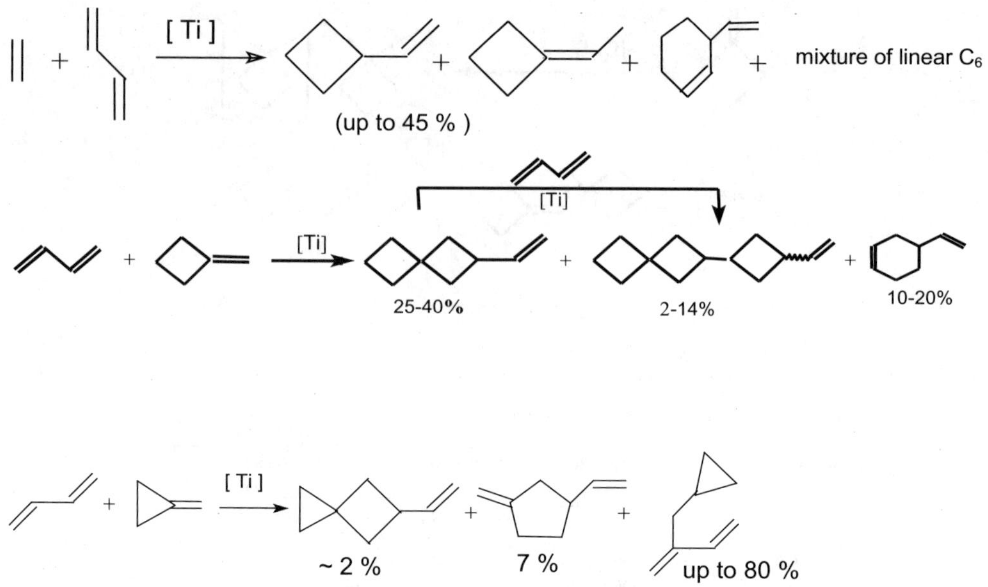

328

[Ti]

(up to 45 %)

[Ti]

[Ti]

25-40%

2-14%

10-20%

+ mixture of linear C$_6$

[Ti]

~ 2 %

7 %

up to 80 %

Scheme 6

A disadvantage of cyclodimerization with participation of ethylene is in rather high ethylene pressure (>100 atm) needed for the active formation of 4-membered ring. Lower pressure leads to more intensive [4+2]$_\pi$-cyclization resulting in vinylcyclohexane and linear codimerization, resulting in linear hexadienes.

Recently we have tried MCB and methylenecyclopropane as starting olefins for this reaction. The cross-cycloaddition of MCB with butadiene proceeded smoothly at 70-90°C at 1 atm pressure. Probably the presence of some strain in semi-cyclic double bond (~ 1,8 kkal/mol) is a reason for its increased activity [16]. As a rule the product of secondary cycloaddition of butadiene to vinylspiroheptane was also indicated in the reaction mixtures (60-90°C, C$_p$TiCl$_2$Br – RMgX). The reaction between methylenecyclopropane and butadiene in the presence of the same catalyst proceeded in other manner yielding mainly the products of linear codimerization and [3+2]-cycloaddition as a result of 3-membered ring opening (Scheme 6).

Allene, light hydrocarbon of petrochemical origin, has long been known (from the beginning of the last century) as a contributor of cyclobutane structures [23,24]. Allene [2+2]$_\pi$-cyclodimerization can be realized both thermally and catalytically. As a result, complicated mixtures of dimethylenecyclobutanes and higher cyclobutane products are obtained. Efforts of many chemists having an interest to cyclobutane hydrocarbons were focused on an increase of selectivity in this reaction [24]. The best results have been achieved by one of us: 489°C, continual process, tube reactor, 75% yield of dimethylenecylobutanes [25,26].

Unfortunately, 1,3-dimethylenecyclobytane suitable for metathesis is a minor product of allene [2+2]$_\pi$-cyclodimerization. 1,2-Isomer contains conjugated double bonds of 100% cis-like configuration. They are very suitable for formation of stable complexes with transition metals of catalyst systems. It leads to poisoning of catalysts. All our attempts to involve 1,2-dimethylenecyclobutane into metathesis were unsuccessful.

Probably catalytic cyclodimerization of allene is a promising approach which can provide an increased content of needed 1,3-isomer.

Recently it was shown with participation of one of us that [2+2]-cycloaddition of allene and functional olefins (such as acrylonitryl) is a real way to MCB [27-29]. We have managed to prepare and involve into metathesis 1,3-dimethylecyclobutane. It was isolated from allene pyrolisis dimer fraction with help of a pricise fractionation in a 100 plate column filled with copper rings (together with Drs K.Makovetsky and A. Yurkovetsky) [11,30]. This diolefine of 99% concentration turned out to be an active substrate for ADMET and cometathesis with MCB in the presence of our permanently used rhenium-on-alumina catalysts. A number of 4-membered rings can be regulated by reaction conditions (reaction time) and the presence of the second substrate, such as MCB. This allowed to prepare a unique set of unsaturated and saturated cyclobutane hydrocarbons containing two or three 4-membered cycles.

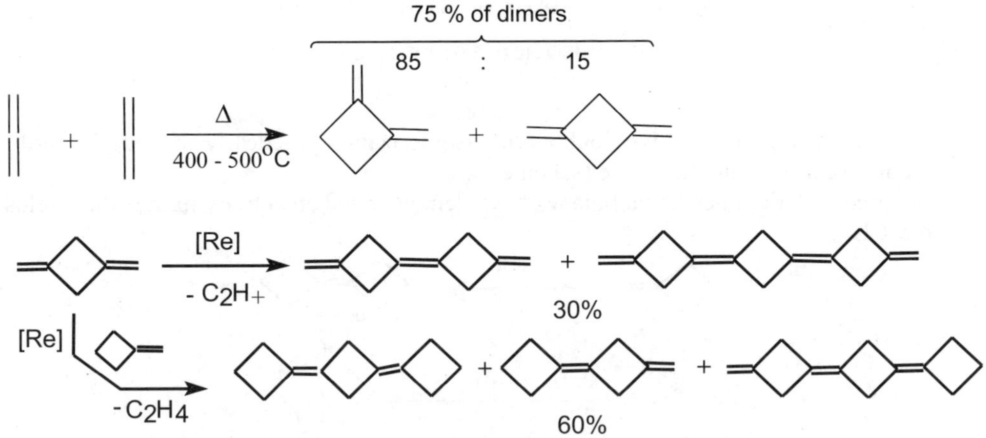

Scheme 7

We have realized an effective shift of terminal double bonds in vinylcyclobutane hydrocarbons to semi-cyclic position in the presence of typical anionic and Ti-containing catalysts. Anionic catalyst, tert-butoxypotassium (t-BuOK), turned out to be the best one

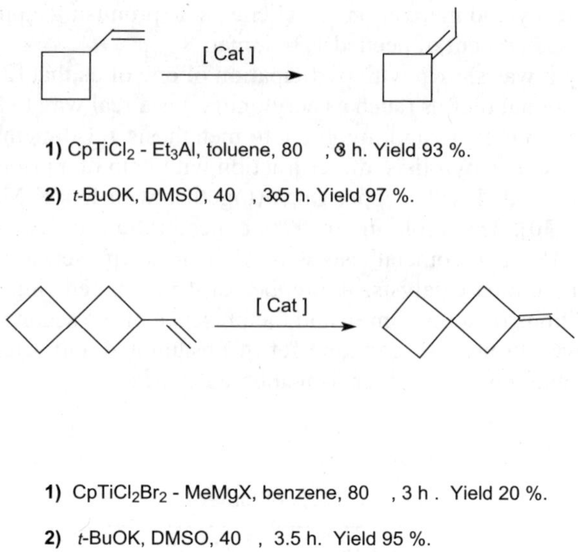

[Cat]

1) CpTiCl$_2$ - Et$_3$Al, toluene, 80 , 3 h. Yield 93 %.

2) t-BuOK, DMSO, 40 , 3–5 h. Yield 97 %.

[Cat]

1) CpTiCl$_2$Br$_2$ - MeMgX, benzene, 80 , 3 h . Yield 20 %.

2) t-BuOK, DMSO, 40 , 3.5 h. Yield 95 %.

Scheme 8

providing a smooth selective double bond isomerization practically without the further double bond shift into the cycle (scheme 8).

The prepared alkylidenecyclobutanes have demonstrated an activity in metathesis close to MCB.

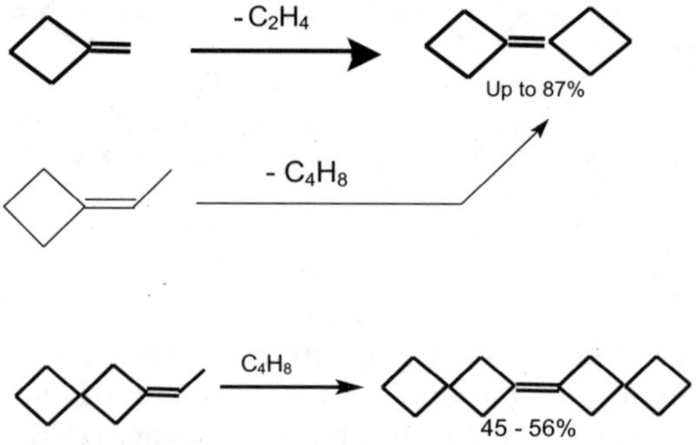

- C$_2$H$_4$

Up to 87%

- C$_4$H$_8$

C$_4$H$_8$

45 - 56%

Scheme 9

In the case of ethylidenecyclobutane (or MCB) a technique including "fractionation' in the course of the reaction, which is a continual removal of ethylene or butene-2 as well as dicyclobutyl products from catalytic zone, was used. Vinylspiroheptane is too heavy for this technique. Its metathesis was carried out by the regular way in liquid phase providing a removal of only butene-2. That is why the yield of the target dispirane was not so high. It is important that highly strained spiro-systems didn`t break up in the course of metathesis in the regular reaction conditions (Re$_2$O$_7$/Al$_2$O$_3$-PbEt$_4$, 20-45°C). An increase of reaction temperature gave a rise of by-products formed as a result of spiro-system destruction.

So, the shift of double bond from terminal to semi-cyclic position according to the scheme 8 and subsequent metathesis accompanied by elimination of non-cyclic bonds according to the scheme 9 allows to prepare bi- and polycyclobutanes containing only one non-cyclic bond (that is the shortest bridge) between 4-membered cycles. Vinylcyclobutane derivatives doesn't present any problem at metathesis. Practically they work like linear α-olefins. Spiro-systems were kept unchanged at 20-45°C (Scheme 10).

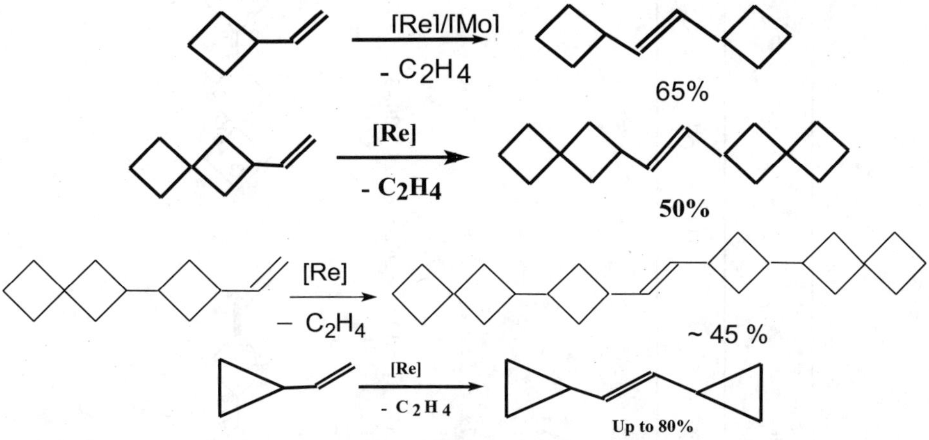

Scheme 10

Vinylcyclopropane has shown less activity but the yields of dicyclopropylethylene were substantially increased using "rectification" technique, that is continual removal of ethylene and dicyclopropylethylene from catalytic zone. In the Tables 2 and 3 some characteristics of polycyclobutanes first prepared by us are displayed.

TABLE 2. NMR ^{1}H data for some polycyclobutane hydrocarbon obtained by olefin metathesis and [2+2]-cycloaddition

Compound	NMR ^{1}H, δ (ppm)
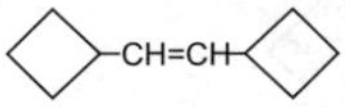 ◇-CH=CH-◇	1..9 m (12H, CH$_2$ cycl), 2.86 m (2H, CH cycl), 5.41 m (2H, CH=).
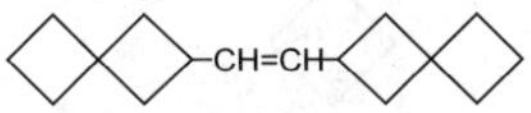 ◇◇-CH=CH-◇◇	1..9 m (20H, CH$_2$ cycl), 2.7 m (2H, CH cycl), 5.4 m (2H, CH=).
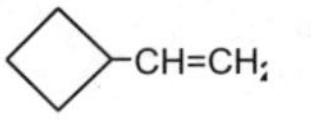 ◇-CH=CH$_2$	1.6-2.3 m (6H, CH$_2$ cycl), 2.9 m (1H, CH cycl), 4.8-5.2 m (2H,=CH$_2$), 6.0 m (1H,CH).
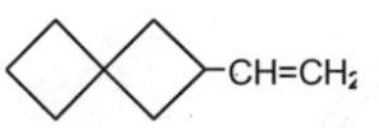 ◇◇-CH=CH$_2$	1.9-2.8 m (10H, CH$_2$ cycl), 2.77 m (1H, CH cycl), 4.78-5.1 m (2H,CH= cycl), 5.87 d (1H, CH=).

As a result of this study we propose a flexible technologically promising synthetic strategy allowing preparation of highly strained hydrocarbons containing desired number of 4-membered rings with various aliphatic bridges. The majority of synthesized hydrocarbons is difficult or even impossible-to-prepare by any other existing methods. The starting materials are accessible simple olefins such as ethylene, allene, butadiene and MCB of petrochemical origin (see the general scheme 11).

TABLE 3. NMR 13C data for some polycyclobutane hydrocarbons obtained
by [2+2]$_\pi$-cycloaddition.

Com- pound	NMR-^{13}C (*ppm*)/mult.												
	C^1	C^2	C^3	C^4	C^5	C^6	C^7	C^8	C^9	C^{10}	C^{11}	C^{12}	C^{13}
1	112.70 / d	143..38 / t	34.13 / d	41.01 / s	36.26 / t	35.48 / t	41.55 / 2t	17.21 / t					
2	13.81 /q	116.86 /d	135.95 /s	44.31 / t	42.37 / t	39.89 / s	35.60 / 2t	16.83 / t					
3	111.80 / t	153.14 / d	36.39 / d	37.92 / t		34.31 / d	34.47 / d	35.79 / t	35.01 / t	39.63 / s	31.40 / t		16.40 / t
4	111.53 / t	143.41 / d	36.45 /d	38.47 / t		34.69 / d	34.34 / d	35.85 / t	35.12 / t	40.35 /s	30.43 / t		16.40 / t

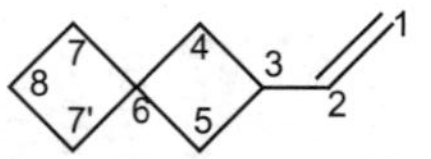

(1)

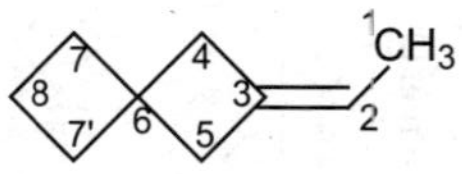

(2)

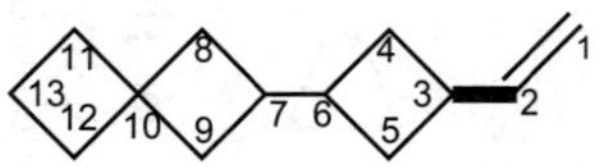

(3)

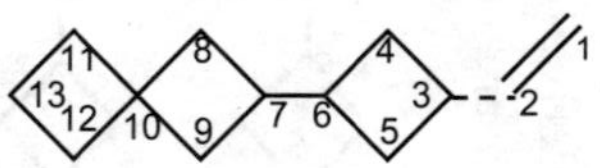

(4)

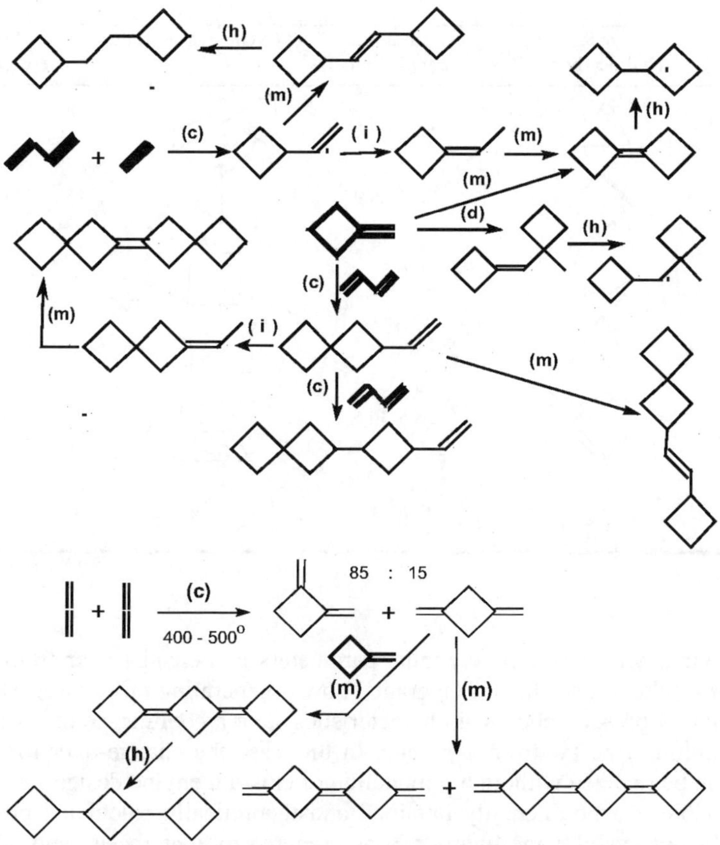

Scheme 11

3. Application of cyclobutane hydrocarbons as effective propellants.

So, our long-standing efforts have led us to elaboration of a common flexible catalytic strategy allowing to synthesize a great variety of highly strained bi- and polycyclobutane hydrocarbons on the basis of accessible petrochemical olefins. It is metathesis that provides a doubling and even multiplying a number of 4-membered rings in hydrocarbon molecules.

Since the middle of the last century strained cyclic hydrocarbons have been considered as potential highly effective liquid propellants containing enhanced energy storage in the molecules.

In the table 4 the strain energy values for cyclic hydrocarbons of different type are displayed showing that energy storage in 4-membered rings is rather high [31-33]. It is clear that energy released by combustion of a strained hydrocarbons in a rocket engine chamber has to be higher than in the case of strain free hydrocarbons entered into rocket kerosene.

TABLE 4.   Strain energy in carbocycles

Cyclic Hydrocarbon	Strain energy(kkal/mol)	Cyclic hydrocarbon	Strain energy(kkal/mol)
	27.6-28.1		63.1
	26.2-26.9		63.9-65.0
	6.5		53.0-53.6
	52.6-54.5		18.5-19.0
	28.5-30.6		95.0
	41.0-41.7		28.8

The obvious way to improve engine parameters is a change-over from kerosene type fuel for synthetic ones having upgraded power-generating properties. At the same time the standard physico-chemical characteristics such as boiling point, viscosity, density, etc. should not be far from kerosene. In this case the change-over for such synthetic fuels can be realized without a remodeling of existing engine design.

In the table 5 experimentally obtained and theoretically calculated physico-chemical properties of cyclobutane hydrocarbons prepared by metathesis and related reactions are displayed. The important criteria of synthetic propellant effectiveness is ΔS_{sp} – an increase of a specific impulse relative to the regular kerosene (I_{sp}-specific impulse [kg/kg/sec]=[sec]; $I_{sp,s}$- specific impulse of a synthetic propellant [sec]; $I_{sp,k}$ – specific impulse for kerosene [sec]; $\Delta I_{sp}=I_{sp,s}$ - $I_{sp.k}$ [sec] [34].

Two parameters play role as contributors to the I_{sp} value: enthalpy of hydrocarbon formation (that is the strain in cyclic fragments) and mass H-contribution in the common hydrocarbon formula, that also should be high. These parameters on frequent occasions are in conflict to one another. An increase of 4-membered ring number in molecules leads to a hydrogen depletion decreasing the I_{sp}. Therefore, a number of strained rings in propellant molecules should be optimal. For example, spiro-fragments enlarge enthalpy value, but lead to a reduction of H-contribution. As a result an increase of I_{sp} value remains conservative in spite of high strain energy.

Analysis of physico-chemical properties demonstrates that cyclobutane hydrocarbons, synthesized by olefin metathesis provide I_{sp} value as 4,7 – 5,9 sec and densities - 0,83 – 0,95 kg/l.

TABLE 5. Experimental and calculated physico-chemical properties of cyclobutane hydrocarbons synthesised by metathesis and related reactions

Formula	Structure	Mass H − contribution H_f (%)	Entalphy of formation E_f (kkal/kg)	Density d_4^{20} kg/l	Increase of Specific impulse ΔI_{sp} [sec]	B.p., °C	M.p., °C
C_8H_{14}		12.805	80	0.83	5.85	134	-56
$C_{10}H_{16}$	*)	11.838	232	0.89	6.1	-	-
$C_{10}H_{18}$		13.124	-30	0.85	3.75	168-173	-129
$C_{11}H_{18}$	*)	12.075	166	0.89	5.5	-	-
$C_{12}H_{20}$		12.27	110	0.89	5.1	187-191	-47
$C_{14}H_{22}$		11.65	220	0.85	5.4	-	--73
$C_{15}H_{26}$		12.70	- 5.0	0.90	4.15	235-245	-105
$C_{16}H_{26}$		12.01	140	0.92	4.75		-
Kerosene "N" **)		13.9	-407	0.833	0.00		195-270

*) - Calculated data.
* *) – V.P. Glushko .Termodynamic parameters of Chemical fuels. Reference book. USSI of Sciences Moscow. 1965.

We can argue that dicyclobutyl obtained by metathesis in combination with hydrogenation is the best candidate for real application as a component of liquid propellants. Metathesis way to formation of needed bi- (or three) cyclic hydrocarbons is preferable in comparison to di- (or tri) merization way. At oligomerization the structure of dimer (or trimer) copies the monomer structure retaining energetically non-profitable C-C bonds unchanged. At metathesis these bonds are removed in the form of ethylene or butene-2, enriching the aimed hydrocarbon by highly strained cycles. (Scheme 12).

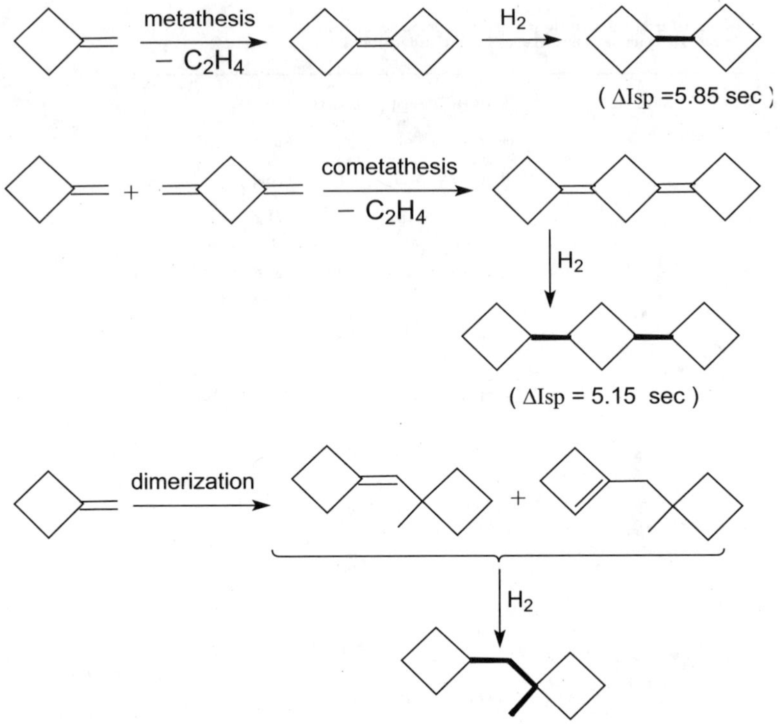

Scheme 12

In the Fig. 1   there are two curves demonstrating the increase of specific impulse (Δ I_{sp}) obtained experimentally in a real rocket engine chamber, as a result of change-over of regular kerosene for dicyclobutyl. The latter in combination with liquid oxygen provides the increase of I_{sp} value for rocket engines of RD-107 type by 2% [4]. For an engine power of high and medium scale the most profitable ratio of oxygen to hydrocarbon is in the range 2,4-3,0. This provides the most increase of specific impulse for rocket engines of this type.

The temperature of dicyclobutyl in the tank should be in the range from +50 up to $-$ 50°C [4].

So, dicyclobutyl is a convenient   effective propellant component for liquid rocket engines providing increased engine thrust. The prospects for its industrial production are in economic and technological problems.

4.   Some technological aspects of dicyclobutyl synthesis.

Briefly, possible technological process combining olefin metatheses of MCB followed by and hydrogenation of metathesis product, looks like follow according to our Russian Patent of 2000 [15].

Metathesis unit includes a reaction apparatus containing a column with heterogeneous catalyst (for example Re_2O_7/Al_2O_7-$PbEt_4$) and fractionation column. The catalyst is activated by oxygen and nitrogen flows at high temperature and than it is promoted. Catalyst column is connected with fractionation one having a still filled by a starting olefin located at the bottom of the column. Heating of column is realized by hot water. Olefin vapors arrive at catalytic column going through deflegmator, ethylene removes at the top, simultaneously. Liquid reaction products and unconverted olefin drain down into the fractionation column. In the course of the reaction boiling point of reaction mixture is increased at the cost of a rise of a heavy metathesis product (dicyclobutylidene, in the case of MCB metathesis) in the bottom still. The process performs up to practically full olefin conversion due to continuous removal of both metathesis products from catalyst reaction zone (98% conversion in the case of MCB, Tab. 6)

TABLE 6. Synthesis of dicyclobutyl from methylencyclobutane [15]

N	Metathesis catalyst	T, °C	Solvent at the bedining of metathesis	Hydroge-nation catalyst	Dicyclobutyl-idene Hydrogenation		Solvent at the bedining of hydrogenation	Dicyclobutyl Yield Calculated on Methylenecyclobutane,%
					T, °C	P,ата		
1.	$Re_2O_7/Al_2O_3-PbEt_4$	40	-	Ni/CrO_3	20-140	60-80	-	78,5
2.	$Re_2O_7/Al_2O_3-PbEt_4$	60	-	Ni/CrO_3	20-140	60-80	-	75,2
3.	$Re_2O_7.WO_3/Al_2O_3-SnMe_4$	100	-	Ni/CrO_3	20-140	60-80		70,8
4.	$Re_2O_7.MoO_3/Al_2O_3-SnBu_4$	60	-	Ni/kiselgur	20-140	60-80	-	70,2
5.	$Re_2O_7/Al_2O_3-PbEt_4$	40	-	Ni/CrO_3	20-140	60-80	Hexane	83,5
6.	$Re_2O_7.MoO_3/Al_2O_3-SnBu_4$	60	Hexane	Ni/CrO_3	20-140	60-80	-	80.4
7.	$Re_2O_7/Al_2O_3-PbEt_4$	40	Pentane	Ni/CrO_3	20-140	60-80	Pentane	93.0
8.	$Re_2O_7.MoO_3/Al_2O_3-SnBu_4$	60	Hexane	Ni/CrO_3	20-140	60-80	Pentane	85,2
9.	$Re_2O_7.WO_3/Al_2O_3-SnMe_4$	100	Heptane	Ni/kiselgur	20-140	60-80	Hexane	80,3
10.	$Re_2O_7/Al_2O_3-PbEt_4$	60	Hexane	Ni/kiselgur	20-140	60-80	Hexane	89,1
11.	$Re_2O_7/Al_2O_3-PbEt_4$	60	Hexane	Ni/CrO_3	20-140	60-80	Hexane	92.0
12.	$Re_2O_7.WO_3/Al_2O_3-SnMe_4$	100	Heptane	Ni/kiselgur	20-140	60-80	Tetra-decane	71.2
13.	$Re_2O_7/Al_2O_3-PbEt_4$	40	Pentane	Ni/CrO_3	20-180	60-80	Hexane	65,5
14.	$Re_2O_7/Al_2O_3-PbEt_4$	40	Pentane	Ni/CrO_3	20-140	60-100	Hexane	60,8
15.	$Re_2O_7/Al_2O_3-PbEt_4$	40	Pentane	Ni/CrO_3	20-180	60-100	Hexane	52,3

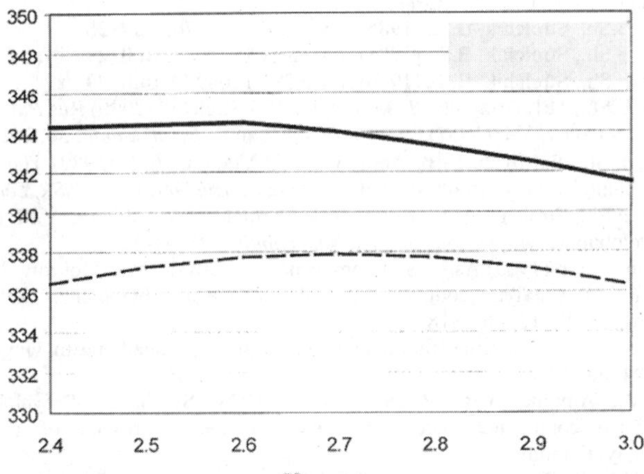

Figure 1. Energy parameters of propellants in a rocket engine [4]
——— Dicyclobutyl
— — Kerosene "N"

The unit works in a periodic mode of operation. Reaction products arrives to further processing, hydrogenation without additional fractionation due to high selectivity of the metathesis.

The temperature in catalyst zone is defined by boiling point value of raw materials and metathesis product mixtures. It should not be too high to avoid deactivation of rhenium containing catalysts. This limits a range of starting olefins by hydrocarbons of C_5-C_7 content.

5. References

1. Vinogradov, M.G. and Zinenkov, A.V. (1996) Chemistry of methylencyclobutane, *Uspehi khimii* **65(2),**140-155.
2. Bullups, W.E.,Gross,J.H., Smith, C.V. (1973) A synthesis of (±)-grandizol,   *J.Am.Chem.Soc.* **95,** 3438-3439.
3. Mori, K. (1998) *Comprevensive Natural Chemistry*, **8**, Pergamon Press.
4. Katorgin, B.I., Chelkis, F.Yu., Finkelshtein, E.Sh., Chernykh, S.P. (1998) *Rus.   Pat.* 2146334.
5. Anufriev, V.S., Bakulina, G.V., Zefirov, N.S., Palulin V.A., Finkelshtein E.Sh. (2000) About fundamental research program promising propellant of LRE, *Third international aerospace congress IAC`2000, Moscow*, Abst.,172. Finkelshtein, E.Sh, Gringolts, M.L., Portnykh, E.B., Anufriev, V.S., Chernykh, S.P., Strelchik, B.S., Chelkis, F.Yu. (2003) Synthesis of effective propellants on the basis of olefin metathesis as a key reactions. *27 –th Academic conference on astronautics*. Moscow, Anstr. 38-39.
6. Schwendeman, D.P., Kutal,C. (1977) Transition metal photoassisted valence isomerization of norbornadiene. An attractive energy-storage reaction, Inorg. *Chem.*16, 719-721.
7. Safonov, V.I., Vasileva, V.P.(1975) About the reaction of methylenecyclobutane formation in the isoprene synthesis from dioxane, *Neftehimiya,* **15**, 77-81.
8. Vostrikova, V.N., Moiseeva, T.P., Grigorjev, A.A., Chernykh, S.P. (1979)   Isolation of methylenecyclobutane from by-products of isoprene production, *Prom. SC*, 6, 11-13.
9. Finkelshtein, E.Sh., Strelchik, B.S., Zaikin, V.G., Vdovin, V.M. (1975) Catalytic synthesis of bicyclic hydrocarbons on the basis of alkylene- and vinylcycloalkanes, *Neftehimiya,*15, 667-672.

10. Finkelshtein, E.Sh., Strelchik, B.S., Portnykh, E.B., Vdovin, V.M., Nametkin, N.S. (1977) Disproportionation of olefins, including small carbocycles, *Dokl.Akad. Nauk SSSR*, **232**,1322-1325.
11. Finkelshtein E.Sh., Bykov, V.I., Portnykh, E.B. (1992) The olefin metathesis reaction – a versatile tool for fine organic synthesis, *J. Mol,Catal.*, **76**, 33-52.
12. Finkelshtein, E.Sh., Strelchik, B.S., (1985) *A.S. SSSR* 1165676, Bull. **25**
13. Finkelshtein, E.Sh., Strelchik, B.S., (1985) *A.S. SSSR* 1171087, Bull. **29**
14. Finkelshtein, E.Sh., Strelchik, B.S., (1986) *A.S. SSSR* 1264973, Bull. **39**
15. Finkelshtein, E.Sh., M.L.Gringolts, E.B.Portnykh, B.S.Strelchik (2000) Rus.*Pat.*, 2175312, Bull.**30**.
16. Zefirov, N.S., Sokolov, V.I. (1967) Strained double bond, *Uspehi khimii*, **36**, 241-268
17. Finkelshtein, E.Sh., Portnykh, E.B., Antipova, I.V., Vdovin, V.M. (1989) The lenthening of α–olefin hydrocarbon chain by its cometathesis with allylsilenes, *Izv.Akad.Nauk SSSR, Ser.Khim.*, 1358.
18. Finkelshtein. E.Sh., Strelchik, B.S., Kotov, S.B., Portnykh, E.B., Vdovin, V.M. (1985) Cometathesis of methylenecyclobutane with various olefins, *Neftekhimiya*, **25**, 41-47.
19. Lefebr, G., Shoven, I. (1973) Aspects of homogeneous catalysis, Mir, Moscow, 158-251.
20. Su, A.C.L. (1979) Catalytic codimerization of ethylene and butadiene, Adv. Organometal. Chem. Academic Press, N.Y., **17**, 269-318.
21. Cannel L.G. (1972) Cyclodimerization of ethylene and 1,3-butadiene to vinylcyclobutane, J. Amer. Chem.Soc., **94**, 6867-6869.
22. Vdovin, V.M., Amerik, A.B., Poletaev, V.A. (1978) Synthesis of cyclobutane derivatives by codimerization of conjugated dienes with ethylene in the the presence of catalysts on the basis of cyclopentadienyl titanium
23. Lebedev, S.V., (1913) Research in the field of polimerization. Diethylene hydrocarbons. Allene polymerization, *Gurnal Russkogo Khimicheskogo Obshestva (GRKhO)*, **45**,1357-1391.
24. Taber, A.M., Mushina, E.A., Krencel, B.A., (1987) *Allene hydrocarbons*. Nauka, Moscow, 87-116.
25. Chernykh, S.P., Rudenkov, A.U. (1982) Termal dimerization of allene, *Khim. Prom.,***6** ,335-336.
26. Chernykh, S.P., Taber, A.M. (1981) Synthesis of dimethylencyclobutanes. *Neftepererabotka i neftekhimija*, **7**, 48-50.
27. Menshikov, V.A., (2000) *Rus.Pat.* № 2174505, Bull. **28**
28. Menshikov, V.A., (2000) Rus.Pat. № 2175962, Bull. **32**
29. Lange,C.A., Tur, M.N., Salamatova, T.A., Chernykh, S.P. (2001) Application for the *Rus.Pat.* №20000128849 .
30. Yurkovetskii,A., (1986) Polymerization of dimethylenecyclobutanes. Dissertation.
31. Feruson, L.N. (1970) Ring strain and reactivity of alicycles , *J.Chem.Educ.*, **47**, 46-53.
32. Schleger, P.R., Williams, J.E., Braucham, K.R. (1970) The evaluation of strain in hydrocarbons, *J.Am.Chem.Soc.*,**92**, 2377-2388.
33. Boatz, J.A., Gordon, M.S., Hildebrandt, R.L. (1988) Structure and bonding in cycloalkanes and monosilacycloalkanes, *J.Amer.Chem.Soc.*, **110**, 352-358.
34. Alemasov, V.E. (1969) *Theory of rocket engines*, Mashinostroenie, Moscow

DIRECTIONAL SYNTHESIS OF MEMBRANE MATERIALS BY ROMP OF SILYL-SUBSTITUTED NORBORNADIENES AND NORBORNENES

J.V.ROGAN, M.L.GRINGOLTS, N.V.USHAKOV, YU.P. YAMPOLSKII, E.SH.FINKELSHTEIN
Topchiev Institute of Petrochemical Synthesis, Russian Academy of Sciences, 29, Leninskii prospect, 119991 Moscow, Russia

1. Introduction

A clarification of correlation between chemical structure of polymers and their gas transport parameters is an important task in membrane science, because it promises to solve the problem of directional construction of materials having high gas permeability and selectivity. In this respect ROMP of silylnorbornenes is a suitable subject of investigation. It allows preparing series of polymers bearing regularly changed substituents located at the desired places in the polymer backbones of a different microstructure.

Earlier we have synthesized the ROMP polynorbornenes having F-, Cl-, silyl-, alkyl and Ge- substituents [1-6]. It was shown that poly(trimethylsilyl norbornene) has the highest gas permeability. Positive effect of trimethylsilyl group on this parameter has been observed earlier by examples of the different type polymers - poly(vinyltrimethylsilane) and poly(trimethylsilylpropyne) [7]. Therefor we decided to study influence of the number of trimethylsilyl groups and their location at backbone of poly(cyclopentylenevinylene)s and poly(cyclopentenylenevinylene)s on the gas transport parameters.

2. Results and Discussion

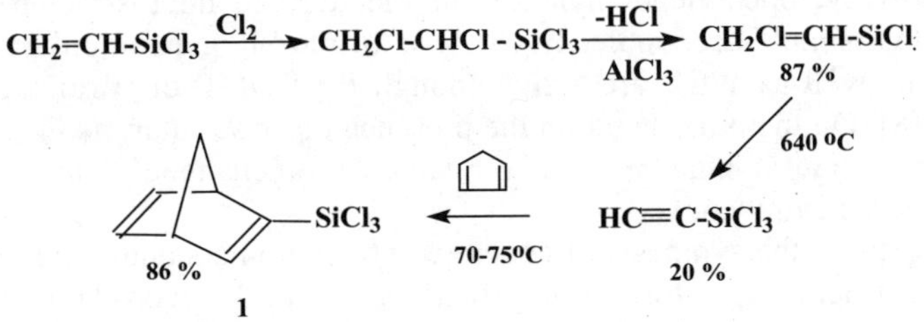

$$CH_2=CH-SiCl_3 \xrightarrow{Cl_2} CH_2Cl-CHCl-SiCl_3 \xrightarrow[AlCl_3]{-HCl} CH_2Cl=CH-SiCl \quad 87\%$$

$$\xrightarrow{640\ ^oC} HC\equiv C-SiCl_3 \quad 20\%$$

86%   **1** $-SiCl_3$ $\xleftarrow{70-75^oC}$

Scheme 1

The synthesis of norbornadienyl-silyl derivatives was undertaken by the Diels-Alder reaction between cyclopentadiene and ethynyltrichlorosilane. The latter has been prepared from accessible vinyltrichlorosilane according to the Scheme 1:

Y. Imamoglu and L. Bencze (eds.), Novel Metathesis Chemistry: Well-Defined Initiator Systems for Specialty Chemical Synthesis, Tailored Polymers and Advanced Material Applications, 341–349.
© 2003 *Kluwer Academic Publishers. Printed in the Netherlands.*

342

Subsequent alkylation of **1** by means of the Grignard reagents has led to the desirable norbornadienes **2,3,4** (Scheme 2).

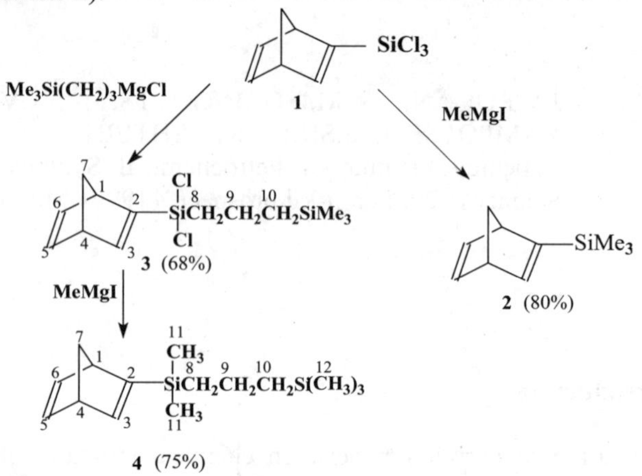

Scheme 2

Endo-exo-5,6-bis-(trimethylsilyl)-bicyclo[2.2.1]hept-2-ene (**6**) has been prepared from cyclopentadiene and *trans*-1,2-di(trichlorosilyl)ethylene by the Diels-Alder reaction followed by full methylation of all Si-Cl bonds with the yield of **6** equal to 72% (Scheme 3).

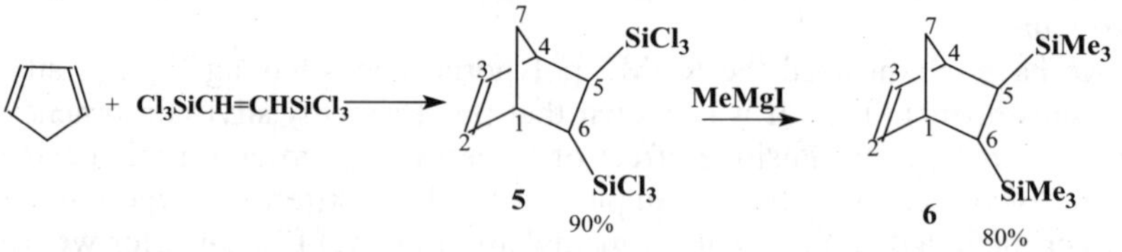

Scheme 3

Norbornene and its derivatives are highly strained structures that can be readily polymerized by ring-opening schemes, especially by ROMP. This reaction proceeds in the presence of a wide range of catalysts including ones, which are not active with respect of majority of monocycloolefins. That is why choosing of the catalysts for effective ROMP of norbornenes doesn't make any problem, if there is no special need to prepare poly(cyclopentylenevinylene)s of the definite microstructure. Earlier we have shown that simple accessible Ru-catalysts of the "1st generation" such as RuCl$_3$, RuCl$_2$(PPh$_3$) as well as WCl$_6$ are active enough for ROMP of various Si-containing norbornenes [8]. On the other hand, in the presence of these catalysts the Si-containing side groups are intact and retained unchanged in the target poly(cyclopentylenevinylene)s.

This work directs to the synthesis of concrete new monomers and polymers of desired chemical structure for clarification of correlation between poly(cyclopentylenevinylene) structures (the presence, number and location of Me$_3$Si - side groups) and gas transport parameters of the respective polymer films. The main part of our earlier investigations in the framework of this problem published in [1-7] has been realized using Ru- and W- catalysts mentioned above. It was logical to use the same catalysts in the present work.

Thus, ROMP of synthesized silylnorbornadienes and norbornenes (Scheme 4) has been successfully undertaken in the presence of Ru- and W-containing catalysts such as

RuCl$_3$·3H$_2$O, RuCl$_2$(PPh$_3$)$_3$, and WCl$_6$ – 1,1,3,3-tetramethyl-1,3-disilacyclobutane(TMSB).

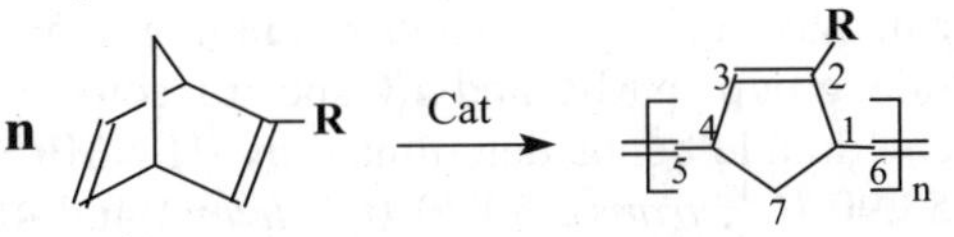

R = -SiMe$_3$(**7**), -SiMe$_2$(CH$_2$)$_3$SiMe$_3$(**8**), -SiCl$_3$(**9**)

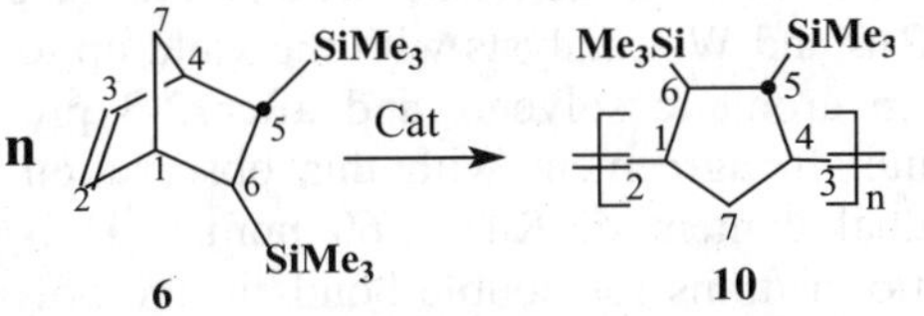

Scheme 4

The results of ROMP are presented in the Table 1. Rather high catalyst concentrations were used to avoid too high polymer molecular weight that could makes film casting a difficult problem. Only W-containing catalysts were shown to reveal activity in the ROMP of norbornadienes **1,2,4**. Incapability of norbornadiene substrates to ROMP in the presence of Ru-containing catalysts is probably explained by di-*endo* chelation of these molecules to metal carbene centre alike unsubstituted norbornadiene [9]. ROMP of highly reactive **1** proceeds with the formation of a polymer soluble in the reaction mixture. Its methoxylation in methanol led to a cross-linked insoluble polymer.

TABLE 1. ROMP of silylsubstituted norbornenes and norbornadienes

Monomer (M)	Catalyst	[M]/[cat] mol/mol	T, °C	τ, h	Yield %	M$_w$	PDI	T$_g$, °C	Cis, %
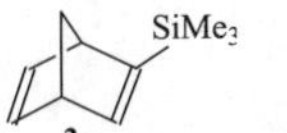 **2** (SiMe$_3$)		70	16	0,5	97	51100	2,9	108	
	WCl$_6$/TMSB	250	16	2,5	70	549000	3,1	-	42
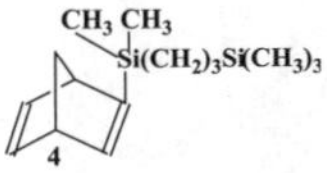 **4** (SiMe$_2$(CH$_2$)$_3$Si(CH$_3$)$_3$)	WCl$_6$/TMSB	92	25	24	40	318700	3,0	5	43
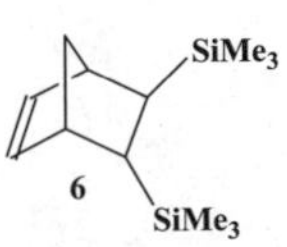 **6** (SiMe$_3$, SiMe$_3$)	RuCl$_3$·3H$_2$O	40	75	12	98	732900	1.4	167	6
	RuCl$_2$(PPh$_3$)$_3$	42	60	6	62	464400	5.3	154	9
	WCl$_6$/TMSB	50	20	24	93	291400	3,1	170	54
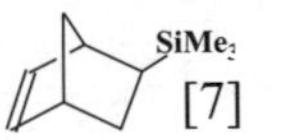 **[7]** (SiMe$_3$)	WCl$_6$/TMSB	1000	20	2	99.9	*	*	113	-
	RuCl$_3$·3H$_2$O	800	60	15	84	**	**	101	-

*-[η]=4 dl/g; **- [η] = 6 dl/g

The characteristics of the polymer **7** prepared in this work were identical to those described in the literature [10]. In the presence of WCl_6/TMSB the new monomer **4** was successfully polymerized giving previously unknown product containing two silicon atoms in the side group. NMR and IR spectra confirmed its structure. The cis/trans ratio of vinylene double bonds determined by ^1H NMR according to [10] and by ^{13}C NMR (54.06, 53.90 (C^1, *trans*), 50.70 (C^4, *trans*) and 48.15, 47.90 (C^1, *cis*), 45.84 (C^4, *cis*),) was 43/57.

Unlike the norbornadienes studied, the norbornene derivative **6** readily polymerizes in the presence of both Ru- and W- catalysts with the yield up to 98%. A novel polymer **10** is easily soluble in aromatic solvents and after 2-3 precipitations in methanol became perfectly white. In agreement with this observation, experimental analysis indicated a low residual content of Ru - 60 ppm. ^1H NMR spectra allowed a determination of a ratio cis/trans for double bonds in the polymer **10** on the basis of signals of the allylic protons (C^1H and C^4H, see experimental part).

Investigation of gas permeation parameters of the polymers **7** and **10** and our previous results can provide an opportunity to examine the following effects of the modification of polymer structure: the appearance of the double bond in the five-membered cycle: comparison of **7** with poly(trimethylsilyl norbornene) (PSNB) studied earlier [7]; the introduction of the second $Si(CH_3)_3$ group in the repeat unit: comparison of **10** with PSNB; the location and number $Si(CH_3)_3$ group: comparison of **10** with polymers PSNB and poly[dimethylsilyl(trimethylsilylmethyl) norbornene] (PDSNB) studied in [2].

Permeability coefficients and selectivity of gas permeation are presented in Table 2. It can be concluded that introduction of a double bond in the cyclopentylene ring in **7** does not led to a further increase in gas permeability in comparison with PSNB. The permeability coefficients are somewhat decreased for the most gases. In **7** double bonds are included in five-membered ring. Because of it the appearance of an additional double bond in the main chain doesn't result in increases in the glass transition temperature as can be seen from the T_g values of PSNB and **7** (Table 2).

TABLE 2. Permeability coefficients P (Barrer) and selectivity of gas permeation $\alpha=P_i/P_j$ of silylsubstituted polynorbornenes and polynorbornadiene.

Polymer	T_g (°C)	P, Barrer						$\alpha=P_i/P_j$				
		H_2	O_2	N_2	CO_2	CH_4	C_2H_6	H_2/N_2	H_2/CH_4	O_2/N_2	CO_2/CH_4	CH_4/C_2H_6
[7] (SiMe$_3$)	31	21	2.8	1.5	15.4	2.5	1.4	14.4	8.6	1.9	6.3	1.8
[7] PSNB	113	140	30	7.2	89	17	7.0	19.4	8.2	4.2	5.2	2.4
Me$_3$Si SiMe$_3$, 10	167	375	95	25	445	45	30	15	8.3	3.8	9.9	1.5
Me$_3$SiCH$_2$SiMe$_2$ [2] PDSNB	24	73	16	3.7	67	8.5	10.0	19.7	8.6	4.3	7.9	8.5
SiMe$_3$, 7	108	81	20	4.9	64	8.5	4.5	16.5	9.5	4.1	7.5	1.9

Entirely different result is observed for **10**, the polymer containing two adjacent Si(CH$_3$)$_3$ groups, prepared on RuCl$_3$·3H$_2$O. As has been mentioned in Introduction, gas permeation properties of numerous polymers of different classes (vinylic, acetylenic, etc) containing silyl groups (SiR$_3$) have been studied. In all the cases the SiR$_3$ group appeared only once in the repeat units. If substituents contained more than one Si atom within the same longer side group the observed permeability was rather modest (e.g. 16 Barrer for permeability coefficient of oxygen in PDSNB [2]). The polymer **10** presents a very rare example of design with two Si(CH$_3$)$_3$ that attached to different sites within a repeat unit. The data of the Table 2 indicate that its permeability is relatively high: this polymer is characterized by an increase in the permeability coefficients by a factor 3-5 in average as compared with PSNB containing only one Si(CH$_3$)$_3$ group. Introduction of the second Si(CH$_3$)$_3$ group in the polynorbornene main chain is accompanied by approximately the same increase in permeability as for the first group Si(CH$_3$)$_3$.

Usually, a trade-off between permeability and permselectivity is observed, that is materials with larger permeability are distinguished by smaller permselectivity and vice versa. Opposite trend is observed due to introduction of the second Si(CH$_3$)$_3$ group in the structure of **10**: a significant increase in permeability with almost no reduction of permselectivity is observed; in fact, the latter quantity even slightly increases for CO$_2$/CH$_4$, what was not observed for other norbornene polymers. On the other hand, another polynorbornene derivative **PDSNB** containing two silicon atoms in the same side group (Si(CH$_3$)$_2$CH$_2$Si(CH$_3$)$_3$) did not reveal such an effect. It is likely related with self-plasticization effect caused by long side group: T_g of this polymer decreased down to 24°C, whereas the glass transition temperature of PSNB is in the range 101-113°C [7] . The same conclusion is apparently true for polymerization product **8** of the present study containing even longer substituent: owing to this, the polymer is rubbery at room temperature (T_g=5°C, Tab.1).

3. Experimental

3.1. MEASUREMENTS

^{1}H and ^{13}C NMR-spectra were recorded in CDCl$_3$ solutions at 300 and 75,47 MHz, respectively using a Bruker MSL-300 spectrometer (CDCl$_3$ was used as an internal standard). IR spectra were obtained with a Specord M-82 spectrometer on KBr plates. Gel-permeation chromatography (GPC) analysis of the polymers was performed on a Waters system with a differential refractometer (Chromatopack Microgel-5, toluene as the eluent, flow rate 1 ml/min). Molecular mass and polydispersity were calculated by standard procedure relative to monodispersed poly(styrene) standards. Differential scanning calorimetry (DSC) was performed on Mettler TA 4000 system at a heating rate 20 °C/min. Chromato-mass-spectrometric analysis was carried out using a Kratos MS-25 RF-instrument.

A mass spectrometric apparatus with MI-1309 instrument was used in permeability measurements. The details of the technique were the same as described in [2]. Gas permeation rate was measured at 22±2°C with an upstream pressure in the range of 0.1 to 1 atm, and a downstream pressure about 10^{-3} mm Hg. In this instrument, time-lag method permitted the simultaneous determination of permeability coefficients (P).

The residual metal content in the polymers was determined by the method of Inductively Coupled Plasma Atomic Emission (ICP-AE).

3.2. MATERIALS

As catalysts, commercial RuCl$_3$·3H$_2$O, RuCl$_2$(PPh$_3$)$_3$ were used without preliminary purification. WCl$_6$ was sublimated and used as 0,01 M solution in absolute toluene. 1,2-bis(trichlorosilyl)ethene [12] were prepared according to published procedures. Solvents were purified and distilled before their use by standard methods.

3.2.1. Monomer preparation

The methods of monomer synthesis are disclosed below. All the procedures were carried out under the dry argon.
2-Trichlorosilylbicyclo[2.2.1]hepta-2,5-diene(**1**) and *2-trimethylsilylbicyclo[2.2.1]hepta-2,5-diene*(**2**) were prepared according to published procedures [11]. Their spectral properties are in agreement with the literature data.
2-dichlorolsilyl(trimethylsilylpropyl)bicyclo[2.2.1]hepta-2,5-dien (**3**)
To a stirred diethyl ether solution of ClMg(CH$_2$)$_3$Si(CH$_3$)$_3$ prepared from 0,125 mol Mg and 0,1 mol Cl(CH$_2$)$_3$Si(CH$_3$)$_3$, 20 g (0,0879 mol) **1** in 20 ml dry diethyl ether was slowly added at (-)70°C. Then temperature was slowly elevated, and the mixture was kept overnight at room temperature. After heating the mixture under reflux for 1.5 h, precipitated MgCl$_2$ was filtered out and washed with diethyl ether. Filtrate was

concentrated and fractionated under reduced pressure to yield 17,8 g (0,0584 mol) (68%) of a colorless liquid (b.p. 97-101°C, 0.25 mmHg). ^{1}H NMR (CDCl$_3$) δ 7.49 (d, C^{3}H), 6.79 (s, C^{5}H), 6.69 (s, C^{6}H), 3.90 (s, C^{4}H), 3.78 (s, C^{1}H), 2.00 (m, C^{7}H$_2$), 1.62 - 1.48 (m, C^{9}H$_2$), 1.24 (t, C^{8}H$_2$), 0.64 (t, C^{10}H$_2$), 0.011 (s, Si(CH$_3$)$_3$). IR (KBr): 1296, 1248 (Si – C), 864, 840, 776, 696 (Si –C), 552, 532 cm^{-1} (Si-Cl). MS (70eV), C$_{13}$H$_{22}$Cl$_2$Si$_2$, Calc.mass= 305: m/z = 304/306/308 (M$^{+\cdot}$), 289/291/293 (M$^+$ - CH$_3$), 223/227/225 (M$^+$ -CH$_3$ - C$_5$H$_6$), 196/198 (M$^+$ - Me$_3$Si - Cl), 73 ((CH$_3$)$_3$Si), 66 [C$_5$H$_6$]$^{+\cdot}$.

2-Dimethylsilyl(trimethylsilylpropyl)bicyclo[2.2.1]hepta-2,5-diene(**4**)

was prepared by methylation of 14 g (0.046 mol) of **3** with 2,2 mol MeMgI in dry diethyl ether at room temperature for 1 h and under reflux for 1 h. Diethyl ether was distilled off, and the product obtained was diluted with anhydrous hexane, ClMgI was filtered out and a residue was distilled under reduced pressure. The yield of **4** was 75% (b.p. 95,5-97,5 °C/ 2 mmHg). ^{1}H NMR (CDCl$_3$) δ 7.04 (d, C^{3}H), 6.68 (s, C^{5}H, C^{6}H), 3.73 (s, C^{4}H), 3.64- 3.60 (m, C^{1}H), 1.92 - 1.82 (m, C^{7}H$_2$), 1.28-1.41 (m, C^{9}H$_2$), 0.52-0.70 (m, C^{8}H$_2$, C^{10}H$_2$), 0.11 (s, C^{11}H$_3$), 0.08 (s, C^{11}H$_3$), 0.02 (s, C^{12}H$_9$). IR(KBr): 1296, 1248 (Si – C), 864, 832, 760, 696 cm^{-1} (Si – C). MS (70eV), C$_{15}$H$_{28}$Si$_2$, Calc.mass=264: m/z = 264(M$^+$), 249(M$^+$ - CH$_3$), 198 (M$^+$ - C$_5$H$_6$), 183 (M$^+$- CH$_3$-C$_5$H$_6$), 150 (M$^+$- Me$_3$SiCH$_2$CH$_2$CH), 149 (M$^+$- Me$_3$SiCH$_2$CH$_2$CH$_2$), 83 (M$^+$- Me$_3$Si(CH$_2$)$_3$ - C$_5$H$_6$), 73 [(CH$_3$)$_3$Si]$^+$, 66 [C$_5$H$_6$]$^{+\cdot}$.

Endo-exo-5,6-bis(trichlorosilyl)bicyclo[2.2.1]hept-2-ene (**5**).

6.4 g (97.5 mmol) of cyclopentadiene was added dropwise to 19.2 g (65 mmol) of *trans*-1,2-bis(trichlorosilyl)ethene for 1 h at 50°C. Then temperature was gradually raised to 90°C, and the mixture was stirred at this temperature for 3 h. Distillation of resulting mixture under reduced pressure afforded 21.1 g of **5** (yield 90%); b.p. 69-71°C/ 0.05 mmHg. ^{1}H NMR (CDCl$_3$) δ 6.36 – 6.27 (m, C^{2}H or C^{3}H), 6.27 – 6.18 (m, C^{2}H or C^{3}H), 3.50 (s, C^{1}H or C^{4}H), 3.39 (s, C^{1}H or C^{4}H), 2.27 –2.18 (m, C^{6}H *exo*), 1.65 – 1.53 (m, C^{7}H$_2$), 1.51 – 1.44 (m, C^{5}H *endo*). ^{13}C NMR (CDCl$_3$) δ 135.1, 130.5 (C^2, C^3), 45.2(C^7), 43.9, 42.4 (C^1, C^4), 32.9, 32.2 (C^5, C^6).

Endo-exo-5,6-bis(trimethylsilyl)bicyclo[2.2.1]hept-2-ene (**6**).

To a stirred solution of CH$_3$MgI in diethyl ether (prepared from 1.6 mol Mg and 1.6 mol CH$_3$I) under dry argon 55.3 g (153 mmol) of **5** in 45 ml of dry diethyl ether was slowly added under reflux for 2 h. Then the mixture was stirred under reflux for 1,5 h. Diethyl ether was distilled off, and the product obtained was diluted by 250 ml of anhydrous hexane. Upper layer was removed. Residue product was poured carefully in water and was acidified by a little portion of NH$_4$Cl water solution. Upper layer was removed again. After removing the solvent the residue was distilled under reduced pressure. The yield of **6** was 28.8 g (80 %), b.p. 81.0-81.5°C/3mmHg. ^{1}H NMR (CDCl$_3$) δ 6.1 – 6.03 (m, 1H, C^{2}H or C^{3}H), 5.93 – 5.85 (m, 1H, C^{2}H or C^{3}H), 3.02(s, 1H, C^{1}H or C^{4}H), 2.80(s, 1H, C^{1}H or C^{4}H), 1.25 – 1.16 (m, 1H, C^{6}H *exo*), 1.12 – 1.02 (m, 2H,C^{7}H$_2$), 0.46 – 0.39(m, 1H, C^{5}H *endo*), 0.04 (s, 9H, SiMe$_3$ *exo*), -0.06 (s, 9H, SiMe$_3$ *endo*). ^{13}C NMR (CDCl$_3$) δ 137.8 132.8 (C^2, C^3), 49.6, 46.5, 44.9 (C^1, C^4 C^7), 27.8 (C^5, C^6), -0.3 (SiMe$_3$ *exo)* and -0.9 (SiMe$_3$ *endo*).

3.3. POLYMERIZATION PROCEDURE

The polymerization in the presence of RuCl$_3$·3H$_2$O and RuCl$_2$(PPh$_3$)$_3$ was carried out using the procedure previously described [6].

The polymerization in the presence of WCl₆/1,1,3,3-tetramethyl-1,3-disilacyclobutane (TMSB) was performed under dry argon in a reaction vessel equipped with magnetic stirring bar. TMSB was added to 0.01 M toluene solution of WCl_6 (TMSB/ WCl_6 = 3mol/mol). The color of the solution changed from deep blue to deep purple. After 5-10 minutes, the monomer was added. The polymers were isolated using standard procedures.

The ^1H and ^{13}C NMR spectra were assigned as follows:

Polymer **8** from *2-Dimethylsilyl(trimethylsilylpropyl)bicyclo[2.2.1]hepta-2,5-diene*: ^1H NMR (CDCl₃) δ 5.82 (broad s, 1H, C^3H), 5.60 – 4.95 (broad m, 2H, C^5H, C^6H), 3.65 (broad s, 1H, C^4H), 3.27 (broad s, 1H, C^1H), 2.33 (broad s, 1H, C^7H, *syn*), 1.33 (broad s, 3H, C^7H, *anti*, C^9H_2), 0.8 – 0.47(broad m, 4H, C^8H_2 and $C^{10}H_2$), 0.05 (s, 6H, Si(CH₃)₂), - 0.01 (s, 9H, Si(CH₃)₃). ^{13}C NMR (CDCl₃) δ 147.4 (s, C^2), 145.6 (m, C^3), 134.6, 134.3, 133.7, 133.1, 132.2 (C^5, C^6), 54.1, 53.9 (C^4, *trans*), 50.7 (C^1, *trans*), 48.2, 47.9 (C^4, *cis*), 45.8 (C^1, *cis*), 41.2, 40.7, 40.2, 36.1 (C^7), 21.4, 20.2, 18.4 (C^8, C^9, C^{10}), - 1.5, -2.4, -2.6 (Si(CH₃)₂, Si(CH₃)₃)

Polymer **10** from *endo-exo-5,6-bis(trimethylsilyl)bicyclo[2.2.1]hept-2-ene*: ^1H NMR (CDCl₃) δ 5.6 – 5.08 (m, 2H, C^2H, C^3H), 3.01, 2.86 (broad s, 2H, C^4H, C^1H, cis), 2.65, 2.50 (broad s, 2H, C^4H, C^1H, trans), 1.74 (broad s, 1H, C^6H, exo), 1.42 – 1.0 (m, 2H, C^7H_2), 0.92 (broad s, 1H, C^5H, endo), - 0.003 (broad s, 18H, 2 SiMe₃). ^{13}C NMR (CDCl₃) δ 135.8, 135.2, 134.1, 134.0, 132.2, 132.2, 131.8, 130.7 (C^2, C^3), 48.3, 48.2, 47.3, 47.1, 45.3, 45.2, 44.8, 44.2, 43.9, 43.1, 42.8, 41.8, 41.8, 39.3 (C^1, C^4 C^7), 33.6, 32.7, 32.0, 31.9, 31.7, 29.6 (C^5, C^6), 0.1, -2.2 (Si(CH₃)₃).

4. CONCLUSIONS

Novel 2-dimethylsilyl(trimethylsilylpropyl)bicyclo[2.2.1]hepta-2,5-diene and endo-exo-5,6-bis(trimethylsilyl)bicyclo[2.2.1]hept-2-ene have been polymerized according to the ROMP scheme in the presence of $RuCl_3 \cdot 3H_2O$, $RuCl_2(PPh_3)_3$, and WCl_6/tetramethyldisilacyclobutane as catalysts. Highly molecular mass silyl-substituted poly(cyclopentylenevinylene)s and poly(cyclopentenylenevinylene)s have been prepared. Transport properties of novel cyclo-linear glassy polymers were studied. The additional double bond in the cyclopentyl ring doesn't improve gas transport parameters. On the other hand the introduction of the second $Si(CH_3)_3$ side group in the backbone chain results in strong increase in gas permeability.

5. REFERENCES

1. Finkelshtein, E.Sh., Makovetskii, K.L., Yampolskii, Yu.P., Ostrovskaya, I.Ya., Portnykh, E.B., Kaliuzhnyi, N.E., Plate, N.A. (1990) A new organosilicon polymer with high parameters of permeability, *Vysokomolek.Soedin.*, (B)**32**, N 9, 643-644.
2. Finkelshtein, E.Sh., Makovetskii, K.L., Yampolskii, Yu.P., Portnykh, E.B., Ostrovskaya, I.Ya., Kaliuzhnyi, N.E., Pritula, N.A., Golberg, A.I., Yatsenko, M.S., Plate, N.A. (1991) Ring-opening metathesis polymerization of norbornenes with organosilicon substituents. Gas permeability of polymers, *Makromol.Chem.*, **192**, N1, 1-9.
3. Makovetskii, K.L., Finkelshtein, E.Sh., Ostrovskaya, I.Ya., Portnykh, E.B., Gorbacheva, L.I., Golberg, A.I., Ushakov, N.V., Yampolskii, Yu.P. (1992) Ring-opening metathesis polymerization of substituted norbornenes, *J.Mol.Catal.*, **76**, 107-121.

4. Finkelshtein, E.Sh., Bespalova, N.B., Portnykh, E.B., Makovetskii, K.L., Ostrovskaya, I.Ya., Shishatsky, S.M., Yampolskii, Yu.P., Plate, N.A., Kaliuzhnyi, N.E. (1993) Synthesis and gas permeability of polymers of norbornene halogen derivatives, *Vysokomolek.Soedin.*, (B)**35**, N 5, 489-494.

5. Finkelshtein, E.Sh., Portnykh, E.B., Ushakov, N.V., Gringolts, M.L., Fedorova, G.K., Plate, N.A. (1994) Synthesis of polymers containing carbazolyl-substituted norbornene derivative, *Makromol.Rapid Commun.*, **15**, 155-159.

6. Finkelshtein, E.Sh., Portnykh, E.B., Ushakov, N.V., Gringolts, M.L., Yampolskii, Yu.P. (1997) Synthesis of a new polymer - polycyclopentylenevinylene bearing Ge-containing group, *Makromol.Chem.Phys.*, **198**, 1085-1090.

7. Finkelshtein, E.Sh., Portnykh, E.B., Makovetskii, K.L., Ostrovskaya, I.Ya., Bespalova, N.B., Yampolskii, Yu.P. (1998) Synthesis of membrane materials by ROMP of norbornenes, in Y.Imamoglu (ed.), *Metathesis Polymerization of Olefins and Polymerization of Alkynes,* Kluwer Academic Publishers, Dordrecht, pp. 189-199.

8. Finkelshtein, E.Sh. (1998) Olefin metathesis in organosilicon chemistry. in Y.Imamoglu (ed.), *Metathesis Polymerization of Olefins and Polymerization of Alkynes,* Kluwer Academic Publishers, Dordrecht, pp. 201-224.

9. Ivin, K.J., Mol, J.C. (1997) *Olefin Metathesis and Metathesis Polymerization*, Academic Press, London p. 317.

10. Stonich, D.A., Weber, W.P. (1991) Synthesis and characterization of poly[(2-trimethylsilyl-2-cyclopentene-1,4-diyl)vinylene], *Polym. Bull.*, **26**, 493-497.

11. Cunico, R.F.(1971*)* The Diels-Alder Reaction of α,β-Unsaturated Trihalosilanes with Cyclopentadiene, *J .Org. Chem.* **36,** 929-932.

12. Sheludjakov, V.D., Zhun, V.I., Lakhtin, V.G., Scherbinin, V.V., Chernishev, E.A. (1983) Interaction of tri- and tetrachloroethylene with trichlorosilane, *Zh.Obschei Chimii* **53,** 1192-1193.

SUBJECT INDEX